MW00844047

CONVERSION TABLE
FOR
THE INTERNATIONAL SYSTEM OF UNITS (SI SYSTEM)

To convert from	to	Conversion factor	To convert from	to	Conversion factor
A/dm^2	A/ft^2	9.290 30	kW	CV or Ps	1.359 62
A/ft^2	A/dm^2	0.107 639	kW	hp	1.341 02
Btu	kj	1.055 06	lb	kg	0.453 592
Btu/h	W	0.293 071	lbf	N	4.448 22
cal	J	4.186 8	lbf/in^2	kN/m^2	6.894 76
cm^2	in^2	0.155 000	$10^3 \times lbf/in^2$	hbar	0.689 476
cm^3	in^3	0.061 023	litres (dm^3)	UK gal	0.219 969
CV	kW	0.735 499	litres (dm^3)	US gal	0.264 172
ft	m	0.304 8	m	ft	3.280 84
ft^2	m^2	0.092 903	m	yd	1.093 61
ft^3	m^3	0.028 316	m^2	ft^2	10.763 9
ftlbf	J	1.355 82	m^2	yd^2	1.195 99
g	oz(av)	0.035 274	m^3	ft^3	35.314 7
g	oz(Troy)	0.032 150	m^3	yd^3	1.307 95
g/1	oz/UKgal	0.160 359	miles	km	1.609 34
g/1	oz/USgal	0.133 526	mm	in	0.039 370
hbar	kfg/mm^2	1.019 72	N	lbf	0.224 809
hbar	$10^3 x lbf/in^2$	1.450 38	oz (av)	g	28.349 5
hbar	$UKtonf/in^2$	0.647 490	oz (Troy)	g	31.103 5
hp	kW	0.745 700	oz (av)/UK gal	g/1	6.236 03
in	mm	25.4	oz (av)/US gal	g/1	7.489 15
in^2	cm^2	6.451 6	PS	kW	0.735 499
in^3	cm^3	16.387 1	t	UK ton	0.984 207
J	cal	0.238 846	UK gal	litres	4.546 09
J	ftlbf	0.737 562	UK ton	t	1.016 05
J	kgf m	0.101 972	UK tonf	kN	9.964 02
kg	lb	2.204 62	UK tonf/in^2	hbar	1.544 43
kgf m	J	9.806 65	US gal	litres	3.785 41
kfg/mm^2	hbar	0.980 665	W	Btu/h	3.412 14
kJ	Btu	0.947 817	yd	m	0.914 4
km	miles	0.621 371	yd^2	m^2	0.836 127
kN	UK tonf	0.100 361	yd^3	m^3	0.764 555
kN/m^2	lbf/in^2	0.145 038			

PERIODIC SYSTEM OF ELEMENTS

IA	IIA	IIIA	IVA	VA	VIA	VIIA	VIII			IB	IIB	IIIB	IVB	VB	VIB	VIIB	0
1 H 1.008																	2 He 4.003
3 Li 6.940	4 Be 9.013											5 B 10.82	6 C 12.010	7 N 14.008	8 O 16.000	9 F 19.00	10 Ne 20.183
11 Na 22.997	12 Mg 24.32											13 Al 26.98	14 Si 28.06	15 P 30.98	16 S 32.066	17 Cl 35.457	18 A 39.944
19 K 39.10	20 Ca 40.08	21 Sc 45.10	22 Ti 47.90	23 V 50.95	24 Cr 52.01	25 Mn 54.94	26 Fe 55.85	27 Co 58.94	28 Ni 58.69	29 Cu 63.54	30 Zn 65.38	31 Ga 69.72	32 Ge 72.60	33 As 74.91	34 Se 78.96	35 Br 79.916	36 Kr 83.8
37 Rb 85.48	38 Sr 87.63	39 Y 88.92	40 Zr 91.22	41 Cb 92.91	42 Mo 95.95	43 Tc (99)	44 Ru 101.1	45 Rh 102.91	46 Pd 106.4	47 Ag 107.88	48 Cd 112.41	49 In 114.82	50 Sn 118.70	51 Sb 121.76	52 Te 127.61	53 I 126.92	54 Xe 131.3
55 Cs 132.91	56 Ba 137.36	57 *La 138.92	72 Hf 178.5	73 Ta 180.88	74 W 183.86	75 Re 186.31	76 Os 190.2	77 Ir 193.1	78 Pt 195.09	79 Au 197.2	80 Hg 200.61	81 Tl 204.39	82 Pb 207.21	83 Bi 209.0	84 Po 210.0	85 At (210)	86 Rn 222
87 Fr (223)	88 Ra 226.05	89 •Ac 227.0															

TRANSITION METALS — REFRACTORY METALS — PRECIOUS METALS

METALLIC BONDING — COVALENT BONDING TENDENCY — COVALENT (NON-METALLIC) BONDING

RARE GASES

* LANTHANUM SERIES

58 Ce 140.13	59 Pr 140.92	60 Nd 144.27	61 Pm (147)	62 Sm 150.43	63 Eu 152.0	64 Gd 157.26	65 Tb 158.93	66 Dy 162.46	67 Ho 164.94	68 Er 167.2	69 Tm 169.4	70 Yb 173.04	71 Lu 174.99

• ACTINIUM SERIES

90 Th 232.12	91 Pa 231.05	92 U 238.07	93 Np (237)	94 Pu 239.11	95 Am (241)	96 Cm (242)	97 Bk (243)	98 Cf (245)	99 Es (253)	100 Fm (254)	101 Md (256)	102 No (254)	103 Lw (259)

SUPERALLOYS II

SUPERALLOYS II

Edited by

CHESTER T. SIMS
Rensselaer Polytechnic Institute
Troy, New York

NORMAN S. STOLOFF
Rensselaer Polytechnic Institute
Troy, New York

WILLIAM C. HAGEL
Arbormet Ltd.
Ann Arbor, Michigan

A Wiley-Interscience Publication

JOHN WILEY & SONS

New York • Chichester • Brisbane • Toronto • Singapore

A NOTE TO THE READER
This book has been electronically reproduced from
digital information stored at John Wiley & Sons, Inc.
We are pleased that the use of this new technology
will enable us to keep works of enduring scholarly
value in print as long as there is a reasonable demand
for them. The content of this book is identical to
previous printings.

Copyright © 1987 by John Wiley & Sons, Inc.

All rights reserved. Published simultaneously in Canada.

Reproduction or translation of any part of this work
beyond that permitted by Section 107 or 108 of the
1976 United States Copyright Act without the permission
of the copyright owner is unlawful. Requests for
permission or further information should be addressed to
the Permissions Department, John Wiley & Sons, Inc.

Library of Congress Cataloging in Publication Data:
Superalloys II.

   "A Wiley-Interscience publication."
   Includes bibliographies.
   1. Heat resistant alloys.   I. Sims, Chester T.
(Chester Thomas), 1923–   .   II. Stoloff, N. S.
III. Hagel, William C.   IV. Title: Superalloys
II.

TN700.S85   1987   671   86-32564
ISBN 0-471-01147-9

Printed in the United States of America

10 9 8 7 6 5

CONTRIBUTORS

Adrian M. Beltran
Turbine Technology Laboratory
General Electric Company
Schenectady, NY

Earl E. Brown
Pratt & Whitney Division
United Technologies, Inc.
East Hartford, CT

William L. Chambers
Gas Turbine Division
General Electric Company
Schenectady, NY

David S. Chang
Materials & Process Technology Laboratory
Aircraft Engine Business Group
General Electric Company
Evendale, OH

Wilford H. Couts, Jr.
Research Laboratories
Wyman-Gordon Company
North Grafton, MA

Willard P. Danesi
Garrett Turbine Engine Company
Phoenix, AZ

Robert L. Dreshfield
National Aeronautics & Space Administration
Lewis Research Center
Cleveland, OH

David N. Duhl
Materials Engineering & Research Laboratory
Pratt & Whitney Division
United Technologies, Inc.
East Hartford, CT

Steven Floreen
Knolls Atomic Power Laboratory
General Electric Company
Schenectady, NY

William R. Freeman, Jr.
Howmet Turbine Components Corporation
Greenwich, CT

T. P. Gabb
National Aeronautics & Space Administration
Lewis Research Center
Cleveland, OH

C. S. Giggins
Materials Engineering & Research Laboratory
Pratt & Whitney Aircraft
East Hartford, CT

Edward Goldman
Materials & Process Technology Laboratory
Aircraft Engine Business Group
General Electric Company
Evendale, OH

Timothy E. Howson
Wyman-Gordon Company
North Grafton, MA

George S. Hoppin III
Garrett Turbine Engine Company
Phoenix, AZ

Louis W. Lherbier
Universal-Cyclops Steel Corporation
Bridgeville, PA

Gerald M. Meier
Department of Materials Engineering
University of Pittsburgh
Pittsburgh, PA

Harold E. Miller
Gas Turbine Division
General Electric Company
Schenectady, NY

Robert V. Miner
National Aeronautics & Space Administration
Lewis Research Center
Cleveland, OH

Donald R. Muzyka
Cabot Refractory Metals
Reading, PA

Fred S. Pettit
Department of Materials Engineering
University of Pittsburgh
Pittsburgh, PA

Steven Reichman
Research Laboratories
Wyman-Gordon Company
North Grafton, MA

Earl W. Ross
Materials and Process Technology Laboratory
Aircraft Engine Business Group
General Electric Company
Cincinnati, OH

Chester T. Sims
Materials Engineering Department
Rensselaer Polytechnic Institute
Troy, NY

James L. Smialek
National Aeronautics & Space Administration
Lewis Research Center
Cleveland, OH

Norman S. Stoloff
Materials Engineering Department
Rensselaer Polytechnic Institute
Troy, NY

John H. Wood
Gas Turbine Division
General Electric Company
Schenectady, NY

William Yeniscavich
207 Colonial Court
Lynchburg, VA

FOREWORD

The first successful flights of jet engine–powered airplanes (in World War II, by the German and British military) were made with materials-limited engines of relatively modest performance. As they advanced, jet engines continued to be materials oriented. Nonetheless, examination of *materials progress* since 1942 shows a spectacular series of developments that permitted uninterrupted increases in temperature and operating stress. The developments were both process- and alloy-oriented, and often a combination of the two. As a result the net 800-lb thrust of the 1942 Whittle engine has risen to the level of 65,000 lb—a factor of 80 in a little over 40 years.

Initially, cobalt-base alloys emerged as the leaders for blade manufacture, while iron-base alloys served for lower temperature requirements, disks, for example. From more or less improved conventional practice, wrought alloys, such as S-816, gave way to the coarse-grained precision-cast cobalt-base alloys. Then industry learned how to control the grain size and structure, designers learned how to live with less-than-desired ductilities, and operating temperatures climbed to 815 °C (1500 °F). Precision castings then and now continue to play a commanding role in the superalloy world.

There were parallel developments in Ni-base systems, the valuable, flexible, and now dominant γ/γ'-strengthened alloys. Here it took the process development of vacuum metallurgy to make possible the production of strong "high alloy" compositions by controlling the impurity levels. Then still higher alloy contents leading to greater strength and temperature potential were realized through the development of specialty remelting technologies, of which vacuum arc remelting is the most outstanding.

These developments required unparalleled efforts by research and development groups to demonstrate and evaluate the roles of alloy composition and structure, to utilize the benefit of purity levels previously considered unattainable, and to develop advanced techniques to further modify the structures and the chemistries to solve special problems. Ultimately, this led to the exciting developments of directionally solidified and single-crystal blades, the latter reaching engine application only very recently.

Throughout this period the concern among metallurgists, designers, and manufacturers was always that the nickel-base and cobalt-base alloys ultimately would have to be replaced with higher-melting alloy systems, the refractory metals. This is hardly surprising when one realizes that increased alloying tends to produce lower-melting alloys; here were alloys being used at higher and higher fractions of their melting temperatures!

At first, major efforts were made with alloys of molybdenum and columbium (niobium). These were without success for the then-planned operating temperatures and anticipated lifetimes, but they may still hold promise for temperatures above about 1100 °C (2000 °F) if suitable coatings can be found. Excellent strength levels were realized and some promising coatings were developed, but expected lifetimes were not realized. Later, chromium-base alloys looked to be a natural, but ultimately were not successful because of brittleness problems.

We must also mention the early trials with cermets, and the first of a series of ceramic-age developments from 1950 onwards, both of which produced interesting solid structures, but still no acceptable applications in the superalloy competition. The austenitic superalloys remained dominant.

With the advent of rapid solidification processing, alloys of still more complexity are being developed and studied, now with the advantage of even closer control over impurity segregation and structure of desired phases. Further, production of superfine grain sizes and structures in the powder metallurgy area makes superplasticity easy to achieve and utilize. Nominally cast alloys such as IN-100 and Mar-M 509 are made very strong at low and intermediate temperatures and are easily formable into complex shapes, including near-net-shapes. In the 1960s, who would ever have predicted that IN-100, a casting alloy, could be made to be superplastic and a candidate for disk applications at about 650–700 °C (1200–1300 °F)? Superplastic structures can be expected to have a major impact on superalloy technology.

Finally, we are beginning to see significant applications of ODS (oxide-dispersion-strengthened) alloys, again using a blend of processes and alloying techniques developed over the intervening years. Mechanical alloying, and now the use of RS (rapid solidification; fine, fully alloyed powders), will permit use of ODS nickel-base and cobalt-base alloys to temperatures in excess of 1100 °C (2000 °F).

Use at 1100 °C (2000 °F) and above for alloys melting under 1400 °C (2550 °F)? Use in excess of 80% of the absolute melting temperature? Yes, that time has arrived. Even higher fractions of the melting point may be achieved with metal-matrix composites.

In summary, the extremely effective interplay of alloying processes with alloy compositions and structures, coupled with excellent supporting scientific studies of structures, properties, and stability have given the superalloys an engineering position never dreamed of by their early proponents!

This book, *Superalloys II*, is a very important part of utilizing and perpetuating this industrial success, especially at a time when alternative alloys and materials are being sought but have not yet emerged. The messages in *Superalloys II* should also serve as a model for some of these new materials being studied to replace or supersede the superalloys.

NICHOLAS J. GRANT

Professor, Materials Science & Engineering
Massachusetts Institute of Technology

PREFACE

Many definitions of superalloys have appeared over the years. In our view, superalloys are alloys based on Group VIIIA-base elements developed for elevated-temperature service, which demonstrate combined mechanical strength and surface stability. Progress in superalloy development has made possible the advent of the modern jet engine, with progressively higher thrust-to-weight ratios. Superalloys similarly play a vital role in industrial gas turbines, coal conversion plants, and other applications involving high temperatures and severe environmental conditions.

The Superalloys, published in 1972, was the first comprehensive book on the subject and quickly became the standard reference work in the field; it was adopted as a text in some university courses even though not written primarily as a textbook. Twenty chapters contributed by some 28 authorities in the field were supplemented by appendices of phase diagrams, mechanical properties, and chemical compositions of many commercial superalloys. The success of the book notwithstanding, it became clear several years ago that progress in superalloy melting, alloy development, and processing techniques had been so extensive that a revision was needed. Although several excellent books have been published in recent years, none as comprehensive in coverage as *The Superalloys* appeared.

The editors, therefore, decided in early 1985 to prepare a new book following essentially the same format as the original volume but with special emphasis on new developments—the growth of powder metallurgy, the commanding advent of directionally solidified and single-crystal superalloys, and so on. Although many chapters have been totally rewritten (in some cases by different authors than in the original volume), some chapters have just been extensively updated (e.g., "Fundamentals of Strengthening," "Nickel-Base Alloys") or left nearly unchanged (e.g., Joining). With current superalloys operating at temperatures approaching 90% of their absolute melting points, further advances in gas turbine technology are expected to come from new materials; for example, ceramics, refractory metals (columbium, or niobium), composites, and intermetallic compounds. Therefore, a chapter on alternative materials is included. Conversely, we have dropped the chapter on machining and the chapter on chromium. A new chapter on design has been included for a more rounded picture of the superalloys field for the benefit of materials engineers.

As in the original volume, each physical metallurgical chapter aims to provide both the scientific and technical background needed to understand the superalloy systems topic covered. The rapid, continuing advent of new processing techniques is highlighted by devoting one chapter solely to directional solidification and another to wrought alloys.

The international readership for which this book is intended required the use of dual units (English and SI) throughout the text and in most figures and tables. English units are listed first in deference to current practice in the superalloy industry in the United States. We hope that overseas and/or academic readers will not find this arrangement too distracting. Further, we point out that although the authors of our chapters all are from the United States, there is no lack of vigorous research, development, and applications of superalloys in Japan, Western Europe, and the Soviet Union. In particular, the COST 50 and COST 501 programs of the European community may be cited as examples of active European work in this field. We suggest that *Superalloys II* be utilized in conjunction with conference volumes on superalloys arising from the regular Seven Springs Meetings in the United States and the COST volumes issued at conferences in Belgium to completely characterize the status of superalloys in industry and commerce.

To improve readability, mathematical formulations have been kept to a minimum. Nevertheless, we believe that this book will continue to be particularly useful as a text in college courses and in intensive short courses on high-temperature alloys. A list of registered alloy trademarks is given in Appendix C.

The editors are particularly grateful to the contributing authors for the time and effort they have devoted to their respective chapters. We appreciate also the contributions of several colleagues who have assisted by proofreading, commenting on text, and contributing information to the authors.

CHESTER T. SIMS
NORMAN S. STOLOFF
WILLIAM C. HAGEL

Troy, New York
Ann Arbor, Michigan
October 1986

CONTENTS

SUPERALLOYS II

PART ONE

INTRODUCTION

Chapter 1

Superalloys: Genesis
and Character

CHESTER T. SIMS

Rensselaer Polytechnic Institute, Troy, New York

Superalloys are a class of materials around which it is difficult to create an exact boundary. However, the definition utilized in *The Superalloys*[1] some 14 years ago has proven reasonably acceptable: "A superalloy is an alloy developed for elevated-temperature service, usually based on group VIIIA elements, where relatively severe mechanical stressing is encountered, and where high surface stability is frequently required."

Superalloys are divided into three classes; nickel-base superalloys, cobalt-base superalloys, and iron-base superalloys. In addition, a major subgroup, those that have metallurgical characteristics similar to nickel-base alloys but contain relatively large iron contents, are called nickel–iron superalloys.

Superalloys are utilized at a higher proportion of their actual melting point than any other class of broadly commercial metallurgical materials. Superalloys, which have made much of our very high temperature engineering technology possible, are the materials leading edge of the gas turbines that drive jet aircraft. In turn, these engines have been the prime driving force for the existence of superalloys. However, in addition to aircraft, marine, industrial, and vehicular gas turbines, superalloys see service in space vehicles, rocket engines, nuclear reactors, submarines, steam power plants, and petrochemical equipment. Many (perhaps 15–20%) have been developed for utilization in corrosion-resistant applications. While this book is intended principally to treat the topic of superalloys for high-temperature applications, much of the information herein can be applied to the corrosion-service group; they will be covered briefly later.

This chapter is an attempt to tell the story of these alloys. It is intended in part to be a technically and scientifically useful chronology-based analysis of their behavior and their manufacture and will include some evaluation of the significant property factors and of other external forces that have generated these unusual and essential materials. It also will give some feeling for their applications and economics. But above all, it is intended as introduction and background for the rest of the chapters.

BACKGROUND AND UTILIZATION

The Machines

Throughout history, humans have developed mechanical devices to satisfy their needs. Hundreds of years ago, perhaps beginning with observations of the power of rising warm air (Fig. 1a), it became apparent that efficiency in performing useful work was related to use of high temperatures. This then led thermodynamically to the Brayton cycle, a basic physical tenet that holds that higher use temperatures (accompanied by lower heat rejection temperatures) result in more efficient operation (see Chapter 2).

Brayton's concept was applied to rotating engines, and relatively advanced steam turbines began to appear in the 1800s. In the early 1900s, gas turbines were used for power generation in Europe; the first successful one, designed by Norway's Aegidius Elling[2] is shown in Fig. 1b. Furthermore, inventive man was moving fast around the turn of the century, and the gasoline engine and propellor-driven winged flight were being developed, essentially in parallel with turbine engine power.

Within the first decade of engine development, it was perceived that airplanes needed some device to provide their internal combustion engines with pressurized air–fuel mixtures because of the lower air pressure at flight altitudes. Work progressed in Europe and the United States. One such effort between Sanford Moss of Cornell

Fig. 1. Genesis of the gas turbine engine. (a) The first gas turbine. Useful for household chores, after Bishop Gibbons's "Mathematical Magick," 1648.[7] (b) Rotor of the first commercially successful gas turbine. Designed by Aegidius Elling, 1923.[2]

Fig. 2. Aircraft utilization of the gas turbine engine. (a) First turbojet aircraft, German He-178 with Von Ohain HeS3B 1100-lb thrust engine, flown August 27, 1939. (b) Whittle engine. English W-1 powered Gloster meteor in 1941.

University, the General Electric Company, and the U.S. Army culminated in the aircraft turbosupercharger. This invention spurred a continuing improvement in metal alloys, eventually creating a leading position in high-temperature metallurgy for the United States.

Then a scientific and technological phenomenon of immense importance occurred. Advances in aerodynamic theory caused a radical change in the thinking of designers in England, Germany, and Italy; (a) it was realized that turbulent drag wasted two-thirds of the power applied to conventionally driven aircraft, (b) Prandtl's airfoil theory involving the lift concept was applied to axial compressors and turbines, and (c) it became understood that the supersonic forces at the ends of propellers would not allow airplanes to be driven much above 400 miles/h. These three factors together led to a technology paradigm, the concept of jet-engine-powered aircraft. It was revolution, not evolution.[3-6]

Jet aircraft were thrust into public awareness with the 1937 flight of Hans von Ohain's turbine engine Heinkel in Germany (Fig. 2a) and an independent development, the 1939 flight of Whittle's engine (Fig. 2b) in England. Key events are summarized in Fig. 3.[7]

With this new technology it was obvious to the designers that progress would be made through still higher temperatures and that new materials were needed to do the job. Since that time, progress in jet propulsion and in industrial gas turbines has been a growing engineering technology of immense importance. It also has been inexorably dependent on high-temperature alloy capability. The technology area for superalloy performance is as defined by the gas turbine engine: its disk (or wheels); its airfoils*; its combustion chambers; its many structurals (see Chapter 2 and below, Fig. 9).

The Metal

Metallurgy progressed from the age of copper and iron to that of stronger and more corrosion-resistant alloys. In the 1910–1915 era, austenitic stainless steel was "discovered" and developed. Important here is that the gamma [face-centered-cubic (FCC)] austenitic stainless steel field eventually became the fertile phase structure

---

* Called *blades* in an aircraft engine and *buckets* in industrial turbines.

from which superalloys developed, although at that time alloy development for turbosuperchargers was along conventional lines, through the strengthening of ferritic steels.

In 1929, Bedford and Pilling and virtually simultaneously, Merica added small amounts of titanium and aluminum to the by-then well-known FCC "80/20" nickel–chromium alloy. Significant creep strengthening occurred, and superalloys, too, were off the ground, chronologically synchronized with the jet engine paradigm.[8] Then in England, the United States, and Germany, the Edisonian experimenters of the 1930s succeeded in creating strong alloys, built on the austenitic nickel-base solid solution containing chromium (called γ), carbides, and a fine precipitated phase. However, even after almost 10 years, the vital coherent phase present as FCC γ' had not yet been identified visually.[9] Carbide-strengthened austenitic cobalt-base alloys developed competitively at this time since they were more readily cast to complex shapes.

The parallel development of jet engine technology added to the demand for stronger and stronger austenitic alloys since it was realized that the potential of the new engine was (relatively) unlimited.

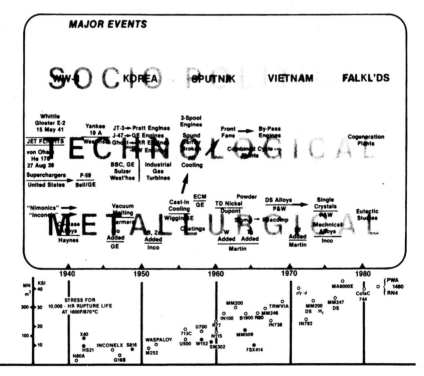

Fig. 3. The inception of superalloys to the present, set against the advent of superalloys of increasing capability.[7]

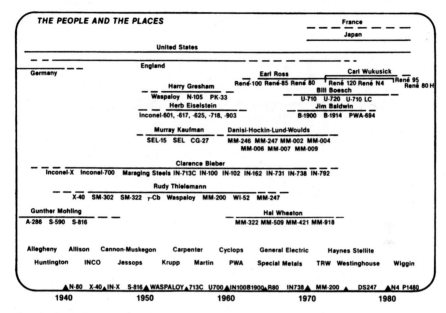

Fig. 4. The countries, companies, and people in the free world responsible for the core generation of superalloy development. Circa 1984.[7]

From the 1940s and World War II the story of superalloys is one of further improvement through the invention of new alloy compositions and new processes. First military jet engines continued as the drive, but then the power and transportation industries needed gas turbines for electricity, gas-line pumping, and other prime-mover applications; these turbines often required superalloys with different characteristics. Alloy development virtually exploded in the fifties and sixties, as did process development in the seventies and eighties.

The alloy development and the process development in superalloys is a story generated by people and by companies, principally in the United States and England. Figure 4[7] identifies the more prominent inventors (illustrating the invention of three commercially utilized alloys) and alloy companies.

DEVELOPMENT OF THE TECHNOLOGY

Chemical Composition

Superalloy ownership and use right patterns are based principally on their patented chemical compositions. Composition is the core of the engineering specification, giving the physical and legal statement of the solid matter. Figure 5 gives a broad perspective of the trends in chemical composition over time. Since composition is

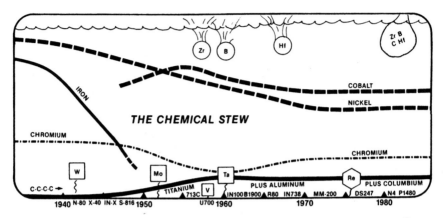

Fig. 5. A qualitatively comparative view of trends in superalloy chemical composition.[7]

very complex, the figure is not to exact chemical content scale (it does not "add to 100"); it attempts only to identify broad qualitative trends.

Prior to the early 1930s, the alloys were built only on an iron or a nickel base, with sufficient chromium present for oxidation resistance. Then, the small amounts of aluminum, titanium, and/or columbium produced coherent creep-resistant γ'. Sometimes hidden embrittling enemies, such as colonies of $M_{23}C_6$, also occurred. Typical alloys in this action were Rex 78, K42B, Nimonic 75 and 80, and Inconel X.

It can be seen that in the 1930s, iron generally disappeared as an alloy base in favor of nickel and cobalt since they stabilized the stronger FCC structure. Chromium, always a major alloying element contributing oxidation resistance to the systems, was perceived as hobbling strength in the 1960s. However, overreduction of chromium led the unwary into "hot corrosion" problems (e.g., Inco-713C), resulting in a more carefully balanced use of that element (e.g., IN-738). Of course, aluminum, titanium, and columbium, added to form γ', have never been present in large amounts. Added excessively, they too, can cause structural problems by a variety of mechanisms. Aluminum's role as the primary γ' former and as an important protective oxide-forming element makes it the most vital of the three.

In the late 1940s, it was found that additions of molybdenum (first in M-252) created significant additional strengthening through solid solution and carbide effects. Soon, other refractory elements, tungsten, columbium, tantalum, and now rhenium, were utilized to this end. Hafnium is involved in a complex set of reactions with γ', carbides, and the matrix.

Carbon, of course, has always been a complexing agent. Matrix carbides act as point strengtheners in several solid-state reactions. Carbides (and zirconium and boron) also have salutary grain boundary effects. However, in the present cortege of single-crystal alloys, carbon, zirconium, and boron are not generally needed because there are no grain boundaries.

Thus, superalloys have been through a score of years (1950–1970) when increasing amounts of many elements were added for specific mechanical and chemical effects; generally, a maximum level was found followed by removal of some elements in the 1980s due to advancing process developments. Table 1 compares the composition of two alloys invented in the 1930s with several that are popular today. However, it is important to note that most of the early alloys are not extinct: Nimonic 80A, Inconel X, and X-40 are specified today for many critical components where their properties are still appropriate.

Table 1 indicates that austenitic superalloys are *very* complex. A broad balance of alloy base (Ni + Co at ~50%), surface stability addition (Cr at ~10–15%), and γ' formers (Al + Ti, ~4–8%) was reached about three decades ago and now vary only slightly. Changes in lesser elements are now the major "action," connoting the subtle depth of understanding.

As in all classic equilibrium metallurgy, the alloy chemical composition defines the solid phases present. The phases in turn create a visible microstructure. Thus, the *chemical composition, phase constitution*, and *microstructure* define superalloys in the physical sense. Phase constitution is discussed next.

Phase Constitution

Figure 6 shows the most significant physical phases (identified over the last 50 years) that have created such unique strengthening in superalloys. Some of the phases observed to be deleterious to behavior also are identified. All phases, of course, are potentially interactive with each other and the alloy matrix. At highest operating conditions, superalloy structural metal is a white-hot, chemically dynamic entity of *constantly changing* solid-state phases, just a few degrees below its melting point.

The alloy matrix always consists of the close-packed FCC austenite. Figure 7 illustrates the FCC field as seen from three useful perspectives, a simple ternary, a typical quaternary, and a polar phase diagram. The austenite evolves from the small FCC field in the iron–chromium system expanded by nickel or cobalt; in most cases the iron is eventually eliminated. Thus, superalloys are the daughters of stainless steel. (Selected variations of these diagrams are in Appendix A.) Solution strengthening of the matrix is a major contributor to alloy mechanical capability.

The major strengthening phases in the austenitic matrix were carbides until the 1929–1930 period, when the coherent, cubic gamma prime (i.e., γ', or Ni_3Al) was created in the austenite. For instance, γ' was in both the English Nimonic 80 and the German Tinidur. It is an unusual strengthening phase in that it is very similar to the matrix in composition, it can be very widely alloyed, *its yield strength increases with temperature*, and it has good oxidation resistance on its own. Then in the early 1950s, Eiselstein's development of IN-718 revealed another unique phase, γ''; while γ' is characterized by a simple FCC structure, γ'' is BCT (body-centered-tetragonal) that is, two cubes stacked.

The carbides now present in both nickel and cobalt alloys primarily are of the $M_{23}C_6$ and M_6C types, readily heat treatable, whereas the slowly decomposing MC

Table 1. Chemical Compositions: A Comparison of Early Superalloys and Those Used 50 Years Later[a]

		Fe	Ni	Co	Cr	Al	Ti	Ta	Mo	W	Hf	Zr	C	Other
1935	Rex-78 Wrought Fe-base bucket alloy	60.0	18	—	14	—	0.6	—	4.0	—	—	—	0.01	0.015 B, 4Cu
	K 42 B Wrought Ni–Co–Fe base blade alloy	13.0	43	22.0	18	0.2	2.1	—	—	—	—	—	0.05	
1985	FSX-414 Cast Co-base vane alloy	—	10	52.5	29	—	—	—	—	7.5	—	—	0.25	
	CM SX-2 Cast Ni-base single-crystal alloy	—	66.5	4.6	8.0	5.6	0.9	5.8	0.6	7.9	0.1	0.01	0.005	10 ppm S, N, O
	MA 6000E Wrought Ni-base ODS blade alloy	—	70	—	15	4.5	2.5	2.0	2.0	4.0	—	0.15	0.05	0.01 B 1.1 Y$_2$O$_3$
	IN-718 Forged Ni-Fe-base wheel alloy	18.5	52.5	—	19	0.5	0.9	—	3.0	—	—	—	0.04	5 Cb, 0.005 B

[a] REX-78 bladed the Whittle engine (from ref. 7).

10

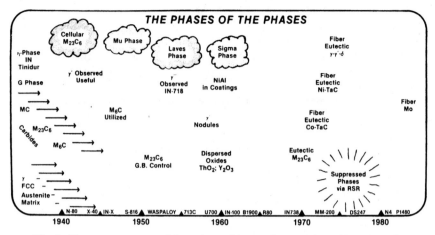

Fig. 6. Discovery occurrence of the major useful superalloys and the problem phases.[7]

type are utilized as a sump for carbon release throughout the life of the alloy. (Details of the γ' carbide behavior are in chapters devoted to the individual alloy classes.)

As alloying efforts exceeded the ability of the austenite matrix to hold phases, undesirable compounds such as mu, sigma, and Laves occurred. A task of the alloy metallurgist is to avoid these, perhaps by use of phase control tools (Chapter 8). More recently, advanced "phase diagram metallurgy" has reached a new level of

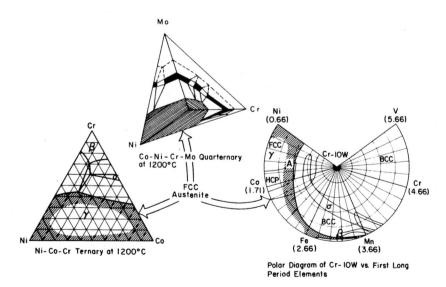

Fig. 7. Phase diagrams illustrating the FCC γ' field; basis of austenitic superalloys.

understanding and practice in the development of experimental eutectic superalloys strengthened by constitutionally directed eutectic lamellae formed by freezing from the melt.

Knowledge of the phases occurring in superalloys and understanding and control of their generation and reactions is the centerpiece of alloy composition development and process development.

The Microstructure

As chemical composition establishes the phases, the phases in turn create the microstructure. As mentioned, the earliest superalloy metallographers did not see the tiny, coherent precipitate particles of γ' that were making their alloys strong and useful. It was not until the advent of electron microscopy in the 1950s that we began truly to understand the visual (i.e., physical) relationship between the phases and the extreme complexity of superalloy phases and alloy behavior and still later that the interactions between dislocations and the γ/γ' alloy system began to be understood.

Figure 8 is a drafted panoramic at about $10,000\times$ (original reduced approximately 4:1) showing the 50-year development of nickel superalloy appearance, characterizing the microstructures that have made the alloys increasingly strong over time, yet retaining usable ductility, as in the upper two-thirds of the picture. The lower part

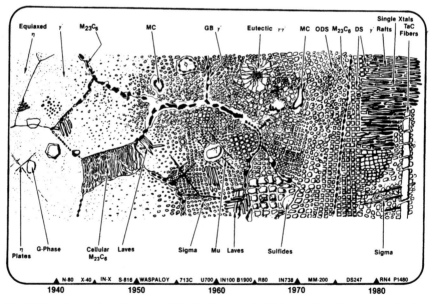

Fig. 8. The microstructure. Panorama of the development of nickel superalloy microstructure showing both useful and deleterious phases.[7]

of the figure includes some of the phases that have been found to cause brittleness, have lowered strength, or create other problems.

From the early 1930s until the 1950s the alloys were filled with increasing "structure" to make them stronger. In the 1950s, packing with strengthening elements accelerated but then led to significant problems created by embrittling phases, such as σ and Laves. The 1950s also saw the generation of very complex grain boundaries, with carbides engloved in γ', creating a dispersion-strengthened layer bonding the grains together. By 1970, the effects of hafnium had been discovered, and the γ'-engloved carbide structure was less essential. Hafnium contorts the grain boundary to create strength and ductility in a more purely mechanical fashion as well as generating additional γ'.[10]

Directional solidification (DS) processing then created aligned grain structures, aligned grain boundaries, and even aligned strengthening filaments (such as TaC), shown in Figure 8 as they appeared in the seventies and eighties. Finally, we see the aligned homogeneous single-crystal (SX, or SC) structures. Recently, through heat treatment, transverse plates of γ' have been created in single crystals, which gives still further strengthening. (Because DS and SX alloys are so important, Chapter 7 is solely devoted to the subject.)

Cobalt superalloy microstructures (Chapter 5) are somewhat less complex than nickel alloys. For creep strength cobalt alloys depend primarily on solid-solution strengthening and interaction between hard carbides and alloy imperfections such as dislocations and stacking faults. Cobalt alloys do not form γ', but metallurgists tend to utilize varying combinations of the carbides (e.g., MC, M_6C, and $M_{23}C_6$) to attempt similar results. The iron-base alloys of the 1930s were similar to the cobalt alloys. The nickel/high-iron alloys (Chapter 6), however, are *more* complex, forming both γ' and γ'', and can be thought of as nickel-base alloys heavily diluted with iron. Thus, superalloy metallurgists have developed and utilized a series of strengthening reactions that create a complex interactive structure of finely divided phases of unparalleled complexity.

DEVELOPMENT OF BEHAVIOR

Physical Properties

Some primary physical properties of the three Group VIIIA elements are given in Table 2.

Several factors stand out. For instance, the characteristic FCC structure is close packed. This is the best atom arrangement to provide strength to very high fractions of the melting point. (Co becomes FCC at high temperatures.) Density, which varies around 0.3 lb/in. depending on base and alloying elements, is a critical property in aircraft engines. For instance, a 10% reduction of alloy density from 0.310 to 0.280 can increase disk life by a factor of 3 or, alternately, significantly reduce disk weight.[11]

The low thermal expansion coefficient (α) of nickel- and cobalt-base alloys (compared to iron) helps in the use of components to close tolerances to achieve

Table 2. Some Physical Properties of Superalloy Base Elements[a]

	Crystal Structure	Melting Point		Density		Expansion Coefficient[b]		Thermal Conductivity[b]	
		°F	°C	lb/in.^{3}	g/cm^3	°F × 10^{-6}	°C × 10^{-6}	Btu/ft^2/hr/°F/in.	cal/cm^2/s/°C/cm
Co	HCP	2723	1493	0.32	8.9	7.0	12.4	464	0.215
Ni	FCC	2647	1452	0.32	8.9	7.4	13.3	610	0.165
Fe	BCC	2798	1535	0.28	7.87	6.7	11.7	493	0.175

[a] From ref. 11.
[b] At room temperature.

maximum efficiency. High thermal conductivity helps in the cooling of turbine hot-stage parts.

Mechanical Properties

The prime reason for the existence of superalloys is their outstanding strength over the temperature ranges at which gas turbine components operate. Their close-packed FCC lattice has the capability to maintain relatively high and reliable tensile, rupture, creep, and thermomechanical fatigue properties to homologous temperatures that are much higher than for equivalent body-centered-cubic (BCC) systems. Contributing factors are the high modulus of the FCC lattice, its many slip systems, and its low diffusivity for secondary elements. Further, of utmost importance are the broad solubility of secondary elements in the austenitic matrix and their physiochemical characteristics enabling precipitation of intermetallic compounds (such as γ' and γ'') for strength. Strengthening effects also can be obtained by solid-solution hardening, carbide precipitation and grain boundary control, directional solidification, and single-crystal generation. The great capability of nickel–cobalt–iron austenite to be usefully strengthened cannot be underestimated.

A comparison of alloy systems is given in Fig. 9. Some refractory metal alloys are stronger at higher temperatures, but they have never been capable of service in

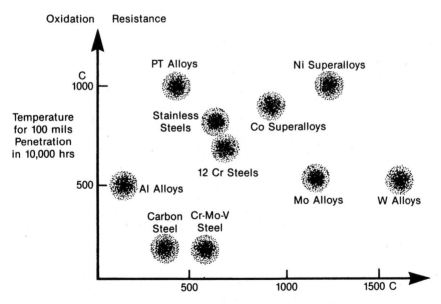

Fig. 9. Relative capabilities of alloy systems.

oxidizing conditions. The *combination* of oxidation resistance and strength is unparalleled. Although platinum group alloys are stable in oxidizing atmospheres, they have little strength. It is, of course, a major objective of this text to create an understanding of the factors that have created this strength superiority of superalloys.

By 1950 leading superalloy metallurgists were well versed in many of these strengthening techniques. Subjected to incessant demands to create continuing increases in tensile and creep rupture properties for higher and higher temperatures, they responded by adding refractory metals in increasing amounts for carbide and solution effects. More aluminum and titanium were added to create more γ', and chromium was decreased to increase the perceived allowable amount of γ'. Strength indeed increased, as shown by the tensile and the stress rupture properties of successively issued alloys. Interest concentrated on nickel-base alloys, since (because γ') they could be made stronger and are more oxidation resistant than cobalt alloys.

As is usual in metallurgy, the increase in strength was generally accompanied by a concomitant decrease in ductility (Fig. 10). By the 1960s widely used alloys such as IN-100, René-100, and B-1900 were testing the limits of acceptable ductility.

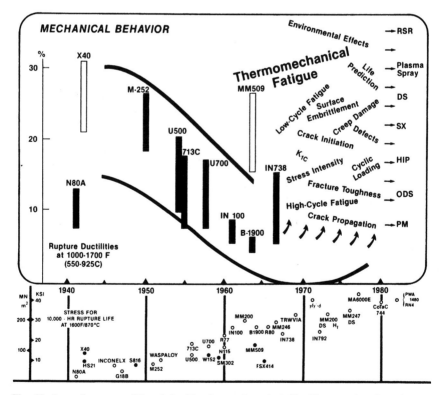

Fig. 10. Superalloy rupture life and ductility change chronologically. Illustrates the effect of intense attention to study of failure.[7]

Metallurgists joked that the elongations reported from rupture or creep-bar tests were principally a function of capability to fit the fracture faces together for after-test length measurements. Still, the alloys grew stronger and hung together and turbines performed.

Then the roof fell in. Sigma phase formed, generating heats of unacceptable IN-100. That emphasized the problems of μ and Laves in other alloys. The IN-100 problem was fixed (Chapter 8), but the problem now was just around the corner for many old and new alloys.

Then, more universally, a host of related mechanical difficulties seemed to saturate "the business," virtually all provoked in some part by low ductility. Metallurgists and designers battled these problems, each concerned with some critical limiting property for their specific application. Optimizing composition and heat treatment, for instance, often had salutory corrective effects. But the entropy available for "fixing" was draining away. Things were coming to a head again.

By the middle seventies it became apparent that the *general* failure mechanism at high temperature for most HPT (high-pressure turbine) components was that of *thermomechanical fatigue*, a combination of ductility limitations in mechanical behavior, often including effects from surface attack. It most succinctly shows up as low fracture toughness. Superalloys had now entered an age in which the measure of success was K_{IC}, the measure of fracture toughness.

Since it was not acceptable in the long run to recover ductility by retrogression to earlier, weaker nickel alloys (or to choose cobalt alloys that had not suffered this problem to the same degree), superalloy metallurgists turned to developments in processing. Led by directional solidification and powder metallurgy, such new processes strove to maintain or improve strength properties while concomitantly producing acceptable ductility. Processing in detail, of course, is discussed extensively in this book, with focus on Chapters 7 and 14 through 18. The provocative effects of new process developments on rectifying the thermomechanical fatigue problem will become evident.

Present Design Requirements.   Figure 11 provides a link directly to gas turbines by showing the upper limits of the stress rupture capability of the three basic classes of alloys, superimposed on which are the design requirements (in general, of course) for aircraft engine disks, rotating blades, and static nozzle guide vanes. These are the three most critical applications for superalloys in turbines.

Surface Stability

High-Temperature Oxidation and Hot Corrosion.   Protection of superalloys from oxidation and corrosion by the aggressive atmosphere in which they serve is equal in importance to generation and utilization of their high strength, although it attracts marginally less attention. A long look to the past suggests there have been three distinct eras related to surface protection (Fig. 12).

In the first decades of this century superalloy operation was at moderate temperatures [approximately 700 °C (1500 °F)], and the chromium level that came naturally from

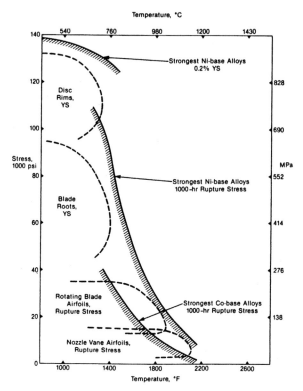

Fig. 11. Comparison of widely-used cast nickel- and cobalt-base alloys at turbine blade and vane operating temperatures and stresses (adapted from ref. 11).

stainless steels (~16–25%) was quite sufficient to do an acceptable protection job. Further, it protected the alloys against the then relatively unknown problem of hot corrosion. However, as temperatures rose, greater oxidation resistance was needed under the pressure of greater oxidation attack (Chapter 11). The potential of aluminum (which replaces Cr_2O_3 with more stable Al_2O_3) was observed, and its level in the alloys increased. Chromium was lowered from 18 to 15% and even to 10%. Oxidation behavior improved. Aluminum, up to 5% or so, was doing the job, and of course, it formed more γ' for higher strength. Two gains at once!

However, some of these new alloys suffered from "hot corrosion" (Chapter XII), a problem previously experienced and handled in industrial turbines. Hot corrosion, a mode of enhanced oxidation that destroys protective oxide layers, results mainly from sodium and sulfur in the fuel and the gas stream. Further, in the late sixties ingestion of seawater spray into helicopter engines in Vietnam wrought havoc in low-chromium turbine blades. The absolute necessity of alloying to *balance* oxidation and corrosion resistance was dramatized.

As temperatures continued to increase, it became clearer that the alloy changes needed to protect against both oxidation and corrosion also often ran counter to alloying trends for higher strength. Increasing chromium and lowering aluminum lowered γ' solution temperatures and thus lowered strength. Engineers turned to coating of superalloys (Chapter 13) to obtain required surface protection without significantly degrading mechanical properties of the underlying blade or bucket alloy. This, in turn, led to the (current) period of "enhanced alumina" where carefully balanced coating alloys (based on nickel, iron, or cobalt with chromium, aluminum, and other active elements) create extremely oxidation- and/or corrosion-resistant protective skins of alloyed aluminum oxide. In today's technology virtually all superalloy load-bearing parts used at very high temperatures under dynamic conditions are coated. But it is also noteworthy that SX alloys, by their nature devoid of grain boundaries, often show a new and unusual level of high surface stability without coating.

At present the advent of TBCs (thermal barrier coatings) brings another dimension. These are thick oxide coatings (such as Y_2O_3-stabilized ZrO_2) that can lower metal alloy surface temperature by reducing the heat flux, thereby enhancing the effects of air cooling and promoting life and reliability. TBCs have been used principally in combustion chambers, but their future use on nozzles and blades is even more promising.

To summarize, surface protection of superalloys from gas stream oxidation–corrosion has been shown to be just as essential as developing greater strength. In fact, the coating–substrate alloy combination now long used in superalloys can rightfully be called the first true *composite* hot-stage turbine part.

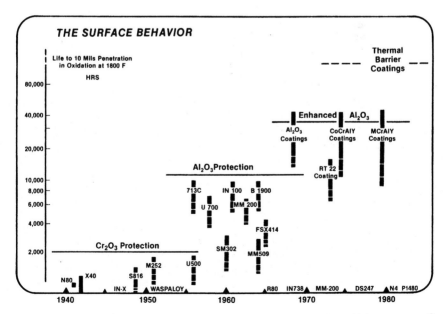

Fig. 12. Advancing steps in the protection of superalloys against oxidation at high temperatures.[7]

Corrosion-Resistant Superalloys.†   Since the early 1972 publication of *The Superalloys*, major growth has occurred in the production and use of weldable corrosion-resistant superalloys for service in reactors, chemical equipment, and the like.

Wet scrubbers, acid-handling equipment, reaction vessels, nozzles, agitators, and so on can, of course, present a wide spectrum of corrosion problems. To reduce overall costs, plant designers are beginning to recognize the value of using a qualified corrosion-resistant superalloy to obtain relatively trouble-free long-time service rather than suffering along with numerous repairs with less expensive, but more degradable stainless steels.

Compositionally, the superalloys are nickel base with about 20% chromium, major amounts of molybdenum and/or tungsten, and negligible amounts of aluminum and titanium since very high temperature strength is not a prime goal. The low γ' former contents are required for ease in welding plate, sheet, and turbine components that are prepared in accord with the same high-quality melting and hot-working practices given high-temperature γ'-strengthened superalloys. The presence of molybdenum and/or tungsten provides some degree of solid-solution strengthening and much higher "wet" corrosion resistance to a variety of environments than with chromium alone. Commercial examples are Hastelloy B-2 (28% Mo), Hastelloy C-276 (16% Mo), Hastelloy C-22 (13% Mo), Inconel 625 (9% Mo), and Hastelloy G-30 (5% Mo and 2.5% W).

Roughly one-fifth to one-third (depending on estimating source) of U.S. production of all superalloys (50–80 million pounds a year) currently goes into these alloys for primarily corrosion-resistant applications. With further stringent demand placed on the metals, chemicals, petrochemicals, glass, paper pulp, and antipollution industries, this market fraction might rise to one-half.

For the past 10 years producers of these corrosion-resistant superalloys have been actively compiling corrosion case histories using both field and laboratory data on problems served. Selecting specific alloys to assure maximum service is best handled by direct communication with the producers.

The objective of *Superalloys II* is to treat superalloys as they are utilized by high-temperature heat engines, with emphasis on gas turbines. The increasing use pattern for lower temperature corrosion service (with corresponding changes in chemical composition emphasis) is considered a separate development. In time, another text devoted to corrosion-resistant, rather than high-temperature, superalloys may be warranted.

DEVELOPMENT OF PROCESSING

Processing always has been an even-handed partner with alloy development in superalloy technology. Processing seems to have developed in three principal stages. Superalloys were first used only in the wrought condition, as sheet or forged from bar stock and machined. Then, evolving from dental technology, it was found that

† This section was supplied by W. C. Hagel.

investment casting was very effective in creating complex shapes for hot-stage parts. Thus, in the 1940s and 1950s the first question for a new blade or vane often was "cast versus wrought?" When a wrought alloy would appear superior for a component, through design changes and perhaps a change in chemical composition, a cast alloy would emerge as stronger and more economically viable. Then the pattern would reverse, and a cast alloy would seem superior.

The advent of vacuum melting around 1950 generated a second major phase in processing technology. It benefited both forged products and castings. Vacuum melting removed undesirable alloy impurities that had hamstrung alloy advances in the thirties and forties; it permitted additional and closer control of reactive strengthening and oxidation-resistant elements; total alloy chemistry was improved, and complex cast shapes were possible (Fig. 13). Vacuum melting was the most important process development made for superalloys in the first 30 years. In fact, the invention of vacuum melting by Falih N. Darmara may well have been the most significant single development in superalloys.

Fig. 13. A contemporary investment cast aircraft engine blade (left) and industrial gas turbine bucket (right).

During the 1950s and into the 1960s alloy development flourished. "Processing" seemed to be taking a breather while the benefits of vacuum melting were absorbed. Then, by the 1960s and 1970s a virtual explosion of new process developments occurred. By the mid-1970s this third-era technological thrust surpassed alloy development as the prime forward drive to create superalloys of greater capability.

The processing "horn of plenty" created is shown in Fig. 14. It attempts to illustrate the prime avenues of activity as one step forward led to another. Perhaps two can be singled out. In the 1960s Frank VerSnyder and associates at Pratt and Whitney and TRW developed directional solidification (DS) for airfoils. This technology base generated new directionally solidified (DS) alloys, single crystals (SX), and DS eutectics. While the future of eutectics is yet unsure, cast DS and SX alloys are now a permanent part of superalloy commercial products. High-pressure turbine and materials selection in dozens of turbine lines take advantage of these strong, oxidation-resistant and thermal-fatigue-resistant alloys.

In parallel, oxide dispersion strengthening (ODS) was invented by Anders et al. at DuPont. A product of powder metallurgy technology, ODS features very fine grain structures and superplastic properties. Combined with strengthening, wrought ODS "mechanical alloys" can generate usable creep rupture strength to within 90% of melting temperatures. However, the requirement of extensive mechanical working during processing (they cannot be investment cast) limits somewhat their applicability.

When problems of phase embrittlement, hot corrosion, and general ductility limits arose, processing came to the rescue in a wave of ingenious developments.

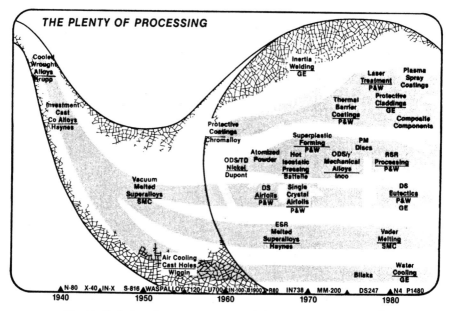

Fig. 14. Explosion of process discoveries and developments in superalloys.[7]

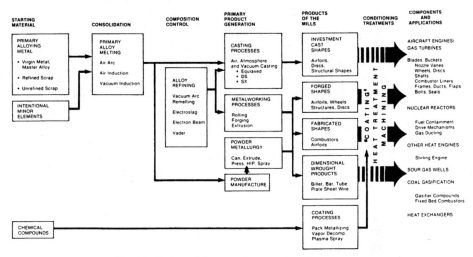

Fig. 15. A view of the superalloy manufacturing process.

They were based on improved knowledge of the science of solids, new and innovative tools to do the processing itself, and advanced study tools. Now, complex and subtle processing consists not only of "forging" and "casting" but also of filtering molten metal, controlling heat removal to make single crystals, atomization to make powder alloys, extremely rapid cooling to suppress or create new structures, and isothermal forging to control metal flow with the greatest of care.

But the pendulum keeps swinging. This surge of "new processing" has reopened opportunities for alloying. Present alloying inventions, however, are clearly of a more delicate nature than previously, such as subtle control of sub–grain boundary elements in single-crystal alloys.

All of the processes described above, of course, have some relationship to each other. Together, they create a rational industrial manufacturing pattern to make component configurations in superalloys that are bolted, welded, or fitted together as an aircraft engine, industrial gas turbine, Stirling engine, reactor components, or other. In an attempt to collate the whole, Fig. 15 gives a total view of the superalloy manufacturing process.

Summarizing, it is clear that after years of improvements in superalloys driven by alloying changes, a flood of process inventions has taken over the task of creating still better superalloys.. Interaction of both processing and alloy modification enhances the advantages of both, but in the mid-1980s we are clearly in the *age of processing*.

THE DISASTERS

Progress in creating greater mechanical capability at higher temperatures in corrosive atmospheres has been outstanding for superalloys. However, as in other technologies, such progress does not occur without setbacks and failures. By mastering such failures, the base is created for a new advance. A few words about "disasters," some mentioned previously, give perspective.

In the 1940s it was found that cobalt-base alloys (i.e., Vitallium) could suffer from overalloying with carbon and that this uncontrolled age hardening severely reduced serviceability.[12] This was corrected by more care in alloying with carbon and with the carbide-forming elements, and by control with improved heat treatments; this led to the first successful investment cast bucket alloy (X-40) for the first U.S.-developed aircraft engine.

A similar broad difficulty arose in the 1960s when chromium reduction in nickel alloys to improve creep strength led to increased oxidation and hot-corrosion attack. This resulted in significantly shortened life for alloys in many industrial gas turbine applications as well as in aircraft engines subjected to a salt-containing atmosphere. An improved balance of chromium, aluminum, and titanium together with coatings solved the problem.

The intense alloy development pressure in the 1950s so packed nickel systems with strengtheners that "disaster" came in the form of deleterious service-incurred precipitation of platelike phases. These hard σ or μ plates resulted in premature alloy cracking and reduced creep and rupture performance. Compositional control by PHACOMP (a computer program) solved that problem. (PHACOMP was probably the first direct application of solid-state electron theory to alloy science utilizing a computer-driven compositional analysis approach.) These hard phases and their control are discussed in Chapter 8.

In the 1950s a shortage of cobalt caused a shift to nickel-base alloys. The world promptly forgot the lesson, and the same thing happened in the late 1970s, when Communist forces invading the Congo triggered precipitation of a series of events more involving inefficient production, poor maintenance, and corruption than mortar shells. Enormous costs were incurred; during the 1979–1980 period this author has estimated that in the United States alone businessmen paid a *premium* of 0.5 billion dollars per year for cobalt. Furthermore, parallel shortages of other elements such as tantalum and molybdenum developed with concomitant cost effects. Now (apparently) proper attention is being paid to reducing the need for "strategic" elements, to recycling scrap, and to government stockpiles.

The general "position" of superalloys as societies' premier high-temperature air-breathing structural materials has been challenged twice. First in the 1950s an effort to develop refractory metal alloys occurred. However, it was not possible to protect the refractory alloys from surface attack, and the effort died. This approach is now beginning to recur; much has been learned in the last 25 years. Recently an extremely heavily funded effort has been mounted to develop ceramics for turbine and other high-temperature service. The vigor in this drive has never been paralleled. However, after nearly 10 years of effort there are yet no ceramic materials acceptable for engine use; recent major tests of ceramics were a disaster, and attention is turning to "ceramic composites."

Much more viably, intermetallic compounds springing from superalloys (Ni_3Al, NiAl, Ti_3Al, TiAl, and others) are now being studied. Their low density, high

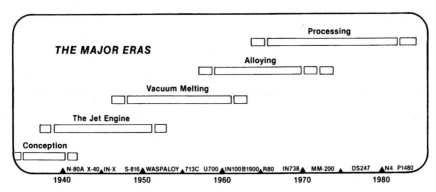

Fig. 16. An overview of the dominant technology eras through the development of superalloys.

modulus, and good oxidation resistance are provocative. Brittleness, leading to difficulties in manufacture and in service, are their limitations. Time will tell.

Thus, we see that some severe problems have arisen in superalloy technology. However, these "disasters" always have been solved by application of metallurgical and process developments that have led to still greater superalloy service capability.

SUMMARY: THE MAJOR ERAS

Figure 16 illustrates the dominant factors in superalloy development. Following superalloy conception from austenitic stainless steels, the demand of the concurrent paradigm of the jet engine obviously provided the initial technical impulse.

Without question, Damara's invention of vacuum melting in the late 1940s provided the process metallurgy leverage for superalloy composition development by introducing a new era of alloy cleanliness and alloying element freedom. This, of course, is now followed by the dominance of processing developments.

The inescapable conclusion is that superalloy metallurgists have invented and developed alloys more complex and used at higher fractions of their melting point than any other comparable group in metallurgy. Superalloys have been the indispensable materials partner to mechanical design for one of the world's foremost "high-technology" products, jet engines and gas turbines.

REFERENCES

1. C. T. Sims and W. Hagel (eds.), *The Superalloys*, Wiley, New York, 1972.
2. J. Mowill, personal communication, February 1984.
3. W. T. Griffiths, "The Problem of High-Temperature Alloys for Gas Turbines," 739th Royal Aeronautical Society Lecture, London, October 1947.
4. J. F. Hanieski, *Technol. Cult.*, **14**(4), 547 (1973).

5. R. Schlaifer and S. D. Heron, "Development of Aircraft Engines and Fuels," lecture at Harvard University, Cambridge, MA, Maxwell Reprint Co., Elmsford, NY, 1950.
6. A. Williams, "The 20th Century," in *A History of Technology*, Parts I and II, McGraw-Hill, New York, 1965.
7. C. T. Sims, *Superalloys 1984*, TMS-AIME, Warrendale, PA, 1984, p. 399.
8. N. P. Allen, "A Summary of the Development of Creep-Resisting Alloys," Symposium on High Temperature Steels and Alloys for Gas Turbines, The Iron and Steel Institute, London, July 1952.
9. C. R. Austin and H. D. Nickol, *J. Iron Steel Inst.*, **137**, 177 (1938).
10. B. Kear, J. Doherty, and A. Giamei. *J. Met.*, **23**, 59 (November 1971).
11. R. W. Fawley, in *The Superalloys*, C. Sims and W. Hagel (eds.), Wiley, New York, 1972, p. 3.
12. F. S. Badger, *J. Met.*, **10**, 512 (August 1958).

Chapter 2

Gas Turbine Design
and Superalloys

HAROLD E. MILLER and WILLIAM L. CHAMBERS

Gas Turbine Division, General Electric Company, Schenectady, New York

Superalloy development responds to the need for materials with creep and fatigue resistance at high temperatures. Historically, these needs have been most acute in aircraft jet engines and other gas turbines, although applications exist in heat exchangers and high-performance heat engines operating with other thermodynamic cycles. In this chapter, the economic implications of increased temperatures in heat engines are described to illustrate the leverage realized through increased efficiency or output permitted by the application of superalloy materials, despite their higher cost. Jet engine and industrial gas turbine high-temperature components are described together with their failure modes, and necessary material data for life calculation are provided.

HEAT ENGINES: MOTIVATION FOR DEVELOPING THE SUPERALLOYS

Among heat engines actively employed today, many are cyclic in regard to the energy content such as the Otto and Diesel cycles. The steady-flow cycles are the Rankine (steam) cycle and the Brayton (gas turbine) cycle and their variations. The Otto, Diesel, and Brayton cycles are internal combustion cycles wherein the fuel is burned in the working fluid; thus, the highest cycle temperature is not achieved through a heat transfer process. This peak temperature is, however, dependent on the material capabilities of the parts in contact with the hot fluid. In the gas turbine using the Brayton cycle, the combustion and turbine parts are in contact with a "continuously" hot working fluid, whereas in the Otto and Diesel cycles, the fluid is intermittently hot and cool. Thus, the peak temperature of the Otto and Diesel cycles can be stoichiometric, whereas the gas turbine can only approach stoichiometric temperatures as the material capabilities enable it. In this chapter, we will concentrate on the gas turbine.

In the history of the development of gas turbines, when goals were either high efficiency or high work output for a given size device, the designers moved to higher operating temperatures. This is the fundamental reason for the development and application of the superalloys and their continued improvement.

Figures 1 and 2 represent the gas turbine in its power generation (or mechanical drive) configuration and its aircraft propulsion configuration, respectively. The stations noted are

1. inlet conditions (a),
2. compressor discharge (b),
3. turbine inlet (c),
4. turbine outlet (d), and
5. jet nozzle outlet (d').

From a thermodynamic standpoint, the differences between the two are, first, that not all of the energy is extracted as shaft work in the aircraft jet engine turbine (only what is required to drive the compressor and accessories) and second that the jet engine converts the remaining energy to thrust work by accelerating the working fluid in a jet nozzle. The temperature–entropy diagrams of these two systems appear in Fig. 3.

Line ab represents the compression process. If this process were completed at perfect efficiency, its line would be vertical (constant entropy). Line bc represents the combustion process, or the heat addition portion of the cycle; it reflects a process that occurs at constant pressure. Both lines cd and cd' represent the expansion of the working fluid and energy extraction. Line cd is specifically turbine work. As

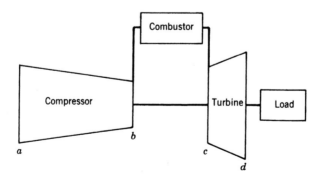

Fig. 1. Industrial gas turbine diagram.

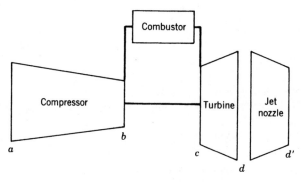

Fig. 2. Jet aircraft engine diagram.

is the case with the compressor, a vertical line would reflect 100% component efficiency. The segment dd' in the aircraft engine case is the work extracted by the nozzle and converted to thrust. Lines da and $d'a$ show the cooling process that occurs at constant pressure.

Note that the heating and cooling lines grow further apart at higher relative entropy points. Hence, the temperature difference between point c and either point d or d' is greater for otherwise identical cycles that feature higher temperatures at point c. Enthalpy, the energy in the fluid that can be converted to work, is a direct function of temperature in constant-pressure processes; therefore, the enthalpy difference between points c and d' (d for shaft drive turbines) increases as higher values of temperature at point c are reached. In other words, the further to the right the cd line is, the more energy there is to be extracted by the turbine (and jet nozzle); the higher the temperature at point c, the more power is produced by the

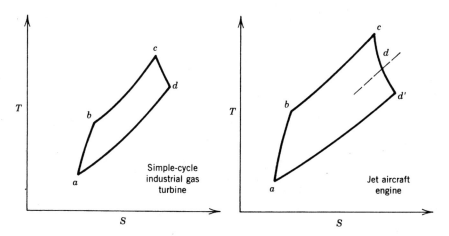

Fig. 3. Temperature–entropy diagrams for Brayton cycle engines.

gas turbine per pound per second of flow (this is called specific power). In pursuing higher thrust without increasing the weight or size of a jet engine, turbine inlet temperatures continue to be increased. In pursuing higher horsepower ratings for given sizes of industrial turbines, turbine inlet temperatures are increased. Figure 4 shows how power is related to turbine inlet temperature.

Thermal efficiency is also related to turbine inlet temperature but not as directly as specific power. In fact, it can be shown that when component inefficiencies are ignored and perfect gas with fixed specific heat is assumed, the thermal efficiency of a simple cycle gas turbine is a function only of the pressure ratio. In reality, this is only approximately the case, and thermal efficiency in a simple cycle gas turbine becomes a weak function of turbine inlet temperature. One can intuitively appreciate the benefit of higher turbine inlet temperature to thermal efficiency by noting the relative significance of the change of entropy in the compression and expansion processes to the shape of the cycle diagram as the lines *ab* and *cd* are separated. Figure 5 shows the increase in thermal efficiency due to turbine inlet temperature for one particular set of component efficiencies and other cycle parameters as noted.

Combined Cycles

A more dramatic effect of turbine inlet temperature on thermal efficiency can be observed when the Brayton cycle is combined with other cycles (e.g., Rankine)

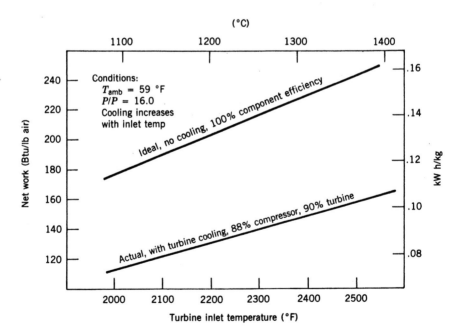

Fig. 4. Simple cycle network.

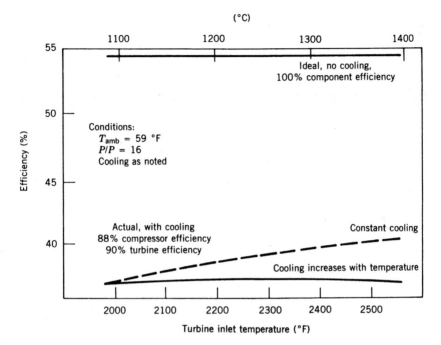

Fig. 5. Simple cycle efficiency.

that convert the heat available at point *d* to additional useful work. The primary gas turbine cannot convert the energy in this relatively high temperature air to work since there is no remaining pressure differential to drive additional turbine stages. The hot exhaust gas can be used in waste heat boilers that produce steam that is in turn expanded through steam turbines via the Rankine cycle. The added Rankine cycle becomes a bottoming cycle to the Brayton cycle. The impact on the topping cycle is limited to a slight increase in backpressure (which has minimal efficiency, power, and exit temperature effect). Increasing the temperature at point *c* will increase the temperature at point *d*, with at worst no effect on the efficiency of the topping cycle. The increase in temperature at *d* can be converted to more work by the boiler and steam turbine. The combined cycle thermal efficiency gains due to increased gas turbine inlet temperature are quite dramatic, as can be seen in Fig. 6.

Temperature Leverage

Superalloy applications continue to increase in spite of their high cost. This is due to the considerable economic leverage that results from the relatively small region of an aircraft or power plant that is exposed to the highest temperatures of the thermodynamic cycle. In a modern aircraft engine or gas turbine, for example, about 0.09 lb of superalloy is used in the combustion and turbine section per kilowatt of power produced. As can be seen from Fig. 4, if the turbine inlet temperature

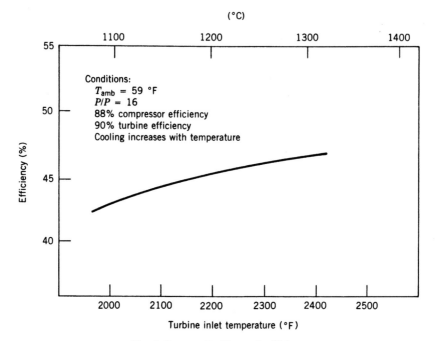

Fig. 6. Brayton–Rankine cycle efficiency.

could be increased 100 °F (56 °C), power output could be increased 4%. If a gas turbine is sold for $100/KW, an uprate of 100 °F could result in an increased revenue to the manufacturer of $4/KW of the original rating. If the 100 °F increase in turbine inlet temperature was due solely to the improvement in the hot-section material, the break-even added alloy cost would be about $44/lb.

A similar example of alloy leverage can be drawn from a case involving a combined Brayton–Rankine cycle power plant. About 2% of the cost of the turbines in such a power plant is due to the superalloys used. From a recent study of electric power generation,[1] one finds that the levelized average carrying charge (reflecting the cost of the plant) is on the order of 7.5 mils/kw h). A fuel cost of 75 mils/ kw h is reasonable. Figure 6 shows that efficiency could be increased 2.25% by increasing turbine inlet temperature 100 °F. At these fuel and plant costs, a break-even situation would exist if a plant cost increase of 22.5% permitted this temperature increase. If only hot-section material changes were required, the alloy cost for the hot-section materials could be increased 11 times.

Such leverage is anticipated to increase as the business of engine design becomes more competitive and as the concern over the scarcity of fuel resources becomes more acute.

GAS TURBINE COMPONENTS AND ENVIRONMENT

The components of the hot section of a modern aircraft engine are shown in Fig. 7. Following is a description of the combustor, rotating, and stationary blading.

Combustors and Transition Pieces

The combustor is the location of the highest gas temperatures, in excess of 3000 °F (1650 °C). The combustor shown in Fig. 7 is an annular type. The inner and outer combustor liners partially enclose an annular space concentric with the engine centerline. Compressor exit air passes through this space, mixes with the fuel, and the mixture is ignited. The fuel is introduced through nozzles at the upstream end of the combustor. Once the fuel–air mixture is ignited by a spark plug, combustion continues until the fuel is shut off. The engine thrust is adjusted principally by regulating the flow of fuel to the combustor. By the time the extremely hot gas reaches the first stage of turbine stationary airfoils (called vanes or nozzles), it has mixed with excess compressor discharge air cooling and dilution flows and enters the turbine at temperatures ranging from 1700 °F (950 °C) in first-generation gas turbines and first aircraft engines to over 2700 °F (1500 °C) in some advanced machines. The combustor shown is an annular in-line design fabricated of machined rings of superalloy material. The thicker sections that occur regularly along both the inner and outer wall contain cooling holes through which compressor discharge air is forced. The convection cooling plus the film of relatively cool air thus formed protect the combustor material from the hot gas. Differences between metal temperature and flame temperature may well exceed 1500 °F (850 °C). Thermal radiation from the flame to the cooler combustor is a significant source of heat. Thermal barrier coatings that may be applied to the inside of combustion liners and transition ducts provide an insulating layer and a reflective coating.

Earlier aircraft engines and many industrial gas turbines have multiple-"can"-type combustors. Each can-type combustion chamber is a cylinder, the walls of which are similar to the walls of the annular aircraft engine combustor. Single combustors are employed as well as various multiple-chamber arrangements. Large industrial turbines with multiple combustors have the combustors arranged in a circle about and more or less parallel to the engine centerline. The cross-sectional change from the discharge of each of the circular combustors to the annulus of the first stage of turbine blading is accomplished by transition ducts of superalloy sheet material similar to that of the combustors but less aggressively cooled since the gas here is cooler.

Throughout the history of combustor design, engineers have been dealing with large areas of thin metal in hot environments. The combustion process can produce periodic pressure variation. High-cycle fatigue (HCF) problems can result from this combination. To avoid high-cycle fatigue, combustor designers have sought to stiffen and restrain their structures to avoid low natural frequencies that could be stimulated by the combustion process.

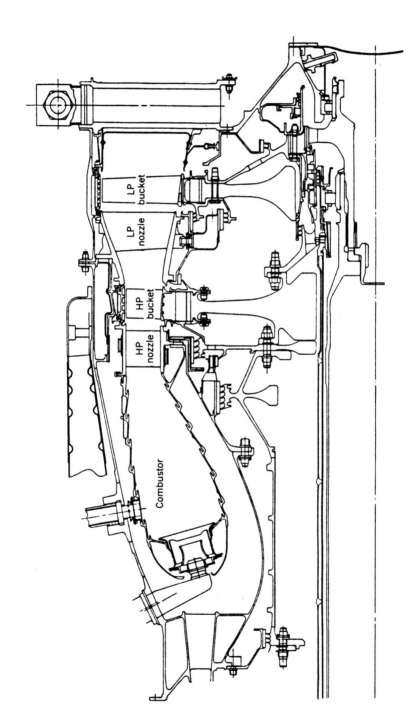

Fig. 7. Combustion and turbine section of F404 engine (Courtesy GE Co.).

Since the combustor contains the hottest gases in the turbine, it must withstand significant temperature changes caused by the starting and shutting down of the turbine. Mechanical restraints, and, in fact, cooling features, inhibit the free thermal expansion of combustor structures, resulting in thermally-induced strain and low-cycle fatigue (LCF). Combustor mechanical designers are faced with the task of balancing their design so as to feature sufficient stiffness to avoid high-cycle fatigue and yet be free enough to expand sufficiently to have acceptable low-cycle fatigue life.

Creep also is a problem. Air pressure outside the combustor is above the pressure inside. Since the surface area of the combustor and associated ductwork is relatively large, the small pressure differential acting over the large area can produce creep in thin sections. Thus, material demands for combustion and transition ducting include workability, weldability, ductility for fatigue resistance, and moderate creep strength with temperature capability to 1400–2000 °F (780–1100 °C), depending on the application. Desirable qualities also include a low coefficient of thermal expansion and wear resistance. Thermal barrier coatings must adhere well to the combustor or transition piece material, have similar expansion characteristics, low conductivity, and high reflectivity.

Nozzles

The first-stage turbine stationary blades, or nozzles, are located at the discharge of the combustor and accelerate the hot working fluid and turn it so as to enter the following rotor stage at the proper angle. The first-stage nozzles are subjected to the highest gas velocity in the engine. Here, the gas temperature is reduced from flame temperature only by mixing with that compressor discharge air introduced specifically for dilution and cooling; in subsequent stages, the temperature of the working fluid will be reduced by work extraction. Because of this environment, the first-stage nozzle metal requires aggressive cooling. The high-pressure turbine nozzle shown in Fig. 7 is a segmented assembly that is bolted to the combustor. It is cooled by a combination of convection, impingement, and film cooling.

Since the gas entering the first-stage nozzle can regularly be above the melting temperature of structural metals, cooling is a necessity. Cooling to a uniform temperature over the entire nozzle structure, although a goal of the designer, is not practical due to a variety of reasons. As a result, temperature differentials cause thermal stresses that in turn cause low-cycle fatigue and fatigue cracking.

Aerodynamic loading as well as the pressure loading of the cooling air itself can, in some instances, combine to produce creep deformation at very high metal temperatures. The deformation, by changing the shape of the nozzle, degrades the efficiency of the turbine and eventually warrants the replacement of the nozzle. Creep can lead to cracking and reduction in load-carrying capability as well. Required first-stage nozzle material capabilities include properties that combine to reduce low-cycle fatigue, low coefficient of expansion, high ductility, and high conductivity. Also desirable are creep strength, weldability (repairability), and machinability.

By virtue of their location in the engine, later-stage nozzles cannot be supported at both the inner and outer sidewall, as can the first-stage nozzle. For this reason, stresses due to aerodynamic loading are higher, and the designer must guard against creep in the outer sidewall and the vane; permanent downstream deflection at the inner diameter can result. Primary material requirements for later stages must include creep resistance as well as the requirements noted for first-stage nozzles.

Blades

The rotating blades of the turbine, or buckets, convert the kinetic energy of the hot gas exiting the nozzles to shaft horsepower used to drive the compressor and load devices. The blades consist of an airfoil section in the gas path, a dovetail joint connecting the blade to the turbine disk, and often a shank between the airfoil and the dovetail, allowing the dovetail to run at a lower temperature than the root of the airfoil. Here the shape transition between the airfoil root and the dovetail is accomplished. Mechanical damping devices are located between adjacent blade shanks. Some bucket designs feature tip shrouds that have an aerodynamic benefit and also raise natural vibratory frequencies. The blades in Fig. 7 are cooled utilizing convection and film techniques.

The total temperature of the gas relative to the blade is lower than that relative to the preceding nozzles as a result of the velocity of the airfoil relative to the gas stream and addition of cooler air into the gas path. This cooler air is that provided for cooling and the result of parasitic leakage into the gas path.

Due to rotation, the blade is subject to centrifugal stresses. The centrifugal force acting on a unit mass at the blade's midspan is 13,000–90,000 times that of gravity. Midspan airfoil centrifugal stresses range from 10,000 psi in industrial first-stage blades to 40,000 psi (28.2 kg/mm^2) at the airfoil root of aggressively cooled aircraft blades and the last stage of industrial gas turbines. Stresses of approximately 25,000 psi (17.6 kg/mm^2) occur in the last stages of aircraft engine fan turbines. The desire to extract the maximum energy from the working fluid in the turbine of industrial machines results in relatively larger annulus areas in the last stage compared with aircraft turbines. For this reason, last-stage root stresses tend to be higher in industrial machines. This combination of stress and temperature result in creep being the primary concern in the design of turbine blades. Blade material selection generally results in the application of an alloy with one of the best creep resistance capabilities.

As the blades move with the rotating turbine disk, they pass through the wake of nozzles, combustors, and struts. The periodic variation of force on the bucket that results can create high-cycle fatigue. To avoid this, the designer shapes the bucket to avoid natural frequencies in resonance with such stimuli insofar as is possible. Frequently avoiding all known stimuli over the entire operating speed range is impossible, forcing the designer to apply damping devices or to avoid only the most damaging resonance situations.

Military aircraft engines designed over the past two decades and second-generation industrial gas turbines have employed air-cooled blades in at least the first stage. Lowering of the average section temperature of the bucket airfoil increases the component's creep life. As is the case with the nozzles, it is practically impossible to cool an airfoil uniformly. Hot spots occur at the trailing and leading edges. Cold spots occur at interior points where the cooling is most active, such as on the perimeter of cooling holes and on ribs between cavities and bucket with serpentine cooling passages. The necessity for starting and stopping the engine means that thermally induced strain will vary with time, and low-cycle fatigue results. More rapid warm-up needs and greater cooling-to-gas stream temperature differentials characteristic of aircraft engines (particularly military) have led to thermomechanical fatigue as the controlling failure mode in aircraft blades. This is usually manifest as simple cracking initiating at transverse grain boundaries in equiaxed alloys. It has led to the rapid acceptance of directionally solidified and single-crystal blading in aircraft.

The air admitted to engines and fuel, particularly in helicopter engines and land-based turbines, contains levels of sodium, potassium, vanadium, and lead that enter into the sulfidation process and thus cause corrosion of blade airfoils. Concentrations as low as a few parts per million have resulted in the destruction of turbine blading in very few hours.

To solve this problem, materials development efforts continue on two fronts. Alloys are being modified to enhance resistance to attack, and coatings are being developed which can be applied to protect the bucket alloy from attack. The designers' concern does not end here, since coatings have been known to reduce fatigue resistance. In assembling a list of candidate bucket materials, only those with good hot corrosion resistance or those that can be used in conjunction with adequate coating systems are considered for industrial turbines. For aircraft engines, similar attention is paid to oxidation.

Blade material requirements include corrosion and excitation resistance or the existence of a good protective coating system and fatigue and creep strength. Also desirable are tensile strength, fatigue strength, and toughness. Castability is currently a requirement. Machining is accomplished via grinding and electrochemical and electrical discharge processes; hence, single-point machinability is not generally required.

Disks

Turbine disks, to which the blades are attached by dovetail joints, are subjected to the radial pull of the blades and body forces resulting from the centrifugal forces associated with rotation. Additional stress results from transient and steady-state temperature variations within the disks. The temperature environment of the disks is determined by the cooling and sealing air used inboard of the gas path and by whatever working fluid manages to infiltrate the spaces upstream and downstream of the disk rim. As a practical matter, the temperature is more nearly, and not significantly higher than, compressor discharge temperature. Materials suitable for application below 1200 °F (670 °C) are generally chosen. Alloy steels are commonly applied in industrial turbines; IN-718 and similar alloys are found in aircraft engines.

Disk failures result in the release of pieces from the turbine at high velocities; hence, avoidance of disk burst is the rotor designer's overriding concern. Disks have been burst by overspeed wherein the ultimate stress is approached by the average tangential stress. (Ductile failure occurs when average tangential stress reaches a large fraction of the ultimate stress; tests show values below 0.9.) Failures have occurred that are characterized by brittle fracture of the turbine disk. Good fracture toughness, low crack growth rate, and inspectability are prime requisites for turbine disk materials. A low coefficient of linear thermal expansion is desirable so as to minimize the thermal contribution to stress.

MATERIAL BEHAVIOR MODELS

The design engineer produces two types of output, the component design, represented by drawings and specifications, and a prediction of the component's performance. It is this later work that involves the greatest effort and most extensive collaboration with the materials development community. The aspect of component performance germane to this discussion is component durability. To predict component life, the engineer must determine, through calculation, test, or communication, the stress, temperature, environment chemistry, and material behavior data. Material data can be both failure mode related and non-failure mode related. The later type consists of the material properties of Young's modulus, shear modulus, Poisson's ratio, coefficient of linear expansion, conductivity, emissivity, and density. These properties are needed for stress, strain, and temperature calculation. Failure-related data consists of corrosion data, creep and creep rupture data, high- and low-cycle fatigue curves, fracture mechanics data, yield, and ultimate strength. From the discussion preceding, one finds that the time- and cycle-related failure mechanisms become those of greatest concern to the designer of combustors, blades, and nozzles.

Creep

Although many excellent books and papers have been written concerning the physics of creep, one of the more informative and useful techniques described in recent times is the deformation mechanism map concept developed by Ashby.[2] There are six recognized independent ways that a polycrystalline material can be deformed and still retain its structure. First, there is the defect-free flow that occurs when the theoretical shear strength is exceeded. The other five require defects. Dislocations are the source for two forms of plastic flow, dislocation glide and dislocation creep. Point defect movement causes two independent kinds of different flow, through grains and around grain boundaries. The sixth mechanism is caused by twining and is generally of little importance to the engineers. Fig. 8 demonstrates the mechanism fields for pure nickel. It succinctly summarizes the concept. The mechanism fields are plotted on a map using normalized tensile stress (stress divided by shear modulus) versus homologous temperature (temperature divided by the melting point). The various field boundaries are obtained by equating the constitutive equation for each

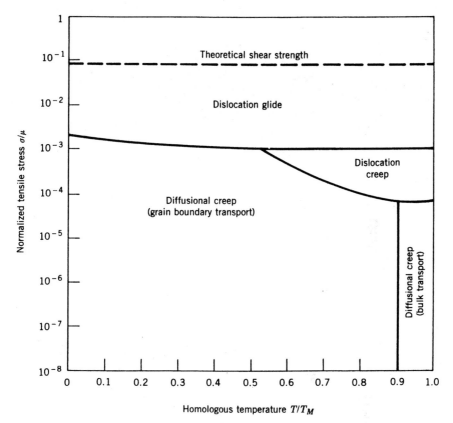

Fig. 8. Mechanism field for pure nickel with 32 μm grain size (adapted from ref. 2).

mechanism. Fig. 9 demonstrates the power of the concept for MAR-M 200, a turbine blade material, as taken from Gittus.[3] In this example, minimum strain rates are plotted as parameters with the strengthening effects of grain size clearly demonstrated. A stress–temperature region of turbine operation is also shown. For a complete discussion of the physics of creep and the deformation mechanism map concept see Ashby[2] and Gittus.[3]

Creep Models

The parametric method is probably the most practical means of communicating material data. In 1952, Larson and Miller[4] first introduced the concept of a time–temperature parameter of the form $T(C + \log t) = \text{const}$ for a given stress. Here T is absolute temperature, t is time, and C is a material constant. This parameter is used to express stress as a function of temperature and time to rupture or some creep strain of interest. The expression is valuable since it yields a single curve for

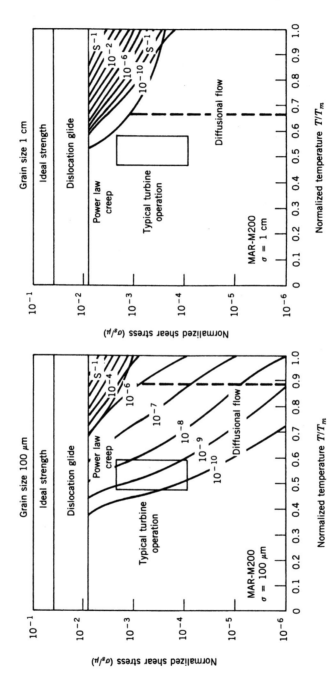

Fig. 9. Maps of Mar-M200 With grain sizes of 100 μm and 1 cm (adapted from ref. 2).

the characteristic strain level or rupture life. Such a curve, often referred to as the "master curve," is generally created by using data from short-term tests at elevated-temperature combinations for extended service lives. This sounds simple enough, although it can be dangerous to use such extrapolations as design tools. The use of short-term data is not generally justified if one is to believe Ashby's deformation mechanism maps. Short-term tests at elevated temperatures will be conducted in a range of temperature where a dominating creep mechanism is different from that operating over the life of a long-life component. This was not obvious when Larson and Miller published, but within a year Manson and Haferd[5] published results of a study based on data from a number of well-characterized materials that demonstrated the Larson–Miller form was inadequate for accurate long-range extrapolation and that a different parametric form was needed to account for the nonlinearity observed in the data when plotted as log t versus $1/T$ for constant stress.

Others found parametric forms that better described data and were considered more appropriate for their particular needs. Some prominent examples are Manson and Succop,[6] Orr, Sherby, and Dorn,[7] and Goldhoff and Hahn.[8] The behavior of some of those models are demonstrated by schematics in Fig. 10. The equations mentioned are summarized in Table 1.

Special forms evolved almost two decades after Larson and Miller first published. Of particular note is one discussed by Manson and Ensign[9]:

$$\log t + AP \log t + P = G \qquad (1)$$

where

$$
\begin{aligned}
P &= \text{function of temperature} \\
G &= \text{function of stress} \\
A &= \text{material constant} \\
t &= \text{time}
\end{aligned}
$$

Equation (1) has the particular property that if the material behaves in the manner peculiar to the special forms in Table 1, equation (1) defaults into that form.

The deformation mechanism map concept by Ashby and the multiplicity of time–temperature parameters should make it clear to the practicing engineer the possible limitations of the time–temperature parametric forms for extrapolation. One can assume that much data collected by means of elevated temperature and stress tests are out of the operating and design condition range of a component designed for long operation life. The problem is that such data are obtained subjected to a creep mechanism that the operating component will not experience. It is not safe to assume that extrapolations based on such data are conservative. The minimum-commitment method approach by Manson resulting in equation (1) and similar approaches of others[9] are attempts to make an equation sufficiently general such that for a given stress, time, and temperature combination, the state of strain can be accurately predicted.

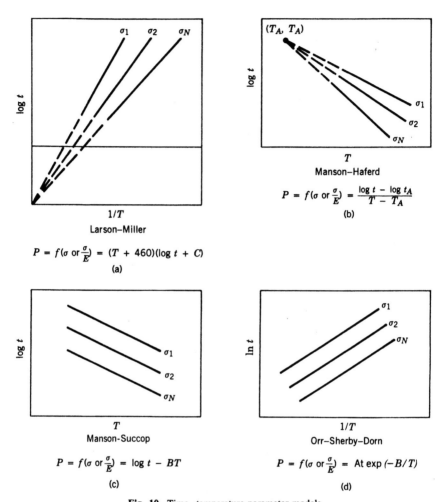

Fig. 10. Time–temperature parameter models.

Fatigue

Fatigue (discussed in detail in Chapter 10) is the decreased resistance of a material to repeated cycles of stress or strain. The terms low-cycle fatigue and high-cycle fatigue have been introduced in preceding paragraphs. Engineers have created these terms and differentiated between them by number of cycles, by stress level relative to yield, and by fracture appearance. From the point of view of the turbine component designer, a more practical demarcation is that low-cycle fatigue results from (usually) thermomechanical strains applied once per start or once per load change; high-cycle fatigue results from strains applied alternately once or more per revolution of the rotor.

Low-Cycle Fatigue

To calculate low-cycle fatigue life, the designer of turbine components needs a model of material behavior that relates salient conditions to the number of operating cycles withstood without rendering a component useless. The first of such models developed from solid-state physics yielded extremely optimistic results when compared with tests. To improve agreement, the concept of everpresent microcracks and a material property defined as surface energy per unit crack area was developed and applied generally to fracture by Griffith[10] and born out in experiments with brittle materials. Although this foundation of fracture mechanics was laid down in 1920, most fatigue life estimation done since has been based on empirical relationships between cyclic loading magnitude and cycles to failure for particular materials of interest.

Early fatigue data took the form of $S-N$ (number of cycles to failure as a function of the level of alternating stress) curves for specimen separation at constant test temperature. Such curves served well enough in predicting the cycles to separation of other specimens subjected to uniform stress at uniform temperatures. Improvements followed that considered pseudostress (the stress calculated using a perfectly elastic material deformation model) rather than actual stress, plastic deformation, and total strain as independent variables. The effect of mean stress was considered, and data became available that showed a line on the $S-N$ plot for crack initiation rather than specimen separation.

A bilinear logarithmic relationship between plastic strain reversal and cycles to failure was observed in the early 1950s independently by Manson and Coffin.[11,12] An approach for generating $S-N$ curves with a minimum of data was proposed later by Manson.[13] The independent variable chosen here was total strain range. This approach, called the method for universal slopes, combines the plastic Coffin–Mason function form with the elastic Basquin function form to describe the total strain-versus-cycles-to-failure plot. Often, material selections and preliminary life calculations are based on behavior models thus developed.

Table 1. Popular Time–Temperature Parameters for Representing Creep Rupture Data

Parameter Name	Equation
Larson–Miller	$T(C + \log t)$
Manson–Haferd	$(T - T_A)/(\log t - \log t_A)$
Orr–Sherby–Dorn	$t \exp(-\Delta H/RT)$
Goldhoff–Sherby	$(\log t - \log t_A)/(1/T - 1/T_A)$
Manson–Succop	$\log t - BT$

where

T	=	temperature
B, C, T_A, t_A	=	material constants
R	=	universal gas constant
t	=	time

The Coffin–Manson fatigue model and the universal slopes method developed by Manson were based largely on low-strength, high-ductility metals. Nickel-based superalloys used in blades are high-strength, low-ductility alloys and are employed at high temperatures where under thermomechanical loads they must resist creep and fatigue in addition to chemically hostile environments. In these situations, the blind use of these models to estimate fatigue life is not recommended: actual, relevant, LCF data must be employed.

Relating solid-state physics and laboratory test data to gas turbine component life prediction is complicated by numerous factors. Consider one location where low-cycle fatigue cracks form, the leading edge of a cooled turbine blade at a location remote from the root and tip. A simple engine mission consisting of a start, acceleration, loading, unloading, and shutdown will have a turbine inlet temperature and engine speed history (Fig. 11), creating a corresponding temperature history shown in Fig. 12. The three histories shown are significant, since for simple cooling systems it can be shown that the spanwise strain at the surface of an airfoil is a function of the difference between the temperature at the location of interest and the airfoil cross-sectional average temperature. This is not the case for thin-walled airfoils, but it is instructive insofar as local strain is a strong function of some pair of temperatures that typically respond differently to rates of change of environment temperature.

Figure 13 shows a strain history for a blade leading edge. Figure 14 shows the strain and temperature cross-plotted. On the axes of Fig. 14 a typical laboratory test would be plotted as a vertical line. The lack of even a passing resemblance between the actual strain–temperature cycle and that of the test specimens should concern the designer and those responsible for alloy characterization as well.

Fortunately, the work of Ostergren,[14] Russell,[15] and others has made significant steps toward correlating uniaxial, isothermal fatigue data with real and idealized engine component cycles. Aspects of the stress or strain versus temperature loop that have been studied for correlation are total strain range, maximum stress level, stress level at steady-state running conditions, time at steady-state conditions, temperatures corresponding to maximum strain, maximum temperature, and others. Although correlating methodologies have been proposed, their proponents uniformly caution against casual application. Correlations have been quite good for particular bucket locations and particular engine cycles but depend on correctly identifying the particular micromechanism active in the fatigue of a particular alloy and cycle. Accordingly, the state of low-cycle fatigue crack initiation life prediction for turbine blading now is that a significant improvement in accuracy is still achieved by modeling the actual component local strain and temperature histories in a laboratory specimen having a geometry similar to the component of interest.

Current life prediction activity also depends heavily on fracture mechanics techniques, especially when dealing with the higher strength superalloys as used in blades. Standard crack initiation, crack growth, and critical crack testing are performed on alloys of interest. The results can be applied over regions of varying stress levels and in three-dimensional solids. The superimposition of creep and environmental effects complicates fracture mechanics applications as it does conventional crack

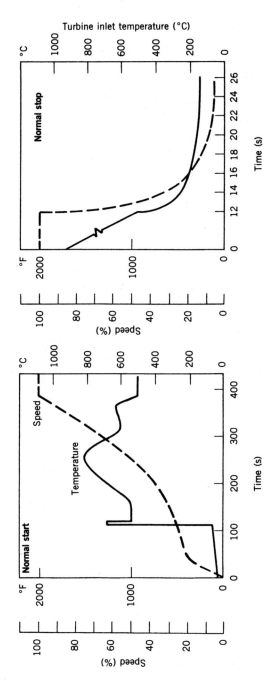

Fig. 11. Normal start and shutdown characteristics of an industrial gas turbine. (Temperatures are for turbine inlet.)

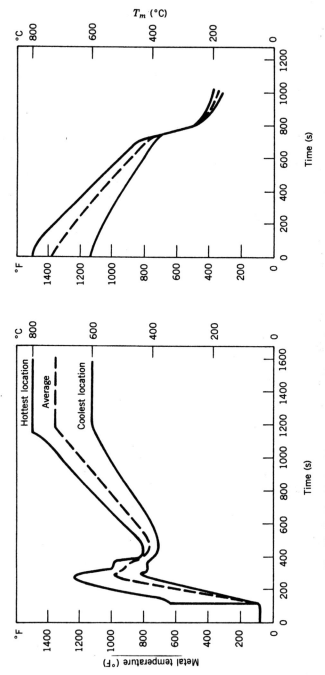

Fig. 12. Turbine blade midspan temperature histories for start-up and shutdown cycle. (Temperatures are average metal temperatures.)

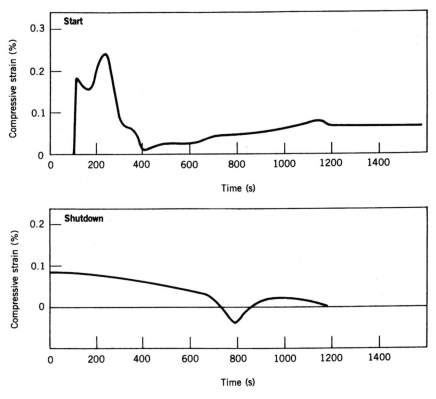

Fig. 13. Strain history of a blade leading edge.

initiation models. Fracture mechanics techniques are also applied to high-cycle fatigue and will be discussed later.

Isothermal fatigue test results continue to be used to rank candidate alloys, but life prediction must involve further testing, the modeling of the component, the duty cycle, and the environment.

High-Cycle Fatigue

High-cycle fatigue was first generally recognized in turbines in the early 1920s. Since a high-cycle fatigue in blades and combustion components invariably involves a vibratory resonance situation, it is the designer's first task to determine the natural frequencies of components, particularly blades and combustors, and the second to determine stimuli and damping and then to calculate the resulting stresses. Because of the complex shapes of combustion components and turbine blades, the calculation of frequencies is not simple. For blades, complex beam theory computer programs or finite-element analyses are employed for calculation of frequencies and mode shapes and for calculating local stress per unit damping and stimulus. Besides the

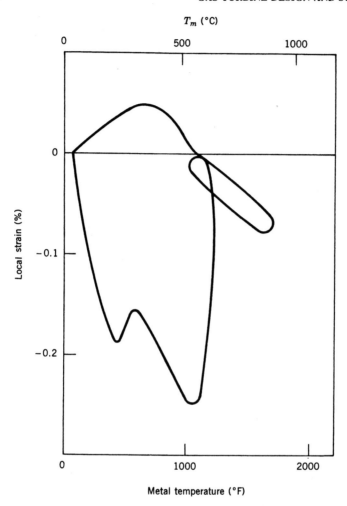

Fig. 14. Strain temperature plot for blade leading edge.

requirements associated with temperature calculation, the designer needs to know the material density, Young's modulus, and Poisson's ratio. The damping significant to the calculation is provided by mechanical features such as bars or pins that bear against the shank portions of adjacent blades in a stage. Their effect is determined by test.

In steam turbines, per-revolution stimulus can be very severe due to the employment of partial arc steam admission. This is an economic necessity in large steam turbines but is not used in gas turbines. Stimulus in gas turbine engines comes from the aerodynamic wakes of upstream nozzles and struts, from the backpressure of downstream nozzles and struts, and from the discrete ("can"-type) combustion chambers when used.

The result of frequency calculation and investigation of stimuli is generally displayed in a Campbell diagram (Fig. 15). The designer endeavors to avoid resonance with a known stimulus at speeds where the turbine is expected to run for long periods. This is not always possible. When faced with the choice, the designer chooses to avoid these resonances that this stress analysis indicates produce the highest relative stresses. Damping features are designed and provided to mitigate the effect of the modes of greatest concern. In any event, alternating stresses are calculated or measured for each mode.

Early research revealed the importance of the superimposed mean stress on high-cycle fatigue life. Material data are commonly displayed in what has become known as a modified Goodman diagram (Fig. 16). Component life is determined by the placement of the calculated mean and alternating stress on the diagram and interpolating the number of cycles to failure.

High-cycle fatigue analysis involves the application of fracture mechanics in establishing allowable defect sizes for components. In low-cycle fatigue situations

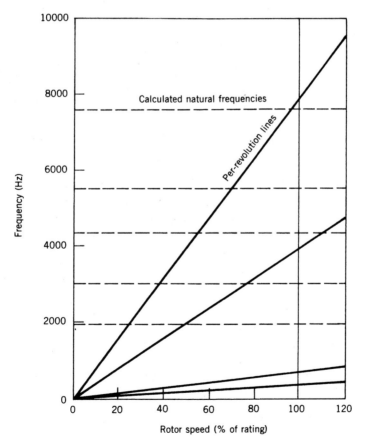

Fig. 15. Campbell diagram of a turbine blade.

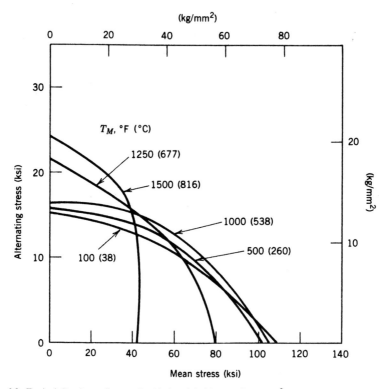

Fig. 16. Typical Goodman diagram for blade (nickel-base) alloy at 10^9 cycles (adapted from ref. 16).

fracture mechanics techniques can be used to predict the growth of a crack and allow the establishment of inspection or retirement intervals. In these situations the useful life of a part may go beyond the time to crack initiation. In high-cycle fatigue, however, any crack that will propagate at all under the stress reversals associated with the number of revolutions of the engine can be assumed to grow to fracture in a short time.

Allowable Flaw Size

The procedure for establishing allowable defect size is begun by determining the size of defect that cannot be grown by the anticipated level of vibratory stress. This defect size is then taken to be the upper limit for the growth of a crack in the number of engine start cycles required by the turbine specifications. By determining what initial flaw size will grow to this size in the given life of the turbine, one sets the allowable initial flaw size. The task does not stop here, however. Statistical methods and an understanding of nondestructive inspection techniques are employed to adjust the acceptable initial flaw specification to ensure a low rate of acceptance of unacceptable flaws.

Corrosion

Corrosion is one of the major reasons for the replacement of industrial turbine blades. Corrosion contributes to failures driven by other primary mechanisms, mainly LCF, as well as causing failures alone. Corrosion failures occur when the aerodynamic shape or tip clearance of the gas path component is changed to the point where turbine efficiency and output deterioration necessitate the component's replacement, when the cooling circuit is breached, or when the load-carrying capability of the component is compromised. Blades have historically been more susceptible to hot corrosion than nozzles and combustion components due to general use of nickel-base alloys. In some applications, the corrosion life of blades is reached when substantial creep and fatigue life remain. The exact mechanism of hot corrosion continues to be debated (Chapter 12), but it is generally agreed that the alkali salt Na_2SO_4 or a similar compound is a prerequisite. It is formed from sodium (potassium behaves similarly) in the fuel or borne by the inlet air and from sulfur, which is in ample supply in fuels and in the air. The oxides of vanadium and the sulfates or oxides of lead are other species that cause corrosion.

Controlling sulfur has not been effective in controlling corrosion, but by careful fuel control, inlet filtration, and the use of inhibitors, corrosion can be reduced. Alloys have been developed and modified to improve corrosion resistance, and coatings are now used on aircraft and industrial turbines which significantly increase blade life (Fig. 17). Blade configuration can be varied to improve corrosion life as well. The wall thickness remaining between the bucket exterior and a cooling feature can be optimized to result in a best combination of surface temperature and allowable wastage.[16] To perform such optimization and to predict component life, an accurate model of corrosion progression is necessary.

Until recently, a simple model of corrosion wastage was thought to exist wherein corrosion rate (mils of metal lost from a surface per hour) was considered an increasing function of the amount of sodium, potassium, vanadium, or lead present in the environment (from fuel and air) and the temperature of the gas path metal surface. Two observations changed this. First, instances of aggressive attack have been noted in cooled portions of blades and nozzles, and second, corrosion pitting was observed on last-stage blades operating at temperatures below that considered the threshold temperature for hot corrosion. Currently, cooling is thought to accelerate corrosion by condensing molten salts on the airfoil surfaces. The phenomenon noted on last-stage blades has given rise to a study of "low-temperature hot corrosion," which is due to a different mechanism than the classic hot corrosion noted at metal temperatures over 1600 °F (900 °C). Liquids that form the combination of Na_2SO_4 and the sulfates of nickel and cobalt which are stable at temperatures below 1400 °F (800 °C) are believed to cause low-temperature hot corrosion.

A useful model of corrosion must include several independent variables: metal surface temperature, gas temperature, coating chemistry and thickness, allowable wastage, metal chemistry, and contaminant level in the combustion products. Such models are developed by turbine manufacturers through elaborate series of material tests wherein these independent variables are examined for effect on corrosion. Since the application is largely design specific and the expense involved in acquiring

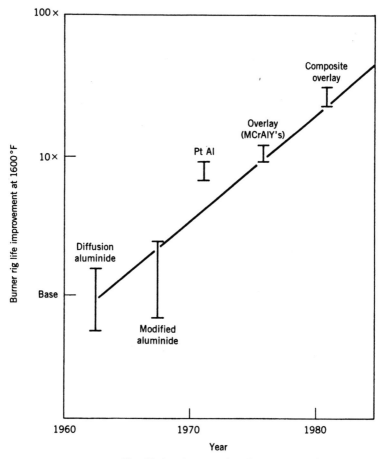

Fig. 17. Development of coatings.

data is high, the models are proprietary. Accurate models take the form of volumes of numerical data or sets of curves of wastage versus time for specific combinations of independent variables.

MULTIAXIAL STRESS STATES AND ANISOTROPY

In cooled nozzles and blades particularly, the state of stress at critical locations is more complex than in creep and fatigue specimens. Generally speaking, only uniaxial data are available, and the component designer must predict service life for the biaxial or triaxial stress state with uniaxial data. Stress analysis techniques for complex configurations are becoming commonplace; hence, determining stress state and magnitude is less a problem than finding an accurate material behavior model.

Multiaxial Stress State of Cooled Airfoils

The temperature-related stress states of leading edges of blades with concentric circular cooling passages have been analyzed using a closed-form technique.[17] The state of stress at the outer surface stagnation point determined from this analysis is biaxial, with the surfacewise stress (sigma-s), and the spanwise stress (sigma-z) normal to each other. The mutually perpendicular stress normal to the surface (sigma-n) is zero (Fig. 18). The spanwise and surfacewise stresses are principal stresses and are equal in magnitude and sense. Furthermore, upon computing the Huber–Mises–Hencky effective stress, we find it equal to the magnitude of either of the two nonzero principal stresses. Superimposed on the stress resulting from the temperature gradient is that due to centrifugal force. Most cooled nozzles and combustion liners and blades having film, impingement, and serpentine cooling systems have more complicated states of stress that require sophisticated techniques to assess accurately.

In practice, designers apply uniaxial creep and fatigue data to components with complex stress states but not without experimental support. For example, airfoil out-of-plane stress in turbine blades and nozzles is used for creep life calculation.[17]

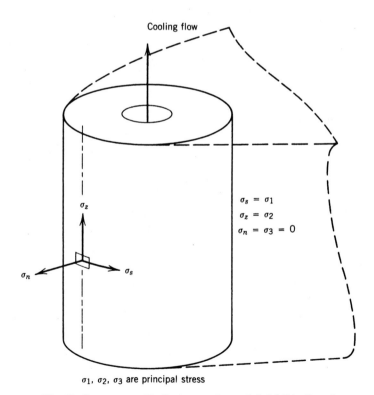

Cooling flow

$$\sigma_s = \sigma_1$$
$$\sigma_z = \sigma_2$$
$$\sigma_n = \sigma_3 = 0$$

σ_1, σ_2, σ_3 are principal stress

Fig. 18. Stress state on idealized convection-cooled airfoil leading edge.

The justifying assumption is that the in-plane stresses are entirely due to temperature rather than load and will gradually relax. Out-of-plane local stress is used in calculating the instantaneous out-of-plane local creep strain rate, which is in turn used to compute stress redistribution, a new local creep strain rate, and so on, eventually resulting in a life calculation of the component. Field experience provides the significant body of justifying data.

The prediction of crack initiation in low-cycle or high-cycle fatigue is thought to be best related to effective stress. In actual components, the magnitude of principal stress is not significantly different from that of effective stress; hence, the verification of the merit of using effective rather than principal stresses is not complete. The orientation of cracks is normal to the maximum tensile principal stress. Crack growth is generally agreed to be related to the maximum reversing tensile principal stress. Alloys not exhibiting such behavior require revised analytical procedures and data models.

Anisotropic Materials

Over the past two decades directionally solidified and single-crystal components have been under development and are now employed in turbines (Chapter 12). These materials are characterized by a lack of grain boundaries perpendicular to the direction of loading and favorable grain orientation. Such components can be found in the latest aircraft engine designs and have the potential for life improvements of nearly 10 : 1 with comparable fatigue life improvements.

These anisotropic materials require special analysis. Blade natural frequencies must be calculated with analytical tools that consider anisotropy. Finite-element codes regularly can be used for frequency and stress calculation for materials with different moduli in three orthogonal directions.

Grain orientation is important also to fatigue life calculation. A predominantly γ' alloy such as a nickel-base superalloy has the best fatigue properties in the $\langle 100 \rangle$ and $\langle 010 \rangle$ directions. In a directionally solidified bucket, this axis is oriented radially, in line with the maximum net tensile principal stress. Fortunately, the other nonzero principal stress is in the plane of the $\langle 100 \rangle$ and $\langle 010 \rangle$ directions, which are equally strong. The weak $\langle 011 \rangle$ and similar directions are oriented with lower tensile stresses. Orientation of the $\langle 100 \rangle$ and $\langle 010 \rangle$ axes with the airfoil cross-sectional axes could become a significant parameter in determining fatigue life.

STATISTICAL REPRESENTATION OF MATERIAL BEHAVIOR

It is to the mutual benefit of the turbine manufacturer and user to be able accurately to predict the life of hot-section components. Failures that require the shutting down of the turbine are second in significance only to those resulting in the escape of failed components or fragments from the engine. Among such failures are blade creep and fatigue failures wherein a significant portion of the blade comes adrift in the gas path, in turn causing the failure of other gas path components. The prediction of such failures is the prediction of the failure of the weakest blade in a stage.

Creep and low-cycle fatigue failures appear to be normally distributed when the histograms are plotted against time or cycles on a logarithmic scale, as indicated in Fig. 19. The characteristics of the lognormal distribution are such that 0.15% of the population will fail before the time or cycles on the population passes the −3 standard deviation point. A hypothetical engine containing 300 blades, all designed to the same life, has about an even chance of containing a blade with worse than −3 sigma properties. Hence, −3 sigma creep life is a good estimator of average engine creep life. Similarly, −3 sigma fatigue life approximates average engine fatigue life.

An acceptable failure density in the turbine's life is typically less than 50%; values like 1% are more appropriate. Fig. 20 shows the histogram for material failure superimposed on that for turbine failure for a machine containing 300 blades. For the purpose of this example variables in operating environments are not considered. They would widen the shape of the bell curves. In this example the 1% machine failure density point is separated from the average material failure point by about four standard deviations. What this means to the turbine bucket designer is that material data are required that allow the life calculation of this worst blade in 100 turbines (30,000 blades). Even if this is not carried out rigorously in all design practices, a similar logic is present. It may be the custom to design with −3 sigma material properties and a temperature margin representing a combination of worst cases. In any event, the magnitude of the standard deviation is important.

Consider creep data expressed with a standard deviation of a magnitude equivalent to 40 °F (or 2.4 : 1 in life, or 10% in stress, they are approximately equivalent). The design temperature limit would be perhaps 160 °F (4 sigma) below the temperature where an average specimen would fail at the design stress in the design life. Alloys

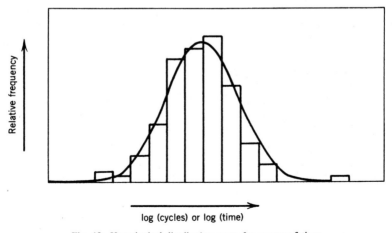

Fig. 19. Hypothetical distribution curve for creep or fatigue.

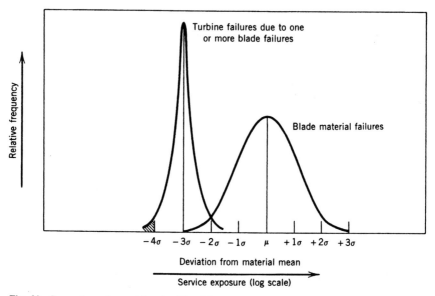

Fig. 20. Comparison of material and turbine failure distributions for hypothetical 300-blade turbine.

have been developed and adopted that have promised a 30–50 °F improvement in average creep capability. A 25% improvement in central tendency would allow an identical uprate.

SUMMARY

Manufacturers of heat engines are driven by the economics of the marketplace to increase the ratio of the value of their product to cost. Although the cost of superalloy development, characterization, and production may be high, the value added to the engines in which they are implemented is disproportionately higher. Designers will continue to be required to raise the stresses and metal temperatures of combustor and turbine components. The ultimate users of engines will expect no decrease in reliability and no increase in maintenance costs. In addition to the demand for higher temperature capability in superalloys, there is an increasing challenge to characterize more accurately alloy behavior. Then the best use can be made of today's superalloys, and with similar data, the future application of material and metallurgical advances is assured.

REFERENCES

1. R. Zeren, "Technical Assessment Guide," EPRI P-2410-SR, EPRI, Palo Alto, CA, May 1982.
2. M. Ashby, *Acta Met.*, **20**, 887 (July, 1972).

3. J. Gittus, *Creep, Viscoelasticity and Creep Fracture in Solids*, Halsted, Wiley, New York 1975.

4. F. Larson and J. Miller, *Trans. ASME*, **74**, 765 (1952).

5. S. Manson and A. Haferd, "A Linear Time Relationship for Extrapolation of Creep and Stress-Rupture Data," NACA TN 2890 1953.

6. S. Manson and G. Soccup, "Stress Rupture Properties of Inconel 700 and Correlation on the Basis of Several Time Temperature Parameters," ASTM STP 174, 1956.

7. R. Orr et al., *Trans. ASM*, **46**, 113 (1953).

8. R. Goldhoff and G. Hahn, "Correlation and Extrapolation of Creep Rupture Data of Several Steels and Superalloys Using Time-Temperature Parameters," ASM Publications, D-8-100, ASM, Metals Park, 199, 1968.

9. S. Manson and C. Ensign, *Trans. ASME, J. Eng. Mat. Technol.*, Goldhoff Issue: Materials Elevated Temperatures (October 1979).

10. A. Griffith, *Philos. Trans. Roy. Soc. Lond.*, **221A**, 163 (1920).

11. S. Manson, "Behavior of Materials Under Conditions of Thermal Stress," NACA TN-1933, 1953.

12. L. Coffin, *Trans. ASME*, **76**, 931 (August 1954).

13. S. Manson, *Int. J. Fract. Mechan.*, **2**(1) (March 1966).

14. W. Ostergren, "Correlation of Hold Time Effects of Elevated Temperature Modified Damage Function, 1978 ASME-MPC Symposium on Creep-Fatigue Interaction, MPC-3, ASME, p. 1979 (1976).

15. E. Russell, in *Proceedings of the Minnowbrook Conference on Life Prediction for High Temperature Turbine Materials, Blue Mountain Lake, NY, October 1985*, to be published by EPRI, Palo Alto, CA.

16. R. Kunkel, "High Reliability Gas Turbine Combined-Cycle Development Program: Phase 1", AP-1681, Vol. 1 (January, 1981), p. 1187.

17. S. S. Manson, *Thermal Stress and Low-Cycle Fatigue*, McGraw-Hill, New York, 1966.

PART TWO

BASIC ALLOY SYSTEMS

Chapter 3

Fundamentals of Strengthening

NORMAN S. STOLOFF

Rensselaer Polytechnic Institute, Troy, New York

The purpose of this chapter is to review the status of our current understanding of mechanisms contributing to the strength of austenitic superalloys. To approach this task, we shall discuss the mechanisms of strengthening of the matrix–austenite phase as well as the means by which precipitated phases, usually γ' [Ni$_3$(Al,Ti)] but sometimes η (Ni$_3$Ti) or γ'' [Ni$_3$(Cb,Al,Ti)], contribute to strength and creep or fatigue resistance. Under certain circumstances iron–nickel-base and cobalt-base alloys are hardened by precipitation of either carbides or intermetallic compounds; however, since the most striking hardening effects are achieved in nickel-base alloys, most of our discussion is concerned with the latter. Dispersion strengthening by hard, incoherent particles such as oxides is included in this review; for a detailed account of the preparation of dispersion strengthened alloys, see Chapter 17. Factors affecting fatigue resistance are discussed in Chapter 10.

Commercial cobalt-, iron-, and nickel-base superalloys always contain substantial alloying additions in solid solution to provide strength, creep resistance, or resistance to surface degradation. In addition, the nickel-base alloys contain elements that, after suitable heat treatment or thermo-mechanical processing, result in the formation of small coherent particles of an intermetallic compound. Therefore, typical nickel-base superalloys are variations of an austenitic nickel–chromium–tungsten (or molybdenum) matrix, further hardened by coherent particles of γ' (Ni$_3$Al,Ti) with optional additions of cobalt, columbium, tantalum, zirconium, boron, hafnium, carbon, and iron. The newer single-crystal superalloys do not require grain boundary strengthening elements so that boron, carbon, zirconium, and hafnium are eliminated.

Most alloying elements partition to some degree between both the matrix and the precipitate so that both primary phases generally are highly alloyed. Since the complexity of the alloy compositions is bound to produce a complicated pattern of hardening behavior, there has been a proliferation of theories to account for the high strength thus developed.

The strength of commercial superalloys arises from a combination of hardening mechanisms, including contributions from solid-solution elements, particles, and grain boundaries. In addition, thermomechanical processing is sometimes utilized to provide strengthening through increased dislocation density and the development of a dislocation substructure. Certain alloys benefit also from composite strengthening (e.g., wire-reinforced superalloys and directionally solidified eutectics). In general, strengthening mechanisms are considered to be independent and additive, although there is considerable controversy as to the means of superposing hardening mechanisms. For the purpose of this chapter, we will treat hardening mechanisms as essentially independent. We shall begin with consideration of models for low-temperature, short-time strengthening and then discuss factors affecting creep behavior.

TENSILE BEHAVIOR

Solid-Solution Strengthening

In the case of solid-solution strengthening it is convenient to discuss several theories of yielding in terms of the effects of solutes on various physical or crystallographic properties, for example, lattice parameter and elastic modulus.

Size Misfit

Mott and Nabarro[1] concluded that both solid-solution hardening and precipitation hardening can be accounted for by internal strains generated by inserting either solute atoms or particles in an elastic matrix. According to their model, the yield stress τ for a dilute solid solution is given by

$$\tau = 2G\epsilon c \qquad (1)$$

where G is the shear modulus, ϵ the misfit, and c the concentration of solute atoms. The misfit is produced by the difference Δa between the lattice parameter a_0 of the pure matrix and a, the lattice parameter of the solute atom:

$$\epsilon = \frac{1}{c} \frac{\Delta a}{a_0} \qquad (2)$$

Unfortunately, equation (1) usually overestimates the strengthening obtained with solutes.

While a linear relation between flow stress and lattice parameter change is obeyed for any single solute element in nickel (Fig. 1), Pelloux and Grant[2] have shown

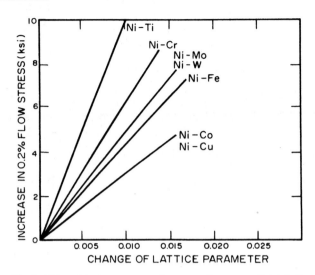

Fig. 1. Effect of lattice parameter changes on flow stress of nickel alloys.[2]

that the change in yield stress for various solutes in nickel is not a single-valued function of the lattice parameter but depends directly on the position of the solute in the periodic table. The term N_v is the number of electron vacancies in the third shell of the first long period. For the same lattice strains the larger the valency difference between solute and solvent, the greater the hardening (Fig. 2). The strengthening influence of alloying elements persists to temperatures at least as high as 1500 °F (815 °C).[2] Fleischer[4] suggests that valency effects may be explained by modulus differences between the various alloys, as discussed in the following section. Alternatively, the effects of valency may be felt through the decrease in stacking fault energy (SFE) of face-centered-cubic (FCC) alloys with increasing electron atom ratio. Beeston et al.[5] have correlated the electron vacancy number N_v with SFE.

The solid-solution elements typically found in the γ phase are likely to include aluminum, iron, titanium, chromium, tungsten, cobalt, and molybdenum. The difference in atomic diameter from that of nickel varies from +1% for cobalt to +13% for tungsten. Decker[3] has shown that for an austenitic phase of the composition shown in Table 1, the strengthening taken from Fig. 1 would be most potent for aluminum, tungsten, molybdenum, and chromium and least effective for cobalt, iron, vanadium, and titanium. The amount of solute, particularly tungsten or molybdenum, added to austenite for strengthening by lattice misfit is generally limited, however, by the instability of the alloy to σ-phase formation (see Chapter 8).

Modulus Misfit

Fleischer's suggestion[4,6] that modulus differences between solute and solvent may give rise to strengthening is based on the argument that extra work is needed to

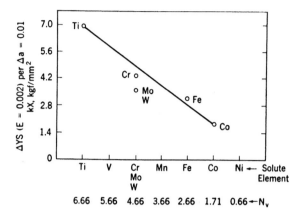

Fig. 2. Effect of valency difference on hardening of nickel alloys. N_v is electron vacancy number of solute.[2,3]

force a dislocation through hard or soft regions in the matrix. The total interaction energy with a screw dislocation is

$$E_G = \frac{G\epsilon_G' b^2 R^3}{6\pi^2 r^2} \tag{3}$$

where

$$\epsilon_G' = \frac{\epsilon_G}{1 + |\epsilon_G|/2} \qquad \epsilon_G = \frac{1}{G}\frac{dG}{dc} \tag{4}$$

Combined Effects of Size and Modulus Misfit

Fleischer[4] concluded that modulus differences and lattice misfit can be incorporated into a single equation, such that the interaction force between solute and dislocation is

$$F = \frac{Gb^2}{120} |\epsilon_G' - \alpha\epsilon| \tag{5}$$

where $\alpha = \pm 16$ for edge dislocations and $\alpha = 3$ for screw dislocations. The sign of the misfit interaction for edge dislocations is negative for solute atoms in the compressive strain field above the slip plane and is positive for solutes in the dilatation field below the slip plane. The smaller interaction force can be excluded.[7]

If L is the mean distance between two solute atoms touched by a dislocation under a stress τ_c, then

$$F = \tau_c bL \tag{6}$$

Friedel[8] estimates $L \sim (6b^3/\tau_c c)^{1/3}$, so that

$$\tau_c = \tau_0 + \frac{G|\epsilon_G'| - \alpha\epsilon|^{3/2}c^{1/2}}{Z} \tag{7}$$

where $Z = 1320$; τ_0 is the critical resolved shear stress of the pure metal.

Labusch[9] extended Fleischer's model to higher concentrations by use of a different type of statistical average of the interaction between solute atoms and dislocations. The resulting expression is

$$\tau_c = \tau_0 + \frac{G\epsilon^{4/3}c^{2/3}}{550} \tag{8}$$

Although Fleischer found some agreement with equation (7) for copper-base alloys, data for ductile gold-, silver-, and copper-base alloy single crystals showed better agreement with $c^{2/3}$, as predicted by equation (8).[7]

Short-Range Order

Concentrated solid solutions are likely to exhibit appreciable short-range order. The energy required to shear a short-range-ordered crystal causes an increase in flow stress of the alloy.

The energy associated with short-range order in a binary solid solution is[10]

$$E_s = NZc(1 - c)va_s \tag{9}$$

where N is the number of atoms in the lattice, Z is the coordination number (equals 12 for FCC lattices), c is the mole fraction of solute, v is the interaction energy $[=V_{AB} - \frac{1}{2}(V_{AA} + V_{BB})$ where V_{AB}, etc., are interaction energies for the various atom pairs], and a_s is the short-range-order coefficient, which may be determined experimentally.

Table 1. Solid-Solution Strengthening of γNi[a]

Solute Atom	Wt % in γ	Change in a_0 (kX)	Change in Flow Stress (ksi)
Co	20	0.011	2.56
Fe	10	0.020	7.96
Cr	20	0.033	22.8
Mo	4	0.035	24.2
W	4	0.038	25.5
V	1.5	0.006	4.55
Al	6	0.025	28.5
Ti	1	0.006	5.69

[a] From ref. 4.

The energy to disrupt short-range order per unit area of slip is

$$E_s = \frac{8}{3^{1/2}} \frac{c(1 - c)va_s}{a^2} \tag{10}$$

Equating this to the work done in moving the dislocation, τb, we obtain for the shear stress to move the dislocation

$$\tau = 16\left(\frac{2}{3}\right)^{1/2} \frac{c(1 - c)va_s}{a^3} \tag{11}$$

Since all terms in equation (11) are temperature independent, short-range order provides an athermal increment to flow stress. However, a_s increases with decreasing annealing temperature. Consequently, the short-range-order component of flow stress is sensitive to thermal history.

Nordheim and Grant[11] have suggested that short-range order exists in nickel–chromium alloys in the neighborhood of 20–25 w/o chromium. This composition range is reached in Hastelloy X and Inconel alloy 625 (22% Cr) and is approached in other γ'-lean alloys such as the early Nimonic series (19.5% Cr). Consequently, short-range-order strengthening may occur in such alloys.[12]

PRECIPITATION HARDENING OF NICKEL-BASE ALLOYS

Properties of $\gamma-\gamma'$ Alloys.

The major contribution to the strength of precipitation-hardened nickel-base superalloys is provided by coherent stable intermetallic compounds such as γ' [Ni$_3$(Al,Ti)] and γ'' [Ni$_3$(Cb,Al,Ti)]. Other phases, for example, borides and carbides, provide little additional strengthening at low temperatures due to the small volume fractions present, although they may provide significant effects on creep rate, rupture life, and rupture strain through their influence on grain boundary properties.

We are concerned with the properties of γ'-strengthened alloys exclusively in this section, with the experimentally determined strength dependent on such diverse factors as

1. volume fraction f of γ',
2. radius r_0 of γ',
3. solid-solution strengthening of both γ and γ', and
4. presence of hyperfine γ'.

In view of the critical importance of the properties of the γ' phase, we first discuss those facets of structure and deformation modes that contribute to the strength of $\gamma-\gamma'$ alloys.

Properties of γ' [$Ni_3(Al,Ti)$]

Structure and Slip Systems. Ni_3Al is a superlattice possessing the Cu_3Au ($L1_2$)-type structure, which exhibits long-range order to near its melting point of 2525 °F (1385 °C). It exists over a fairly restricted range of composition, but alloying elements may substitute liberally for either of its constituents to a considerable degree. In particular, most nickel-base alloys are strengthened by a precipitate in which up to 60% of the aluminum can be substituted for by titanium and/or columbium.

Unalloyed Ni_3Al deforms by $\{111\}\langle 110\rangle$ slip at all temperatures, and at temperatures above 750 °F (400 °C) some slip along $\{100\}$ also is observed; at 1292 °F (700 °C) $\{100\}$ slip predominates.[13] Cube slip can be prevented by testing cube-oriented samples, but there is no orientation in which $\{111\}$ slip is precluded. Slip along $\{111\}$ is extremely heterogeneous at low temperature, but at temperatures of 750 °F (400 °C) and above, $\{111\}$ slip is fine and uniform.

Stacking Faults. Three types of stacking faults exist in the $L1_2$ structure[14]:

1. superlattice (intrinsic or extrinsic) faults,
2. antiphase boundary faults, and
3. complex faults.

Superlattice intrinsic (SI) and superlattice extrinsic (SE) faults are produced by shear displacements of type $1/3\langle 112\rangle$ and $1/6\langle 112\rangle$ on $\{111\}$. These control creep of Mar M200 at 1400 °F (760 °C).[14] Antiphase boundary (APB) faults are produced by $a/2\langle 110\rangle$ displacements on $\{111\}$. The energy of the latter should be higher than for SI or SE faults since a portion of the atoms in the APB have incorrect nearest neighbors. The energy of an APB is a sensitive function of crystal orientation and is a minimum for a $\{100\}a/2\langle 110\rangle$ shear-type boundary. Complex faults (CF), which may be regarded as a superposition of APB and SI faults, are related to the $L1_2$ superlattice structure by $\{111\}a/6\langle 112\rangle$ shear displacements.

The large numbers of faults produced in nickel-base alloys hardened by coherent precipitation play a dominant role in deformation. It will be shown in a later section that several precipitation-hardening models predict a very sensitive dependence of critical resolved shear stress (CRSS) upon the fault energy of γ'. Also, creep properties of MAR M200 have been related to the nature of the faults in γ' left by the passage of dislocations through the particles.[14]

Temperature Dependence of Yielding. Both single crystals and polycrystals of unalloyed γ' exhibit a startling, reversible[15] increase in flow stress between −320 °F (−196 °C) and about 1475 °F (800 °C), which is highly dependent on aluminum content,[16] as shown in Fig. 3. While other superlattices exhibit a modest peak in strength over a rather narrow temperature range near T_c, the critical temperature for ordering, such a peak often is connected with a change in the degree of order with temperature.[17] However, several other superlattices, Ni_3Si, Co_3Ti, Ni_3Ge, Ni_3Ga, all of $L1_2$ structure, rise in strength over a temperature range comparable

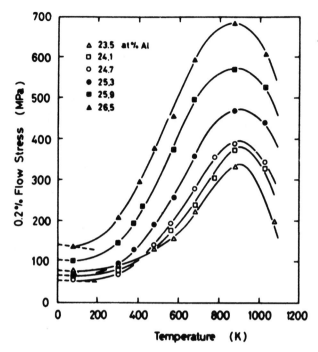

Fig. 3. Rise in flow stress of γ' (Ni₃Al) with temperature at various Al contents.[16]

to that of Ni_3Al.[18] The flow stress in all of these is fully reversible upon changing temperature. While no applications of the anomalous yielding behavior of ordered phases other than Ni_3Al has yet been announced, it is possible that additional γ'-type two-phase alloys may be forthcoming.

The magnitude and temperature position of the peak in flow stress of γ' may be shifted by alloying elements such as titanium, chromium, and columbium (Fig. 4).[13] There is no simple relation between the magnitude of flow stress increase and the change in the temperature of the peak. For example, both chromium and titanium increase the temperature of the flow stress peak, but chromium weakens γ' at low temperatures while titanium strengthens γ'. All substitutional elements in Ni_3Al single crystals (molybdenum, tantalum, columbium, titanium, tungsten) increase the CRSS for (111) [$\bar{1}$01] slip but decrease the CRSS for (001) [$\bar{1}$10] slip relative to binary Ni_3Al.[19]

With this knowledge of the most significant properties of bulk γ', it is now possible to consider the interaction of glide dislocations with γ' in the form of a distributed second phase.

Particle Cutting Models

General Comments. Among the factors that have been suggested to account for observed hardening of austenitic superalloys by coherent particles are the following:

1. coherency strains,
2. differences in elastic moduli between particle and matrix,
3. existence of order in the particles,
4. differences in SFE of particle and matrix,
5. energy to create additional particle–matrix interface, and
6. increases in lattice resistance of particles with temperature.

While several mechanisms may apply to any single system, theoreticians consider only one mechanism at a time and then, if necessary, add the increment in shear stress due to each of the various mechanisms. It now appears, however, that the major factors that contribute to strengthening by γ' in superalloys are coherency strains and the presence of order in the particles. Consequently, our discussion is limited to these mechanisms and to the Orowan dislocation bypass model, which limits the strengthening achieved by the other mechanisms.

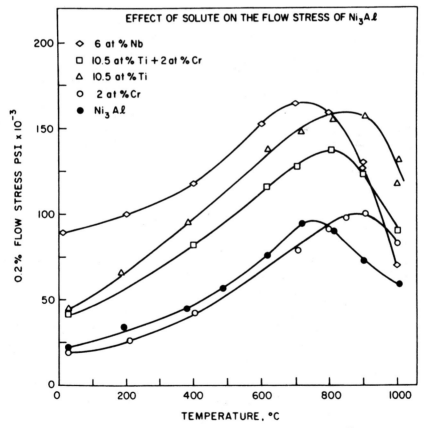

Fig. 4. Shift in flow stress peak of γ' with alloying additions.[13]

Rather than attempt a detailed analysis of the various models proposed to account for order and misfit hardening, it is our intention to dicuss the principles underlying the most pertinent models. The methods used to treat hardening by solutes and by precipitates are basically similar; they depend on calculating the force of interaction between a moving dislocation and whatever obstacles are in its path.

In order to move through a field of dispersed obstacles, a dislocation must bend to an angle ϕ dependent on the obstacle strength, see Fig. 5. For weak obstacles $\phi \rightarrow \pi$, as very little bending is required for the dislocation to escape the obstacle; for strong obstacles $\phi \rightarrow 0$, as the dislocation is forced to almost double back on itself. The number of obstacles per unit length of dislocation line depends on ϕ; if $\phi \simeq \pi$, the number per unit length is found by computing the number that intersects a random line. As ϕ decreases from π, the dislocation sweeps out more area and therefore meets more obstacles, necessitating expressions for obstacle spacing that depend on the applied stress τ. The most commonly used expression is that of Friedel[8]:

$$L' = \left(\frac{2TL_s^2}{\tau b}\right)^{1/3}$$

(12)

where T is the line tension ($\simeq \frac{1}{8}Gb^2$ for edge and $\frac{1}{2}Gb^2$ for screw dislocations) and L_s is the square lattice spacing ($=n^{-1/2}$, where n is the number of particles per unit area of slip plane). To simplify calculations, it is generally assumed that dislocations interact with a random array of obstacles of fixed strength. The limits on L' are

$$L_s \leq L' \leq \frac{4r}{3f}$$

(13)

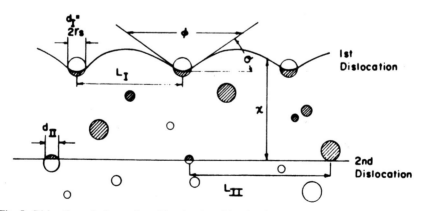

Fig. 5. Dislocation pairs interacting with ordered particles showing effect of bend angle ϕ on obstacle spacing. Shaded areas represent APB.

The upper limit, $4r/3f$, represents the spacing of random obstacles along a straight line.

Order Strengthening. Ham[20] has shown that when one dislocation cuts an ordered particle (neglecting the particle–matrix interfacial energy), the force $\tau_I b$ on the dislocation must balance the APB energy created, which is $2r_s\gamma_0/L_I$, where γ_0 is the APB energy,

$$\tau_I = \frac{2r_s\gamma_0}{L_I b} \tag{14}$$

and r_s is the average radius of particle intersected by a slip band [$= (\frac{2}{3})^{\frac{1}{2}}r_s$ in equation (12) to obtain

$$L_I' = \left(\frac{2T\pi r_s^2}{f\tau_I b}\right)^{1/3} \tag{15}$$

Substituting equation (17) into equation (14), we obtain the stress necessary to force the dislocation through the particle:

$$\tau_I = \frac{\gamma_0^{3/2}}{b}\left(\frac{4fr_s}{\pi T}\right)^{1/2} \tag{16}$$

Therefore, at constant f, τ_I increases with r_s owing to the increasing flexibility of dislocations as they interact with coarser particles. The stress to move edge dislocations is twice that to move screw dislocations because of differences in line tension. An edge dislocation bows out four times as much as a screw dislocation (when $v = \frac{1}{3}$) and, therefore, meets more obstacles. The fraction of dislocation-line-cutting particles, touching APB, is given by $2r_s/L_I$, where $L_I = f(\tau)$ given by equation (15):

$$\frac{2r_s}{L_I} = \left(\frac{4\gamma_0 f r_s}{\pi T}\right)^{1/2} \qquad \frac{\pi T f}{4\gamma_0} \leqslant r_s \leqslant \frac{T}{\gamma_0} \tag{17}$$

The upper limit on this fraction for increasingly bent dislocations is set by the condition $r_s = T/\gamma_0$ for which $L_I = L_s$, that is, just at the point of Orowan bowing:

$$\frac{2r_s}{L_I} \rightarrow \frac{2r_s}{L_s} = \left(\frac{4f}{\pi}\right)^{1/2} \qquad r_s = \frac{T}{\gamma_0} \tag{18}$$

At and above this critical fraction, the stress to shear a particle becomes

$$\tau_I = \frac{\gamma_0}{b}\left(\frac{4f}{\pi}\right)^{1/2} \tag{19}$$

This stress must be less than the Orowan bowing stress (see below) in order for particle shear to continue.

The lower limit of $2r_s/L_I$ [eq. (17)] is given by the value of L_I for a perfectly straight dislocation and corresponds to the case of very small particles; that is,

$$\frac{2r_s}{L_I} = f \qquad r_s \leqslant \frac{\pi T f}{4\gamma_0} \qquad \tau = \frac{f\gamma_0}{b} \tag{20}$$

Let us now consider the case of superlattice dislocation pairs interacting with particles. The calculation follows the principles first elaborated by Gleiter and Hornbogen[21] but utilizes the specific equations developed by Ham[20] and Brown and Ham.[22] As the first dislocation is just shearing the particles (see Fig. 5), the second dislocation is pulled forward by the APB remaining in all particles cut by the first dislocation. Provided that the two dislocations assume the same shape and that the separation x between the two dislocations is sufficiently small but larger than r_s, the second dislocation may lie outside of all the particles. This situation may occur at long aging times. Then, at equilibrium the total forward stress on the second dislocation, τ_{II}, balances the repulsive force between the two dislocations; that is,

$$\tau_{II}b = \frac{Gb^2}{2\pi kx} \tag{21}$$

where τ_{II} is now equal to the applied stress τ ($k = 1 - \nu$ for edge dislocations, $k = 1$ for screw dislocations). The stress on the first dislocation $\tau_I = 2\tau$ due to the presence of the second dislocation. For shear of the particles by the first dislocation to occur [using eq. (18)],

$$\tau_{II} = \frac{\gamma_0}{2b} \frac{2r_s}{L} = \frac{\gamma_0}{2b} \left(\frac{4f}{\pi}\right)^{1/2} = \frac{\gamma_0}{b} \frac{(f)^{1/2}}{\pi} \tag{22}$$

and the applied stress necessary to cut the particles is just half the value computed for shear by single dislocations [eq. (19)].

In general, however, the second dislocation does come into contact with the APB and is nearly straight. The more APB is cut by the second dislocation, the less effective the particles become as obstacles. Then, referring to Fig. 5, the force balances are as follows, neglecting any new particle–matrix interface formed by shear of the particles: on dislocation 1

$$\tau b + \frac{Gb^2}{2\pi kx} - \frac{\gamma_0 d_1}{L_I} = 0 \tag{23}$$

on dislocation 2

$$\tau b + \frac{\gamma_0 d_{II}}{L_{II}} - \frac{Gb^2}{2\pi kx} = 0 \tag{24}$$

Solving equations (23) and (24) simultaneously, we obtain, for the forward stress on the first dislocation,

$$2\tau b + \gamma_0 \frac{d_{II}}{L_{II}} = \frac{\gamma_0 d_I}{L_I} \tag{25}$$

Since the second dislocation is observed to be straight during shear of the particle by the first dislocation, we may substitute equation (20) for d_{II}/L_{II} and equation (17) for d_I/L_I so that

$$2\tau b + \gamma_0 f = \left(\frac{4\gamma_0 f r_s}{\pi T}\right)^{1/2} \gamma_0 \tag{26}$$

leading to the following relation for the applied stress τ:

$$\tau = \frac{\gamma_0}{2b} \left[\left(\frac{4\gamma_0 f r_s}{\pi T}\right)^{1/2} - f \right] \tag{27}$$

If the line tension T is approximated by $\frac{1}{2} G b^2$ (screw dislocations), equation (27) reduces to

$$\tau_c = \frac{\gamma_0}{2b} \left[\left(\frac{8\gamma_0 f r_s}{\pi G b^2}\right)^{1/2} - f \right] \tag{28}$$

Equation (28) cannot hold for r_s approaching zero since τ_c cannot be negative. Nevertheless, the negative intercept, $-\gamma_0 f/2b$, has been used for an alternate computation of APB energy.[23]

The first term of eq. (28), $A\gamma_0^{3/2} f^{1/2} G^{-1/2} b^{-2} r_s^{1/2}$, is similar in form and dependence on particle size to an equation for order strengthening that had been presented earlier by Gleiter and Hornbogen,[21]

$$\tau_c = 0.28\gamma_0^{3/2} f^{1/3} G^{-1/2} b^{-2} r_0^{1/2} \tag{29}$$

except for a different constant and the dependence of volume fraction. The second term of equation (28) may be dropped, however, only when the second dislocation can avoid all particles.[22] The flow stress given by equation (28) would then reduce to one-half of the stress given in equation (16) for single dislocations. The basic features of this model are summarized schematically in Fig. 6 and have been applied successfully to a variety of austenitic superalloys,[20–23] both nickel and iron based. So far as alloy design is concerned, the model emphasizes the role of particle size and antiphase boundary energy on strength in alloys containing a small volume fraction of precipitates.

Copley and Kear[24,25] also modified the Gleiter–Hornbogen theory; the results were applied specifically to the alloy MAR-M 200, which is a high-volume-fraction

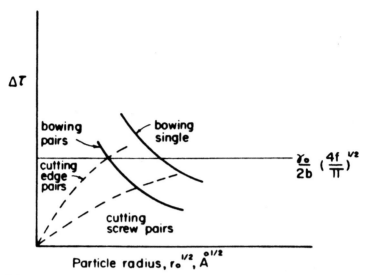

Fig. 6. Schematic age-hardening curves illustrating relation between order strengthening and Orowan bowing as functions of particle size.

γ' alloy. Based on extensive electron microscope examination, the rate-controlling step for plastic deformation was shown to be moving dislocations from γ into γ'. Instead of setting up a force balance with the first dislocation partly through the particles, as in Fig. 5, they contend that the first dislocation wraps around the particle, assuming its curvature, until forced in by the second dislocation. The conditions of static equilibrium for the leading and trailing dislocation of the superlattice pair about to enter a particle are:

for dislocation 1

$$(\tau_c - \tau_p)b + \frac{C}{x} + \frac{T}{r_0} - \gamma_0 = 0 \tag{30}$$

for dislocation 2

$$(\tau_c - \tau_0)b - \frac{C}{x} + \frac{T}{r_0} = 0 \tag{31}$$

where C/x is the force of repulsion between two dislocations, τ_p is the friction stress of particle, and τ_0 is the friction stress of matrix. Here T/r_0 is the line tension force due to a dislocation assuming the curvature of the particle.

Solving equations (30) and (31) simultaneously, one obtains for conditions of static equilibrium at 70 °F (22 °C)

$$\tau_c = \frac{\gamma_0}{2b} - \frac{T}{br_0} + \frac{1}{2}(\tau_0 + \tau_p) \tag{32}$$

where $\gamma_0/2b$ is the stress to constrict the dislocation pair to the point that particle shear begins.

For dynamic conditions the CRSS is predicted from the stress dependence of the plastic strain rate, which is related to a derived dislocation velocity–stress function. A very similar equation is obtained for τ_c,

$$\tau_c = \frac{\gamma_0}{2b} - \frac{T}{br_0} + \frac{k}{2}(\tau_0 + \tau_p) \tag{33}$$

where k is a constant dependent on the dislocation velocity of the crystal and has a value of 0.823 for MAR-M 200 at room temperature.[25] Penetration of small particles is easier than for large particles due to the line tension force. In any case, the major contribution to τ_c at room temperature is provided by the term $\gamma_0/2b$, which represents about 80% of the total flow stress computed for MAR-M 200. Leverant et al.[26] concluded, however, that at high temperatures and high strain rates, where the flow stress of γ' reaches a distinct peak, both the APB energy and the flow stress of γ' are major contributors to τ_c. As in hardening models developed for low-volume-fraction alloys, large particle radii are predicted to increase strength, but the effect will be small since the term T/br_0 in equations (32) and (33) does not make a major contribution to flow stress.

In the most general case, γ_0 should be replaced by Γ, the fault energy for shear of the particle, since faults other than APB-type faults may be produced by shear. For example, particles are sheared by loosely coupled intrinsic–extrinsic fault pairs in MAR-M 200 at 1400 °F (760 °C).[26] In this model the influence of crystal orientation on flow stress is felt throught the variation in the nature of faults produced by different glide mechanisms. Decker[3] has pointed out that based on the critical temperature for ordering of their Ni_3X phase, titanium, columbium, and tantalum in γ' should not appreciably increase APB energy. However, titanium and perhaps tantalum could increase the energy of other fault types. Brown and Ham[22] have analyzed several sets of data to calculate APB energy as a function of alloy content and found that the fault energy may be widely varied (see Table 2). This table is discussed later in conjunction with alloy design principles.

Misfit Strengthening. Early attempts[1] to relate the influence of coherency strains to CRSS failed to explain the dependence of CRSS on particle size. A model has been assumed by Gerold and Haberkorn[31] where the interaction between dislocations and strain plays a dominant role, but particle cutting occurs as a consequence of the interaction.† The calculation is similar in outline to Fleischer's[4,6] for solid-solution hardening and is expected to apply for matrix–particle misfit ϵ of approximately 0.01 in the case of spherical, coherent particles ($\epsilon = a_{ppt} - a_{matrix}/a_{matrix}$).

The increase in flow stress due to interaction of single dislocations with strain fields is given by

---

† Particle cutting must occur on the Mott and Nabarro model [eq. (1)], as applied to precipitates, when the critical particle spacing for maximum strength, $L = b/4\epsilon f$, is reached.

Table 2. Antiphase Boundary Energies for $\gamma-\gamma'$ Alloys

Alloy Composition	APB Energy$^a\gamma_0$, erg/cm^2		Particle Properties		References
	Uncorrected	Corrected^b	r_0 (Å)	f	
Ni-12.7 to 14.0					
a/o Al	153	145	21–45	0.05–0.14	15,22,27
Ni-18.5 a/o Cr-7.5					
Al	104	94	55	0.194	15
Ni-18.8 a/o Cr-6.2					
Al	90	81	45	0.054	28
Fe–Cr–Ni–Al–Ti					
(Ti/Al = 1)	240	—	—	—	29
(Ti/Al = 8)	300	—	—	—	29
Ni-19 a/o Cr–					
14Co–7Mo–					
2Ti–2.3Al	170	—	180	0.3	30
	220	—	50	—	30
Ni-33 w/o Fe–					
16.7Cr–3.2Mo–					
1.6Al–1.1Ti	270	—	—	—	22

^{a}Dependent on f, r_0.
^{b}Corrected for lattice friction within particle.

$$\Delta\tau = \frac{K}{bL''} \tag{34}$$

where K is the maximum repelling force of the strain field of a single particle on a moving dislocation and L'' is the average distance between the force centers. Equation (34) is the analog of equation (6) for solid solutions. As before, the problem is to find appropriate expressions for K and L'': K is found to be equal to or less than T, the line tension of an edge dislocation. For L'', instead of using the Friedel spacing [eq. (12)], the authors use an expression[4] in which the obstacle spacing depends on the bend angle, $\theta = \frac{1}{2}(\pi - \phi)$ (see Fig. 5):

$$L'' = \frac{r_0\pi^{1/2}}{(\theta f)^{1/2}} \qquad \frac{9\pi f}{16} < \theta < \frac{3}{2} \tag{35}$$

The angle to which a dislocation is bent by the force K before escaping the particle is given by

$$2\sin\theta \simeq \frac{K}{2T} \tag{36}$$

The maximum value of K is computed to be

$$K = 4G|\epsilon|br \tag{37}$$

For bend angles smaller than $\frac{9}{16}\pi f$ the dislocation must be treated as a rigid line; for bend angles near $\frac{1}{2}$ the dislocation line is totally flexible and another expression for L'' must be used. Combining equations (34) and (36), the CRSS is obtained:

$$\Delta\tau = AG\epsilon^{3/2} \left(\frac{r_0 f}{b}\right)^{1/2} \qquad \frac{9\pi f}{16} < \frac{3|\epsilon|r_0}{b} < \frac{1}{2} \qquad (38)$$

where $A = 3$ for edge dislocations and $A = 1$ for screw dislocations. This equation predicts that the flow stress should increase slightly more rapidly than ϵ because increasing misfit bends the dislocation more and makes it interact with more regions of adverse stress.

Based on experimental data for copper–cobalt and aluminum–zinc alloys, it was concluded that edge dislocations control the CRSS; also it was concluded that while for small particles ($r_0/b < 20$), which corresponds to $r_0 \simeq b/3|\epsilon|$, the dislocations cut the precipitates, and for larger particles dislocations will bypass by the Orowan mechanism.

Gleiter[32] also has discussed the effect of coherency strain fields on CRSS in a two-phase alloy and, by following the steps outlined above with different assumptions as to the flexibility of dislocation lines and a different averaging procedure for obstacle distribution, has obtained the following relation for flexible edge dislocations:

$$\Delta\tau = 11.8G\epsilon^{3/2}f^{5/6} \left(\frac{r_0}{b}\right)^{1/2} \qquad (39)$$

The main difference between equations (38) and (39) is the dependence of $\Delta\tau$ on the volume fraction. However, better agreement has been found with equation (38) for aluminum–zinc alloys[33] and, as will be discussed below, data for several nickel–aluminum alloys also were in agreement.[30]

Nembach and Neite[34] have extensively reviewed the experimental evidence bearing on lattice misfit effects on the strength of superalloys. It was concluded that there is no convincing experimental proof that misfit affects the flow stress of underaged γ'-hardened alloys and that lattice misfits of the magnitude found in commercial alloys do not make a significant contribution to strength.

Dislocation Bypass Models

Orowan Bowing. All of the dislocation cutting models previously discussed agree that as particles grow beyond a critical size, bypass may occur by bowing, climb or other processes. The Orowan bowing model[35] is generally considered to be most applicable for austenitic superalloys. The increment in flow stress at low temperature due to bowing is given by consideration of the radius of curvature ρ to which a flexible dislocation can be bent by an applied stress τ,

$$\tau b = \frac{T}{\rho} \qquad (40)$$

The minimum value of ρ is half the interparticle spacing L and corresponds to $\theta = \frac{1}{2}\pi$ (see Fig. 5). The line tension T is given approximately by $\frac{1}{2}Gb^2$, but a more exact formulation is[36]

$$T = \frac{Gb^2}{4\pi} \phi' \ln \frac{L}{2b} \qquad (41)$$

where $\phi = \frac{1}{2}[1 + 1/1 - v]$ and L is the edge-to-edge spacing of particles $(=[(\pi/f)^{\frac{1}{2}} - 2]r_s)$, which leads to an expression for the increment in flow stress $\Delta\tau$:

$$\Delta\tau = \frac{Gb}{2\pi L} \phi' \ln \frac{L}{2b} \qquad (42)$$

The effect of increasing volume fraction f for a given particle size is to decrease L, leading to a prediction of increased strength. Greater hardening also should occur as particle size increases; this effect would be enhanced by coherency strains, producing a larger particle diameter in the path of a dislocation.

Grain Boundary Effects

Metals and alloys tested at temperatures below about $0.5T_m$ (T_m = absolute melting temperature) are further strengthened by the resistance of grain boundaries to dislocation motion. The Hall–Petch relation

$$\sigma_y = \sigma_0 + k_y d^{-1/2} \qquad (43)$$

where σ_y is yield stress, σ_0 is a lattice friction stress, d is grain diameter, and k_y is a measure of the grain boundary resistance, demonstrates that significant strengthening can be obtained for fine-grained alloys when k_y is large. Factors tending to increase k_y are solute hardening and difficult cross slip. Therefore, solutes such as cobalt, which lower the stacking fault energy of nickel, are expected to increase the contribution of grain boundaries to yield or flow stresses. For cobalt-base alloys the presence of massive stacking faults reduces the effective slip distance d in equation (43), and strength can be sharply higher.

DISPERSION-STRENGTHENED ALLOYS

While precipitates may be either coherent or noncoherent with the matrix, depending on aging conditions and the crystal structures involved, dispersoids such as oxides are always incoherent. Therefore, only the Orowan mechanism may be applied to strengthening by oxides. In the mechanically alloyed superalloys produced by INCO both precipitates and dispersoids may be found in the same alloy, and hardening mechanisms may be additive provided that coarse, elongated grain structures are produced during processing and retained in service. The ThO_2 or Y_2O_3 particles in the TD–Ni and mechanically alloyed type materials are fine and relatively uniformly dispersed. Size is of the order of 100–500 °Å with interparticle spacings of 500–3000 °Å. The hardening due to these particles must be added to the strengthening effects of grain boundaries and subboundaries as well as those of solid-solution elements. An additional factor is the grain aspect ratio (GAR), the ratio of grain length L to width l. At high temperatures strength varies linearly with GAR, as shown in Fig. 7a[37]:

$$\sigma = \sigma_e + k\left(\frac{L}{l} - 1\right) \qquad (44)$$

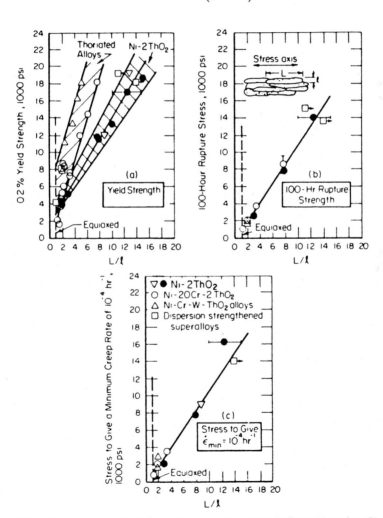

Fig. 7. Effect of grain aspect ratio GAR $= L/l$ on properties of ODS alloys: (a) tension; (b) 100 h rupture strength, (c) minimum creep rate.[38]

where σ_e is the strength of equiaxed material (GAR = 1) and k is defined as the GAR coefficient. Creep and stress rupture behavior also correlates well with GAR (see Figs. 7b and 7c). Later work confirmed the importance of GAR in mechanically alloyed materials such as MA-753[39] and MA-6000.[40] Wilcox and Clauer[37] concluded that when grains are elongated, the GAR effect swamps any effect of grain size alone. The most direct means of achieving coarse, elongated grains is to use an extrusion press both to consolidate powder and to achieve an appropriate structure suitable for subsequent secondary recrystallization.

CRITICAL EVALUATION OF MODELS

The results of aging studies on low-volume-fraction model $\gamma-\gamma'$ alloys are not directly applicable to many modern commercial nickel-base superalloys because of the much higher volume fraction of γ' present in the latter. (However, several low-volume-fraction alloys such as Nimonic 80A and A-286 are still used extensively, and the models are applicable to such alloys.) Also, particle sizes tend to be larger in the superalloys. Another complication is the orientation and strain rate dependence of stress–strain behavior in single-crystal alloys containing a high volume fraction of precipitate.[41]

Each of the models outlined suffers from shortcomings, not the least of which is the fact that the microstructures of nickel-base superalloys are too complex to expect a single mechanism to operate over all ranges of stress and service temperatures. We shall distinguish between alloys in which there is little or no mismatch between γ and γ', that is, nickel–chromium–aluminum type, and high-mismatch alloys, that is, nickel–aluminum–titanium type (see Fig. 8).

Alloys with No Lattice Mismatch. The principal elements of the Brown–Ham model of pairs of dislocations interacting with ordered particles have been directly confirmed in high-voltage electron microscopic experiments on Nimonic PE16.[43] Specifically, the leading dislocation of a pair bows out strongly between γ' precipitates while the trailing dislocation remains nearly straight. The spacing of obstacles along the leading dislocation is in reasonable agreement with equation (12). When there is little or no mismatch, considerable evidence suggests that the volume fraction f is the most significant variable controlling flow stress and creep resistance. The volume fraction of γ' varies from 0.2 in γ'-lean alloys, such as Nimonic 80A, to 0.6 in MAR-M 200 and 713C. Newer alloys possess up to 70% γ'. The flow stress of binary nickel–aluminum aged to peak hardness[15] and of nickel–chromium–aluminum alloys containing fractions of γ' between 0.4 and 0.6 is remarkably insensitive to temperature (see Fig. 9). The yield stress of MAR-M 200 is nearly constant from room temperature to 1380 °F (750 °C).[24] This is due to the particle-cutting mechanism that controls plastic flow.

Beardmore et al.[44] have summarized the influence of the fraction of γ' on the flow stress of a series of nickel–chromium–aluminum alloys. At 1659 °F (900 °C) or more these alloys consisted of strong, approximately 0.5 μm diameter γ' particles

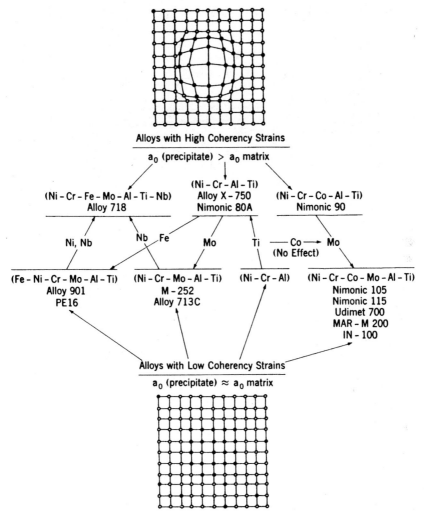

Fig. 8. Classification of nickel-base alloys according to degree of mismatch.[42]

in a weak γ matrix. The flow stress then was a function of f. The strongest alloy at temperatures above 1400 °F (760 °C) contained 100% γ' (see Fig. 9). Alloys with large f are deformed by particle shear and those with small f are deformed by bowing of unpaired dislocations in the FCC matrix.

At approximately 930 °F (500 °C) and below the 0.5-μm γ' particles were weak relative to the γ matrix, which also contained a hyperfine ($r_0 = 38$ Å) precipitate of γ'. The latter formed upon cooling from the aging temperature 1750–1670 °F (950–900 °C). Under these conditions the flow stress was not simply related to f but showed a peak at $f = 0.25$. (Cornwell et al.,[45] on the other hand, have reported a nearly linear increase in yield strength with f in binary nickel–aluminum single

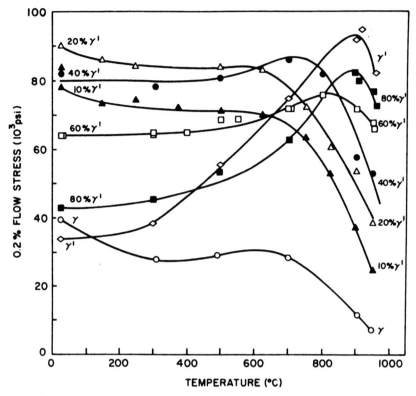

Fig. 9. Temperature dependence of flow stress of Ni–Cr–Al alloys containing various volume fractions of γ'.[44]

crystals, up to $f = 0.6$.[46]) Deformation occurred with large slip offsets, the slip bands becoming more diffuse as f increased or the temperature was raised. Dislocations were paired and stored primarily in the matrix phase. Hyperfine precipitates also have been observed in commercial alloys, for example, Nimonic 115, MAR M 200, and Udimet 700.

Alloys containing a large volume fraction of γ' behave similarly to pure γ' in that the flow stress increases with increasing temperature. If the alloy contains about 50% primary coarse γ' in γ, the strength characteristics are intermediate. Flow strength is moderately high at low temperature; a shallow peak in flow stress is reached near 1290 °F (700 °C), and strength falls off at a temperature somewhat higher than for a leaner alloy (see Fig. 9). Note that an intermediate volume fraction of 20% produces the highest strength at 70 °F (21 °C). Most of the lower temperature yield strength is due to "hyperfine" γ' (50–100 Å diameter).

In spite of the likelihood that at least two mechanisms may be operative in alloys with low mismatch, we now proceed to discuss the individual models in more detail.

Copley and Kear[24] originally had suggested that the drop in flow stress of γ–γ' alloys above 1400 °F (760 °C) is a consequence of a reduction in APB energy due to disordering. However, it has been shown that Ni_3Al does not disorder until at least 2010 °F (1100 °C)[44] and probably not until the melting point.[47] It has been proposed also that local disordering occurs in the vicinity of superlattice dislocation pairs at high temperature.[26] The evidence for this is stated to be an observed increase in separation between the dislocation pairs when viewed in the electron microscope. While local disordering could account for a drop in flow stress at high temperatures, an additional factor also should be considered. When cube slip occurs, the APB fault energy should be lower than under conditions of octahedral slip. Consequently, the flow stresses of particle and matrix may control the strength of the alloy [see eq. (30)], at temperatures where cube slip predominates.

All of the order-strengthening theories in low-volume-fraction γ'-strengthened alloys predict an increase in flow stress with increasing particle size, r_0, for a constant volume fraction. This has been confirmed in Ni-12.7 atom percent (a/o) aluminum alloys.[48] However, other evidence on the effect of particle size has been conflicting; it has been shown for an 18Cr–6.5Al–3.3Cb alloy that an increase in size of γ' from 0.05 to 0.5 μm by varying aging time reduced the room temperature flow stress by about 13%.[49] On the other hand, hardness of nickel–chromium–titanium alloys increases initially with particle size and then decreases (see Fig. 10).[50] Dislocations initially shear the particles; when the particles grow larger, the dislocations shift into a bypass mechanism. Consequently, we conclude that so long as particles are being cut, the flow stress increases with increasing particle size.

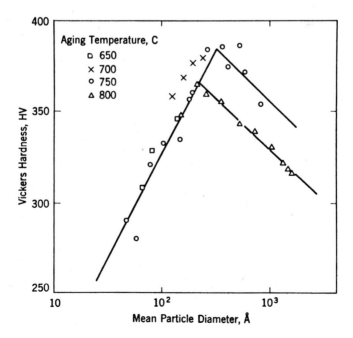

Fig. 10. Effect of particle size on hardness of Ni–Cr–Al–Ti alloys.[50]

For nickel–chromium–aluminum–titanium alloys containing 10–20% γ' the smallest γ' particle size gave optimum creep resistance at 1290 °F (700 °C); also, size was more important than volume fraction in determining creep life. (See Chapter 7 for a discussion of the relation between creep strength and γ' size in single crystals.) Small size is achieved in conjunction with small interparticle spacing, which is about 0.05 μm for optimum creep resistance. Consequently, it may be difficult to produce a particle size and spacing that will simultaneously provide good tensile and creep properties in lean alloys.

The Gleiter–Hornbogen–Ham theories are strictly applicable to low-volume-fraction alloys only but have the advantage of explicitly including f, r_0, and γ_0 in the expressions for flow stress. These models are capable of explaining the observed transition between particle shear and dislocation bowing observed by several investigators as well as the dependence of flow stress on particle size and APB energy. The Copley–Kear model, on the other hand, is applicable only to high-volume-fraction alloys and has only been tested quantitatively with data for MAR-M 200. However, an inverse relation between γ' size and yield strength is noted for cube-oriented, high-volume-fraction single crystals, as predicted by Copley and Kear (see Chapter 7). The major uncertainty involved in the use of all order-hardening theories is that there are no direct means of determining APB energies, and it is difficult to measure precisely the parameters f, r_0, and L', which are so important in applying these models.

Alloys with Lattice Mismatch.   It has been suggested that there is a correlation between the titanium–aluminum ratio of superalloys and strength or creep resistance. However, there is considerable controversy as to the origin of these effects. Phillips[27] and Raynor and Silcock[29] suggest that increasing the titanium–aluminum ratio influences strength through an increase in APB energy from approximately 150 erg/cm^2 (no Ti present) to approximately 240 erg/cm^2 (Ti/Al = 1) and 300 erg.cm^2 (Ti/Al = 8), as shown in Table 2. In this view a difference in lattice parameter between γ and γ' as high as 0.5%, which accompanies high titanium additions, is relatively unimportant as a strengthening mechanism. Rather, mismatch is the driving force in the growth and coalescence of γ' particles. A large mismatch, corresponding to a large interfacial strain energy, may reduce the stability of the γ' precipitate even in the absence of applied stress. Applied stress further lowers the mismatch to stabilize the precipitate, particularly when the stress axis differs from $\langle 111 \rangle$.

Conversely, Decker[3] and Decker and Mihalisin[42] argue that a high mismatch can markedly increase maximum strength by aging. Increasing mismatch from 0.2 to 0.8% doubled peak-aged hardness of several nickel–aluminum ternary alloys, which is in agreement with the theory of Gerold and Haberkorn.[31] Munjal and Ardell[52] found excellent agreement between the Brown–Ham[22] model and experimental results for Ni–12.19a%Al single crystals tested in compression between −320 and 212 °F (−196 and 100 °C) style. Since misfit changes considerably with temperature, and no significant change in $\Delta\tau$ was observed over the same temperature range, it was concluded that the contribution of coherency hardening is negligible in this system. While the relation between coherency strains and low-temperature tensile strength is still in doubt, optimum creep resistance seems to depend on zero mismatch.

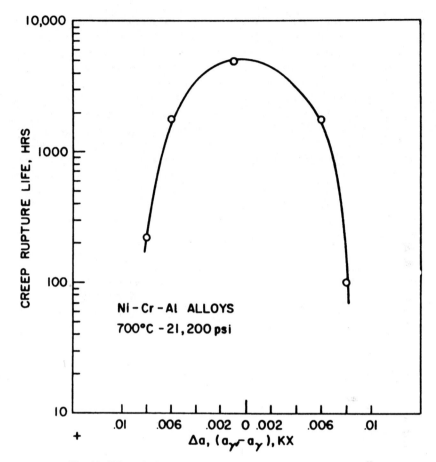

Fig. 11. Effect of mismatch on creep rupture life of Ni–Cr–Al alloys.[53]

As shown in Fig. 11, creep rupture life of nickel–chromium–aluminum alloys tested at 1290 °F (700 °C) and a stress level of 21,200 psi reaches a maximum at zero mismatch.[53] These confirmed[54] results are attributed to high phase stability at low mismatch. However, it is not clear that similar results would be found in stronger alloys or at higher temperatures. In fact, in high-volume-fraction alloys ($f = 0.68$) rupture life increases with increasing mismatch.[55] In any case, the Gerold–Haberkorn theory must be applied at temperatures low enough so that the growth of γ' is not possible.

PRECIPITATION HARDENING OF COBALT-BASE ALLOYS

Preliminary studies of precipitate morphology and growth kinetics of Co_3Ti in binary cobalt–titanium alloys show a close resemblance to binary nickel–aluminum alloys.[56]

Since the temperature variations of the flow stress of Co_3Ti and Ni_3Al are similar,[57] this suggests the possibility of γ'-type strengthening in cobalt alloys. Cobalt-7.5 a/o titanium and Co-10 a/o titanium alloys have been aged to produce, respectively, 22 volume percent (v/o) and 40 v/o of cuboid precipitate.[57] The precipitates possessed a {100} habit plane with a misfit of approximately 1%. The properties of these alloys were compared to those of single-phase Co–2% Ti, corresponding to the matrix composition of the other cobalt alloys. The results are summarized, together with some data for nickel–aluminum alloys and corrected for strain rate differences in the two studies, in Fig. 12. It is apparent that cobalt- and nickel-base alloys behave similarly in that the flow stresses are relatively insensitive to temperature when particles are sheared by paired dislocations. Furthermore, the flow stress of Co-7.5 a/o titanium and Ni-14 a/o aluminum are nearly the same as for Co_3Ti and Ni_3Al, respectively, at 1292 °F (700 °C).

However, upon aging the 7.5 a/o titanium alloy for 50 h at 1292 °F (700 °C), the flow stress dropped rapidly with increasing test temperature above 930 °F (500 °C) as particles redissolved. The flow stress of the aged cobalt-base alloys increased with increasing volume fraction, as is generally observed in nickel-base alloys. In spite of these similarities between cobalt- and nickel-base alloys hardened by intermetallics, no commerical cobalt-base alloys utilizing this mechanism have been developed. Attempts to utilize γ'-type strengthening in cobalt have been unsuccessful owing to the low solution temperature, 1500–1600 °F (815–872 °C),

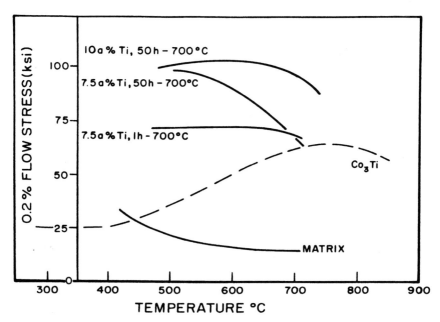

Fig. 12. Effect of temperature on flow stress of aged Co–Ti[56] and Ni–Al[15] alloys illustrating similarities in properties of phases.

of $Ni_3(Al,Ti)$; also, γ' tends to precipitate in cellular form with large mismatch, leading to poor high-temperature properties. Attempts to develop and utilize other A_3B-type precipitates also have been unsuccessful due to low solution temperatures. This has been one of the major elusive goals in superalloy metallurgy; continued work on these systems is warranted.

CREEP MECHANISMS

Primary Creep

Systematic studies of primary creep in austenitic superalloys over a wide temperature range have not been reported. However, Leverant and Kear[58] have investigated the creep mechanisms of MAR-M 200 single crystals at 1400 °F (760 °C). Primary creep strain and rate were markedly orientation sensitive. Slip occurred on {111}, but several Burgers vectors were observed. In particular, $(a/2)\langle 112\rangle$ dislocations were generated and then dissociated into two $(a/3)\langle 112\rangle$ and two $(a/6)\langle 112\rangle$ partials. The $(a/3)\langle 112\rangle$ partials then sheared the γ' particles instead of $(a/2)\langle 110\rangle$ superlattice dislocations, which shear particles during tensile deformation at the same temperature. Thus, creep proceeds by the motion of pairs of superlattice dislocation coupled by intrinsic–extrinsic fault pairs at a rate controlled by diffusion. The latter is necessary since the core of $(a/3)\langle 112\rangle$ dislocations must be altered to obtain the proper shearing sequence. At high strain rates this adjustment cannot occur, and deformation occurs by slip alone.

Larger particles are more effective in restricting primary creep for a constant volume fraction, since dislocations are less able to penetrate the particles owing to line tension considerations. Consequently, closely spaced, large γ' particles are desirable for optimum resistance to primary creep.

Steady-State Creep

Steady-state creep resistance in crystalline, single-phase solids depends on the diffusivity D, stacking-fault energy γ_{SFE}, elastic modulus E, temperature T, and stress σ according to a formula of the form[59,60]

$$\dot{\epsilon} = A\left(\frac{\sigma}{E}\right)^n f(\gamma_{SFE})e^{-Q/RT} \tag{45}$$

where $f(\gamma_{SFE})$ is a function of SFE and Q is the activation energy for creep. On one model $\dot{\epsilon}$ is dependent on $(\gamma_{SFE})^{3.5}$, while another formulation incorporates γ_{SFE} into the stress exponent n such that as γ_{SFE} increases, n decreases.[61] Typical solid-solution alloys reveal an exponent n with values 3–7 and with Q equal to the activation energy for self-diffusion at temperatures above half the melting point. Consequently, high creep strength is favored by solute additions that raise the modulus or lower the SFE and lower the diffusivity. Tungsten and molybdenum serve to raise the modulus and lower the diffusivity of austenitic superalloys, while cobalt is effective in lowering SFE of nickel-base alloys.

When second-phase particles are present, the apparent activation energy for creep is much higher than the activation energy for creep (or self-diffusion) of the matrix. Thus, the activation energy for steady-state creep of MAR-M 200 and other nickel-base superalloys is as high as twice that of unalloyed nickel and considerably higher than for solid-solution alloys of nickel. These discrepancies can be eliminated either by considering the temperature dependence of E[62] or by replacing σ in equation (45) by $\sigma - \sigma_0$, where σ_0 is a friction stress.[63] In either case the activation energy for creep becomes very close to that for self-diffusion. Similar differences between activation energies for creep of a multiphase alloy and the activation energy for self-diffusion of the matrix have been noted for dispersion-strengthened alloys such as TD nickel and $Al-Al_2O_3$. The grain aspect ratio (GAR) seems to play a role in these alloys as Q and n both increase with increasing GAR, although scatter is very large, see Fig. 7.[38] Later work showed that the threshold stress σ_0 for several oxide dispersion strengthened (ODS) alloys increased linearly with GAR.[64] It was suggested that for this type of alloy σ_0 is a better design criterion than a stress to produce a fixed amount of creep in a given time.

Steady-state creep in MAR-M 200 at 1400 °F (760 °C) occurs only after appreciable strain hardening owing to intersecting $\{111\}$ $\langle 112 \rangle$ bands and the development of a substructure during primary creep. Dislocation networks form at $\gamma-\gamma'$ interfaces, thereby limiting the mean free path of gliding dislocations to the order of the particle size. These networks reduce the rate of recovery, leading to a low creep rate. The

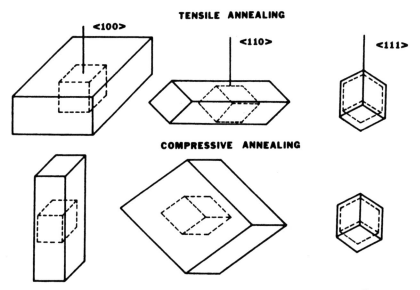

Fig. 13. Morphology of γ' as influenced by stress state during annealing.[65]

observation that ⟨112⟩ slip is responsible for particle shear suggests that crystal orientations with a low Schmid factor for ⟨112⟩ slip are desirable for good creep resistance. Supporting this conclusion is the observation that single crystals with a ⟨111⟩ tensile axis have unusually long creep lives.[58] However, work is needed at temperatures closer to actual service temperatures.

Influence of γ′ Morphology

The morphology of γ′ in nickel-base alloys can be modified by annealing under stress (see Fig. 13).[65] In ⟨100⟩ and ⟨110⟩ orientations both plates and rods of γ′ may be generated depending on the sense of the applied stress. Tensile annealing produces γ′ plates (rafts) for the ⟨100⟩ orientation, while compressive annealing causes rods to form. In the ⟨110⟩ orientation the opposite occurs, while ⟨111⟩-oriented crystals show no change in morphology under tension or compression. The sign of the lattice misfit also influences stress-coarsening behavior; the results described above are for alloys with negative misfit. Morphological changes in γ′ can affect yielding behavior of U-700 crystals.[66] The yield strength of ⟨100⟩ crystals is increased by rod or plate formation, with plates providing the greater effect to 1400 °F (760 °C) (see Fig. 14). At higher temperatures morphology has little effect on strength. However, in creep rupture tests a substantial improvement in properties of ⟨100⟩ crystals has been reported for a nickel–aluminum–molybdenum–tantalum alloy.[67] Specimens in the solution-treated condition (air cooled) exhibit lower steady-state creep rates and longer rupture lives than material given a standard heat treatment.

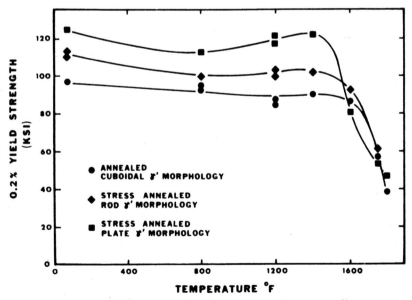

Fig. 14. Effect of γ′ morphology on flow stress of Udimet 700.[66]

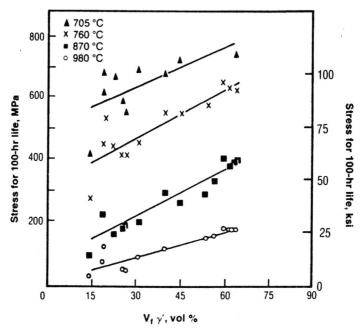

Fig. 15. Influence of volume fraction V_f of γ' on rupture strength of Ni-base alloys.[6]

A prestrain under creep conditions leads to still further improvement in properties due to the formation of γ' plates or rafts during primary creep. The molybdenum content is critical, with the creep strength maximized at the solubility limit of molybdenum in γ.[68] In summary, optimum strengthening due to γ' rafting in Ni–Al–Mo–X alloys is achieved in alloys with γ saturated in molybdenum and that exhibit large negative γ/γ' misfit. It is not yet clear whether these results can be practically utilized in alloys intended for long-time service.

Volume Fraction Effects. Decker[3] has reported a linear increase in stress for 100 h rupture life with f at temperatures between 705 and 980 °C (see Fig. 15). The range of f was 0.15–0.60. Similarly, Jackson et al.[70] have shown a sharp increase in short-term creep rupture life of columnar-grained MAR-M 200 at 1800 °F (982 °C) with increasing quantities of fine γ'. However fine γ' is not expected to be stable for long at 1800 °F (982 °C), so that the applicability of these results to long-term service applications is questionable. The volume fraction of γ' can be increased in cast alloys by increasing solution treatment temperatures in the range 2160–2260 °F (1187–1250 °C). For the newly developed single-crystal superalloys, grain-boundary-strengthening elements (C, B, Hf, and Zr) are not present, thereby raising the alloy melting point and permitting still higher solution treatment temperatures and higher subsequent volume fractions of particles after aging.

SUMMARY: STRENGTHENING MECHANISMS AND ALLOY DESIGN

Table 3 summarizes the various models of low-temperature strengthening that appear to be directly applicable to austenitic superalloys. For solid solutions the critical parameters are solute concentration and differences in moduli and atomic radii between solute and matrix. The precipitation of coherent ordered particles offers a potent increment to the strength of austenitic iron and nickel-base matrices but has not proven feasible for cobalt-base alloys. The critical precipitate parameters are volume fraction, radius, and antiphase boundary energy; in some cases particle–matrix misfit also is important, particularly at levels of 1% or more. The latter factor controls the strength of IN-718 and IN-901, which are hardened primarily by misfitting γ'' (Ni_3Nb). Donochie[46] has pointed out that for lower temperature applications where tensile properties are the critical factors a duplex γ' size often is preferred in order to disperse slip and reduce notch sensitivity. The actual γ' distribution and volume fraction will be further influenced by section size and by preservice coating heat treatments. Therefore, compositional control and selection of initial heat-treating cycles are not sufficient to ensure optimum microstructures for service.

The strength of complex nickel superalloys can be analyzed in terms of the basic strengthening mechanisms operative in binary nickel–aluminum alloys modified by the partitioning of alloying elements to the γ and γ' phases to influence particle-coarsening kinetics, antiphase boundary energy, and misfit.

A summary of general alloy design principles appears in Table 4. We shall now consider some of these points individually. Analysis of several sets of experiments on austenitic alloys in terms of an order-hardening model, as was summarized in Table 2, showed a marked effect of alloy content on APB energy.† Virtually all superalloys contain both chromium and titanium; yet they produce opposite effects on APB energy. It should be the objective of alloy design to increase γ_0 to the greatest possible extent. Gleiter and Hornbogen[21] have reported evidence for a change in ordering parameters with particle size in nickel–chromium–aluminum alloys so that it is possible for strength to change with aging time and temperature from this factor alone. Alternatively, if the mechanism of particle shear changes with temperature, the fault produced by shear must be taken into account in assessing strength.[14,24,25].

Unfortunately, those factors favoring high strength at low temperatures are not necessarily favorable for either good creep rupture strength or fatigue resistance. While a high volume fraction of γ' undoubtedly is desirable for high yield and tensile strengths (e.g., see Fig. 15), the creep strength of pure γ' is very poor, and relative high cycle fatigue resistance seems to be lowered with increasing volume fraction of particles. Methods of predicting the optimum particle size and lattice mismatches for alloy design are in doubt. Models of particle strengthening in low-volume-fraction alloys suggest that the stress required to shear particles increases

---

† The question of whether APB and misfit strengthening are additive has not yet been resolved.

Table 3. Summary of Hardening Mechanisms[a]

Author	Nature of Obstacles	Total Flow Stress	Conditions
		Solid Solutions	
Mott–Nabarro[1]	Misfitting atom or precipitate	$2G\epsilon c$	$L \geqslant \dfrac{b}{4|\epsilon|f}$
Fleischer[4,6]	Misfitting atom, modulus	$\tau_0 + \dfrac{G(\epsilon'_G - \alpha\epsilon)^{3/2}\, c^{1/2}}{760}$	
Flinn[10]	Short-range order	$\tau_0 + \dfrac{16(\frac{2}{3})^{1/2}\, c\,(1 - c)\,v\alpha}{a^3}$	
		Precipitates	
Copley–Kear[24]	Coherent, ordered, $\epsilon = 0$	$\dfrac{\gamma_0}{2b} - \dfrac{T}{br_0} + \dfrac{(\tau_0 + \tau p)}{2}$	$f \sim 0.6$
Gleiter–Hornbogen[21]	Coherent, ordered, $\epsilon = 0$	$\tau_0 + \dfrac{0.28\gamma_0^{3/2} r_0^{1/2} f^{1/3}}{b^2 G^{1/2}}$	

Brown–Ham[22]	Coherent, ordered, $\epsilon = 0$	$\tau_0 + \dfrac{v_0}{2b}\left[\left(\dfrac{4\gamma_0 r_s f}{\pi T}\right)^{1/2} - f\right]$	$\dfrac{\pi T f}{4\gamma_0} < r_s < \dfrac{T}{\gamma_0}$
Brown–Ham[22]	Coherent, ordered, $\epsilon = 0$	$\tau_0 + \dfrac{\gamma_0}{2b}\left[\left(\dfrac{4f}{\pi}\right)^{1/2} - f\right]$	$r_s > \dfrac{T}{\gamma}$
Gerold–Haberkorn[31]	Coherent, ordered, $\epsilon \neq 0$	$\tau_0 + 3G\epsilon^{3/2}\left(\dfrac{r_0 f}{b}\right)^{1/2}$	Edge dislocation, $\dfrac{9\pi f}{16} < \dfrac{3|\epsilon|r_0}{b} < \dfrac{1}{2}$
Gerold–Haberkorn[31]	Coherent, ordered, $\epsilon \neq 0$	$\tau_0 + G\epsilon^{3/2}\left(\dfrac{r_0 f}{b}\right)^{1/2}$	Screw dislocation, $\dfrac{9\pi f}{26} < |\epsilon|r_0 < \dfrac{1}{2}$
Gleiter[32]	Coherent, ordered, $\epsilon \neq 0$	$\tau_0 + \dfrac{11.8 G\epsilon^{3/2} f^{5/6} r_0^{1/2}}{b^{1/2}}$	
Orowan[35]	Hard particles	$\tau_0 + \dfrac{Gb}{2\pi L}\phi'\ln\dfrac{L}{2b}$	$\dfrac{r_0}{b} > 30$ or incoherent ppt, $\phi' = \frac{1}{2}[1 + (1 - v)^{-1}]$

$^a\tau_0$ = flow stress of matrix without obstacles, r_0 = particle radius, $r_s = (2/3)^{1/2}r_0$, T = line tension, ϵ = misfit; see text for definiton of other terms.

Table 4. Summary of Alloy Design Principles

Low-temperature strength
    Solutes: large lattice and modulus misfits, induce SRO, lower SFE
    Precipitates: coherent, large, high APB energy, large misfit
    Grain size: small
High-temperature strength
    Solutes: large lattice and modulus misfits, SRO, low SFE
    Precipitates: coherent, fine and hyperfine, high APB energy, low misfit
    Grain size: small at $T < 0.5T_m$, large at $T > 0.5T_m$
Creep resistance
    Crystal structure: close packed, stable to T_m
    Solutes: high modulus, slow diffusivity of matrix
    Precipitates: incoherent, fine and hyperfine, large volume fraction, high fault energy,
        low misfit
    Dispersoids: large volume fraction, stable, high GAR
    Grain size: large, columnar or single crystal
    Fibers: large volume fraction, stable

with increasing particle size until the particles grow so large that bypass by dislocation bowing becomes possible. Nevertheless, there are no similar models relating creep resistance to particle size. Similarly, large mismatch appears to be favorable for low-temperature strength but is decidedly harmful for good creep resistance, except perhaps at intermediate temperatures; at 1300 °F (704 °C) the rupture life of three superalloys was reported to *increase* with mismatch.[55] This apparent conflict is easily reconciled, however, if one considers that creep resistance is improved by any factor that increase the stability of the precipitated phase. With regard to grain size effects, there is no doubt that fine grain size is beneficial to low-temperature strength. While it is generally assumed that creep strength is favored by coarse grain sizes, there are no definitive data in the literature to support this view. Some work by Gibbons and Hopkins[69] supports the view that large grains are advantageous for creep resistance in nickel–chromium–titanium–aluminum alloys, but volume fraction and mismatch were not held constant in this work.

    In view of the apparent significance of mismatch in alloy stability at high temperatures, it is necessary to consider the best means of controlling it in austenitic alloys.[3] Partitioning of solute elements between γ and γ' is the best means of controlling mismatch. Titanium and columbium partition to γ' and increase its lattice parameter; chromium, molybdenum, and iron tend to partition to γ, resulting in expansion of this phase (the effect will be small for Cr). Tantalum should behave similarly to columbium and tungsten similarly to molybdenum. Cobalt substitutes primarily in γ and has little effect on lattice parameter. To approach zero mismatch, secondary elements that when added partition to γ' should be balanced by those that partition preferentially to γ.

    Changes in lattice parameter of γ due to removal of molybdenum and tungsten from solution can occur either through precipitation or transformation of carbides or by the formation of σ, μ, and other TCP (topologically close-packed) phases, see Chapter 7. Consequently, alloys that exhibit little or no mismatch prior to service may develop considerable mismatch during exposure to high temperatures, leading to loss of creep resistance. Also, the coefficient of thermal expansion of γ is greater than that of γ' so that it is desirable to produce an alloy in which the room temperature lattice parameter of γ' is somewhat greater than that of γ in order to achieve low mismatch at operating temperature.

REFERENCES

1. N. F. Mott and F. R. N. Nabarro, *Rep. Conf. Strength Sol.* Phys. Soc., 1–9 (1948).
2. R. M. N. Pelloux and N. J. Grant, *Trans. Met. Soc. AIME*, **218**, 232 (1960).
3. R. F. Decker, *Proc. Steel Strength. Mech. Symp.* Chemax Molybdenum Company, Greenwich, Connecticut *Zurich, May 5–6*, **1**: 147, 1964.
4. R. L. Fleischer, *Acta Met.*, **11**, 203 (1963).
5. B. E. P. Beeston, I. L. Dillamore, and R. E. Smallman, *Met. Sci. J.*, **2**, 12 (1968).
6. R. L. Fleischer, *The Strengthening of Metals*, Reinhold, New York, 1964, p. 93.
7. P. Jax, P. Kratochvil, and P. Haasen, *Acta Met.*, **18**, 237 (1970).
8. J. Friedel, *Dislocations*, Pergamon, Oxford, 1964.
9. R. Labusch, Acta Met., **20**, 917 (1972).
10. P. A. Flinn, *Acta Met.*, **6**, 631–635 (1958).
11. R. Nordheim and N. J. Grant, *J. Inst. Met.*, **82**, 440 (1954).
12. A. Akhtar and E. Teghtsoonian, *Met. Trans.*, **2**, 2757 (1971).
13. P. H. Thornton, R. G. Davies, and T. L. Johnston, *Met. Trans.*, **1**, 207 (1970).
14. B. H. Kear, G. R. Leverant, and J. M. Oblak, *Trans. ASM*, **62**, 639 (1969).
15. R. G. Davies and N. S. Stoloff, *TMS-AIME*, **233**, 714 (1965).
16. O. Noguchi, Y. Oya, and T. Suzuki, *Met. Trans. A*, **12A**, 1647 (1981).
17. N. S. Stoloff and R. G. Davies, *Prog. Mat. Sci.*, **13**(1), 3 (1966).
18. D. M. Wee, O. Noguchi, Y. Oya, and T. Suzuki, *Trans. Jap. Inst. Met.*, **21**, 237 (1980).
19. L. R. Curwick, Ph.D. Thesis, University of Minnesota, 1972.
20. R. K. Ham, *Ordered Alloys: Structural Applications and Physical Metallurgy*, Claitors, Baton Rouge, LA, 1970, p. 365.
21. H. Gleiter and E. Hornbogen, *Mat. Sci. Eng.*, **2**, 285 (1968).
22. L. M. Brown and R. K. Ham, *Strengthening Methods in Crystals*, Elsevier, Amsterdam, 1971, p. 9.
23. V. Marteus and E. Nembach, *Acta Met.*, **23**, 149 (1975).
24. S. M. Copley and B. H. Kear, *TMS-AIME*, **239**, 977 (1967).
25. S. M. Copley and B. H. Kear, *TMS-AIME*, **239**, 984 (1967).
26. G. R. Leverant, M. Gell, and S. W. Hopkins, *Proc. Sec. Int. Conf. Strength Met. Alloys*, **3**, 1141 (1970).
27. V. A. Phillips, *Scr. Met.*, **2**, 147 (1968).
28. H. Gleiter and E. Hornbogen, *Phys. Stat. Sol.*, **12**, 251 (1965).
29. D. Raynor and J. M. Silcock, *Met. Sci. J.*, **4**, 121 (1970); see also ref. 22.
30. J. L. Castagne, A. Pineare, and M. Sidzingre, *C. R. Acad. Sci.*, **C263**, 1465 (1966).
31. V. Gerold and H. Haberkorn, *Phys. Stat. Sol.*, **16**, 675 (1966).
32. H. Gleiter, *Z. Angew. Phys.*, **23**(2), 108 (1967).
33. V. Gerold, *Acta Met.*, **16**, 823 (1968).
34. E. Nembach and G. Neite, *Prog Mat Sci* (in press).
35. E. Orowan, *Symposium on Internal Stresses in Metals*, Institute of Metals, London, 1948, pp. 451–453.
36. A. Kelly and R. B. Nicholson, *Prog. Mat. Sci.*, **10**(3), 151 (1963).

37. B. Wilcox and A. H. Clauer, in *The Superalloys*, C. T. Sims and W. C. Hagel (eds.), Wiley, New York, 1972.

38. B. A. Wilcox and A. H. Clauer, *Oxide Dispersion Strengthening*, Gordon & Breach, New York, 1968, p. 323.

39. J. P. Morse and J. S. Benjamin, in *New Trends in Materials Processing*, ASM, Metals Park, OH, 1976, p. 165.

40. E. Arzt and R. F. Singer, in *Superalloys 1984*, M. Gell, C. S. Kortovich, R. H. Bricknell, W. B. Kent, J. F. Radavich (eds.) TMS-AIME, New York, 1984, p. 367.

41. R. R. Jensen and J. K. Tien, in *Metallurgical Treatises*, J. K. Tien, J. F. Elliott (eds.) TMS-AIME, Warrendale, PA, 1981, p. 529.

42. R. F. Decker and J. R. Mihalisin, *Trans. ASM*, **62**, 481 (1969).

43. E. Nembach, K. Suzuki, M. Ichihara, and S. Takeuchi, *Philos. Mag. A*, **51**, 607 (1985).

44. P. Beardmore, R. G. Davies, and T. L. Johnston, *TMS-AIME*, **245**, 1537 (1969).

45. L. R. Cornwell, J. D. Embury, and G. R. Purdy, in *Ordered Alloys: Structural Applications and Physical Metallurgy*, B. H. Kear et al. (eds.), Claitors, Baton Rouge, LA, 1970, p. 387.

46. M. Donachie, *Source Book on Superalloys*, ASM, Metals Park, OH, 1984, p. 7.

47. C. L. Corey and B. Lisowsky, *TMS-AIME*, **239**, 239 (1967).

48. V. A. Phillips, *Philos. Mag.*, **16**, 117 (1967).

49. R. G. Davies and T. L. Johnston, *Ordered Alloys: Structural Applications and Physical Metallurgy*, B. H. Kear et al. (eds.), Claitors, Baton Rouge, LA, 1970, p. 447.

50. W. J. Mitchell, *Z. Metallkd.*, **57**, 586 (1966).

51. Y. G. Sorokina and S. A. Yuganova, *Met. Sci. Heat Treat.*, 456 (1968).

52. V. Munjal and A. J. Ardell, *Acta Met.*, **23**, 513 (1975).

53. I. L. Mirkin and O. D. Kancheev, *Met. Sci. Heat Treat.*, (1,2), 10 (1967).

54. G. N. Maniar and J. E. Bridge, *Met. Trans.*, **2**, 95 (1971).

55. C. C. Law and M. J. Blackburn, *Met. Trans.*, *A*, **11A**, 495 (1980).

56. M. N. Thompson and J. W. Edington, *Proc. 2nd Int. Conf. Strength Met. Alloys*, **3**, 1150 (1970).

57. P. H. Thornton and R. G. Davies, *Met. Trans.*, **1**, 549 (1970).

58. G. R. Leverant and B. H. Kear, *Met. Trans.*, **1**, 491 (1970).

59. O. D. Sherby and P. M. Burke, *Prog. Mat. Sci.*, **13**, 325 (1967).

60. A. K. Mukherjee, J. E. Bird, and J. E. Dorn, *Trans. ASM*, **62**, 155 (1969).

61. A. K. Mukherjee in *Treatise on Materials Science and Technology*, Vol. 6, *Plastic Deformation of Metals*, R. J. Arsenault (ed.), Academic, New York, 1975, p. 163.

62. M. Malu and J. K. Tien, *Scripta Met.*, **9**, 1117 (1975).

63. K. R. Williams and B. Wilshire, *Met. Sci. J.*, **7**, 176 (1973).

64. J. D. Whittenberger, *Met. Trans. A*, **8A**, 1155 (1977).

65. J. K. Tien and S. M. Copley, *Met. Trans.*, **2**, 543 (1971).

66. J. K. Tien and R. P. Gamble, *Met. Trans.*, **3**, 2157 (1972).

67. D. D. Pearson, B. H. Kear, and F. D. Lemkey, in *Creep Fracture of Engineering Materials and Structures*, Pineridge, 1981, p. 213.

68. D. D. Pearson, private communication.

69. T. B. Gibbons and B. E. Hopkins, *Met. Sci. J.*, **5**, 233 (1971).

70. J. J. Jackson, M. J. Donachie, R. J. Henricks, and M. Gell, *Met. Trans. A*, **8A**, 1615 (1977).

Chapter 4

Nickel-Base Alloys

EARL W. ROSS and CHESTER T. SIMS

Aircraft Engine Business Group, General Electric Company, Cincinnati, Ohio, and Rensselaer Polytechnic Institute, Troy, New York

Nickel-base alloys are the most complex, the most widely used for the hottest parts, and to many metallurgists, the most fascinating of all superalloys. Their use extends to the highest homologous temperature of any common alloy system, and they currently comprise over 50% of the weight of advanced aircraft engines. The physical metallurgy is complex, subtle, and sophisticated, yet the relationship of properties to structures in these systems is certainly the most best known of all materials for use in the 1200–2000 °F (650–1100 °C) range.

From 1940 to 1965 the properties given the most attention for applications such as aircraft engine blades were high-temperature tensile strength, creep rupture strength to 5000 h, and oxidation resistance. Conversely, industrial gas turbine designers required blade alloys with known longer time creep rupture properties and good hot-corrosion resistance. Aircraft engines for longer periods of service and industrial gas turbines for peaking electrical power generation needs now require materials with a combination of the previously noted properties coupled with excellent high- and low-cycle thermal fatigue resistance. Thus, aircraft engines for advanced transport systems aimed at 20,000–50,000 h life and the industrial turbines aimed at up to 100,000 h life require consideration of many factors to ensure high performance and reliability.

To deal with these problems, new tools and new metallurgical strategy are being developed, and a study of the fundamental characteristics of austenitic solid-state systems is occurring at a sharply increasing rate. Nickel superalloy metallurgy is showing significant scientific enlightenment after years of existence as an art. Meanwhile, competitive materials such as ceramics and refractory metals are being evaluated for future advanced application but have yet to prove their reliability.[1]

This chapter discusses the physical metallurgy of nickel superalloys from the classical standpoint of "enlightened empiricism" and to generate guiding relationships between composition, structure, and properties. Fundamentals of strengthening are discussed in Chapter 3.

CHEMICAL COMPOSITION

The composition of many types of nickel-base alloys are listed in Appendix B. At first glance the complexity of their compositions appears to defy logic. At least 12–13 important elemental constituents are included and carefully controlled; in addition, "tramp" elements such as silicon, phosphorus, sulfur, oxygen, and nitrogen also must be controlled through appropriate melting practice. Trace elements such as selenium, thallium, tellurium, lead, and bismuth must be held to very small (ppm) levels in critical parts; this is accomplished by careful selection of raw materials coupled with optimum melting practice. However, it should be obvious that most of the nickel alloys contain 10–20% chromium, up to about 8% aluminum and titanium, 5–10% cobalt, and small amounts of boron, zirconium, and carbon. Optional common additions are molybdenum, tungsten, columbium, tantalum, and hafnium.

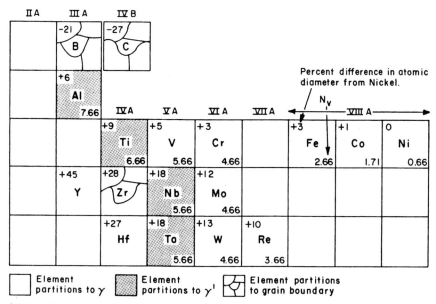

Fig. 1. Elements important in the constitution of nickel-base alloys.[2]

Table 1. Elements in Superalloys

Elements	Ni	Co	Fe	Cr	Mo, W	Cb, Ta, Ti	Al	C, B, Zr, Hf
Matrix class	X	X	X	X	X			
γ' class						X	X	
Grain boundary class								X
Carbide subclass				X	X	X		
Oxide scale subclass				X			X	
Examples[a]								
René 77[b]	58.4	15	—	14.6	4.2 Mo	3.4 Ti	4.3	0.07 C, 0.016B
MAR-M 200 H	58.9	10	1.5	9.0	12.5 W	2.0 Ti	5.0	0.15 C, 0.015 B, 2.0 Hf[c]

[a] In weight percent.
[b] René 77 is similar to U-700/Astroloy; René 77 is phase controlled to be σ free after long-time exposure.
[c] The role of hafnium can be very complex; it could be equally justified as a γ' class element here.

It is, of course, a primary task of this chapter to explain the reasons for the presence of these elements. Thus, an orderly initial view can be discerned from Fig. 1, which illustrates that the alloying elements do tend to be grouped with some commonality in the periodic system.

The first class consists of elements that prefer and make up the face-centered-cubic (FCC) austenitic (γ) matrix. These are from Groups V, VI, and VII and include nickel, cobalt, iron, chromium, molybdenum, and tungsten. The second class of elements partition to and make up the γ' precipitate Ni_3X. These elements are from Groups III, IV, and V and include aluminum, titanium, columbium, tantalum, and hafnium. Boron, carbon, and zirconium make up a third class of elements that tend to segregate to grain boundaries. These elements are from Groups II, III, and IV and are very odd sized in atomic diameter.

As shown in Table 1, two subclassifications are beyond these three major classifications. One includes the carbide formers: chromium, molybdenum, tungsten, columbium, tantalum, and titanium. The second subclassification comprises the chromium and aluminum oxide formers, which develop adherent diffusion-resistant oxides to protect the alloys from the environment.

STRUCTURE AND MICROSTRUCTURE

Figure 2 illustrates how the microstructure (*following service*) of these alloys has developed[3] with time. The major phases present in these and other nickel superalloys are as follows:

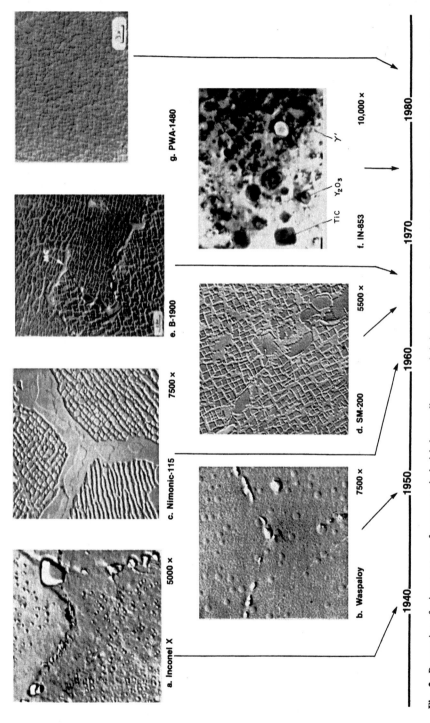

Fig. 2. Progression of microstructures for some typical nickel superalloys over their invention span. Sources: (1) General Electric; (b) Special Metals; (c) Henry Wiggin; (d) Martin Metals; (e,g) Pratt and Whitney; (f) INCO.

a. Inconel X   5000×

b. Waspaloy   7500×

c. Nimonic-115   7500×

d. SM-200   5500×

e. B-1900

f. IN-853   10,000×

g. PWA-1480

1. *Gamma Matrix* (γ). The continuous matrix is an FCC nickel-base austenitic phase called gamma that usually contains a high percentage of solid-solution elements such as cobalt, chromium, molybdenum, and tungsten.

2. *Gamma Prime* (γ'). Aluminum and titanium, for example, are added in amounts and mutual proportions to precipitate high volume fractions of FCC γ', which invariably precipitates coherently with the austenitic γ matrix.

3. *Carbides*. Carbon, added at levels of about 0.05–0.2%, combines with reactive and refractory elements such as titanium, tantalum, and hafnium to form MC carbides. During heat treatment and service these begin to decompose and generate lower carbides such as $M_{23}C_6$ and M_6C, which tend to populate the grain boundaries.

4. *Grain Boundary* γ'. For the stronger alloys, heat treatments and service exposure generate a film of γ' along the grain boundaries; this is believed to improve rupture properties.

5. *Borides*. Occur as infrequent grain boundary particles.

6. *TCP-Type Phases*. † Under certain conditions, platelike phases such as σ, μ, and Laves form; this can result in lowered rupture strength and ductility. These phases are discussed in detail in Chapter 9.

CONSTITUTION, STRUCTURE, AND REACTIONS OF PHASES

An important first principle is that all superalloys exposed to high temperatures are chemically dynamic structures. The phases present are constantly reacting and interacting. We only observe a nickel-base alloy structure temporarily at room temperature to record its appearance and to analyze it. The very complex high-temperature solid-state reactions prevent defining chemical equations of state (with appropriate activation energies) to categorize the systems.

THE GAMMA MATRIX (γ)

Although nickel alone is not endowed with a distinctly high modulus of elasticity or low diffusivity (two factors that promote creep rupture resistance), the γ matrix is favored by most gas turbine designers for the most severe temperature and time excursions. It is remarkable that some of these alloys can be utilized at $0.9T_M$ (melting point) and for times up to 100,000 h at somewhat lower temperatures. The basic reasons for this endurance must be attributable to the following.

1. The high tolerance of nickel for alloying without phase instability owing to its nearly filled third electron shell[2].

---

† TCP is topologically close packed.

2. The tendency, with chromium additions, to form Cr_2O_3-rich protective scales having low cation vacancy content, thereby restricting the diffusion rate of metallic elements outward, and oxygen, nitrogen, sulfur, and other aggressive atmospheric elements inward[2].

3. The additional tendency, at high temperatures, to form Al_2O_3-rich scales with exceptional resistance to oxidation.

Phase Relations

The γ in these alloys consists principally of nickel, cobalt, chromium, and refractory metals such as molybdenum or tungsten. The γ can be considered as a phase common to a group of quaternary phase diagrams extending outward from the nickel–cobalt binary tieline, as shown in Fig. 3. One can observe obvious similarities in the quaternaries, particularly with regard to the band of compounds separating the FCC austenitic quaternary space from the body-centered-cubic (BCC) quaternary space (Cr–Mo tieline). This band contains the hard TCP phases, which must be constitutionally avoided.

Brewer[5] has described these phase relations in planar form while applying the Engel correlation between electronic configuration and crystal structure. This work was then used as the basis for a further development by Tarr and Marshall,[6] who plotted Brewer's phase relation data on polar coordinates. Examples of diagrams of this type for a few elements of interest in superalloy metallurgy have been plotted and are included in Appendix A; they are called polar phase diagrams. Study of the polar phase diagrams can give additional insight into general phase relations.

In polar diagrams the effect of electron vacancy [electron hole number (N_v)] is evident. Lines of constant electron–atom ratio (e/a) for alloys should spiral outward in a counterclockwise fashion. The close relationship between e/a and intermetallic compounds formed becomes obvious, since σ, μ, and Laves, as well as the important γ field boundary, have a tendency to follow a locus of relatively constant e/a. The alloy matrix of all iron, nickel, or cobalt superalloys stems from the FCC γ field with most of the compositions residing fairly close to the field limit.

Solid-Solution Strengthening

From phase analyses of complex nickel-base superalloys[7–9] the common solid-solution elements in γ are cobalt, iron, chromium, molybdenum, tungsten, titanium, and aluminum. As seen in Fig. 1, these elements differ from nickel by 1–13% in atomic diameter and 1–7% in N_v. As reported previously,[2] hardening can be related to atomic diameter oversize as measured by lattice expansion.

Some effect also may result from lowering of stacking-fault energy by the alloying elements to make cross-slip more difficult in γ. The N_v can be correlated with stacking-fault measurement.[10,11]

The strengthening potency of solid-solution elements can be estimated; consider a highly alloyed superalloy with γ of atom percent composition as follows:

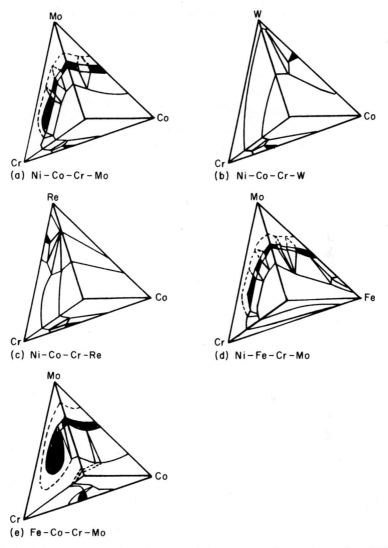

Fig. 3. Approximate quaternary phase diagrams of high-temperature alloy matrices at about 2200 °F (1200 °C).[3] Nickel corner at facing apex.

Co	Fe	Cr	Mo	W	V	Al	Ti
20	10	20	4	4	1.5	6	1

The change in lattice constants (kX) for binary addition to the nickel content of this matrix would be:

Co	Fe	Cr	Mo	W	V	Al	Ti
0.011	0.020	0.033	0.035	0.038	0.006	0.025	0.006

The change in flow stress at room temperature (in kgf/mm^2) is estimated[10,11] to be:

Co	Fe	Cr	Mo	W	V	Al	Ti
1.8	5.6	16	17	18	3.4	20	4

Aluminum, usually noted only as a precipitation strengthener, rates as a potent solid-solution strengthener. Tungsten, molybdenum, and chromium also contribute strongly, whereas iron, titanium, cobalt, and vanadium serve only as weak solid-solution strengtheners.

These effects persist to high temperatures, but above $0.6T_M$, the range of high-temperature creep, γ strengthening is diffusion dependent. The slow-diffusing elements molybdenum and tungsten would be expected to be the most potent hardeners. An additional beneficial effect on diffusion has been shown[12]; in a Ni–22%Cr–2.8%Ti–3.1%Al alloy the presence of molybdenum and tungsten lowered the diffusivity of titanium and chromium at 1650 °F (900 °C).

GAMMA PRIME PHASES (γ')

The precipitation of FCC A_3B compounds, γ' in superalloys is a most fortunate event. The nickel atom is incompressible owing to its $3d$ electron state. Thus, a high-nickel matrix favors precipitation of γ', which requires little size change (experience has shown that FCC alloys require at least 25% nickel). More complex phases that require atomic size changes are avoided. Matrices with higher electron hole characteristics (N_v), such as iron, favor these latter undesirable phases. Also, the compatibility of the γ' FCC crystal structure and lattice constant (approximately 0.1% mismatch) with γ allows homogeneous nucleation of a precipitate with low surface energy and extraordinary long-time stability. Coherency between γ' and γ is maintained by tetragonal distortion.

In the γ'-type A_3B compound relatively electronegative elements such as nickel, cobalt, or iron compose the A, and more electropositive elements such as aluminum, titanium, tantalum, or columbium compose the B. Typically, in a nickel-base alloy, γ' is $(Ni,Co)_3(Al,Ti)$ with nickel and aluminum dominating, although it is common to add at least as much titanium as aluminum. However, substitutions for A and B are more complex than this and are discussed in more detail later.

The γ' is a unique intermetallic phase. It contributes remarkable strengthening by dislocation interaction to force, bypassing or particle cutting to the γ–γ' alloy. More remarkably, the strength of γ' increases as temperature increases. Furthermore, the inherent ductility of γ' prevents it from being a source of fracture. This is also in direct contrast to the severe embrittlement created by the formation of the brittle σ phase. Details of the mechanisms by which γ' strengthens are explored in Chapter 3.

Microstructure

The γ' was first observed as a spherical precipitate and then as cubes; its shape was later found to be related to matrix-lattice mismatch. Hagel and Beattie[13] observed that γ' occurs as spheres at 0–0.2% lattice mismatch, becomes cubes at mismatches around 0.5%–1.0%, and then becomes plates at mismatches above about 1.25%. Several illustrations of typical γ' structures found in nickel superalloys such as AF-1753,† IN-100, and others are shown in Fig. 4; many other variations in size and shape can be found as well.

Composition Relations

The schematic isothermal ternary section of elements with nickel and aluminum of Fig. 5 points out how elements substitute and partition in γ'. Cobalt, with its horizontal phase field, substitutes for nickel. Titanium, columbium, tantalum, and hafnium would substitute for aluminum positions in the ordered structure, as demonstrated by a phase running diagonally from Ni_3Al to Ni_3X. Molybdenum, chromium, and iron would substitute for both nickel and aluminum positions since their phase fields are intermediate between the above two extremes.

Much of this has been confirmed by work on commercial alloys. For instance, chromium was identified by Decker and Bieber[17] to be present in the γ' of Inconel 713C. While its atom size permits substitution on either side of A_3B, it has been commonly assumed to be committed for nickel. Mihalisin and Pasquine[7] determined the composition of extracted γ' from 713C to be

$$(Ni_{0.980}Cr_{0.016}Mo_{0.004})_3(Al_{0.714}Cb_{0.099}Ti_{0.048}Mo_{0.038}Cr_{0.103})$$

showing chromium mostly on the aluminum side. When cobalt is present, it substitutes mostly for nickel, and the chromium shifts left as in IN-731. (IN-731 is an experimental alloy; its composition is given in Table 2):

$$(Ni_{0.884}Co_{0.070}Cr_{0.032}Mo_{0.088}V_{0.003})_3Al_{0.632}Ti_{0.347}V_{0.013}Cr_{0.006}Mo_{0.002}).$$

Until recently, of the refractory elements molybdenum, tungsten, columbium, and tantalum, only columbium was believed to enter γ' extensively. In fact, in certain nickel–chromium–iron superalloys, such as Inconel 718, high additions of columbium lead to precipitation of Ni_3Cb. In γ' the columbium combines with aluminum and titanium to increase the volume of Ni_3X precipitate and may possibly increase the solution temperature of γ' so that its strengthening effect carries to higher temperatures.

Since early analysis of γ' indicated little tendency for molybdenum and tungsten to enter its lattice, these elements were felt to participate only in carbide reactions and solid-solution strengthening. However, further work has indicated this may not

† AF-1753: 16.3Cr–9.5Fe–7.2Co–3.2Ti–1.9Al–8.4W–1.6Mo–0.24C, balance Ni.

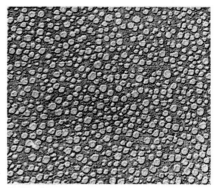

a. Cubical and Trigonal γ' in NASA IIb
S-R Tested at 1900F (1040C)
2725X. Kent[15]

b. Typical Spherical and Cooling γ'
in S-R Tested U500. 5,450X.

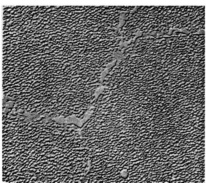

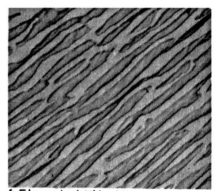

e. Very Fine γ' in AF-1753. S-R
Tested at 1350F (735C). 4,100X

f. Elongated γ' in Alloy 713 C
Tested at 1500F (815 C)[2]

Fig. 4. Illustrations of γ' morphology in nickel-base alloys. Various conditions and heat treatments (S-R = stress rupture).

be the case.[18] For example (Table 2), γ' removed from MAR-M 200, nominally containing about 4.0 a/o tungsten and 0.6 a/o columbium, contains 3.2 a/o tungsten and 0.5 a/o columbium in the γ'. Clearly, considerable tungsten is present in the MAR-M 200 γ'. Other data in Table 2 show similar effects. Although not listed in this table, tantalum also is a strong participant in γ'; it is particularly important in single-crystal alloys.

Molybdenum has been found by Loomis et al.[9] to dissolve extensively in γ' of titanium-free alloys and to a lesser extent in γ' of alloys with a high titanium–aluminum ratio (Table 2). Molybdenum raised the lattice parameter, solvus temperature, and weight fraction of γ' in proportion to the molybdenum content of the γ'.

c. Typical Cubical γ' in IN-100 S-R
Tested at 1500 F (815 C) 13,625 X
Mihalisin[14]

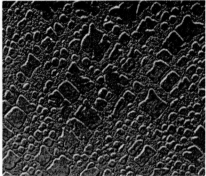

d. Fine, Medium and Coarse γ' in
Heated-treated IN-738. 5,450 X

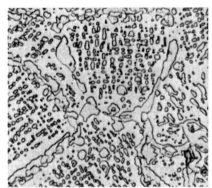

g. Condensed γ' in U700 Aged 180
Days at 1900F (1040C). 545 X

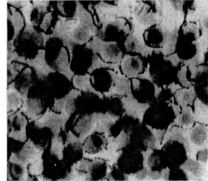

h. Dislocations Decorating γ' in U500
Tested at 1450F (790 C). Transmission, 4,800 X. Phillips[16]

Fig. 4. (*continued*)

Substitution of neighboring elements for nickel can decrease A_3B solubility. In Fig. 6 it can be seen that substitution of cobalt for nickel reduces low-temperature solubility of the nickel–chromium matrix for aluminum and titanium. In fact, chromium, iron, and cobalt can all be added to increase the volumetric percentage of γ' at a given aluminum plus titanium level. However, this effect must not be confused with effects on strength; it is common knowledge that in complex commercial alloys, decreasing chromium and iron increase net alloy strength. Conversely, use of cobalt in wrought superalloys to facilitate hot working at high aluminum plus titanium contents may be related to increased solubility for γ' above 2000 °F (1100 °C).

Table 2. Analysis of γ' in Nickel-Base Alloys[18]

Alloy	Component Analyzed	Composition a/o									Remarks
		Ni	Co	Cr	Al	Ti	Mo	W	Cb(V)	C	
Mar-M 200	Av. alloy	59.6	10.95	10.24	10.97	2.47	—	4.02	0.64	0.74	Nominal composition. Separated and analyzed[a].
	γ'	64.19	6.71	4.02	18.01	3.42	—	3.16	0.49	—	
Udimet 500	Av. alloy	49.62	18.11	20.84	6.22	2.33	2.5	—	—	0.37	Wet chemical analysis. 180 days at 1900 °F (1050 °C), separately analyzed. No stress.
	γ'	50.0	4.0	1.6	—	4.0	1.3	—	—	—	
Udimet 520	Av. alloy	55.35	11.63	20.87	4.45	3.58	3.57	0.31	—	0.24	Nominal composition. 90 days at 1900 °F (1040 °C) microprobe. No stress.
	GB γ'[b]	73.89	7.22	4.31	—	9.49	1.8	0.31	—	—	
	γ'[c]	68.8	6.92	4.08	—	13.87	0.64	0.3	—	—	Same as above.
Alloy 713C	Av. alloy	68.77	—	13.96	12.03	0.92	2.58	—	1.24	0.5	Analyzed composition as cast.[7] 4 hr at 1900 °F (1040 °C)[2].
	γ'	73.36	—	3.73	17.95	1.26	1.21	—	2.49	0.0	
IN-731X	Av. alloy	61.56	8.91	9.97	11.22	5.3	1.41	—	(0.9V)	0.72	Analyzed composition as cast.[7] 737.7 hr at 1800 °F (980 °C).
	γ'	65.55	5.17	2.30	15.28	10.62	0.50	—	(0.6V)	0.0	
Ni-Cr-Al-Mo	Av. alloy	69.9	—	13.42	12.09	—	4.46	—	—	0.02	Wet chemical analysis[9]. Separated and analyzed after 112 hr at 1700 °F (927 °C).
	γ'	75.2	—	4.4	17.1	—	3.3	—	—	—	
Ni-Cr-Ti-Al-Mo	Av. alloy	74.4	—	14.25	2.21	4.14	4.84	—	—	0.03	Wet chemical analysis[9]. Separated and analyzed after 112 hr at 1700 °F (972 °C).
	γ'[b]	76.0	—	1.1	8.6	13.5	0.79	—	—	—	

[a] B. H. Kear, personal communication.

[b] GB = grain boundary.

[c] I = intergranular.

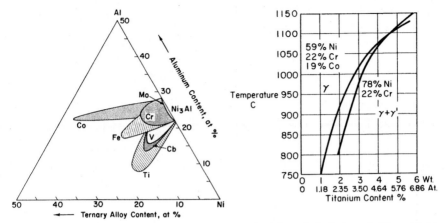

Fig. 5. Semischematic presentation of Ni₃Al solid-solution field at approximately 2100 °F (1100 °C) for various alloys.[2]

Fig. 6. Effect of substituting cobalt in a nickel-base system on the solvus. Ti : Al ratio = 1:1.[19]

γ′ Dispersion Stability

On thermal exposure above about $0.6T_M$ the γ′ ripens (increases in size) at a significant rate, facilitating dislocation bypassing. Thus, measures that minimize ripening will help retain long-time creep resistance. Fleetwood[20] has applied Wagner's theory of Ostwald ripening to γ′:

$$h^3 = \frac{64\gamma_e D C_e V_m^2 t}{9RT} \tag{1}$$

where

$$t = \text{time}$$

$$\gamma_e = \text{specific } \gamma' - \gamma \text{ interfacial free energy}$$

$$D = \text{coefficient of diffusion of } \gamma' \text{ solutes in } \gamma$$

$$C_e = \text{equilibrium molar concentration of } \gamma' \text{ solute in } \gamma$$

$$V_m = \text{molar volume of } \gamma'$$

$$R = \text{gas constant}$$

$$h = \text{particle size}$$

The significant changes with composition come in the terms γ_e, C_e, and D. Fleetwood found that the ripening rate of γ' in nickel–chromium–titanium–aluminum alloys decreased as chromium increased from 10 to 37%. This arose partly from reduced C_e but also partly from reduction of coherency strains and the consequent reduction of $\gamma_e D$. Increased coherency strains by a higher titanium–aluminum ratio increased ripening rate.[21]

The coarsening of γ' is retarded significantly by cobalt, molybdenum, or a combined addition of molybdenum and tungsten. Increasing columbium from 2 to 5% markedly reduces the coarsening rate despite the concurrent increase in coherency strains (22). Columbium partitions almost completely to γ', yielding low C_e and low D. These reductions in C_e and D from columbium were more influential than the increased coherency strains with increasing titanium–aluminum ratio. No effect of boron and zirconium, however, has been found on the growth rate of γ' in a nickel–chromium–aluminum alloy.[23]

Creep strain has little effect on ripening with 33 or less volume percent of γ'.[22-24] Obviously, the stress on γ' at 50 vol. % in Fig. 4f has accelerated the generation of γ' "rafting"; this has been particularly observed in single crystals after stressed exposure.

Studies of cyclic overheating of superalloys[24] showed that fine γ' dispersions were automatically restored at the normal service temperature. However, the loss of creep resistance during γ' ripening was very dependent on volume percent γ'. M-252 (low-volume-percent γ') weakens faster than Inconel 700 (high-volume-percent γ'), and sensitivity of flow stress to changes in γ' particle size was much greater in the leaner alloys. Thus, the available options to retard ripening would be (1) increase the volume percent γ' and (2) add high partitioning, slow-diffusing elements such as columbium and tantalum to the γ'.

Transformations of γ' to Eta and Ni$_3$Cb

The γ' containing only aluminum or sufficiently high aluminum cannot transform to other Ni$_3$X compounds but can transform when the titanium and/or columbium and/or tantalum content is sufficiently high. Pearson and Hume-Rothery[25] related stability of the Ni$_3$X compounds to size factor. Ranking them in the decreasing order of stability yields Ni$_3$Al, Ni$_3$Ti, and Ni$_3$Cb (or Ta). Although the equilibrium diagrams predict otherwise, the aluminum in Ni$_3$Al can be replaced by titanium, columbium, or tantalum, leaving the possibility of metastable γ'.

A-286 is an example of a commercial wrought iron–nickel γ'-strengthened superalloy that contains 2.1% titanium and 0.3% aluminum. The Ni$_3$(Ti, Al) γ' in A-286 is metastable and will transform from the strengthening cubic γ' to a weakening platelike hexagonal eta (η) phase on exposure to temperatures above 1200 °F. This limits A-286 to 1200 °F maximum usage.

The titanium-rich metastable γ' transforming to η (Ni$_3$Ti hexagonal close packed) also occurred in a Ni–22%Cr alloy when the titanium–aluminum ratio was raised to 5 : 1[26]; tungsten retarded the transformation in such an alloy.[27] The effect has also been noted at ratios of 3 : 1. Some intergranular η developed in Incoloy 901

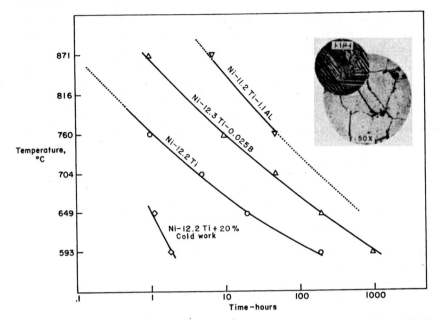

Fig. 7. Effect of cold work and boron and aluminum additions on time for the start of cellular precipitation of η.[29]

after 2850 h at 1350 °F (730 °C), but no undue deterioration in strength or ductility was noted.[28] When η precipitates as cells at grain boundaries, notch stress rupture strength is reduced; when precipitated intergranularly, in Widmanstätten fashion, strength, but not ductility, is reduced.[2] Several means are available to retard the γ' → η reaction. Trace levels of boron are commonly added to nickel-base superalloys; equilibrium segregation of boron to grain boundaries retards the nucleation of cells,[29] with an accompanying increase in notched stress rupture strength. This effect, plus the retarding effect of aluminum and the accelerating effect of cold work is illustrated in Fig. 7. This factor and others are particularly closely related to nickel–iron alloys; detailed studies on effects and mechanisms for these alloys have been conducted,[30,31,33,34] as discussed in Chapter 6.

CARBIDES

The role of carbides in superalloys is complex. First, carbides appear to prefer grain boundaries as location sites in nickel alloys, while in cobalt and iron superalloys and other alloy matrices of higher N_v, intragranular sites are common. Early investigators noted detrimental effects on ductility from certain grain boundary carbide morphologies and took the logical step of reducing carbon to very low levels. However, further studies of this variable uncovered sharply reduced creep life and ductility with less than 0.03% carbon in Nimonic 80A[35] and Udimet 500.[36]

Thus, opinions seem to vary as to whether carbides are to be tolerated or are essential in superalloy grain boundaries. However, most investigators today feel carbides exert a beneficial effect on rupture strength at high temperature. In addition, it is quite clear that carbide morphology can influence ductility and that carbides can influence chemical stability of the matrix through the removal of reacting elements. Therefore, understanding of the desirable chemistry, class, and morphology of carbides is critical to the alloy designer in the selection of compositions and heat treatments.

Classes of Carbides and Typical Morphologies

The common nickel-base alloy carbides are MC, $M_{23}C_6$, and M_6C (Table 3). MC usually takes a coarse random cubic or script morphology. $M_{23}C_6$ shows a marked tendency for grain boundaries; it usually occurs as irregular discontinuous blocky particles, although plates and regular geometric forms have been observed. M_6C can also precipitate in blocky form in grain boundaries and more rarely in a Widmanstätten intragranular morphology, as seen in B-1900. Although data are insufficient for precise correlation, it is apparent that continuous and/or denuded grain boundary $M_{23}C_6$ and Widmanstätten M_6C are to be avoided for best ductility and rupture life. Structures of some typical and nontypical carbide forms are illustrated in Fig. 8.

MC carbides usually form in superalloys during freezing; they occur as discrete particles (Fig. 8) distributed heterogeneously throughout the alloy, both in intergranular and transgranular positions, often interdendritically. Little or no orientation with the alloy matrix has been noted. MC carbides are a major source of carbon for the alloy to use later during heat treatment and service.

These primary carbides are FCC. Their dense, closely packed structures are very strong, and, if pure, they are some of the most stable compounds in nature. They

Table 3. Carbides and TCP Phases[a] Identified in Nickel Superalloys

Alloy	Mo + W, a/o	Carbides			TCP Phases	
IN-100	1.7	MC	$M_{23}C_6$	—	σ	—
Nimonic-115	2.0	MC	$M_{23}C_6$	—	σ	—
U-500	2.3	MC	$M_{23}C_6$	—	σ	—
U-700	2.4	MC	$M_{23}C_6$	—	σ	—
Alloy 713C	2.6	MC	$M_{23}C_6$	—	σ	—
B-1900	3.5	MC	$M_{23}C_6$	M_6C	σ	μ
AF-1753	3.7	MC	$M_{23}C_6$	M_6C	—	μ
AF2-1DA	3.7	MC	$M_{23}C_6$	M_6C	—	μ
Mar-M 200	4.0	MC	$M_{23}C_6$	M_6C	—	μ
René 41	6.0	MC	$M_{23}C_6^b$	M_6C	—	μ
M-252	6.1	MC	$M_{23}C_6^b$	M_6C	—	μ

[a] The TCP phases observed generally have been found in abnormal heats of these alloys; modern chemistry control practices effectively eliminate occurrence in most cases.
[b] $M_{23}C_6$ found only infrequently.

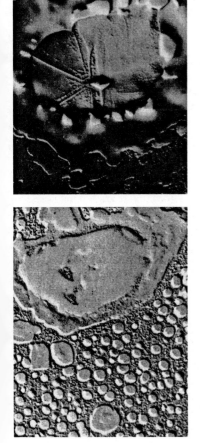

a. Typical MC Particle in γ' Strengthened Alloy; Degeneration Commenced. 4,900X

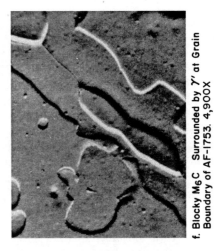

c. Cellular $M_{23}C_6$ Formed in Nimonic 80A at 1200F (650C). 4,900X . Hagel and Beattie

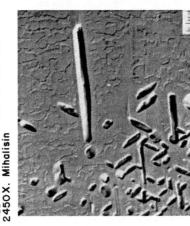

b. Degenerated MC (Diamond) Surrounded by $M_{23}C_6$ Particles and Matrix in IN-100. 2450X. Mihalisin

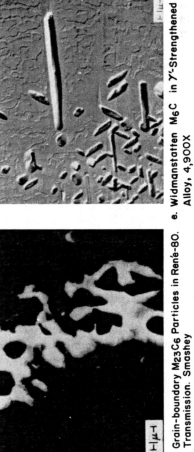

d. Grain-boundary $M_{23}C_6$ Particles in René-80. Transmission. Smashey

e. Widmanstatten M_6C in γ'-Strengthened Alloy, 4,900X

f. Blocky M_6C Surrounded by γ' at Grain Boundary of AF-1753. 4,900X

Fig. 8. Examples of carbides in nickel-base alloys.

113

occur from simple combination of carbon with reactive and refractory metals, classically possessing a formula such as TiC or TaC. The preferred order of formation in superalloys for these carbides is HfC, TaC, CbC, and TiC in order of decreasing stability. This order is not the same as dictated by thermodynamics, which is HfC, TiC, TaC, and CbC.

Obviously, M atoms can substitute for each other, as in (Ti,Cb)C. However, the less reactive elements, principally molybdenum and tungsten, also can substitute in these carbides. For instance, one will find (Ti,Mo)C in U-500, M-252, and René 77. In the latter alloy it has been analyzed to be $(Ti_{0.8}Mo_{0.2})C$ plus traces of nickel and chromium. It is reasonably certain that the change in stability order cited above is due to the molybdenum or tungsten substitution, which so weakens the binding forces in MC carbides that degeneration reactions, discussed below, can occur. This leads typically to formation of $M_{23}C_6$ and M_6C-type carbides as the more stable compounds in the alloys after heat treatments and/or service.

Additions of columbium and tantalum tend to counteract this effect. Modern alloys with high columbium and tantalum contents contain MC carbides that do not break down easily during solution treatment at, for instance, 2200–2300 °F (1200–1260 °C). One such carbide, from MAR-M 200, has been analyzed[39] to have an approximate composition of $(Ti_{0.53}Cb_{0.31}W_{0.16})$ C. Lund (personal communication) has categorized MC carbides as being of essentially two types. MC(1) is a single monatomic low-parameter FCC compound characterizing those commonly found in most cast or wrought superalloys. It can be, and usually is, readily decomposed in the carbide-degenerated reactions described later [Eq. (5) and (6)]. However, if an alloy contains hafnium and is held at high temperature for considerable time, FCC MC(3) can form; MC(3) is very finely divided and has not been observed to decompose.

$M_{23}C_6$ carbides are profuse in alloys with moderate to high chromium content. They form during lower temperature heat treatment and service, that is, 1400–1800 °F (760–980 °C), both from degeneration of MC carbides and from soluble carbon residual in the alloy matrix. Although usually forming at grain boundaries (Fig. 8c), they occasionally occur along twin lines, stacking faults, and at twin ends (the "zipper" structure). In MAR-M 200,[39] it was found that $M_{23}C_6$ precipitated intergranularly as platelets parallel to (110) planes of the austenite and that the $M_{23}C_6$ may be initially coherent. $M_{23}C_6$ carbides have a complex cubic structure, which, if the carbon atoms were removed, would closely approximate the structure of the TCP σ phase. In fact, coherency between $M_{23}C_6$ and σ is high; σ plates often nucleate on $M_{23}C_6$ particles.

When tungsten or molybdenum are present, the approximate composition is $Cr_{21}(Mo,W)_2C_6$ and generally appears rigidly fixed as such. However, it also has been shown that considerable nickel can substitute in the carbide, and it is also suspected that small portions of cobalt or iron could substitute for chromium. Other refractory elements can locate in the "Mo,W" position, as shown by analysis in cobalt-base alloys.

$M_{23}C_6$ carbides have a significant effect on nickel alloy properties. Their critical location at grain boundaries promotes a significant effect on rupture strength, apparently through inhibition of grain boundary sliding. Eventually, however, rupture failure can initiate either by fracture of these same grain boundary $M_{23}C_6$ particles or by decohesion of the $M_{23}C_6$ interface. In certain alloys cellular structures of $M_{23}C_6$ have been experienced (see Fig. 2) but can be avoided by heat treatment and chemistry control. Cellular $M_{23}C_6$ has shown to initiate premature rupture failures.

Figure 8b is an interesting example found by Mihalisin[14] of $M_{23}C_6$ particles forming from a degenerating MC particle in 713C. The only remaining MC appears to be the typical diamond in the center; it is surrounded by matrix, replacing the originally large MC particle. At the original MC surface $M_{23}C_6$ particles have formed in a ring and are encased in γ', another reaction product.

M_6C carbides have a complex cubic structure and form at slightly higher temperatures, 1500–1800 °F (815–980 °C), than $M_{23}C_6$. They are a similar carbide to $M_{23}C_6$ but are formed when the molybdenum and/or tungsten content is high, more than 6–8 a/o. M_6C forms as well as $M_{23}C_6$ in MAR-M 200, B-1900, René 80, René 41, and AF-1753. Typical formulas for M_6C are $(Ni,Co)_3Mo_3C$ and $(Ni,Co)_2W_4 6$.

Other data suggest even a wider range of composition for M_6C, with formulas ranging from approximately M_3C to $M_{13}C$, depending on alloy matrix content. In a study of Hastelloy X,[40] M_6C type carbides were analyzed to range from

$$M_{2.48}C = (Mo_{0.91}Ni_{0.90}Cr_{0.50}Fe_{0.17})C$$

to

$$M_{13.25})C = (Mo_{6.34}Ni_{5.73}Cr_{0.69}Fe_{0.49})C$$

Thus, M_6C carbides are formed when molybdenum or tungsten acts to replace chromium in other carbides; unlike more rigid $M_{23}C_6$, the composition can vary widely. Since M_6C carbides are stable at higher temperatures than $M_{23}C_6$ carbides, M_6C is more beneficial as a grain boundary precipitate to control grain size in processing wrought alloys.

Carbide Reactions

A major source of carbon in most nickel-base superalloys below 1800 °F (980 °C) is the high-temperature carbide MC. However, during heat treatment and service, MC decomposes slowly, yielding carbon, which permeates the alloy and triggers a number of important reactions.

The dominating carbide reaction in many alloys is believed to be the formation of $M_{23}C_6$ by the following reaction:

$$MC + \gamma \rightarrow M_{23}C_6 + \gamma' \qquad (2)$$

or

$$(Ti,Mo)C + (Ni,Cr,Al,Ti) \rightarrow Cr_{21}Mo_2C_6 + Ni_3(Al,Ti) \qquad (3)$$

This equation cannot be balanced with thermodynamic exactness on the basis of present evidence, but the reaction was assumed by metallographic observations of phase transformations at grain boundaries by Sims[4] and Phillips.[41] Reaction 2 or 3 begins to occur at about 1800 °F (980 °C) and has been observed as low as approximately 1400 °F (760 °C). In isolated circumstances it has been found reversible. M_6C can form in a similar manner,

$$MC + \gamma \rightarrow M_6C + \gamma' \tag{4}$$

or

$$(Ti,Mo)C + (Ni,Co,Al,Ti) \rightarrow Mo_3(Ni,Co)_3C + Ni_3(Al,Ti) \tag{5}$$

Further, M_6C and $M_{23}C_6$ interact, forming one from the other,

$$M_6C + M' \rightarrow M_{23}C_6 + M'' \tag{6}$$

or

$$Mo_3(Ni,Co)_3C + Cr \leftrightarrows Cr_{21}Mo_2C_6 + (Ni,Co,Mo) \tag{7}$$

depending on the alloy. For instance, René 41 and M-252 can be heat treated to generate MC and M_6C initially; long-time exposure then causes conversion of M_6C to $M_{23}C_6$. Conversely, in MAR-M 200,[39] M_6C can be formed from $M_{23}C_6$. The type of refractory metal atoms present may well control the reaction.

These reactions yield the lower carbides in various locations but most usually at grain boundaries. Perhaps the most beneficial reaction, and that controlled in many heat treatments, is reaction (2) or (3). Both the blocky carbides and the γ' produced are important. As discussed previously, it is believed that the carbides inhibit grain boundary sliding; in any case the γ' generated by this reaction engloves the carbides and the grain boundary in a relatively ductile, creep-resistant layer.

In certain early alloys $M_{23}C_6$ was found in cellular (Figs. 2 and 8) rather than blocky morphology; ductility is reduced sharply, so this structure must be avoided. However, alloys that generate profuse γ' at grain boundaries appear resistant to this phenomenon; grain boundary γ' also is believed to play a key role in blocking growth of $M_{23}C_6$ cells.

Chromium is depleted from the matrix in forming $M_{23}C_6$. This raises the solubility for γ' near the grain boundary, which can leave denuded zones, as shown in Inconel X-750 by Raymond[32] utilizing Huey tests.

Carbon, of course, is also in solution; at temperatures of 1100–1400 °F (595–760 °C) its solubility has been exceeded in cooling. Examples of very fine $M_{23}C_6$ precipitating directly on stacking faults and other standard defects have been seen:

$$\gamma_1 \rightarrow M_{23}C_6 + \gamma_2 \tag{8}$$

$$(Ni,Co,Cr,Mo,C) \rightarrow (Cr_{21}Mo_2)C_6 + (Ni,Co) \qquad (9)$$

Mihalisin[42] has suggested that carbon was slowly depleted through the following sequence on observing a series of experimental alloys:

$$TiC \rightarrow M_7C_3 \rightarrow Cr_{23}C_6 \rightarrow \sigma \qquad (10)$$

BORIDES

Boron is generally present to the extent of 50–500 ppm in superalloys; it is an essential ingredient. It goes to grain boundaries where it blocks the onset of grain boundary tearing under creep rupture loading. In U-700, for instance,[38] more than 1200 ppm boron reacts to form two types of M_3B_2 borides depending on thermal history; one is approximately $(Mo_{0.48}Ti_{0.07}Cr_{0.39}Ni_{0.03}Co_{0.03})_3B_2$ and the other is $(Mo_{0.31}Ti_{0.07}Cr_{0.49}Ni_{0.06}Co_{0.07})_3B_2$. Borides are hard refractory particles observed at grain boundaries with shapes varying from blocky to half-moon in appearance; they act as a supply of boron for the grain boundary.

TCP PHASES

In certain alloys where composition has not been carefully controlled, undesirable phases can form either during heat treatment or more commonly during service (see Fig. 2 and Chapter 8). These precipitates, called TCP phases,[45] are characterized as composed of close-packed layers of atoms forming in "basket weave" nets aligned with the octahedral planes of the FCC matrix. Generally detrimental, they may appear as thin linear plates, often nucleating on grain boundary carbides. Those commonly found in nickel alloys are σ and μ.

Although a TCP phase such as σ often seems to grow directly through γ' particles, it is logical that it must form from the nickel alloy matrix, since γ' and most carbides precipitate first. Its formula is $(Cr,Mo)_x(Ni,Co)_y$, where x and y can vary from 1 to 7. Formation in a cast commercial superalloy was confirmed by Wlodek[46]; previously this phase had often been thought to be Ni_3Ti.

Sigma has a specific and detrimental effect on alloy properties. Its physical hardness and platelike morphology is an excellent source for crack initiation and propagation, leading to low-temperature brittle failure, just as for sigmatized ferritic stainless steels. However, of perhaps more grave concern is the effect on elevated-temperature rupture strength. Sigma contains a high refractory metal content sapped from the γ matrix of the superalloy causing loss of solution strengthening. Also, high-temperature rupture fracture can occur along σ plates (intersigmatic fracture) rather than the normal intergranular fracture, resulting in severe loss in rupture life. This was first shown by Ross for IN-100; excessive formation caused a 1500 °F, 40 ksi rupture bar to fail in 967 h versus an expected life of 8000 h.[47,65] These and related problems are discussed in detail in Chapter 8. Platelike μ also can form, but less is known concerning its detrimental effects.

Carbides and TCP Phases

An interesting relationship exists between $M_{23}C_6$–M_6C-type carbides and the σ and μ TCP phases in alloys containing molybdenum and tungsten. As the molybdenum and tungsten content increase to above approximately 7 w/o, a shift from σ to μ formation will occur in TCP-prone heats, particularly when the molybdenum content is high (Table 3). It is noteworthy that the $M_{23}C_6$ crystal structure is very similar to σ, and the M_6C structure is similar to μ. These effects also can be related to the quaternary phase space position of the residual austenite composition.[50]

PHASE STABILITY AND TEMPERATURE

A stability map of the variation of abundance of phases with temperature is often obtained by solutioning the alloy, exposing it for a long time at a variety of lower temperatures, and then extracting and measuring the abundance of the phases present. A map was first published[48] for René 41 and then later by Collins[43,49] for Udimet 700, IN-100, B-1900, AF2-IDA, and René 41. Some of these results are reported in Fig. 9. It should be noted that γ' is not always reported in these studies.

The degeneration of MC carbides to $M_{23}C_6$ (and sometimes M_6C) is obvious. However, the René 41 diagram is probably incorrect with respect to MC since most observers believe there is an abundance of MC after exposure.

GRAIN AND GRAIN BOUNDARY EFFECTS

Grain Size

The strength of superalloys is very dependent on grain size and its relation to component thickness. Richards[51] found that rupture life and creep resistance increased as component-thickness-to-grain-size ratio increased. With a wrought superalloy, provided the ratio was kept constant, life and creep resistance increased with grain size. Cast superalloys show the same dependence of life and creep resistance on thickness-to-grain-size ratio.

These conditions can be serious when large grains occur in thin section. Thin sections usually exhibit reduced creep rupture resistance; the thinner the section, the lower the rupture strength compared to thick sections.

In modern cast superalloys control of grain size is vital. A balance must be struck to avoid excessively fine grains, which decrease creep and rupture strength, and excessively large grains, which lower tensile strength (but conversely have good rupture strength).

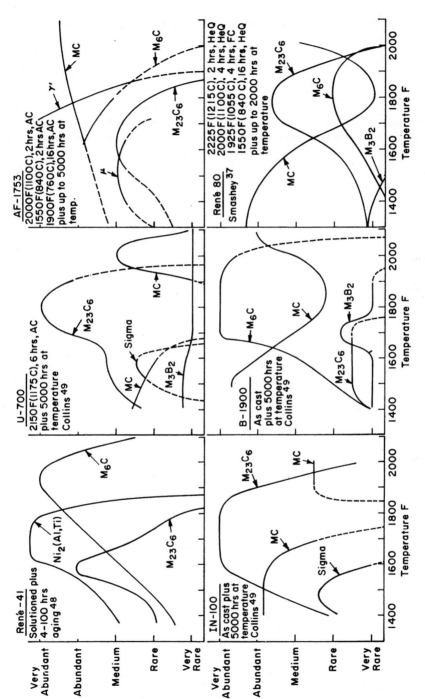

Fig. 9. Phase concentration maps for wrought (top) and cast (bottom) nickel superalloys.

119

Grain Boundary Chemistry

One of the most remarkable alloying effects in superalloys is the enhancement of creep properties from minute additions of boron and zirconium[52-55] (Table 4). Boron and zirconium can increase life 13 times, elongation 7 times, rupture stress 1.9 times, and n (stress dependence of creep rate) from 2.4 to 9. (It is the opinion of some superalloy metallurgists that boron is the key ingredient to these improved properties and that zirconium may play a more important role as a "getter" of deleterious trace/tramp elements.)

Magnesium additions of 0.01–0.05% have resulted in improved properties[2] and forgeability in wrought alloys; this is believed due to the magnesium primarily tying up sulfur, a grain-boundary-weakening element.

Despite these rather clear property effects, mechanisms have resisted clarification. However, it is believed that boron and zirconium segregate to grain boundaries, probably because of their odd size (21–29% oversize or undersize). Since rupture cracks in superalloys propagate along grain boundaries, the importance of this effect becomes clear.

The case of early U-500 (Table 5) is a good example. Boron and zirconium clearly retarded the generation of grain boundary cracking. In U-500 without boron and zirconium, microcracks developed at the end of first-stage creep at 23 h; with boron and zirconium microcracks did not develop until third-stage creep at 214 h.

The sharp effect of boron and zirconium on the exponential dependence, n, of strain rate on stress in second-stage creep (Table 4) and the lack of influence on primary creep are of interest also. When γ' denuded zones were fully developed in second-stage creep without boron and zirconium, $n = 2.4$, which is characteristic of solid-solution alloys. With a few denuded zones and coated carbides, $n = 9$, which is typical of γ'-hardened alloys. However, recent work[66] has revealed a significant "gettering" mechanism; zirconium has been found to form $Zr_4C_2S_2$, significantly reducing elemental sulfur at the grain boundaries.

Boron and zirconium benefit γ'-free alloys, cobalt alloys, and stainless steels, so that the "denuded zone" or "coated carbide" rationale cannot be universal for all alloys. Although there are exceptions, some evidence also exists for reduction

Table 4. Effect of Boron and Zirconium on Creep of Udimet 500 at 1600 °F (870 °C)[54]

Alloy	ϵ in Primary Creep	Life (hr)	Elongation (% in 4 D)	σ for $\dot{\epsilon}_{ss}$ of 0.004%/ hr (psi $\times 10^3$)	n^a
Base	0.002	50	2	17	2.4
+0.19 %Zr	0.002	140	6	23	4
+0.009% B	0.002	400	8	28	7
+0.009% B + 0.01% Zr	0.002	647	14	32	9

$^a n = \log \dot{\epsilon}_{ss}/\log \sigma$, where σ = stress over range of 20,000–30,000 psi and $\dot{\epsilon}_{ss}$ = minimum second stage creep rate.

Table 5. Effect of Boron and Zirconium on Grain-Boundary Stability of Udimet 500 at 1600 °F (870 °C)[54]

	After $\epsilon = 0.012$ in 200 hr			After 200 hr, No Stress[a]
	Denuded Grain Boundaries[b]	Microcracks[b]	γ' Carbide Nodules[b]	γ' Carbide Nodules[b]
Base	264	314	418	230
+0.19% Zr	127	78	175	90
+0.009% B	60	30	63	60
+0.009% B +0.01% Zr	23	2	20	20

[a] No denuded grain boundaries or microcracks in any alloys without stress.
[b] Number of features detected at 1000D in 5 mm^2.

of carbide precipitation in grain boundaries and shunting of carbon into the grains by boron.[56-59] It has also been reported that magnesium served this purpose in a nickel–chromium–titanium–aluminum alloy by generating intragranular MC.[2]

A general mechanism for the above effects is that odd-size atoms segregate to grain boundaries, filling vacancies and reducing grain boundary diffusion. This is consistent with the findings of Tien and Gamble[60] on the formation of denuded zones by Herring–Nabarro-type diffusion. If this is generally true, boron and zirconium are effective not only at austenitic boundaries but also at γ carbide and γ' carbide boundaries.

HAFNIUM

The first two decades of cast nickel-base alloy development for the aircraft gas turbine industry concentrated on increasing elevated-temperature creep resistance. By the mid-1960s the most creep-resistant alloys used for turbine airfoils were operating routinely at temperatures approaching 85% of their incipient melting point, but ductility in some alloys had dropped to levels of only a few percent; these alloys had irregular tertiary creep behavior. There also were severe limitations on the production of airfoils by investment casting, the best method of producing complex internal geometries necessary for air cooling. Failures were predominantly intergranular, so the challenge was to provide for accommodation of localized plastic strain without degrading the high creep resistance provided within the bulk of the grain.

An addition of hafnium was found to be very effective in solving the above problem by promoting a more viable grain boundary structure. Hafnium has high solubility in γ' relative to γ, strengthens the γ', and is an extremely active carbide former. Also, although generally accepted theory holds that a controlled amount of $M_{23}C_6/M_6C$ particles are advantageous at grain boundaries to inhibit sliding, an extensive, interconnected series of such particles permits rapid crack propagation.

Excessive boundary $M_{23}C_6/M_6C$ can be inhibited by hafnium, which reacts with the carbon freed by MC breakdown and results in a secondary, stable MC (predominately HfC) precipitated as finely divided random particles.[65] Further, during solidification, strong partitioning of hafnium to the γ' results in a convoluted γ'–γ structure in the grain boundary area that inhibits rapid crack propagation. Also, hafnium has an added advantage of increasing the oxidation resistance of the base alloy. However, its high reactivity causes an added "adventure," creating difficult (but controllable) problems during ingot melting and component processing.

Grain Boundary Structure

MC degeneration leading to grain boundaries abundant with $M_{23}C_6$ engloved in γ' is discussed in connection with other subjects. In some alloys, particularly "leaner" compositions such as Inconel X-750 and Nimonic 80A, the effect has not been found. In these alloys a γ layer (*depleted* of γ') often appeared next to the grain boundaries. This is caused by diffusion of chromium to form grain boundary carbides; the zone thus depleted of chromium has increased solubility for nickel and aluminum, causing γ' to disappear. The appearance of such a grain boundary in X-750 is shown in Fig. 2a.

The concentration of γ' at grain boundaries in stronger superalloys provides a significantly better combination of strength and ductility than the surrounding components. Engloving the hard grain boundary carbides in an environment that allows "controlled" slip inhibits the onset of intergranular fracture, which can result in outstanding rupture life. It also appears that excessive development of these effects can be troublesome; overgeneration of a grain boundary γ' film in U-700 and Nimonic 115 can lead to tensile and notch brittleness. Figure 2 shows, however, that some cast alloys do not have significant englovement of grain boundaries with γ'; yet they also demonstrate good strength and ductility. Ultimately, however, the grain boundary is always the site of creep rupture fracture initiation.

These effects were first illustrated by Decker and Freeman[23] and then were summarized by Smashey[61] in a study of the behavior of René 80. Figure 10 illustrates a well-developed grain boundary in René 80 that has finally initiated fracture (cracking) during rupture testing. The cast alloy, given a four-part heat treatment, was tested at 1800 °F (980 °C) at 15 ksi to rupture failure at 1236 h. Figure 10 shows grain boundary fissuring near the failure. The $M_{23}C_6$ carbides lining the boundary are illustrated with a surrounding γ' englovement. The carbides obviously initiated fracture. The electron microstructure shows the function of γ' in promoting cross-slip, delaying the onset of fracture.

HEAT TREATMENT

The heat treatment of nickel-base alloys is, to a large degree, an art; knowledge of constitution, phase stability, structural effects, and properties needed is utilized to properly heat treat these alloys. General effects and specific examples for heat treatment are discussed first for wrought and then for cast alloys.

Grain Boundary Effects

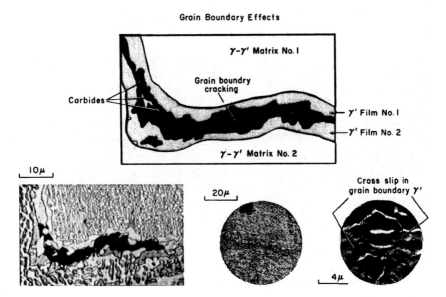

Fig. 10. Flow and fracture behavior in the grain boundary of René 80.[61]

Wrought Alloys

In the solution-treated condition wrought alloys consist mainly of the matrix and MC carbides. Solutioning temperature is usually in the range of 1900–2250 °F (1040–1230 °C) and prepares the matrix for uniform precipitation of γ' on subsequent aging. For some alloys, such as René 41, solutioning at approximately 1950 °F (1070 °C) also causes formation of the M_6C carbide, which can decrease other subsequent carbide reactions. $M_{23}C_6$, however, does not normally form during solution treatment.

A series of ages are then given to precipitate and develop the major strengthening phases. For creep rupture strength γ' is precipitated in the range of 1500–2000 °F (840–1100 °C). Subsequently, an age at about 1400 °F (760 °C) will complete its development.

Figure 11 (top) illustrates the structural appearance of U-700/Astroloy given a heat treatment of the type described above and exposed in service. Nimonic 115 would appear identical; alloys such as Waspaloy and U-500 would be similar. Note both the coarse and fine γ' and the well-developed γ' film surrounding the grain boundary $M_{23}C_6$ carbides.

Precipitation of γ'. Studies to establish optimum relationships between heat treatment, γ' morphology, and strengthening are characteristically made on specific alloys. As an example, one such study on U-500 by Murphy[62] is described to illustrate the action and these effects.

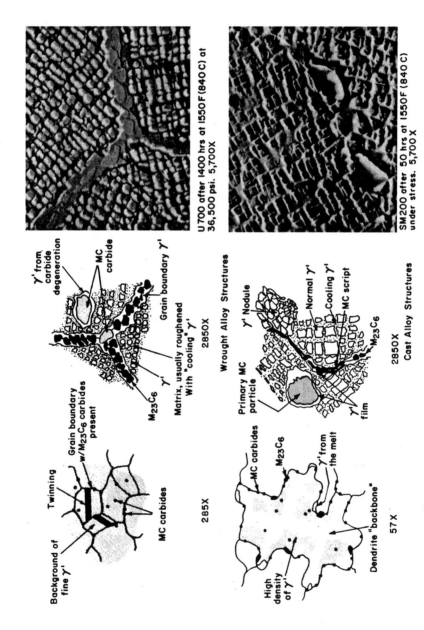

U700 after 1400 hrs at 1550F (840C) at 36,500 psi. 5,700X

SM200 after 50 hrs at 1550F (840C) under stress. 5,700X

γ' from carbide degeneration

MC carbide

Grain boundary γ'

2850X

M23C6

Matrix, usually roughened with "cooling" γ'

Wrought Alloy Structures

Twinning

Grain boundary w/M23C6 carbides present

Background of fine γ'

MC carbides

285X

Normal γ'

Cooling γ'

MC script

γ' Nodule

M23C6

Primary MC particle

γ' film

2850X

Cast Alloy Structures

MC carbides

M23C6

γ' from the melt

High density of γ'

Dendrite "backbone"

57X

Fig. 11. Structures characteristic of wrought (top) and cast (bottom) nickel-base superalloys.

124

Wrought U-500 rupture bars were given the following treatment:

Primary solution	2050 °F (1125 °C), 2 h, air cool
Secondary solution	1975 °F (1085 °C), 2 h, air cool
Primary age	1700 °F (925 °C), 24 h, air cool
Secondary age	1400 °F (750 °C), 16 h, air cool

Analysis of the microstructure showed that 2050 °F (1125 °C) solutioned the γ'; air cooling formed fine γ', most of which redissolved on reheating to 1975 °F (1080 °C). Air cooling from 1975 °F (1080 °C) produced additional γ', a finer and a more effective strengthener than that originally formed by cooling from 2050 °F (1130 °C) since it formed at lower temperatures. Further, the remaining original γ' particles grew slightly. Aging at 1700 °F (925 °C) then produced growth of both previous "cooling" γ' and agglomerated γ' with slight additional precipitation. The final age at 1400 °F (730 °C) generated little further change. The structure (Fig. 4b) has a combination of moderate tensile strength and rupture life well-suited for long-lived industrial turbine airfoils. These or similar principles are applied to virtually all alloys.

Carbides. The alloy designer has used short-time heat treatments to considerable advantage in creating favorable initial types and morphologies of carbides before service. For example, M-252, Nimonic 80A, Nimonic 115, René 41, U-500, U-700, and Waspaloy are wrought superalloys that utilize intermediate carbide heat treatments of 1900–2000 °F (1040–1100 °C) before final γ' aging. Inconel 901, U-500, U-700, and Waspaloy utilize intermediate treatments at 1400–1550 °F (760–850 °C).[2] Two examples follow to illustrate some of the practical carbide-related results of heat treatment.

In René 41 a solution anneal at 2150 °F (1175 °C) dissolves M_6C and makes the alloy susceptible to subsequent rapid precipitation of a continuous grain boundary $M_{23}C_6$ film. Poor ductility and cracking can result, particularly in welding, so the anneal is avoided. A lower temperature solution treatment at 1950–1975 °F (1070–1080 °C) retains the as-worked structure of uniform fine grains with well-dispersed M_6C. Cracking is lessened and ductility improved by the delay in $M_{23}C_6$ formation.

A study of Nimonic 80A has been reported[2] that showed that heat treatment at about 1830–1975 °F (1000–1080 °C) before γ' aging at 1290 °F (700 °C) produced excellent rupture life. At 1975 °F (1080 °C) massive Cr_7C_3 formed at grain boundaries†. This preexisting Cr_7C_3 reduced the initial rate of $Cr_{23}C_6$ precipitation during the 1290 °F (700 °C) γ' age. The benefits of this treatment were rationalized on the basis of a γ'-denuded grain boundary produced by chromium depletion from the Cr_7C_3 precipitation.

---

† Cr_7C_3 carbide has been found in Nimonic 80A.

Heat Treatment and Rupture Properties. Early heat treatments for wrought alloys such as M-252 and Nimonic 80A consisted principally of only a high-temperature solution followed by a low-temperature age. This generated good tensile and short-time rupture properties but did not stabilize the structure sufficiently to produce optimized long-time rupture properties. Heat treatments for these two alloys were studied by Heckman[63] to correct this failing. An additional intermediate-temperature age was added to each respective heat treatment to drive the MC degeneration reaction [eqs. (2) and (3)] forward; grain boundaries of coarse particulate $M_{23}C_6$ engloved in a layer of γ' were developed. In each case rupture strength at low stresses and high P_{LM} (time–temperature parameter) was increased. In effect, the alloy structure was "stabilized." This philosophy was then applied successfully to U-500 and many subsequent alloys. Since U-500 already possessed a four-part heat treatment, the effect was obtained by increasing the temperature of the 1500 °F (845 °C) age to 1700 °F (925 °C) to optimize the carbide degeneration reaction. The results are shown in Fig. 12.

Cast Alloys

The first cast superalloys developed were given a simple heat treatment. The alloy was cooled in its investment mold and then aged at a low temperature such as 1400 °F (760 °C) for about one-half day to ensure full generation of γ'. However, as alloys of a more complex nature have been developed, heat treatments have become more complex. Cast superalloys used in gas turbines are now given exhaustive heat treatments to strengthen γ', improve ductility, homogenize the structure, and achieve a variety of other effects.

Even after "solution" heat treatment [2150–2250 °F (1180–1235 °C)], a cast alloy (e.g., B-1900, René 125) often retains large clusters of eutectic $\gamma–\gamma'$ that had formed during solidification. It is not known that this unsolutioned eutectic harms properties, but it is not believed helpful.

Homogenized or not, the original dendritic freezing pattern of cast alloys is often visibly maintained after service (see Fig. 11, bottom). Dispersed γ' appears as unevenly shaded areas. Dendrite boundaries are far apart and contain some carbides and the previously discussed $\gamma–\gamma'$ eutectic. The "backbone" structure of the dendrite contains concentrations of refractory elements and is usually clearly delineated by etching.

Precipitation of γ'. Broadly, chemical compositions do not vary greatly between wrought and cast alloys. However, cast turbine airfoil alloys developed in the 1970s and 1980s contain additions of tantalum and hafnium, which are not common in the wrought alloys. These additions create some significant effects on carbides, but their effects on γ', that is, on its solution temperature and morphology, are more modest.

René 77 (a σ-free composition based on U-700/Astroloy) heat treatments were investigated by Wood,[64] as shown in Fig. 13. A significant rupture property improvement with heat treatment *B* over *A* and the difference in structure is shown;

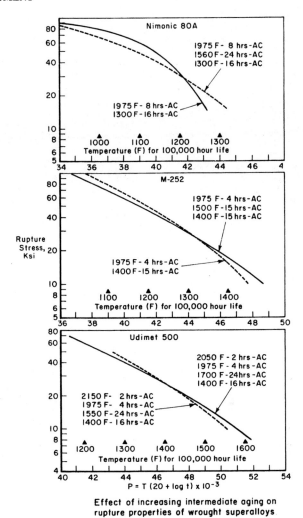

Fig. 12. Effect of increasing intermediate aging on rupture properties of wrought superalloys.[63]

γ' was taken into solution at 2125 °F (1100 °C) by both treatments but began to precipitate at about 2075 °F (1140 °C). The slow cool of treatment A generates a few large γ' nuclei above 1975 °F (1085 °C); considerable quantities of fine γ' is then generated in the 16-h treatment at 1400 °F (760 °C). While treatment A results in excellent tensile ductility, creep rupture is only fair.

Treatment B, rapidly cooled from solution, gives an opportunity for nucleation of large γ' particles but little growth. However, in 4 h at 1975 °F (1085 °C) γ' does grow, and a large number of medium-to-large particles nucleate homogeneously,

Heat Treatment A

2125F(1160C), 2hrs, furnace cool to
1975F (1085C), air cool to R.T.
1400F (760C), 16 hrs, air cool

Heat Treatment B

2125F (1160C), 4 hrs, air cool
1975F (1085C), 4 hrs, air cool
1700F (925C), 24 hrs, air cool
1400F (760C), 16 hrs, air cool

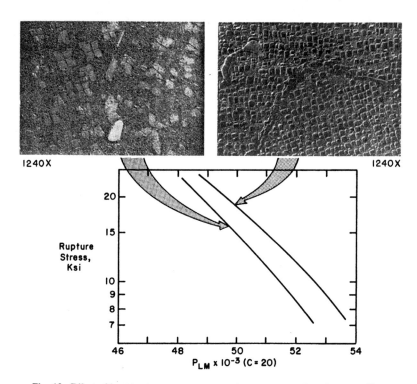

Fig. 13. Effect of heat treatment on structure and rupture properties of René 77.[64]

using up much of the precipitation potential. The "background" γ' generates at 1700 °F (925 °C) and 1400 °F (760 °C), but there is less of it for obvious reasons.

IN-738 is a cast alloy that contains columbium and tantalum in addition to molybdenum and tungsten. The heat treatment currently used is 2050 °F (1125 °C), 2 h air cool; 1550 °F (850 °C), 24 h air cool. As shown by Wood,[64] this heat treatment does not solution the γ'. Thus, large irregular particles are created plus the background of fine γ' generated at 1550 °F (840 °C). Insertion of a 1925 °F (1055 °C) age creates more regularly shaped particles and fine background γ'. When heat treated at 2150 °F (1175 °C), the γ' is more fully solutioned so that the 1975 °F- (1085 °C-) type γ' is dominant. Using 1700 °F (925 °C) as a primary age produces more uniform rounded γ'. (This particular specimen was completely solutioned.) For this alloy, however, the optimum properties result from the duplex γ' structure of the 2050 °F plus 1550 °F heat treatment.

Carbides. The metallurgy discussed for carbides in wrought alloys is also generally true for equiaxed cast alloys. The major change has come with the introduction of Group V refractory metal elements and higher carbon contents in many cast alloys. Although any increased carbon content suggests that a greater potential should exist for carbide reactions during heat treatment and service, the Group V elements columbium and tantalum stabilize MC greatly and reduce or perhaps stop the reaction. Thus, heat treatment is less effective in driving the carbide reaction. For instance, in the René 77 previously discussed, treatment *A* (Fig. 16) will allow modest degeneration of MC to $M_{23}C_6$, but treatment *B* results in a greater degeneration of MC. A heavily developed grain boundary $M_{23}C_6/\gamma'$ structure results (no cellular $M_{23}C_6$) with good rupture life and ductility. The treatment does not, however, improve tensile strength.

IN-738, containing the "big four" (Cb, Ta, Mo, W) refractory elements, differs. Very little $M_{23}C_6$ generates at grain boundaries from 1550 °F (840 °C) to 1925 °F (1055 °C) even though some heat treatments were designed to do this. The γ' film envelopment of $M_{23}C_6$ at grain boundaries is virtually nonexistent since the MC carbides do not readily decompose.

DIRECTIONAL SOLIDIFICATION OF CAST NICKEL–BASE ALLOYS

The controlled solidification of turbine airfoils was introduced in the early sixties[9] using SM-200 (subsequently called MAR-M 200) to directionally solidify the turbine airfoils.† This directional solidification (DS) process caused grains with a significantly lower modulus of elasticity to grow longitudinally in the airfoil. The lack of transverse grain boundaries in the airfoil coupled with the lower modulus resulted in a three- to fivefold improvement in thermal fatigue life over conventionally cast (equiaxed grain) turbine blade alloys such as B-1900, René 80, and MAR-M 247. In the late 1960s approximately 2% hafnium was added to MAR-M 200.[10] Hafnium ductilized the DS grain boundaries, was efficacious from a DS casting viewpoint (prevention of grain boundary cracking), and was very important for improved 1400 °F ductility (particularly transverse). Hafnium has since become a common ingredient in conventionally cast and DS alloys.

The next step in directional solidification was elimination of all grain boundaries.[11] The advantage of this single DS grain over the multigrain DS is that the latter grain boundary ductilizers/strengtheners such as boron, zirconium, and hafnium are not required. These elements significantly reduce the melting point in the alloy; their elimination thus allows 100–200 °F (38–95 °C) higher solution heat treatments of the single crystals. This higher solutioning temperature [2300–2400 °F (1260–1320 °C)] allows more strengtheners in the alloy since all of the γ' can be solutioned. This greater utilization of the γ' has resulted in a 50–100 °F (10–38 °C) improvement in strength capability of the single-crystal alloys over the best multigrain DS or conventionally cast (with equiaxed grain) turbine airfoil alloys.

† See Chapter 7 for detailed description of directional solidification.

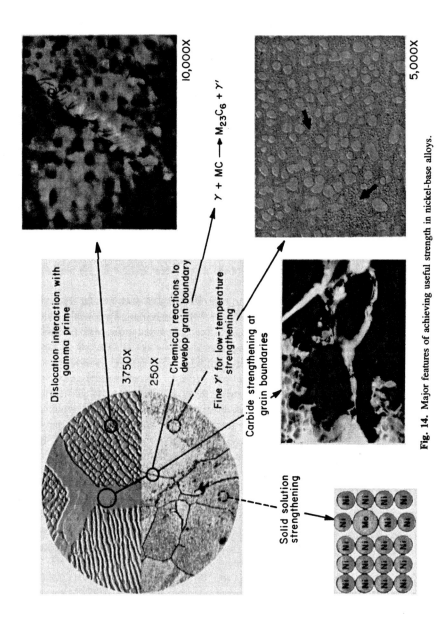

10,000X

5,000X

$\gamma + MC \longrightarrow M_{23}C_6 + \gamma'$

Dislocation interaction with gamma prime

3750X

250X

Chemical reactions to develop grain boundary

Fine γ' for low-temperature strengthening

Carbide strengthening at grain boundaries

Solid solution strengthening

Fig. 14. Major features of achieving useful strength in nickel-base alloys.

DEVELOPMENT OF STRENGTH: A SUMMARY

Knowledge about structure–property relationships in conventional nickel superalloys is extensive, but much is still to be learned. From the relatively empirical view of this chapter, some of the more obvious features, principally microstructural, of nickel alloys leading to good rupture strength are shown in Fig. 14. In more definitive form and for use in understanding of Chapter 3, guidelines for the alloy designer to create strength, adapted from Decker[2], can be summarized as follows:

1. Solid-solution strengthen γ and γ'.
2. Increase volume percent γ'.
3. Increase fault energy of γ'.
4. Increase coherency strains for less than $0.6T_M$.
5. Decrease ripening rate for greater than $0.6T_M$.
6. Minimize formation of η, Ni_3Nb, Laves, σ, and μ phases.
7. Control carbides to prevent denuded zones, $M_{23}C_6$ grain boundary films, and Widmanstätten M_6C for tensile strength.
8. Control carbides and grain boundary γ' to enhance rupture strength.
9. Control component-thickness-to-grain-size ratio.

ACKNOWLEDGMENT

The authors are indebted to Raymond F. Decker, who provided a significant portion of this information through his participation in Chapter 2 of The Superalloys *and to Carl H. Lund who provided the section on hafnium effects.*

REFERENCES

1. D. Coutsouradis, "Meeting Requirements for High Temperature Gas Turbines, A Challenge to Metallurgists," AGARD Propulsion and Energetics Panel Meeting, Florence, September 1970.
2. R. F. Decker, "Strengthening Mechanisms in Nickel-Base Superalloys," Climax Molybdenum Company Symposium, Zurich, May 5–6, 1969.
3. C. T. Sims, ASME Technical Publication 70-GT-24, May 1970.
4. C. T. Sims, *J. Met.* *18*, 1119 (October 1966).
5. L. Brewer, in *High-Strength Materials*, V. F. Zackay (ed.), Wiley, New York, 1965.
6. C. Tarr and J. Marshall, "Phase Relationships in High-Temperature Alloys," AIME Fall Meeting, Chicago, IL, October 30–November 3, 1966.
7. J. Mihalisin and D. Pasquine, "Phase Transformation in Nickel-Base Alloys," International Symposium on Structural Stability in Superalloys, Seven Springs, PA, 1968.
8. O. H. Kriege and J. M. Baris, *Trans. ASM*, **62**, 195 (March 1969).
9. W. Loomis, Ph.D. Thesis, University of Michigan, Ann Arbor, MI, 1969.
10. B. E. P. Beeston, I. L. Dillamore, and R. E. Smallman, *Met. Sci. J.*, **2**, 12 (1960).
11. B. E. P. Beeston and L. France, *J. Inst. Met.*, **96**, 105 (1968).
12. M. V. Pridantsev, *Izv. Acad. Nauk SSSR Met.*, **5**, 115 (1967).
13. W. C. Hagel and H. J. Beattie, Iron and Steel Institute Special Report, London, **64**, p. 98.

14. J. R. Mihalisin, personal communication (May 1967).
15. W. B. Kent, "Mechanical Properties and Structural Characteristics of NASA IIb," AIME Annual Meeting, Cleveland, OH, October 1970.
16. V. A. Phillips, personal communication (1968).
17. R. F. Decker and C. G. Bieber, *Symposium on Electron Metallography*, ASTM STP 262, 1966.
18. C. T. Sims, "The Role of Refractory Metals in Austenitic Superalloys," 6th Plansee Seminar, Reutte, Austria, June 1968.
19. J. Heslop, *Cobalt*, **24**, 128 (1964).
20. M. Fleetwood, personal communication.
21. E. A. Fell, *Metallurgia*, **63**, 157 (1961).
22. W. I. Mitchell, *Z. Metallkd*, **55**, 613 (1964).
23. R. F. Decker and J. W. Freeman, *Trans. AIME*, **218**, 277 (1960).
24. J. P. Rowe and J. W. Freeman, in *Proceedings of the International Conference on Creep*, Institute of Engineers, London, 1963.
25. W. B. Pearson and W. Hume-Rothery, *J. Inst. Met.*, **80**, 641 (1951–1952).
26. W. Betteridge, *The Nimonic Alloys*, Arnold, London, 1959.
27. A. Havalda, *Trans. ASM*, **62**, 581 (1969).
28. C. C. Clark and J. S. Iwanski, *Trans. AIME*, **215**, 648 (1959).
29. J. R. Mihalisin and R. F. Decker, *Trans. AIME*, **218**, 507 (1960).
30. H. L. Eiselstein, ASTM Spec. Tech. Publ. 369, Philadelphia, (1965).
31. F. G. Haynes, *J. Inst. Met.*, **90**, 311 (1961–1962).
32. E. L. Raymond, *Trans. AIME*, **239**, 1415 (1967).
33. W. J. Boesch and H. B. Canada, "Precipitation Reactions and Stability of Ni₃ Cb in Inconel Alloy 718," International Symposium on Structural Stability in Superalloys, Seven Springs, 1968.
34. F. J. Rizzo and J. D. Buzzanell, "Effect of Chemistry Variations on the Structural Stability of Alloy 718," International Symposium on Structural Stability in Superalloys, Seven Springs, PA, 1968.
35. E. A. Fell, W. I. Mitchell, and D. W. Wakeman, *Iron Steel Inst. Spec. Rep.*, **70**, 136 (1969).
36. R. F. Decker and J. W. Freeman, personal communication.
37. R. W. Smashey, personal communication.
38. W. J. Boesch and H. B. Canada, *J. Met.*, **20**, (April 1968).
39. B. J. Piearcey and R. W. Smashey, *Trans. AIME*, **239**, 451 (1967).
40. W. L. Clarke, Jr. and C. W. Titus, "Long-Time Stability of Hastelloy X," ASM Metal Congress, Cleveland, OH, October 1967.
41. V. A. Phillips, personal communication.
42. J. R. Mihalisin, *Trans. AIME*, **239**, 180 (1967).
43. H. E. Collins, "Relative Stability of Carbide and Intermetallic Phases in Nickel-Base Superalloys," International Symposium on Structural Stability in Superalloys, Seven Springs, PA, 1968.
44. H. J. Beattie, Jr. and F. L. VerSnyder, *Trans. ASM*, **49**, 883 (1957).
45. H. J. Beattie, Jr. and W. C. Hagel, *Trans. AIME*, **233**, 277 (February 1965).
46. S. T. Wlodek, *Trans. ASM*, **57**, 111 (1964).
47. E. W. Ross, "Recent Research on IN-100," AIME Annual Meeting, Dallas, TX (February 1963).
48. L. A. Weisenberg and R. J. Morris, *Met. Prog.*, **78**, 70 (1960).
49. H. E. Collins, "Relative Stability of Carbides and Intermetallic Phases in Nickel-Base Superalloys," International Symposium on Structural Stability in Superalloys, Seven Springs, PA, 1968.
50. C. T. Sims, unpublished work.
51. E. G. Richards, *J. Inst. Met.*, **96**, 365 (1968).
52. C. G. Bieber, "The Melting and Hot Rolling of Nickel and Nickel Alloys" in *Metals Handbook*, ASM, Cleveland, OH, 1948.
53. R. W. Koffler, W. J. Pennington, and F. M. Richmond, *Res. and Dev. Dep Rep. No.*, **48** (1956). Universal-Cyclops Steel Corporation, Bridgeville, PA.
54. R. F. Decker, J. P. Rowe, and J. W. Freeman, NACA Technical Note 4049, Washington, DC, June 1957.
55. K. E. Volk and A. W. Franklin, *Z. Metallkd*, **51**, 172 (1960).
56. B. S. Natapov, V. E. Ol'shanetskii, and E. P. Ponomarenko, *Met. Sci and Heat Treat.*, **1**, 11 (1965).

57. E. G. Richards and P. L. Twigg. "Influence of Boron on a Ni-Cr Austenitic Alloy," 11th Creep Colloquium, Saclay, France, 1967.

58. F. C. Hull and R. Stickler, "Effects of N, B, Zr, and V on the Microstructure. Tensile and Creep-Rupture Properties of a Cr-Ni-Mn-Mo Stainless Steel" in *Joint International Conference on Creep*, Institute of Mechanical Engineers, London, 1963.

59. C. Crussard, J. Plateau, and G. Genry, "The Influence of Boron in Austenitic Alloys" in *Joint International Conference on Creep*, Institute of Mechanical Engineers, London 1963.

60. J. K. Tien and R. P. Gamble, *Met. Trans.*, **2,** 1663 (1971).

61. R. W. Smashey, "Effect of Long Time, High Stress Exposures on the Microstructure of René-80 Alloy," AIME Annual Meeting, Cleveland OH, October 1970.

62. H. J. Murphy, C. T. Sims and G. R. Heckman, *Trans AIME*, **239,** 1961–78 (1967).

63. G. R. Heckman, ASME Preprint 67-GT-55, Gas Turbine Conference and Products Show, Houston, TX, March 1967.

64. J. W. Wood, personal communication.

65. E. W. Ross, *J. Met.*, **19,** 12 (December 1967).

66. J. A. Scheibel, C. L. White, and M. H. Yoo, *Met. Trans.*, **16A,** 651 (1985).

Chapter 5

Cobalt-Base Alloys

ADRIAN M. BELTRAN

General Electric Company, Schenectady, New York

The compositional roots of contemporary cobalt-base superalloys stem from the early 1900s, when patents covering the cobalt–chromium and cobalt–chromium–tungsten systems were issued. Consequently, the Stellite alloys of E. Haynes became important industrial materials for cutlery, machine tool, and wear-resistant hardfacing applications.[1] The cobalt–chromium–molybdenum casting alloy Vitallium was developed in the 1930s for dental prosthetics, and its derivative HS-21 soon became an important material for turbocharger and gas turbine applications during the 1940s. Similarly, wrought cobalt–nickel–chromium alloy S-816 was used extensively for both gas turbine blades and vanes during this period. Another key alloy, invented in about 1943 by R. H. Thielemann, was cast cobalt–nickel–chromium–tungsten alloy X-40. This alloy is still used in gas turbine vanes, and it has served extensively as a model for newer generations of cobalt-base superalloys.

The development of vacuum-melted, γ'-strengthened nickel alloys during the 1950–1970 time frame, however, greatly surpassed the capabilities of cobalt alloys, which lack a comparable precipitation hardening mechanism. Cobalt alloys were therefore relegated to a secondary position in the gas turbine industry, which has changed little over the past two decades. This relatively stable marketplace position has been maintained despite the cyclical price and availability problems that have plagued the element cobalt, stemming from its restricted abundance in the mid-African continent. Cast and wrought cobalt alloys continue to be used for the following primary reasons:

1. Cobalt alloys exhibit higher melting temperatures and correspondingly flatter stress rupture curves, providing useful stress capability to a higher absolute temperature than nickel- or iron-base alloys.
2. Cobalt alloys offer superior hot-corrosion resistance to contaminated gas turbine atmospheres due to their higher chromium contents.
3. In general, cobalt alloys exhibit superior thermal fatigue resistance and weldability to nickel alloys.

This chapter will review the physical metallurgical characteristics of cobalt alloys, with a view to correlating the interrelationships between alloy composition, crystallographic phases, microstructure, and physicomechanical properties. Significant advances in superalloy process metallurgy have been achieved in the past decade in the areas of directional solidification, powder metallurgy, and dispersion strengthening, and cobalt alloys have participated modestly in this exciting era. Examples of cobalt alloy development in these fields will be explored for completeness.

CHEMICAL COMPOSITION

The chemical constitution of cobalt alloys is analogous to the general family of stainless steels, and the role of major and minor alloying elements is virtually identical throughout these austenitic alloy systems. Specifically, the key element chromium is added in the range of 20–30 wt. % to impart oxidation and hot-corrosion resistance and some measure of solid-solution strengthening. Where carbide precipitation strengthening is a desirable feature, chromium also plays a strong role through the formation of a series of varying chromium–carbon ratio carbides. Since the cobalt–chromium binary system exhibits a stable sigma phase at about 58 at.% chromium, higher chromium levels must be avoided.

Carbon is clearly critical to those casting alloys formulated for the highest creep rupture strength levels, since carbide strengthening is the primary precipitation hardening mechanism utilized in cobalt alloy systems. The control of carbon is critical for tensile and rupture strength and ductility since it has been shown that a nonlinear increase in strength occurs over the range of about 0.3–0.6 wt. % carbon. Conversely, ductility decreases over this range. More importantly, ductility will decrease noticeably as a consequence of secondary carbide precipitation due to service exposures in the temperature range 1200 °F (650 °C) to approximately 1700 °F (927 °C). Carbides are also important in the simpler wrought alloys (C < 0.15 wt. %) to control grain size during processing, heat treatment, and subsequent service.

The refractory elements tungsten and molybdenum are utilized as the major solid-solution strengtheners for both wrought and cast cobalt alloys, while those of lower solubility such as tantalum, columbium, zirconium, and hafnium are generally more effective in a carbide-forming role. Tungsten additions of up to 11 wt. % in cast WI-52 and 15 wt. % in wrought L-605 are used in conventional alloys. However, a family of Co–25W–1Zr–1Ti–0.5C alloys was developed for low-vacuum space environment applications.[2] Chromium was omitted since it was not required for oxidation resistance and it exhibits high volatility at high temperatures. Tantalum has been successfully utilized as a replacement for tungsten in high-temperature sheet alloys MM-918 and S-57, where some improvement in oxidation resistance was also demonstrated.

While the majority of contemporary cobalt alloys contain tungsten as the primary solid-solution strengthener, the work-strengthened Multiphase™ alloys rely solely on molybdenum additions of up to 10 wt. %.[3] These will be discussed further subsequently. For cast alloys, Morrow et al.[4] demonstrated that replacing tungsten with atom-equivalent additions of molybdenum improves elevated-temperature tensile and rupture ductility without decreasing strength in alloys such as FSX-414 and MM-509. Additionally, alloy cost and density decrease with little change in expansion coefficient or microstructural characteristics. However, molybdenum additions decrease the solidus and liquidus temperatures slightly and increase the total solidification range, which alter the carbide morphology somewhat and produce additional eutectic carbide.

The refractory element rhenium has been successfully utilized in nickel alloys for solid-solution strengthening; however, its potential in cobalt alloys has not fully been explored. Like tungsten it exhibits extensive solubility in the matrix and increases solidus and liquidus temperatures. A 2% rhenium plus 3% chromium addition to the Co-25W alloys further improved the strength of that alloy system; however, its cost is considered prohibitive.

To enhance the stability of the high-temperature austenitic face-centered-cubic (FCC) cobalt matrix, additions of up to 20 wt. % nickel or iron are used to suppress the transformation to hexagonal-close-packed (HCP) cobalt at low temperatures. The presence of these elements in wrought alloys lowers deformation resistance and benefits workability. Additions are generally limited to 10 wt. % in the cast alloys, since higher levels decrease rupture strength.

The role of the major alloying elements is recapped in Table 1 for representative contemporary wrought and cast cobalt alloys. Of these alloying elements, only tungsten produces a favorable increase in melting temperature (Table 2). It is critical that the solubility limits of the refractory elements not be exceeded, as the precipitation of deleterious intermetallic compounds such as σ and Laves will readily occur with disastrous consequences (see Chapter 9). The ternary cobalt–nickel–chromium and cobalt–chromium–tungsten phase diagrams in Fig. 1 illustrate the relative solubilities of these key elements.

The vast majority of commercially available cobalt alloys are air or argon melted since they are devoid of the highly reactive elements aluminum and titanium, whose presence requires more sophisticated and costly vacuum melting techniques. Silicon and manganese additions are utilized to enhance castibility in terms of alloy fluidity, melt deoxidation practice, and sulfur control. Vacuum melting is required to control the relatively low alloying levels of the strong monocarbide-forming reactive elements zirconium, hafnium, and titanium in contemporary alloys like MM-509. Improvements in tensile and rupture properties of more conventional alloys like X-40 have also resulted from vacuum melting due to lower interstitial levels and "cleaner" material.

Table 1. Function of Alloying Element Groups in Cobalt Superalloys

	Nickel	Chromium	Tungsten	Ti, Zr, Cb, Ta	C
Principal function	Austenite stabilizer	Surface stability + carbide former	Solid-solution strength	MC formers	Carbide formation
Problems[a]	Lowers corrosion resistance	Forms TCP phases	Forms TCP phases	Harms surface stability	Decreases ductility
Examples					
X-40	10	25	7.5	—	0.45
MM-509	10	24	7.0	3.5 Ta, 0.5 Zr, 0.2 Ti	0.60
L-605	10	20	15.0	—	0.10
HS-188	22	22	14.0	—	0.08

[a] When added in excess.

Air-melted alloys, for example, typically exhibit 400 ppm oxygen and 700 ppm nitrogen, whereas vacuum-melted alloys contain less than 100 ppm of these elements.

More recently, electroslag remelting (ESR) was investigated by Nafziger and Lincoln[6] and compared to vacuum arc remelting (VAR). A slight improvement in rupture properties, especially at high stresses, was found for ESR MM-302, MM-509, and X-45 compared to VAR. No significant changes in alloy microstructure or nonmetallic inclusions were noted, although chemical analysis showed a small decrease in the sulfur and phosphorus levels for ESR material.

Aluminum has been added to both wrought and cast cobalt alloys, as represented by sheet alloy S-57 and cast alloy AR-213, respectively. Additions of 5 wt. % aluminum in each of these systems are highly beneficial for oxidation and hot-corrosion resistance, as further exemplified by the cobalt–chromium–aluminum–yttrium coatings in commercial use (see Chapter 13). These alloys are strengthened by a uniform noncoherent precipitate of CoAl that generates properties similar to the carbide-strengthened alloys. CoAl tends to overage above approximately 1400 °F (760 °C); however, refractory element additions of tungsten to alloy AR-215 and tantalum to S-57 stabilize the precipitate to a higher use temperature.

Titanium additions have been utilized in wrought alloys CM-7 and Jetalloy 1650 to generate a uniform coherent precipitate of ordered-FCC $(Co,Ni)_3Ti$ analogous to γ' in nickel alloys. High tensile strengths are achieved up to the temperature stability limit of this phase, that is, about 1300 °F (704 °C). However, titanium levels above about 5 wt. % produce phase instabilities that generate the HCP-Co_3Ti or Co_2Ti-Laves phases.

The incorporation of nitrogen in some air-melted casting alloys, either as an intentional or inadvertent addition, also has a positive although less potent strengthening

Table 2. Effect of 1 wt. % of Alloying Elements on Melting Temperature of Cobalt[a]

Alloying Element	Melting Temperature (°F)
Raise	
Tungsten	+1
Lower	
Nitrogen	−1
Iron	−1
Chromium	−5
Molybdenum	−8
Vanadium	−15
Manganese	−15
Aluminum	−20
Tantalum	−30
Zirconium	−30
Sulfur	−40
Titanium	−65
Columbium	−70
Silicon	−75
Boron	−115
Carbon	−120

[a] From ref. 1.

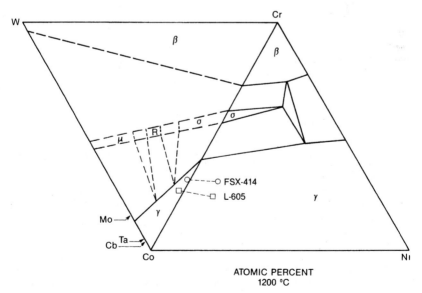

Fig. 1. Schematic illustration of Co–Ni–Cr and Co–Cr–W ternary phase diagrams.[5]

effect similar to carbon through the formation of nitrides and carbonitrides. In general, these are thermodynamically less stable than the carbides and suffer degeneration reactions during service.

Boron is added to cast cobalt alloys to enhance rupture strength and ductility; however, its precise function in the microstructure is usually obscured by the carbides. In nickel alloys boron precipitates at grain boundaries as a molybdenum-rich boride; a similar boride has not been identified in cobalt alloys. Boron levels of typically 0.015 wt. % are used; however, additions of up to 0.1 wt. % have been employed to provide additional strengthening.

Significant improvements in the oxidation resistance of cobalt alloys have been achieved in the past two decades through the addition of the rare-earth elements yttrium and lanthanum in alloys such as cast FSX-418 and wrought HS-188, respectively. Surprisingly, additions of just 0.08–0.15 wt. % promote oxide scale adhesion and reduced oxidation kinetics, especially under thermal cycling conditions, and are particularly effective in stabilizing the Cr_2O_3 oxide and minimizing the formation of $CoCr_2O_4$ spinel and CoO (see Chapter 11).

ALLOY PHASES

Like their nickel and iron alloy counterparts, high-temperature cobalt alloys are a complex chemical and crystallographic composite consisting of an austenitic matrix and a variety of precipitated phases such as carbides and intermetallic compounds belonging to the geometrically close-packed (GCP) and topologically close-packed (TCP) type (electron or size compound) structures. In general, superalloys are not truly stable at service temperatures since a dynamic environment involving stress, temperature, time, and surface–atmosphere interactions is experienced. The interdiffusion of elements between phases, along grain boundaries or to surfaces promotes various solid-state reactions that constantly alter compositional relationships and strongly influence phase stability.

Allotropic Phase Transformation

Pure cobalt exhibits an allotropic phase transformation from the high-temperature γ FCC austenitic crystal structure to the low-temperature ε HCP structure at 783 °F (417 °C). According to Giamei et al.,[7] the reaction is essentially athermal in nature and exhibits reversability during temperature cycling. Upon cooling, the γ-to-ε transformation occurs at 734 °F (390 °C) (the Ms temperature); heating causes a reversion to γ at about 806 °F (430 °C) (the As temperature). Furthermore, the degree of HCP formation is dependent on the alloy impurities and grain size of the starting material, where fine grains and high impurities inhibit the transformation; however, cold work will drive the reaction to completion. The transformation occurs by shear with the following crystallographic relationships between the two phases:

$$\{111\}\alpha\| \{0001\}\epsilon \ : \ \langle 110\rangle\alpha\| \langle 11\bar{2}0\rangle\epsilon$$

The transformation has therefore been classified as martensitic, arising from the mobility of partial dislocations along the close-packed planes.

There is little information available on the effect of the transformation on the mechanical properties of complex cobalt alloys. However, Kamel and Halim[8] examined the properties of pure polycrystalline cobalt in the region of the transformation temperature and found that HCP cobalt has a work-hardening coefficient four times greater than FCC cobalt. The fracture stress decreased and the fracture strain increased with increasing temperature; however, the rate of fracture stress decrease was 10 times greater for FCC cobalt; fracture strains were equivalent. Conversely, the creep rate for the HCP material increased faster with increasing temperature than for FCC cobalt.

Under equilibrium conditions, the addition of alloying elements to cobalt will alter the thermodynamic stability of the HCP and FCC polymorphs by either enlarging or constricting these fields. Simultaneously, these elements will effect the martensitic shear transformation by influencing the Ms and As temperatures. For those elements that stabilize the HCP crystal structure, such as chromium and the refractory elements, an equilibrium transformation via nucleation and growth is possible. In addition, since the refractory elements exhibit limited solubility in both the FCC and HCP crystals, the ε phase terminates in a high-temperature peritectoid reaction isotherm. Hence, metallographic and X-ray diffraction studies of metastable FCC material must be performed with care to prevent a martensitic reversion due to surface deformation during the study. In the case of the elements that stabilize FCC austenite, such as iron, nickel, and manganese, the reaction temperatures are so low that the only mode of transformation possible is martensitic. The effect of individual alloying elements has been strictly treated by Morral[9]; Fig. 2 is a simplified representation relating solubility with the effect of an element on the transformation temperature.

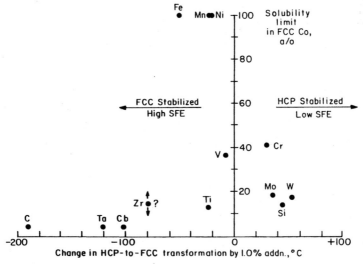

Fig. 2. Effect of alloying additions on the HCP → FCC transformation in cobalt as a function of solubility in FCC cobalt.[9]

Clearly, additions of nickel, iron, and manganese and carbon are favored to stabilize the FCC polymorph. Conversely, it should be noted that chromium and tungsten, which are major alloying additions for corrosion resistance and strength, respectively, are strong HCP stabilizers. The early Stellite alloys did not contain significant additions of the FCC-stabilizing elements, which proved fortuitous for low-temperature applications involving wear, since HCP is favored for easy deformation, and it exhibits a friction coefficient along the basal plane that is less than half that of the FCC phase. The martensitic transformation has also been linked to the absorption of energy in various wear and erosion environment situations to explain the excellent resistance of cobalt alloys in general. This transformation behavior has also been utilized to advantage in the development of the work-strengthened multiphase alloy family.

Stacking Faults

Stacking faults in cobalt alloys are layers of atoms arranged in close-packed array within the FCC austenitic matrix, only the stacking sequence is out of sync with the matrix array. The degree of faulting is obviously a function of alloy composition, temperature, and applied stress or degree of deformation. The alloying elements have an effect on the stacking fault energy in accordance with the data in Fig. 2.

Faulting is of greatest significance within the phase transformation temperature range; mechanical properties are strongly influenced in this regime. Significant strengthening occurs through the interaction of dislocations with the faults. Further strength enhancement is derived from the presence of second-phase particulate that coincidentally occurs during service exposure. Conversely, ductility is of concern since it is likely to reach a minimum in this transition temperature range. Stacking faults in real alloy systems are illustrated in Fig. 3 for as-cast MM-509 and alloy

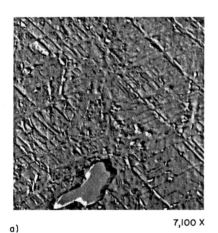 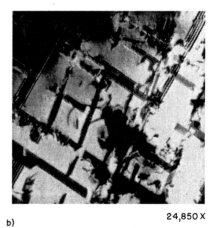

a)   7,100 X   b)   24,850 X

Fig. 3. Stacking faults interacting with carbides can create extensive precipitation in cobalt superalloys. (a) Stacking faults and $M_{23}C_6$ carbides in MM-302 rupture tested at 1600 °F (870 °C). (b) Stacking faults in MM-509. Dark areas are $M_{23}C_6$ carbides at fault intersections.

MM-302 following rupture testing at 1600 °F (871 °C); clearly, carbide precipitation reactions occur in and around the faulted material even at this temperature.

To avoid this potential phase instability in alloys intended for high-temperature applications where cycling to lower temperatures would involve a risk of lower ductility, additions of the FCC austenitic stabilizing elements such as nickel are clearly vital. Nickel sharply raises the stacking fault energy, which increases the difficulty of forming partial dislocations. However, a partial transformation to HCP has been noted in cast cobalt alloys, even in the presence of 10% nickel, following isothermal exposures in the 1200–1400 °F (649–760 °C) temperature range. This demonstrates the powerful alloying influence of chromium and tungsten.

In summary, while the HCP–FCC transformation in cobalt alloys is not well understood and cannot be precisely controlled or utilized to advantage, it is clear that additions of the austenitic stabilizing elements such as nickel are important to the long-term properties and stability of the high-temperature cast and wrought cobalt alloys.

Austenitic Matrix

The genesis of the cobalt-base superalloy family is therefore the FCC austenitic matrix. It is homologous with the austenitic phase of nickel (and iron) systems; this is depicted by the simplified partial cobalt–nickel–chromium ternary system shown in Fig. 1. The residual matrix compositions of L-605 and FSX-414 are plotted on this diagram to illustrate their position relative to the phase boundaries that denote the onset of intermetallic compound precipitation. Of course, the presence of the quaternary element tungsten moves the matrix composition away from the cobalt–nickel–chromium plane, as indicated in the accompanying cobalt–chromium–tungsten diagram. In concert with body-centered-cubic (BCC) chromium, the BCC refractory elements establish a field that is separated from the FCC austenitic corners of nickel and cobalt by a layer of intermediate TCP compounds. Note the binary solubility limits for molybdenum, tantalum, and columbium in cobalt.

It is the start of second-phase precipitation from the alloy matrix that must be anticipated and dealt with in a commercial application to circumvent potential problems in service. However, this diagram illustrates phase relationships at 2192 °F (1200 °C), which is well above the normal service temperature. One can therefore expect a further inward shift of the boundaries with decreasing temperature to reflect decreasing solute solubility. Hence L-605 and FSX-414 lie much closer to the critical boundaries under normal service operating conditions.

The stability of the matrix can be characterized by the electron vacancy theory embodied in the PHACOMP numerical calculation scheme described in Chapter 8. PHACOMP has been utilized far more often and successfully for nickel alloys; however, recent cobalt alloys of the HS-188 and FSX-414 genre have also benefited from this analysis.

In the case of wrought L-605, which is highly alloyed with tungsten (~5 at. %), Laves phase precipitation and a consequent reduction in low-temperature ductility has been blamed on the presence of high silicon levels.[10] Subsequent

PHACOMP analysis and alloy modification successfully produced alloy HS-188 with a higher nickel content, reduced tungsten, and controlled silicon levels. The net effect of these changes was to move the alloy matrix composition away from the phase boundary into an area of stable single-phase composition. Similarly, the high-chromium-content alloys of the FSX-414 type must be monitored to prevent σ-phase formation since they face that particular boundary of the phase diagram.

Carbides

Contemporary cobalt alloys are strengthened primarily by the precipitation of cubic, noncoherent carbide particles. Hence, the carbon content of these alloys is substantially greater than that found in nickel or iron alloys:

Austentic stainless steels	0.02–0.20 w/o C
Nickel-base superalloys (cast)	0.05–0.20 w/o C
Cobalt-base superalloys (cast)	0.25–1.0 w/o C

In general, the carbide-forming elements are drawn from the group illustrated on a periodic system basis by Goldschmidt, as shown in Fig. 4; these are to the left of cobalt since most of the additions are more electronegative and therefore more reactive than cobalt. This is further illustrated in Fig. 5 for an assumed carbon range of 0.1–0.6 w/o; note also that the thermodynamic free energy of carbide formation increases from the left to the right as well.

Precipitation of the carbide phases in cobalt alloys is quite complex, stemming from the relative stabilities and solubilities of the individual types. Coutsouradis[13] documented the solubilities of several carbides in cobalt at 2300 °F (1260 °C) as follows:

WC	22%	V_4C_3	6
Mo_2C	13	CbC	5
Cr_3C_2	12	TaC	3

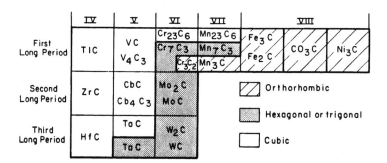

Fig. 4. Relation of carbides to periodic system. Adapted from ref. 11.

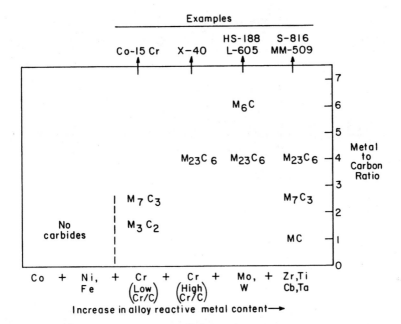

Fig. 5. Effect of adding elements of varying electronegativity on type of carbides formed in cobalt superalloys, carbon assumed 0.1–0.6%.[12]

For purposes of discussion, the carbides can be divided into two groups representing chromium-rich and refractory element-rich carbides, as follows:

M_3C_2, M_7C_3, and $M_{23}C_6$ Carbides.   These are basically chromium carbides containing cobalt, tungsten, or molybdenum in substitution for chromium. Their relationship in terms of the chromium–carbon ratio is illustrated schematically in the phase diagram of Fig. 6. Here M_3C_2 is rhombic and forms a peritectic reaction with chromium; it has been found in some of the early superalloys with lower chromium contents. M_7C_3 is trigonal in structure and forms at low chromium–carbon alloying ratios. It is metastable in X-40, for example, and transforms on aging to $M_{23}C_6$. It has also been identified in slowly cooled MM-509; however, it can be dissolved during solution heat treating. The decomposition reaction from M_7C_3 to $M_{23}C_6$ generates potent secondary carbide strengthening according to the following reactions:

$$23Cr_7C_3 = 7Cr_{23}C_6 + 27C$$

$$6C + 23Cr = Cr_{23}C_6$$

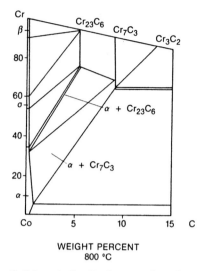

WEIGHT PERCENT
800 °C

Fig. 6. Schematic Co–Cr–C ternary phase diagram.[14]

Electron beam microprobe analysis of $M_{23}C_6$ typically has shown an atom formula of $Cr_{17}Co_4W_2C_6$; hence, significant substitution of cobalt occurs in this phase, supported by the phase diagram (Fig. 6). For investment cast alloys, $M_{23}C_6$ may form as a primary precipitate during solidification. For most commercial alloys it is primarily found as an interdendritic precipitate within the secondary dendrite arms and is the last phase to freeze. This produces a eutectic lathe structure consisting of alternate sheafs of $M_{23}C_6$ and γ matrix; however, its morphology may change with alloy chemistry (see Fig. 7). The schematic representation of eutectic carbide formation is illustrated in Fig. 8.

The primary strengthening role of $M_{23}C_6$ appears to be as a secondary precipitate of fine particles throughout the alloy matrix. This reaction is particularly strong within the 1300–1600 °F (704–871 °C) temperature range. As mentioned previously, these fine $M_{23}C_6$ carbide particles preferentially precipitate along stacking faults and twin boundaries, especially at the lower temperatures. Conversely, this precipitation can have a strong negative influence on low-temperature ductility, especially for casting alloys with carbon levels above about 0.5 w/o.

M_6C and MC Carbides. The refractory-element-rich carbides are utilized for strengthening both wrought and investment cast cobalt alloys. As in nickel alloy systems, the M_6C carbide is generally found in the low-chromium-content alloys with molybdenum and/or tungsten levels exceeding about 4–6 at. %. M_6C generally exhibits excellent temperature stability, which is useful for grain size control during the fabrication of wrought materials.

The M_6C carbides usually occur as $M_3M_3'C$ or $M_4M_3'C$, but the wide solubility limits of this phase allow for even greater compositional variability. One particular example of an analyzed carbide consisted of the following:

$$(Co_{0.45}Cr_{0.3}Ta_{0.15}W_{0.1})_6C$$

The temperature stability of M_6C can, however, be strongly influenced by the alloy composition, as demonstrated by wrought alloys L-605 and HS-188, which undergo transformation of M_6C into $M_{23}C_6$ following 3000-h exposures at 1500–1700 °F (816–927 °C).[16] M_6C may also form as a decomposition product of the MC carbide reaction with the alloy matrix:

$$MC + austenite \rightarrow M_6C$$

that is,

$$TaC + (Co,Ni,Cr,C) \rightarrow (Co,Ni)_4(Cr,Ta)_2C$$

MC carbides are considered to be a major factor in the strengthening of contemporary cobalt alloys especially when properly balanced with $M_{23}C_6$. The strongest MC-forming elements are, as indicated earlier, hafnium, zirconium, tantalum, columbium, and titanium. In casting alloys the MC carbide generally precipitates as discrete blocky particles with regular geometric features, such as diamonds or cubes. Evidence does exist for the precipitation of the most stable of the MC carbides (hafnium and

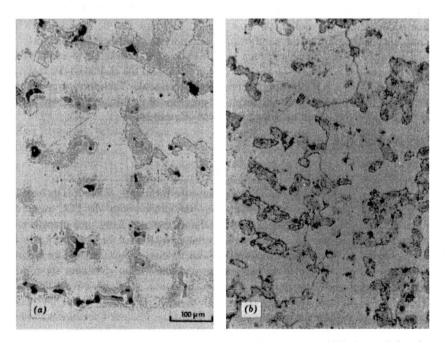

Fig. 7. Alternate solidification morphologies of $M_{23}C_6$–matrix eutectic: (a) thick, slow-cooled section; (b) thin, fast-cooled section.

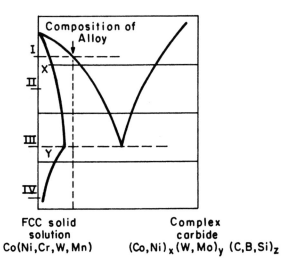

Fig. 8. Schematic representation of eutectic formation in typical cobalt superalloys.[15]

zirconium rich) in the liquid melt as the first phase to form; hence, they migrate to the interstices of the casting dendrites. TaC and CbC, however, usually form in the "Chinese-script" morphology in large-grain-size castings, suggesting precipitation that occurs later in the solidification sequence. This may also be due in part to the pronounced widening of the solidification range by tantalum and columbium.

The MC carbides can, under long service exposures, degenerate to a lower carbide as illustrated above. For alloys high in chromium the displacement reaction from MC to $M_{23}C_6$ is the predominant mode of breakdown. Therefore, an important secondary hardening effect results from the presence of the MC carbide, which acts as a source of significant quantities of the $M_{23}C_6$ carbide.

GCP PHASES

Geometrically close-packed (GCP) phases have the form A_3B, where A is the smaller atom and the phase is an ordered, coherent precipitate within the FCC austenitic matrix. In the case of nickel-base superalloys the γ' phase is the principal strengthening agent and is represented by the general formula $Ni_3(Al,Ti)$. In the highly alloyed compositions of contemporary nickel alloys other elements participate in the ordered phase. Cobalt, iron, and to a small extent chromium may substitute for nickel, whereas chromium and the refractory elements substitute for aluminum and titanium. The γ' solvus temperature may be as high as 2200 °F (1204 °C) for the nickel alloys with the highest volume fraction γ' in use today. Chapter 4 contains additional discussion of the characteristics and behavior of this important phase.

The generation of GCP phases within cobalt alloys is substantially more difficult since the chemical and crystallographic stability is affected by a lattice mismatch that is rarely less than 1%. In addition, the cobalt–aluminum phase diagram does not exhibit a comparable Co_3Al phase, although the Co_3Ti phase does exist in the cobalt–titanium system. During the late 1950s two commercial cobalt alloys were briefly popular, J-1570 and J-1650. These were strengthened by an ordered, coherent γ' precipitate that was stabilized by the high nickel (28 w/o) content; hence, the GCP phase was actually $(Ni,Co)_3Ti$. This is somewhat analogous to the generation of a coherent precipitate in the iron-base alloy A-286, also resulting from the substantial addition of nickel. The rapid development during the 1960s of nickel γ' alloys with superior mechanical properties and temperature capability undoubtedly led to the early demise of J-1570 and J-1650.

During the 1960–1970s a substantial development effort was conducted at the Centre de Researches Metallurgiques (CNRM) in Liege, Belgium, to identify useful

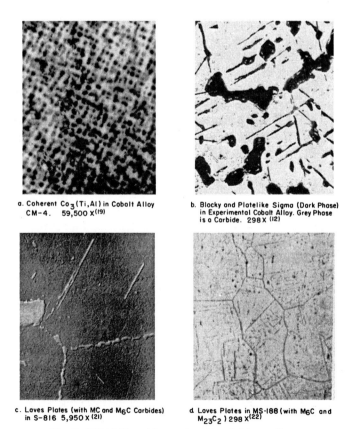

a. Coherent $Co_3(Ti,Al)$ in Cobalt Alloy CM-4. 59,500 X[19]

b. Blocky and Platelike Sigma (Dark Phase) in Experimental Cobalt Alloy. Grey Phase is a Carbide. 298 X[12]

c. Laves Plates (with MC and M_6C Carbides) in S-816 5,950 X[21]

d. Laves Plates in MS-188 (with M_6C and $M_{23}C_2$) 298 X[22]

Fig. 9. TCP, GCP, and Laves phases in cobalt superalloys.

cobalt-base GCP precipitation hardening systems.[17] While it was shown that the cobalt–titanium binary produces a stable and stronger alloy than does nickel–aluminum, the addition of alloying elements to develop other desirable properties caused destabilization of the cobalt–titanium γ' phase. In general, this γ' was stable to approximately 1400 °F (760 °C), but above that temperature a transformation to ordered HCP phase of the Ni_3Ti or Laves A_2B-type phase occurred. These latter phases formed as platelets on matrix {1, 1, 1} planes or stacking faults and signaled a drop in the creep rupture strength of the alloy. The stability of γ' was aided by the addition of both nickel and aluminum as expected, but the alloys developed were not competitive with the available high-strength cast nickel alloys.

The work at CNRM, however, successfully produced an alloy modification of the popular wrought alloy L-605 utilizing small additions of titanium and aluminum and an increase in nickel (to 15 w/o) in order to increase tensile strength and ductility at the lower temperatures.[18] The alloy, identified as CM-7, is carefully balanced to minimize the formation of η and Laves phases in service (Fig. 9a). While the γ' phase is reportedly stable to 1550 °F (843 °C), high stresses will accelerate the transformation to η above about 1475 °F (802 °C).

Considerable effort was also expended in the study of the cobalt–chromium–tantalum system, with particular focus on a Co–8Cr–10Ta alloy.[20] Chromium content was minimized in this alloy because it decreases solubility for tantalum. Aluminum and titanium additions had a minimal effect on hardenability, while the addition of 2.25 wt. % vanadium stabilized the coherent α-Co_3Ta to a higher temperature [1472 °F (800 °C) from 1292 °F (700 °C)]. A 20% nickel addition in conjunction with an increase in tantalum to 15% caused the precipitation of the rhombohedral β-Co_3Ta compound in the matrix {1, 1, 1} planes. Matrix stabilization of the FCC phase was greatly enhanced at this nickel level, and the partly coherent β-Co_3Ta precipitate provided stable hardening to 1652 °F (900 °C).

TCP PHASES

Chapter 8 deals with TCP phases extensively. However, in order to provide a balanced view of cobalt alloy metallurgy, the subject is discussed in abbreviated form here. TCP phases that have been observed in cobalt alloys are sigma (σ), mu (μ), and Laves; pi (π) is a related semicarbide phase also observed. Laves has been observed often in cobalt alloys; it is common in L-605 and found occasionally in S-816 and HS-188 (Figs. 9c and 9d). TCP phases occur when the solubility limit of the austenitic matrix is exceeded primarily by a combination of chromium and refractory element additions.

As in the case of nickel and iron alloys, these phases are of concern because they can cause a loss of strength and ductility at service temperatures as well as a severe loss of low-temperature ductility. In the author's laboratory a significant loss of rupture strength has been found for modified X-45-type alloys when the chromium plus tungsten content exceeds about 37 w/o and σ phase is generated (see Fig. 9b). It has been shown by Wlodek[10] that the α-Co_2W Laves phase degrades the room temperature ductility of L-605 when the alloy is exposed at typical service temperatures such as 1600 °F (871 °C). However, in Wlodek's case the effect of silicon content on Laves phase stability was shown to be a significant factor. As indicated previously, alloy HS-188 was carefully formulated using the PHACOMP calculation tool to balance nickel, tungsten, and silicon levels in order to minimize Laves phase precipitation.

It is worth noting that these phases precipitate in both the accicular and/or blocky morphologies in cobalt alloys. Conversely, reductions in alloy strength may be associated with the removal of the strengthening elements from the matrix. However, the acicular morphology is clearly undesirable from the standpoint of crack initiation and propagation through the microstructure.

Alloy UMCo-50 is a Co–30Cr–20Fe solid-solution alloy developed for high-temperature furnace applications; it is not considered a superalloy per se due to its low rupture strength. However, it is noteworthy because it forms a stable intragranular σ precipitate following lengthy exposures above 1607 °F (875 °C), which favorably enhances strength without decreasing high-temperature ductility.

MICROSTRUCTURE AND HEAT TREATMENT OF ALLOYS

The microstructure of contemporary cobalt alloys is a strong function of the chemical composition, crystallographic phases, and thermomechanical processing history. The nature and morphology of precipitated phases within the alloy structure is likewise a strong determinant of the mechanical properties and structural stability of the alloy system under actual service operating conditions. It is therefore essential to examine the role of alloy microstructure and to characterize microstructural changes as a consequence of heat treatment cycles and service aging exposures.

Wrought cobalt alloys exhibit the simplest microstructures since carbide content is constrained to minimize their effect on workability. HS-188, for example, consists of an intragranular dispersion of blocky M_6C and grain boundary $M_{23}C_6$ carbide particles in the mill-annealed condition (see Fig. 10d). For its primary application in sheet form, a uniform grain size of ASTM 5-6 is optimum for high-temperature creep strength. Klarstrom recently demonstrated[24] that thermomechanical processing (TMP) of thin [0.015 in. (0.4 mm)] sheet can improve the low-strain (<1%) creep strength of HS-188 by developing a strong recrystallized texture. TMP processing consisted of 80% final cold work followed by an anneal at 2250 °F (1232 °C) for 10 min. The major components of the texture were identified as (110) [$\bar{1}$10] and (112) [$\bar{1}$10] with respect to the plane of the sheet and the rolling direction. Transmission electron microscopy revealed that the improvement was due to the combination of enhanced dislocation subboundary formation and the precipitation of carbides on the dislocation structures during creep, as opposed to the dislocation tangles and pileup at carbides experienced by the reference lower strength material.

The wrought GCP-phase-strengthened alloys containing ordered coherent precipitates such as $(Co,Ni)_3Ti$, like Jetalloy-1650 or CM-7, usually contain a small quantity of titanium-rich MC carbide in addition. Viatour et al.[18] demonstrated that

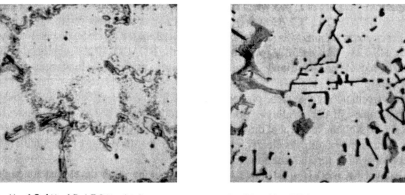

a. X-40/X-45/FSX-414   b. MorM-509

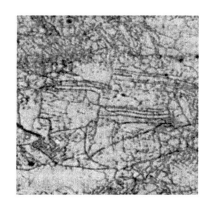

c. SM-302   d. HS-188. 30% Cold-rolled,
                              then annealed 30 min@1800°F
                              (980°C)[22]

Fig. 10. Characteristic structures from prominent cobalt superalloys, 334×.

alloy CM-7 can be solution treated at 2200 °F (1204 °C) and aged at 1470 °F (799 °C) to produce a uniform dispersion of ordered $Co_3(Ti,Al)$. Simulated aging exposures at 1500 °F (816 °C) and above for over 1000 h caused transformation to the ordered η Ni_3Ti-type DO_{24} phase. In addition, coarse particles of β CO_2Ti and Co_2W Laves phases precipitated at grain boundaries near the primary carbides or emanating from η platelets. The time–temperature–transformation diagram for CM-7 is shown in Fig. 11 to illustrate the relatively complex phase behavior of this system.

The work-strengthened cobalt alloys of the Multiphase family are somewhat more complex in microstructure. These alloys are strengthened via strain-induced matrix transformation from γ FCC austenite to ϵ HCP in combination with the precipitation of intermetallic compounds such as Co_2Mo Laves or ordered Co_3Al along FCC–HCP and twin boundaries. Heat treatment cycles are constrained by the required retention of cold work for a given application and level of tensile properties; that is, the transformation temperature must not be exceeded. Recent alloy modifications by Hagan et al.[25] have provided useful mechanical properties to 1300 °F (704 °C), competitive with the popular nickel alloy Waspalloy.

The physical metallurgy of investment-cast cobalt alloys, while less complex than high-strength cast nickel alloys, revolves around the generation and heat treatment control of the carbides. The simplest alloys experience precipitation of the chromium-rich $M_{23}C_6$, M_7C_3, and M_3C_2 carbides only; alloy X-40 and its offspring X-45 and FSX-414 are examples of these. Primary carbide particles precipitate intragranularly during cooling of the casting; these are often coarse, blocky forms with a semirounded morphology. The last liquid to freeze generates the lathlike $M_{23}C_6$ pseudoeutectic carbide structure in the interstices of the dendrite arms and at grain boundaries (Fig. 7). The eutectic is composed of alternating plates of $M_{23}C_6$ and γ matrix; however, the dominant phase in this pseudoeutectic may be either of the two, depending on the chromium–carbon ratio or cooling rate. Additional secondary $M_{23}C_6$ carbide particles may precipitate in thicker casting sections, generally in the immediate neighborhood of the eutectic islands. Long needlelike precipitates, actually comprised of individual carbide particles, may also be produced (Fig. 7b). The lower chromium–carbon ratio carbides M_7C_3 and M_3C_2 are produced in alloys with somewhat lower chromium contents, but these follow the morphologies outlined above.

Heat treatment of the chromium-rich carbides primarily serves to solution and reprecipitate the finer $M_{23}C_6$ particles. Typically, a 2100–2200 °F (1149–1204 °C) solution treatment will dissolve some coarse grain boundary carbides and homogenize the cast structure to some extent. Aging in the 1400–1800 °F (760–982 °C) temperature

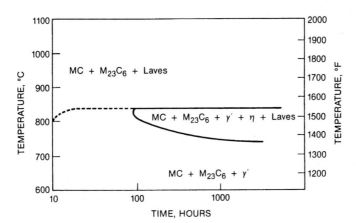

Fig. 11. Schematic time–temperature–transformation diagram of alloy CM-7. Reprinted with permission from P. Viatour, J. M. Drapier, D. Coutsouradis, and L. Habraken, *Cobalt*, June 1971, The Cobalt Development Institute.[18]

range will reprecipitate fine $M_{23}C_6$ particles more uniformly throughout the structure (Fig. 10a). In general, the lowest aging temperatures will produce the finest precipitate with the greatest increase in tensile strength; conversely, ductility decreases. The coarser precipitate is more desirable for high-temperature creep rupture strength and ductility. The eutectic islands are generally not affected, being stable up to the incipient melting temperature of approximately 2425 °F (1330 °C).

Incorporation of the reactive MC-forming elements such as hafnium, tantalum, zirconium, columbium, and titanium causes an immediate shift in the nature and heat treatment response of the precipitates. Alloy MM-509 is an example of an alloy with a balanced distribution of MC and $M_{23}C_6$ carbides, where additions of tantalum, zirconium, and titanium produce a predominantly Chinese-script MC carbide morphology (Fig. 12a). The size and extent of this script is a function of casting section size and cooling rate, where finer carbides are produced at the highest cooling rates in thin sections. Secondary precipitation of the $M_{23}C_6$ and MC carbides may also occur as fine particles; TEM examination has revealed these features more clearly (Fig. 13). With even higher ratios of MC-forming elements to chromium, as in the case of alloy MM-302, very large primary, blocky MC carbides form along with massive Chinese-script carbides (Fig. 10c). Small quantities of M_6C and $M_{23}C_6$ secondary precipitate occur in this alloy, and large blocky eutectic islands readily form.

Heat treatments of MM-509 and similar MC-bearing alloys are largely limited to the solution and reprecipitation of the $M_{23}C_6$ constituent, as in the case of the X-40-type alloys. Since the MC carbides initially form in the melt, they are essentially stable up to the incipient melting point. In general, there is less $M_{23}C_6$ carbide present in these alloys, so the response to heat treatment is substantially less than for other cobalt and nickel alloys. Since solubility is progressively greater for all carbides with increasing temperature, some carbide reprecipitation can indeed be generated. The solution treatment of MM-509 at very high temperatures such as 2325 °F (1274 °C) does remove most of the grain boundary $M_{23}C_6$ and some of the intragranular carbide (Fig. 12b). As expected, this has a beneficial effect on tensile ductility, especially at lower temperatures, and is especially useful for alloy weldability. Subsequent aging at 1700 °F (927 °C) produces general carbide precipitation, as expected (Fig. 12c). Despite the improvement in carbide distribution, tensile ductility falls to near original levels (3–7%) due to the high carbon content. This heat treatment has a slight negative effect on creep rupture strength but increases rupture ductility. It is more useful when applied to the postservice material, where long-term aging has generated significant quantities of fine $M_{23}C_6$ carbides that have a marked negative effect on low-temperature ductility; preservice properties can be generated with this heat treatment.

Directional solidification is a processing technique that has met with considerable success when applied to conventionally cast nickel-base superalloys (see Chapter 7). Vandermousen et al.[28] studied the effect of such processing on the microstructure and properties of representative cobalt alloys, that is, X-40, WI-52, and MM-509. Under experimental conditions that produced dendritic columnar growth at traverse rates of 1.2–12 in./h (3–30 cm/h), phases identical with equiaxed material were

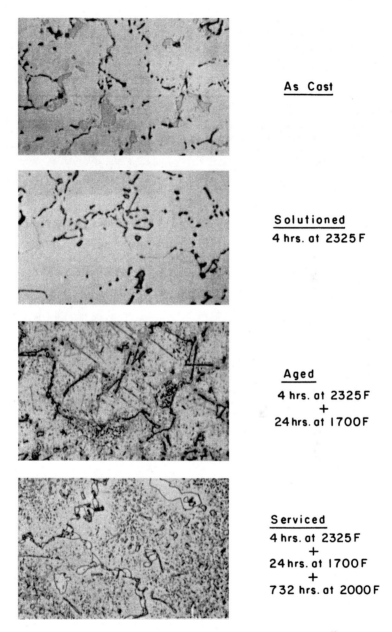

As Cast

Solutioned
4 hrs. at 2325 F

Aged
4 hrs. at 2325 F
+
24 hrs. at 1700 F

Serviced
4 hrs. at 2325 F
+
24 hrs. at 1700 F
+
732 hrs. at 2000 F

Fig. 12. Effect of heat treatment on the cobalt superalloy MM-509, 375×[12].

a. As cast MarM-509, MC carbide and stacking faults. 16,660 X

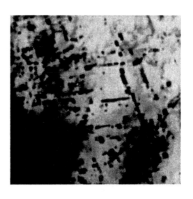

b. Precipitated $M_{23}C_6$ Particles. 21,420X

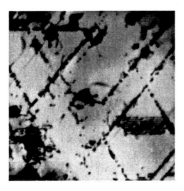

c. Precipitated MC Particles. 21,420 X

d. Large $M_{23}C_6$ Plate Surrounded by Dislocations. 21,420 X

Fig. 13. Structural details of alloy MM-509[26,27].

found. Finer structures were produced with increasing traverse rate, which enhanced the tensile and rupture ductility of all three alloys significantly. Only the rupture strength of X-40 and the thermal fatigue resistance of X-40 and MM-509 were improved by this processing.

The substantial development effort expended during the 1970s to directionally solidify (DS) *in situ* fiber-strengthened composites led to the identification of the M_7C_3 and TaC fiber-strengthened cobalt-base family of materials; these are discussed in greater detail in Chapter 19. The growth of monovariant eutectic TaC–Co,Cr structures is a further example of the thermodynamic stability inherent in cobalt alloy systems. The primary advantages of this material are the significant creep strength and ductility exhibited in the fiber growth direction at very high homologous temperatures. The main disadvantages are the low transverse properties and low-cycle fatigue strength in addition to the economics of DS processing under tight constraints. The commercial viability of these materials has not yet been established.

Another unique family of investment cast alloys employing the precipitation of ordered, noncoherent CoAl particles is exemplified by the AR-213 family of alloys. Solution treating at 2200 °F (1204 °C) followed by aging at 1400 °F (760 °C) produces maximum hardness and presumably strength. Overaging of the precipitate occurs at exposures in excess of 1600 °F (871 °C); hence, the system is more limited in temperature capability than the carbide-strengthened alloys.

Perhaps the most intriguing class of cobalt materials are the powder metallurgy (PM) materials. In general, it has been found that PM processing of carbide-strengthened cobalt alloys produces tensile properties superior to cast alloys up to about 1300 °F (704 °C) due to much finer carbide distribution and grain size (Fig. 14). At higher temperatures rupture properties are adversely affected by the fine grain size and the presence of coarse prior particle boundary (PPB) carbides. To some extent, this situation may be improved by using extruded powder stock, which exhibits a more uniform distribution of carbide due to the breakup of the PPB structure.

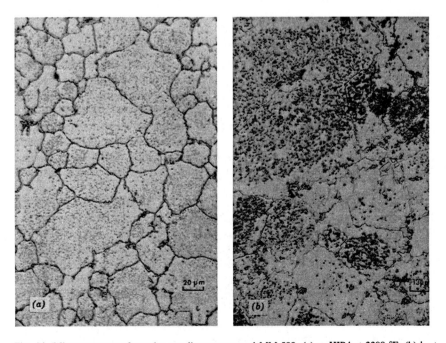

Fig. 14. Microstructures of powder metallurgy processed MM-509: (a) as-HIPd at 2200 °F; (b) heat treated 2250 °F for 4 hr + 1400 °F for 20 h.

More recently[29] it has been shown that HIP heat treatments of consolidated X-40 PM alloy can successfully generate significantly superior creep rupture properties at temperatures between 1200 and 1800 °F (649–982 °C). This is accomplished by solution treating under high argon pressure during HIP at a temperature that approaches or exceeds the incipient melting temperature of the alloy, generally associated with the $M_{23}C_6$ eutectic. The dissolution of the eutectic and the PPB carbides is responsible for accelerated grain growth and reprecipitation of carbides in a more useful morphology. For example, the autoclave heat treatment at 2300 °F (1260 °C) produced complete recrystallization; substantial grain growth occurred in the 2360 °F (1293 °C) HIP treatment. Rupture lives considerably in excess of cast X-40 were achieved at some expense in rupture ductility and possibly fracture toughness. This appears due to the formation of coarse, cellular grain boundary carbide networks at the higher HIP temperatures.

MECHANICAL PROPERTIES AND STRENGTHENING MECHANISMS

This section will review the mechanical property response of cobalt alloys in relation to the composition, phases, and heat treatments described above and broadly compare these to the properties of nickel alloys. Specific mechanical properties have been compiled in Appendix B and the open literature. Finally, the effect of long-time service aging exposures on the microstructure and properties of cobalt alloys is examined.

Mechanical Properties

Wrought solid-solution-strengthened cobalt alloys such as L-605 and HS-188 are superior in creep rupture properties to their nickel alloy counterparts (such as Hastelloy X and IN-617) by up to 100 °F (55 °C) and approach the rupture strength of the lower carbon cast cobalt alloys like X-45 and FSX-414. In addition, they exhibit excellent workability and weldability for gas turbine combustor applications. HS-188 in particular offers outstanding oxidation resistance for these high-temperature components and is not as prone to in-service precipitation of the Laves phases, which reduce ductility in L-605 or the M6C carbide that similarly affects Hastelloy X.

The dominant commercial class of cobalt alloys are the investment cast carbide-strengthened alloys, where tensile strength and rupture strength are essentially a direct function of the carbon content and the resultant volume percent carbides. Compared to nickel alloys, the carbide-strengthened cobalt alloys exhibit a flatter stress rupture parameter curve as a function of temperature (Fig. 15). The lack of a high-temperature ordered, coherent particle-strengthening mechanism produces substantially lower strength at temperatures up to about 1800 °F (982 °C). The greater stability of the carbides, especially M_6C and MC, versus γ' yields a strength advantage at the higher temperatures. This characteristic is the primary reason cobalt alloys are utilized in the lower stress, higher temperature stationary vanes of gas turbines.

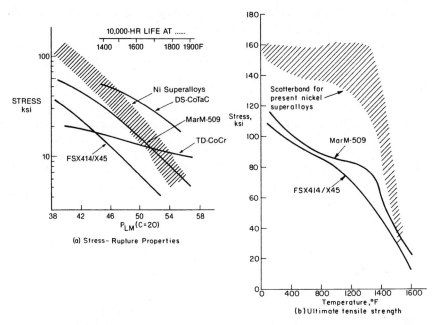

Fig. 15. Mechanical properties of representative cobalt alloys compared to contemporary nickel-base superalloys: (a) stress rupture properties; (b) ultimate tensile strength.

The directionally solidified Co–Cr$_7$C$_3$ and Co–TaC fiber-strengthened materials are substantially stronger in the growth direction than the conventionally cast polycrystalline alloys (Fig. 15). Room temperature tensile strengths in the range of 200–215 ksi (1379–1482 MPa) have been obtained for the Co–Cr$_7$C$_3$ material; conversely, ductilities as low as 1% are not uncommon. Similarly, the creep rupture strength capability is higher than the conventional cobalt alloys, yielding rupture lives 10–20 times greater for temperatures up to 2000 °F (1093 °C).

Oxide-dispersion-strengthened (ODS) alloys exhibit an even flatter parameter rupture curve (Fig. 15) and are superior to all other systems above 2000 °F (1093 °C). However, the difficulty of manufacturing actual hardware components and the loss of parent metal properties in fusion-welded joints are obvious drawbacks.

Strengthening Mechanisms

The strengthening mechanisms utilized in cobalt alloys are principally a careful balance of refractory element solid-solution hardening and carbide precipitation. Both are necessary for high-temperature creep rupture and fatigue strength. The carbides provide strong inhibition to grain boundary sliding and grain growth as well as an impedance to dislocation mobility. In the 1000–1500 °F (538–816 °C) temperature range secondary precipitation of principally fine M$_{23}$C$_6$ carbide particles interacts strongly with stacking faults, yielding preferential precipitation along the faults and at fault intersections with a significant effect on strength and, conversely, ductility. Above approximately 1800 °F (982 °C) grain boundary carbides impede grain boundary sliding. The solid-solution-strengthening role of the refractory elements becomes of greater significance since agglomeration and growth reduces the effectiveness of the intragranular carbides.

The carbon effect is nonlinear above about 0.3 w/o and reaches an effective maximum at 0.5–0.6 w/o in alloys such as X-40 and MM-509. Tensile ductility from room temperature to 1400 °F (760 °C) is inversely proportional to carbon content; rupture ductility above (1400 °F (760 °C) is less affected by carbon content. While carbon additions above 0.6 w/o do generate additional carbides, the morphology of the primary carbides and eutectic islands is mostly ineffective in alloy strengthening. Ductility decreases to very low levels since crack initiation is easier and intercarbidic propagation path length has been decreased.

We have seen that typical solution plus aging heat treatments are relatively ineffective in altering the strength–ductility ratio in high-carbon alloys since the primary MC carbides are so stable. The presence of the $M_{23}C_6$ eutectic islands is a further inhibiting factor for heat treatment response, although HIP treatment under isostatic pressure indeed provides an avenue for further exploration. The $M_{23}C_6$ carbide eutectic morphology would appear to be a near useless form in terms of a strength contribution. Therefore, the trend in recent alloy development has been to balance the carbides in favor of the more stable MC precipitates in order to minimize the primary and eutectic precipitation of $M_{23}C_6$.

The oxide-dispersion-strengthened cobalt alloys are a further manifestation of the high-temperature stability available in cobalt systems.[30] The incorporation of very fine (100–300 Å) inert oxide particles that are thermodynamically stable and nonreactive with the matrix, such as ThO_2 or Y_2O_3, provides significant creep rupture strength to temperatures approaching the melting point of the base. Thermomechanical processing of these materials is a critical factor in achieving the highly textured and high-aspect-ratio grain structure necessary for this property advantage. At lower homologous temperatures the relatively low volume fraction of dispersoid is not an effective inhibitor of dislocation mobility compared to the carbide-strengthened alloys.

Ductility

Woodford and McMahon performed a detailed evaluation of the fracture behavior of as-cast and heat-treated MM-509.[31] They concluded that fracture initiation occurs in the large carbide particles and eutectic islands with the onset of plastic deformation for all conditions. Crack propagation is controlled by the hardness and strength of the matrix. Hence, material aged at 1500 °F (815 °C) for 24 h yields the highest matrix hardness (660 Knoop) due to the very fine secondary precipitation of $M_{23}C_6$ carbides, and fracture proceeds transgranularly with little secondary crack branching and minimal ductility. Conversely, solution treatment at 2354 °F (1290 °C) for 4 h dissolves the $M_{23}C_6$ carbide and softens the matrix (360–370 Knoop). Crack propagation then occurs intragranularly with secondary crack branching, increased fracture of adjacent carbides, and greater ductility. The as-cast condition produces results intermediate to the two heat-treated conditions. Therefore, carbide distribution and spacing, particularly at grain boundaries, is a contributing factor to fracture toughness in cast cobalt alloys.

The secondary precipitation of $M_{23}C_6$ carbides during service exposure in the 1000–1600 °F (538–871 °C) temperature range has, as indicated earlier, a potentially detrimental effect on low-temperature ductility. While not reduced to dead-brittle condition, this loss may be sufficient to cause problems in the handling, repair welding, and machining of serviced components. Figure 16 illustrates this behavior

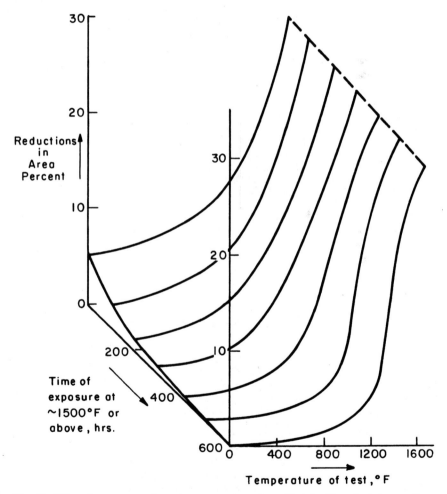

Fig. 16. Effect of exposure at elevated temperatures on the residual ductility of cobalt superalloys. MM-509 used as a guide.

for alloy MM-509 exposed at nominally 1500 °F (816 °C) for various time periods. The source for this aging phenomena is carbon retained in solution from the initially higher preservice aging treatment plus the degeneration of the coarser MC and $M_{23}C_6$ primary precipitates. Stress-accelerated aging, which promotes elemental interdiffusion, also undoubtedly contributes to this behavior. The key to the effect on ductility is the interaction of this carbide precipitation with stacking faults, as discussed previously. Fortunately, this effect is largely reversible via re-solution and aging to restore the normal microstructure prior to additional service exposure.

Woodford and Bricknell[32] recently demonstrated that cast cobalt alloys may be immune to high-temperature environmental embrittlement. Oxidation exposure of many nickel alloys at about 1800 °F (982 °C) for 100 h severely impairs ductility in the temperature range 1292–1652 °F (700–900 °C). Embrittlement may be due in part to the rapid diffusion of oxygen down grain boundaries in association with the grain boundary γ' film. Cobalt alloys MM-509 and FSX-430 suffered slight losses in rupture life following exposures in both air and vacuum, illustrating that the thermal exposure alone has an effect but that environmental embrittlement does not occur in these alloys.

REFERENCES

1. F. R. Morral, **20**, 18 *J. Met.* (July 1968).
2. J. C. Freche, R. L. Ashbrook, and S. J. Klima, *Cobalt*, **20**, 114 (1963).
3. L. A. Pugliese and J. P. Stroup, *Cobalt*, **43**, 80 (June 1963).
4. H. Morrow, W. P. Danesi, and D. L. Sponseller, *Cobalt*, **4**, 93 (1973).
5. S. P. Rideout, W. D. Manly, E. L. Kamen, B. S. Lement, and P. A. Beck, *Trans. AIME*, **191**, 872 (1951).
6. R. H. Nafziger and R. L. Lincoln, *Cobalt*, **4**, 79 (1974).
7. A. Giamei, J. Burma, and E. J. Freise, *Cobalt*, **39**, 88 (June 1968).
8. R. Kamel and K. Halim, *Phys. Stat. Sol.*, **15**, 63 (1966).
9. F. R. Morral, *Cobalt and Cobalt Alloys*, Cobalt Information Center, Columbus, Ohio, 1967.
10. S. T. Wlodek, *Trans. AIME*, **56**, 287 (1964).
11. H. J. Goldschmidt, *J. Iron Steel Inst.*, **160**, 345 (1948).
12. C. T. Sims, *J. Met.*, **21**, 27 (1969).
13. D. Coutsouradis, *J. Int. Appl. Cob., Bruxelles*, **21**, 1–19 (June 1964).
14. W. Köster and F. Spermer, *Arch. Eisenhuttenw.*, **26**, 555–559 (1955).
15. B. Lux and W. Bollmann, *Cobalt*, **42**, 3 (March 1969).
16. R. B. Herchenroeder, 'Haynes Alloy No. 188 Aging Characteristics," International Symposium on Structural Stability in Superalloys, Seven Springs, PA, September 1968.
17. C. Rogister, D. Coutsouradis, and L. Habraken, *Cobalt*, **34**, 3 (March 1967).
18. P. Viatour, J. M. Drapier, D. Coutsouradis, and L. Habraken, *Cobalt*, **51**, 67 (June 1971).
19. J. M. Blaise, P.-Viatour, and J. M. Drapier, *Cobalt*, **49**, 192 (December 1970).
20. J. M. Drapier and D. Coutsouradis, *Cobalt*, **39**, 63 (June 1968).
21. H. J. Beattie, Jr. and W. C. Hagel, *Trans. AIME*, **221**, 28 (1961).
22. R. B. Herchenroeder and W. T. Ebihara, ASM Congress, Detroit, MI, October 1968.
23. E. Hall and S. Algie, *J. Inst. Met.*, **11**, 61 (1966).
24. D. L. Klarstrom, *Superalloys 1980*, ASM, Metals Park, OH, 1980, p. 131.
25. F. C. Hagan, H. W. Antes, M. D. Boldy, and J. S. Slaney, *Superalloys 1984*, TMS-AIME, Warrendale, PA, 1980, p. 623.
26. M. J. Woulds and T. R. Cass, *Cobalt*, **42**, 3 (March 1969).

27. J. M. Drapier, V. LeRoy, C. Dupont, D. Coutsouradis, *Cobalt*, **41**, 199 (December 1974).

28. R. F. Vandermousen, P. Viatour, J. M. Drapier, and D. Coutsouradis, *Cobalt*, **1**, 6 (1974).

29. J. C. Freche and R. L. Ashbrook, *Cobalt*, **2**, 33 (1973).

30. J. M. Drapier, D. Coutsouradis, and L. Habraken, *Cobalt*, **53**, 197 (December 1971).

31. D. A. Woodford and C. J. McMahon, Jr., *Proceedings of the Second International Conference Strength of Metals and Alloys—Asilomar*, ASM, Metals Park, OH, 1970, p. 1067.

32. D. A. Woodford and R. H. Bricknell, *Superalloys 1980*, ASM, Metals Park, OH, 1980, p. 633.

Chapter 6

Nickel–Iron Alloys

E. E. BROWN and DONALD R. MUZYKA

*Pratt & Whitney Aircraft, East Hartford, Connecticut,
and Cabot Corporation, Boston, Massachusetts*

Precipitation-strengthened alloys containing substantial quantities of both nickel and iron form a distinct class of superalloys. They are used in many gas turbine engine and steam turbine components such as blades, disks, shafts, casings, and fasteners and in some automotive engines as valves. This chapter will discuss the nature of the nickel–iron-base superalloys, their chemistry, phases and structures encountered, and the relationship of these factors to properties and other characteristics. For our purpose, nickel–iron-base superalloys include those alloys with an face-centered-cubic (FCC) gamma (austenitic) matrix hardened by precipitation of ordered inter-metallic or carbide precipitates. This class of superalloys is generally characterized by compositions containing approximately 25–60% nickel with additions of iron ranging from 15 to 60%. Major emphasis is placed on the age-hardenable nickel–iron alloys; only a few comments are made concerning the nickel–iron alloys featuring primarily solid-solution strengthening, strain hardening, and/or carbide precipitation as the major strengthening mechanism(s). Some aspects of these alloys are reviewed elsewhere.[1,2]

Emphasis on the examples in this chapter concerns four well-known and widely used nickel–iron alloys: A-286, an iron-rich alloy strengthened by ordered FCC precipitates; Incoloy 901,† a nickel-rich alloy strengthened by ordered FCC precipitates; Inconel 718 and Incoloy 706, nickel-rich alloys strengthened predominantly by

† Referred to subsequently as 901, 718, 706, and 903, respectively.

165

ordered BCT precipitates; and Incoloy 903, a low-expansion, iron-rich alloy strengthened by ordered FCC precipitates.

HISTORICAL

The development of contemporary nickel–iron-base superalloys can be traced to the austenitic iron–base stainless steels. It was initially found that FCC iron–nickel–chromium alloys exhibited age-hardening response through additions of small quantities (less than 2 wt. %) of titanium. These alloys were found to develop high strength levels and retained their properties to remarkably high temperatures.

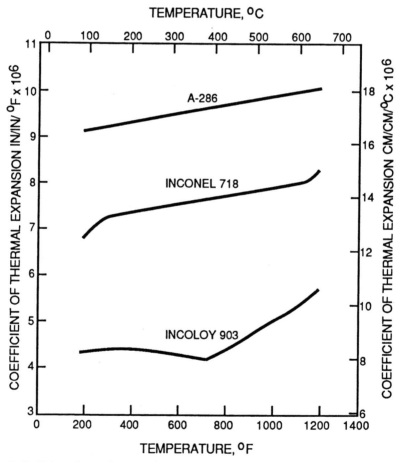

Fig. 1. Coefficient of expansion properties of A-286 and Inconel 718 compared to the low-expansion alloy Incoloy 903.[3]

The more notable alloys to emerge from the early development history include the German alloy Tinidur, the British alloy G18B, and the American alloys 19-9DL and A-286. These alloys were characterized by high iron, low nickel, and relatively low volume fractions of age-hardening precipitate.

From that early beginning the development of improved alloys was limited less by the metallurgists' imagination than by limitations of available melting and forging processes. A notable advancement was the development of vacuum induction melting, which allowed the retention of the more reactive elements such as titanium and aluminum. The evolutionary development from the early alloys to the contemporary alloys followed a general trend toward higher nickel content, lower iron content, increased solid-solution strengthening elements, and higher levels of titanium and aluminum to provide higher volume fractions of the age-hardening precipitate. Incoloy 901 represents a nickel-rich alloy that emerged along this development path.

The identification of columbium as a potent elemental addition for precipitation strengthening has led to the development of contemporary alloys such as 718 and Inconel 706† with mechanical property levels that far exceed those of their predecessors. Alloys such as 903 and Incoloy 909‡ have recently been introduced that combine good elevated-temperature properties with a significantly lower coefficient of thermal expansion (Fig. 1). These alloys are finding application in gas turbine engines for improved clearance control.

ALLOYS AND ALLOY TYPES

The nickel–iron-base alloys derive their high-temperature properties from a combination of alloying and strengthening effects that include solid-solution strengthening, precipitation hardening, and grain boundary strengthening. As such, they have several characteristics in common with respect to their chemical composition. These include (1) an austenitic matrix based on nickel and iron, (2) alloying additions that partition to the austenite for solid-solution strengthening, (3) alloying additions to form the strengthening precipitates (ordered intermetallics, carbides, borides, etc.), and (4) alloying additions that strengthen or otherwise modify grain boundaries. Strengthening mechanisms and their relationship to alloy composition will be discussed subsequently. At this time it would be useful to examine the various alloys of importance to this class of materials and their metallurgical characteristics.

Nickel–iron-base alloys can be grouped into four classes distinguished by composition and strengthening mechanisms. The first group includes those strengthened by ordered FCC gamma prime (γ'). This group can be further subdivided into two groups. The iron-rich subgroup, which includes the earlier alloys such as Tinidur, V-57, and A-286, contain relatively low levels of nickel (25–35 wt. %) and rely predominantly on additions of titanium (less than 2 wt. %) for precipitation hardening.

† Referred to subsequently as 706.

‡ Referred to subsequently as 909.

These alloys exhibit useful mechanical properties up to approximately 1200 °F (650 °C). The nickel-rich subgroup is characterized by significantly higher nickel contents (in excess of 40 wt. %), higher levels of solid-solution strengtheners, and higher volume fractions of strengthening precipitate. Alloys in this group include 901 and Inconel X-750. Relative to the iron-rich alloys, they exhibit improved strength levels and can be used to higher temperatures. The nickel-rich alloys continue to find application in modern gas turbine engines owing to their attractive properties and low cost relative to nickel-base alloys.

The second group of alloys, which includes 718 and 706, is nickel rich and strengthened predominantly by ordered BCT gamma double prime (γ''). The recent discovery of γ'' and the unique property characteristics that it imparts ranks as a very important contribution to nickel–iron-base superalloy metallurgy. The 718 alloy has emerged as the most widely used superalloy of modern times and will be discussed in more detail throughout this chapter. This group of alloys offer exceptionally high property levels from cryogenic temperatures up to 1200 °F (650 °C).

A third group of unique alloys based on the iron-rich iron–nickel–cobalt system strengthened by FCC γ' have been recently developed that combine high strength levels with low coefficient of thermal expansion properties. Incoloy 903 and 909 are the predominant alloys in the group, which derive their low expansion characteristics (Fig. 1) through the elimination of ferrite stabilizers such as chromium and molybdenum.[3] These alloys offer excellent strength characteristics to temperatures of about 1200 °F (650 °C). However, the absence of chromium significantly reduces the oxidation resistance of the materials.

The fourth group of alloys derives major precipitation strength from carbides, nitrides, and/or carbonitrides and includes 16-25-6, HMN, and the CRMD series. These alloys are used to approximately 1500 °F (815 °C). This group also includes alloys with little or no precipitation strengthening available. Hastelloy X and N-155 are typical examples. These alloys are utilized primarily in low-stress applications to approximately 2000 °F (1093 °C), where oxidation resistance is the most significant requirement.

CONSTITUTION, STRUCTURE, AND PHASE RELATIONSHIPS

The nickel–iron-base superalloys all exhibit a number of fundamental similarities. They contain an FCC austenitic matrix that consists of a balance of nickel and iron, contain solid-solution strengthening elements that generate a variety of effects in the overall alloy, and are strengthened by ordered matrix and grain boundary precipitates. This section will focus on the alloying elements of nickel–iron alloys as they affect structure and phase relationships.

Austenitic Matrix

The austenitic matrix of nickel–iron alloys ranges from nickel-lean (less than 35 wt. % Ni) to nickel-rich (>40 wt. % Ni). The ratio of nickel to iron plays a large role in determining the ultimate characteristics of the alloy since it affects alloy cost and useful temperature range. Since most alloys contain relatively low carbon (<0.10%) and relatively large amounts of ferrite stabilizers (e.g., chromium and molybdenum), the minimum level of nickel required to maintain an austenitic matrix is approximately 25 wt. %. The addition of cobalt or other austenitic stabilizers can slightly lower this requirement. The iron-rich matrix lacks the stability of higher nickel matrices, which often limits the quantity of solid-solution and precipitation strengthening elements that can be added. Oxidation and corrosion resistance is also adversely affected as the iron-to-nickel ratio increases. On the positive side, higher iron contents result in lower alloy cost and improved forgeability.

Solid-Solution Strengtheners

The solid-solution strengtheners typically added to nickel–iron-base superalloys include chromium in the range 10–25% and molybdenum in the range 0–9%. Tungsten may be used in place of molybdenum, but the cost and adverse effect on density limits the use. However, the Russian literature[4] cites superior strength for alloys containing tungsten relative to those strengthened by molybdenum. The lattice parameter of cobalt is too close to nickel and iron for cobalt to be an effective solid-solution strengthener. Molybdenum expands and cobalt contracts the lattice of nickel–iron γ matrix as these elements are substituted for iron.[5]

When added to nickel–iron superalloys, chromium and molybdenum partition preferentially to the γ where they impart a number of important effects. Their presence substantially reduces the matrix solubility for the precipitation-strengthening elements (Ti, Al, and Cb). For example, the solubility limit for titanium in Ni–50Fe is about 1.5 wt. %, indicating that for an alloy such as 901 that contains 2 wt. % titanium, little if any precipitation strengthening would occur in the absence of chromium and molybdenum.

Additions of solid-solution strengtheners also impact the characteristics of nickel–iron superalloys as a result of expansion of the lattice parameter of the austenite. This expansion is a measure of the internal strain imposed by the difference in atomic volume between the austenite and the addition. The expansion of matrix lattice parameter influences the degree of mismatch with the coherent strengthening precipitate. In addition, solid-solution strengtheners have been shown to reduce stacking fault energy, thus impeding cross-slip at elevated temperatures.[6] Thus, the solid-solution strengtheners exert a pronounced effect on the efficiency of precipitation hardening.

Chromium provides an additional benefit to nickel–iron alloys by imparting oxidation–corrosion resistance that allows their use in aggressive elevated-temperature environments. Chromium must be present in concentrations sufficient to form a continuous protective oxide film. Studies on 901 have shown that the threshold for formation of a continuous oxide film occurs at approximately 9 wt. % chromium. Nickel–iron alloys typically contain chromium contents well in excess of this level.

The low-expansion alloy 903 contains no chromium and as such exhibits poor oxidation and corrosion resistance. The alloy is susceptible to rust formation at low temperatures and receives little oxidation protection from the NiO + CoO oxide that forms above 900 °F (480 °C). For applications above 900 °F (480 °C) a protective coating is usually required.

Solid-solution strengthening effects can also be attributed to the elements added for precipitation strengthening (Ti, Cb, and Al) owing to their small but finite solubility in the austenite. Irvine et al.[7] have shown that potent solid-solution-strengthening results from the interstitial elements carbon, nitrogen, and boron as well. Carbon and boron are subsequently considered in further detail.

Precipitation Strengtheners

The predominant source of strengthening in nickel–iron-base alloys is that resulting from the precipitation of coherent, ordered A_3B-type compounds. These fall into two categories, ordered FCC γ' and ordered BCT γ''.

The ordered γ' forms in those alloys containing titanium and aluminum as the principal age-hardening constituents. Alloys of this class, which include A-286 and 901, differ significantly from nickel-base superalloys in that they contain a relatively high ratio of titanium to aluminum. The lattice parameter of γ' increases as the titanium-to-aluminum ratio is increased, as does the lattice parameter of austenite as the iron-to-nickel ratio increases. In order to minimize lattice parameter misfit between γ and γ', the nickel–iron alloys necessitate high ratios of titanium to aluminum. When this ratio exceeds approximately 2 : 1 (atomic), a metastable FCC γ' forms that will transform to a stable hexagonal Ni_3Ti (η) phase during elevated-temperature exposure. This transformation affects elevated-temperature stability adversely and will be subsequently discussed in greater detail.

The ordered BCT γ'' forms in those alloys containing columbium (niobium) as the principal age-hardening constituent. Alloys of this class include 718 and 706, which contain 5 and 3 wt. % columbium, respectively. As both alloys also contain smaller amounts of aluminum (0.5 and 0.2 wt. %, respectively) and titanium (0.9 and 1.7 wt. %, respectively), an FCC γ' phase forms and coexists with the γ''.[8–13] For 718 the ratio of γ'' to γ' phases has been determined to be between 2.5 and 4.0.[10] For 706 Raymond et al.[12] have shown that at aluminum levels below 0.2%, the γ'' phase predominates. As aluminum content is increased to 0.5%, the γ' phase predominates. This transition with increasing aluminum content is accompanied by a marked reduction in yield strength. Similar effects have also been reported for 718.[13] These studies indicate that the solubility of columbium in Ni_3Al (γ') is quite high ($\sim$40%) whereas the solubility for aluminum in Ni_3Cb (γ'') is quite low ($\sim$1%), thus explaining the significant impact of small levels of aluminum on the strengthening precipitates in these alloys.[12]

Studies to date on the low-expansion alloy 903 have shown the strengthening precipitate to be exclusively FCC γ'. While this alloy contains 3 wt. % columbium, the aluminum content of 1.0 wt. % is sufficient to stabilize the FCC species.

The γ'' phase has been the subject of numerous investigations.[9–11,14] The structure is ordered DO_{22} and is metastable, transitioning during extended aging times to orthorhombic Ni_3Cb delta (δ) phase.[14] This phase instability occurs during exposure to temperatures in excess of 1200°F (650 °C) and can be related to a degradation in properties of 718 above this temperature.

Elements Providing Other Beneficial Effects

Certain other elements are added intentionally to nickel–iron alloys to improve properties. Boron is an essential alloying element and is added in the range 0.003–0.030% to improve stress rupture properties and hot workability.[14] Zirconium is added for similar reasons and as a carbide former. Studies[15] indicate that the effects of boron and zirconium are related to changes in interfacial energy that favor coalescence and spherodization of secondary phases precipitating at grain boundaries. Such globular, blocky, or spherodized grain boundary particles produce a ductile grain boundary compared to the notch-sensitive conditions associated with continuous films. Boron has been shown to retard the transition of metastable γ' to η by delaying nucleation at grain boundaries.[16]

The Russians add vanadium to nickel–iron-base alloys for improved hot working.[16] Vanadium was originally added to A-286 to improve notch ductility at elevated temperatures.[17] Carbon is used as a deoxidant, as a former of primary MC carbides to refine grain size during hot working, and to form beneficial grain boundary carbides. Manganese and rare-earth elements are sometimes added as deoxidizers. Magnesium has been shown to benefit smooth and notched stress rupture strength and ductility of nickel–iron alloys through modification of grain boundary phases and morphologies.[18,19]

PHYSICAL METALLURGY

A treatment of the physical metallurgy of the iron-base and nickel–iron-base alloys including processing and strengthening effects[1,20] plus others primarily on nickel–base alloys,[6,16–20] have sharply promulgated the understanding of the relationships between composition, processing, microstructure, and properties of superalloys. Aspects of these papers and others that relate to nickel–iron-base alloys are considered here.

Phases

The major strengthening phases in nickel–iron-base superalloys have been previously discussed along with the stable forms of the metastable strengthening phases. Several other phases of importance to these alloys will now be discussed.

Compressibility factors[21] cause most nickel–iron compositions to form phases with abnormally short interatomic distances. Examples are the topologically close-packed (TCP), or A_2B, types, such as σ, μ, X, or Laves.[22,23] Inconel 718 and 901 are susceptible to the formation of Laves phase[24,25] with attendant degradation in mechanical properties. The TCP phases will be covered in greater detail in subsequent chapters.

Carbides form an additional important class of phases that occur in these alloys. All known commercial nickel–iron alloys form MC-type idiomorphic or irregular carbides and/or carbonitrides during solidification. These carbides are not changed drastically during forging, heat treating, or long-time service [<1500 °F (815 °C)] temperatures for nickel–iron alloys.

For alloys containing columbium as a strengthener, the predominant MC carbide is CbC, while titanium-strengthened alloys contain TiC carbides. Other elements such as molybdenum, vanadium, and tantalum may also enter the MC. Some MC-type carbides can be seen as the coarse, irregular particles in Figs. 2 and 3. Carbides with MC stoichiometry may also precipitate at the grain boundaries of nickel–iron-base alloys during processing, heat treatment, or service. Figure 2b shows globular MC formed during the heat treatment of 901. Globular grain boundary MC is generally beneficial to properties and promotes good rupture life and ductility. However, MC carbide films should be avoided during processing or service since they can cause embrittlement.[26]

Several nickel-base superalloys form M_6C upon solidification and/or heat treatment. Few nickel–iron-base alloys contain enough molybdenum to form this phase. However, precipitated M_6C has been identified in the grain boundaries of 718 (which contains 3% Mo) by several investigators.[24,27] A carbide of considerable importance in nickel–iron-base alloys is $M_{23}C_6$. It occurs at grain boundaries during processing, heat treatment, or service. Heat treatments are often designed to promote blocky $M_{23}C_6$ (M is predominantly chromium) in favor of $M_{23}C_6$ films that can form during service. A fine cellular $M_{23}C_6$ formed in A-286 by improper processing has been shown to cause stress rupture notch sensitivity.[28,29]

Strengthening Mechanisms

Strengthening mechanisms in nickel–iron-base alloys closely parallel those for nickel-base alloys (treated in detail in previous chapters). A review of the strengthening mechanisms in nickel–iron-base alloys will be subsequently presented, with major emphasis on those aspects of strengthening that differ from nickel-base alloys. Solid-solution-strengthening elements have been discussed previously; further discussion of their effects will be limited to influence on precipitation strengthening.

γ' Precipitation

The precipitation of ordered FCC γ' ($L1_2$) predominates in nickel–iron-base alloys such as A-286 and 901. Analogous to the nickel-base alloys, it is possible to relate strength of nickel–iron alloys to several (not necessarily additive) effects: antiphase boundary and fault energy of γ'; γ' strength, coherency strains, and volume fraction γ' ($V_f\gamma'$); γ' particle size; and γ–γ' modulus mismatch. A few of these effects are now considered.

Raynor and Silcock[4] have shown for nickel–iron-base superalloys of the A-286 type that at constant $V_f\gamma'$, strength initially increases as γ' particle size increases until peak aging is reached. In this regime strengthening is related to paired dislocations

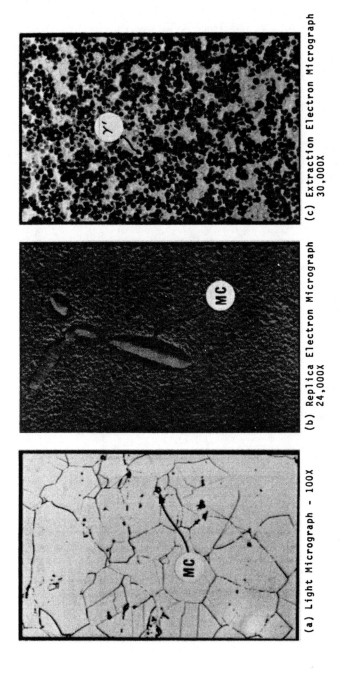

(a) Light Micrograph – 100X

(b) Replica Electron Micrograph 24,000X

(c) Extraction Electron Micrograph 30,000X

Fig. 2. Microstructure of 901, a typical nickel–iron-base superalloy strengthened by γ′. Heat treated: 2000 °F (1100 °C), 2 h water quenched (WQ) + 1425 °F (775 °C), 2 h, air cooled (AC) + 1350 °F (730 °C), 24 h AC.

173

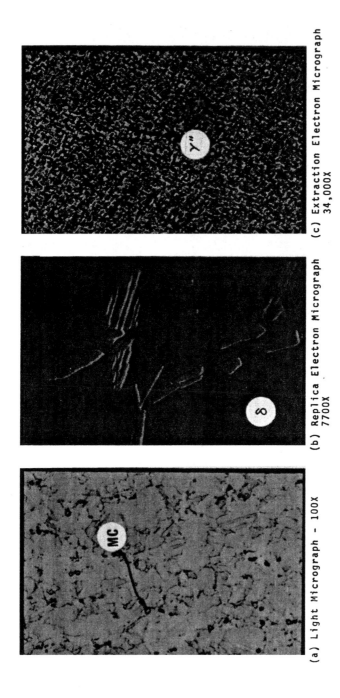

(a) Light Micrograph – 100X

(b) Replica Electron Micrograph 7700X

(c) Extraction Electron Micrograph 34,000X

Fig. 3. Microstructure of 718, a typical nickel–iron-base superalloy strengthened by γ". Heat treated: 1750 °F (955 °C), 1 h AC + 1325 °F (720 °C), 8 h, cool 100 °F (55 °C)/h to 1150 °F (620 °C), 8 h AC.

cutting precipitate particles, alternately creating and annihilating an antiphase boundary in the γ'. This can be seen in Fig. 4a, which shows the arrangement of atoms on the γ' slip plane, illustrating the effect of internal order of the γ' on superdislocation arrangement. A superdislocation consisting of two $a/2<110>$ matrix dislocations are required to restore order to the γ' for either of the three $<110>$ directions the superdislocation moves.[12] Figure 5a shows paired dislocations cutting precipitate particles at peak strength. The first dislocation pair is considerably more bowed than the second, and a few dislocation (Orowan) loops have occurred. After peak aging further particle growth results in decreased strength due to particle bypass by looping around them (Fig. 5b). Maximum strength is generally found to occur in nickel–iron alloys at γ' particle diameters of approximately 100–500 Å.

Work reported by a number of authors[4,27,30] indicates that the γ' is coherent with the matrix but that coherency strains are low and are not a major source of strength in this family of alloys. This would be expected since most of the γ'-strengthened nickel–iron alloys were empirically developed to maximize creep stability, a condition where minimization of γ–γ' mismatch and resultant coherency strains is desirable. (See also Chapter 3.)

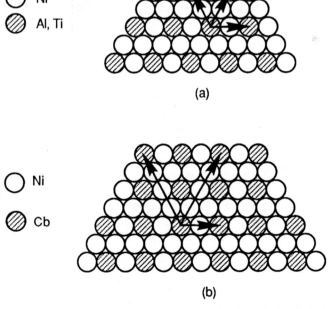

○ Ni

◉ Al, Ti

(a)

○ Ni

◉ Cb

(b)

Fig. 4. Arrangement of atoms on the (111) slip plane illustrating ordering in the principal slip directions for (a) FCC γ' and (b) BCT γ''.[12]

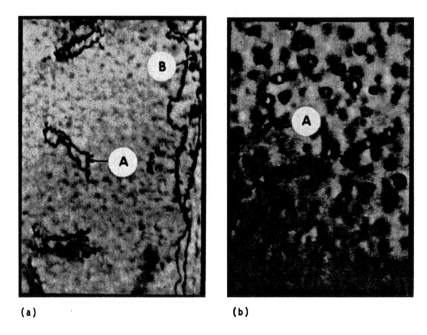

(a)   (b)

Fig. 5. Deformation mechanism in Nimonic PE 16 (Fe–43.5Ni–16.5Cr–3.4Mo–1.2Ti–1.10Al–0.0025B–0.005C). Both solution treated 1800 °F (980 °C)/1 h WQ prior to aging and examined after 0.5–1.0% tensile deformation. Both 60,000×.[4] (a) Aged 455 h at 1290 °F (700 °C). Dislocation (Orowan) loops are indicated at A and B. (b) Aged 200 h at 1380 °F (750 °C). At A a loop encircles two particles.

γ'' Precipitation

The precipitation of ordered BCT γ'' (DO$_{22}$) predominates in nickel–iron alloys strengthened through additions of columbium. Because of the high solubility of columbium in Ni$_3$Al,[12] aluminum contents must also be kept low to favor γ'' precipitation.

Commercial alloys that are known to exhibit γ'' precipitation include 718 and 706. The unusually high strength levels exhibited by these alloys can be attributed to several characteristics of the precipitate, a subject of numerous investigations.[9,11,13,31,32] The unit cell for BCT γ'' is shown schematically in Fig. 6b. When compared to the FCC γ' unit cell (Fig. 6a), the γ'' unit cell can be roughly described as FCC with a somewhat different ordering sequence and a c_0 parameter roughly equal to twice the a_0 parameter of γ'. Measurements of BCT lattice parameters in 718 indicate $c_0 = 7.406$ and $a_0 = 3.624$.[13] The contrast between the ordering sequence for γ' and γ'' can be seen more clearly in Fig. 4, which compares the arrangement of atoms on the slip plane for the FCC and BCT structures. For the BCT structure, in only one of the three slip directions will order be restored as a result of the movement of the two $a/2\langle110\rangle$ matrix dislocations through the

γ'' phase.[12] In the other two directions the superdislocation must consist of four $a/2\langle110\rangle$ matrix dislocations. The moderate limitation of available slip directions is believed to be one of the factors contributing to the relatively high strength levels of γ''-strengthened alloys.

The morphology of the γ'' precipitate in 718 has been reported by several investigators[9,10] as disk shape with an orientation relationship to the matrix described by $(100)\gamma'' \parallel \{100\}\gamma$; $[100]\gamma'' \parallel \langle100\rangle\gamma$. Following commercial heat treatment, the γ'' precipitate has an average disk diameter of 600 Å and a thickness of 50–90 Å.[9]

Numerous investigators[8,9,30,32] have shown that the γ'' precipitate in 718 produces effective hardening as a result of coherency strains between γ and γ'', a strain that has been measured to be 2.86%.[8] Coherency strains may also account for the rapid loss of stability in 718 resulting from exposure to temperatures in excess of 1200 °F (650 °C) since these strains produce a driving force for particle coarsening. It is interesting to note that Inconel 718 represents one of the few alloy compositions developed to maximize strength up to 1200 °F (650 °C) rather than creep rupture properties to higher temperatures.

Inconel 718 is one of the few age-hardenable superalloys that exhibits good weldability by resisting the strain-age cracking phenomenon exhibited by most superalloys. This has been attributed to the sluggishness of the γ'' age-hardening

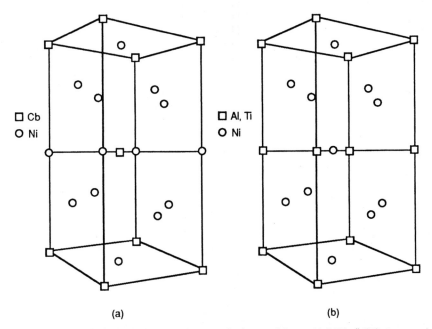

(a)   (b)

Fig. 6. Unit cells illustrating ordering of the strengthening precipitates: (a) BCT γ'' (DO$_{22}$), one unit cell; (b) FCC γ' (L1$_2$), two unit cells.[12]

reaction, which allows relief of weld residual stresses during aging heat treatment prior to nucleation and growth of the γ″ precipitates. It is more likely that the sluggishness of this aging reaction in Inconel 718 is a function of coherency strains rather than an inherent characteristic of the γ″ phase. This is supported by the fact that Incoloy 903, an alloy strengthened by FCC γ′, shows similar weldability to 718 and has also been found to exhibit a high degree of coherency strains.

Overaging

Many nickel-base alloys will overage by the relatively simple mechanism of γ′ particle ripening when exposed to time–temperature conditions in excess of that which produces peak age hardening. The overaging process for nickel–iron alloys is somewhat more complicated owing to the metastable characteristics of the titanium-rich γ′ and columbium-rich γ″ phases. Extended periods of exposure can cause γ′ to transform to η (eta) in alloys such as A-286 and 901 and γ″ to transform to δ (delta) in 718. These transformations are usually associated with a loss in alloy properties. Ripening can be an intermediate step in these transformations, particularly at lower temperatures.

η Phase

The nickel–iron-base superalloys strengthened by γ′ are susceptible to the formation of HCP η. Two modes of formation will be considered: (1) formation during forging/heat treatment and (2) formation during long-time service exposure. There are two forms of η that may occur, intragranular platelets that form (1) by way of a γ′–η transformation, sometimes with a Widmanstatten form, and (2) cellular grain boundary form. Both of these are illustrated in Fig. 7.

Processing (e.g., forging and heat treatment) to form a small amount of η is a useful practice to control microstructure. For many years A-286 has been processed to fine grain sizes by finish forging in the vicinity of the η solvus followed by solution heat treatment below the η solvus [e.g., 1650 °F (900 °C), 2 h, oil quench] followed by aging at 1325 °F (720 °C). This generates a finer grain structure with a better balance of tensile properties and stress rupture ductility than that achievable with higher temperature (η-free) processing. Finer grain sizes that result from this processing approach can result in somewhat lower stress rupture life. Typical data are shown in Table 1. The morphology of η observed after such a process is usually finely distributed intragranular platelets in the vicinity of grain boundaries with the relationship $\{0001\}\eta \parallel \{111\}\gamma$; $\langle 1210 \rangle \eta \parallel \langle 110 \rangle \gamma$.[34]

Cellular η appears in A-286 or other nickel–iron alloys (Fig. 7a) as alternate coherent lamellae of η and γ originating at grain boundaries, with a random orientation relationship with the grain in which the cellular zone is growing.[34,35] The strain energy resulting from γ′ precipitation may contribute to the driving force for cellular η formation. This phenomenon is observed after prolonged exposure in the range 1110–1560 °F (600–850 °C).[27,34]

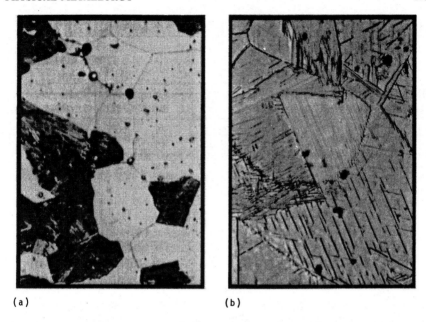

(a)   (b)

Fig. 7. Cellular and platelet (Ni₃Ti) in experimental Fe–Ni base alloys. Both solution treated at 2100 °F (1150 °C) prior to aging. Both 750X. (Photographs courtesy of F. B. Pickering, British Steel Corporation.) (a) Cellular η–Fe–25Ni–15Cr–2.5Ti–3Al alloy aged 10 h at 1290 °F (700 °C). (b) Predominantly platelet η–Fe–25Ni–15Cr–3.75Ti–2.68Co alloy aged 500 h at 1380 °F (750 °C).

Singhal and Martin[36] have observed that (because of interfacial energy considerations) in an Fe–24Ni–21Cr–1.3Ti–0.3Si–0.004C alloy, the cellular η has parallel close-packed directions with those of adjoining grains. This also has been verified[34,37] in other nickel–iron alloys. The coarse interlamellar spacing of the η as well as the loss of the more finely distributed γ' in the matrix resulting from the cellular formation often leads to severe degradation of mechanical properties. However, this point is controversial since other work[28,29] indicates that the ductility loss often attributed to cellular η may be due to an accompanying formation of cellular $M_{23}C_6$ carbide.

The intragranular platelet or Widmanstätten form of η (Fig. 7b) is associated with loss in strength[38] and may result in a loss of ductility. Exposure in the range 1475–1560 °F (800–860 °C) or higher will result in this form of η in A-286.[34,37,39,40] Formation of platelet η is associated with a γ' → η transformation and can be avoided by heat treating above the solvus [above 1675 °F (915 °C) for A-286] and by utilizing the alloy at temperatures below which the γ'–η transformation occurs [less than 1475 °F (800 °C) for A-286].

The exact temperature at which η occurs is a strong function of titanium level and alloy-base chemistry. Increasing titanium levels and warm/cold work strain

Table 1. Effect of Heat Treatment on the Properties of A-286[a]

Heat Treatment[b]	Tensile Properties 70 °F (21 °C)				Stress Rupture 1200 °F (650 °C), 65 ksi		
	0.2% YS[c] (ksi)	UTS[d] (ksi)	EL[e] (%)	RA[f] (%)	Life (h)	EL (%)	RA (%)
A	100	156	24	46	85	10	15
B	108	160	25	46	64	15	20

[a] From ref. 33.
[b] A = 1800 °F (980 °C), 1 h oil quench (OQ) + 1325 °F (720 °C), 16 h, AC. B = 1650 °F (900 °C), 2 h OQ + 1325 °F (720 °C), 16 h, AC.
[c] YS = yield strength.
[d] UTS = ultimate tensile strength.
[e] EL = elongation.
[f] RA = reduction of area.

energy enhances formation of η. Boron is useful in retarding the formation of the cellular form of η but does not appear to affect the Widmanstätten form. Aluminum is useful in preventing both forms of η. The aluminum effect may be related to the fact that hexagonal η has little or no solubility for aluminum. Thus, η can nucleate and grow only as a result of diffusion of aluminum. Aluminum may also decrease γ–γ' mismatch, thus reducing the driving force for the transformation.

δ Phase

Nickel–iron alloys that are strengthened by BCT γ" are susceptible to the formation of orthorhombic Ni_3Cb (δ). The formation of δ in these alloys closely resembles the formation of η in alloys strengthened by γ' although some differences will be noted. The δ phase represents the thermodynamically stable form of the metastable γ". The formation of δ phase has been studied by numerous investigators, primarily in alloy 718.

In 718, δ-phase formation has been shown to occur over the 1200–1800 °F (650–980 °C) temperature range with a platelet morphology. A matching of close-packed planes occurs between the δ and the γ matrix with the following relationships reported[41]: (010)δ ∥ (111)γ; [100]δ ∥ [110]γ. Randomly oriented globular phase can also be observed in grain boundaries of 718, as shown in Fig. 3. The loss of orientation relationship is believed to be a result of the presence or formation of this δ concurrent with forging deformation.

Below 1300 °F (700 °C) the rate of formation of δ is quite slow, requiring hundreds to thousands of hours. Nucleation is usually observed at grain boundaries or columbium-rich MC carbides, and growth occurs at the expense of the γ". A significant acceleration in the rate of δ formation occurs above 1300 °F (700 °C) and is accompanied by a rapid coarsening of γ" up to 1625 °F (885 °C), above which solutioning of the γ" occurs. Formation of δ is most rapid over the 1550–

1750 °F (840–950 °C) range where the Widmanstätten morphology forms profusely in times less than 24 h.

There have been no reports of cellular δ formation in commercial γ″-strengthened alloys following extended elevated temperature exposure.[42,43] Cellular δ has, however, been reported in an experimental Fe–35Ni–15Cr–5Cb–0.08C alloy following aging at 1290 °F (700 °C).[44] Cellular δ has also been reported in experimental Fe–15Cr–Ni–Cb alloys after exposures from 1200 °F (650 °C) to 1380 °F (750 °C) when nickel was greater than 45% and columbium was greater than 5%.[32,41] Cellular δ in these alloys occurred by a mechanism similar to cellular η formation in γ′-strengthened nickel–iron-base alloys.[41] Above approximately 1380 °F (750 °C) these alloys favored formation of intragranular platelets of δ over cellular δ. This is in agreement with η formation in γ′-strengthened alloys where the cellular reaction dominates at lower temperatures and the intragranular reaction dominates at higher temperatures.

In all cases of cellular δ the alloys were not reported to contain inhibiting boron or aluminum. Also, in all cases the alloys were solution treated at 2100 °F (1150 °C) followed by rapid quenching, which differs considerably from the solution treatments used for commercial alloys [e.g., 1750 °F (955 °C), 1 h, air cooled (AC) for 718]. These factors undoubtedly contributed to this apparent anomaly.

The precipitation kinetics and morphology of δ phase in 718 can be dramatically altered as a result of forging at temperatures below the δ solvus temperature of 1825 °F (1000 °C). If sufficient forging deformation is introduced, nucleation of δ will be rapidly accelerated and will occur intragranularly in a uniform manner rather than preferentially at grain boundaries. The resultant δ-phase distribution can be effectively used to control and refine grain size, resulting in optimum tensile properties and stress rupture ductility.[24] Extremely fine grain structures (ASTM 10-13) with exceptional fatigue resistance have been developed by this approach.[45] Studies on 706 also show mechanical property advantages for forging and heat treatment practice which develops globular grain boundary precipitates of δ and η.[46] It is believed that for both 718 and 706 the advantages of globular grain boundary δ and/or η result from a combination of grain size control and inhibition of long-range grain boundary sliding by the grain boundary particles.

The formation of large amounts of δ during long-time service exposure will lead to degradation in properties. The degradation is probably a combined result of the loss of columbium from the matrix plus the γ″ coarsening that accompanies the δ formation. Because the kinetics of this transformation increase rapidly at temperatures exceeding 1200 °F (650 °C), exposure to temperatures in excess of 1200 °F (650 °C) should be avoided. It has been observed[1] that high silicon and columbium and low aluminum promote δ. No specific data have been published on the effects of solid-solution alloy additions on δ formation in nickel–iron alloys.

Minor Phase Formation

Nickel–iron superalloys are more prone to the formation of minor phases such as G, σ, μ and Laves than are the nickel-base superalloys. The formation of these phases generally results in an embrittling effect owing to the brittle nature of the phases. Proper control of chemistry, heat treatment, and service temperatures represent the most effective means of avoiding these phases. Particles of these phases formed during ingot solidification can be effectively eliminated through the use of thermal homogenization cycles and controlled hot working practice.

Increasing columbium, titanium, and silicon in nickel–iron alloys promotes Laves phase, while increasing boron and zirconium minimizes Laves and μ formation. The G phase, a complex cubic nickel titanium, and silicon phase, has been shown to lower the stress rupture life of A-286 but has no significant detrimental effect on other properties.[38] Sigma platelets in nickel–iron superalloys are usually associated with brittle fracture along platelet–matrix interfaces. Sigma phase can be effectively avoided through control of chromium and molybdenum contents.

Since most nickel–iron alloys are used at temperatures below 1400 °F (760 °C), minor phase formation during service is rare. Also, the relatively lean nature of these alloys makes formation of large volume fractions of deleterious phases unlikely.

Precipitation Rates

Time–temperature precipitation (TTP) diagrams can be useful in describing the kinetics of phase precipitation in nickel–iron alloys. These relationships have been thoroughly investigated for 718, and Fig. 8 presents a TTP diagram for this alloy. Caution should be exercised when using TTP diagrams for nickel–iron alloys since the energy state (e.g., grain size, degree of retained forging deformation, and solution heat treatment temperature) can shift the relationships to the left or right.

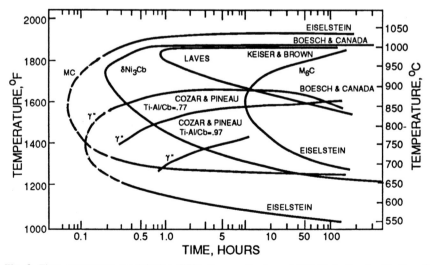

Fig. 8. Time–temperature precipitation (TTP) diagrams for Inconel 718 illustrating the kinetics of formation for MC, M_6C, Laves, δ, and γ″ phases.[11,24,49,50]

From Fig. 8 the time–temperature relationships for phases previously discussed can be readily seen. Note that the commonly used commercial solution heat treatment for 718 [1750 °F (950 °C), 1 h, AC] would be expected to result in the formation of some grain boundary δ phase as shown in Fig. 3. It can also be seen that longer solution heat treatment times should be avoided to prevent the formation of the Laves phase.

It can also be seen from Fig. 8 that a short but finite nucleation time (~10 min) exists for the onset of γ″ precipitation. It is this sluggishness of the age-hardening reaction that accounts for the excellent welding characteristics and freedom from strain-age cracking exhibited by 718. For alloys strengthened by γ′, the precipitation kinetics are sufficiently rapid that it frequently cannot be avoided even by water quenching from the solution treatment temperature.

PROCESSING EFFECTS ON PROPERTIES AND MICROSTRUCTURE

Most nickel–iron superalloys are available in both cast and wrought conditions. For the wrought product large ingots are cast and forged, rolled, or extruded to finish sizes and then heat treated for desired properties. Cast products are used either as cast or after a homogenization or other property-improving heat treatment. Since Chapter 14 discusses in detail the casting metallurgy of superalloys, this chapter deals primarily with the wrought form of the nickel–iron alloys.

Primary Processing

Melting of nickel–iron is generally accomplished using vacuum induction melting (VIM) to minimize the nitrogen and oxygen content of the melt as well as to minimize reaction of oxygen and nitrogen with the more reactive metal additions, such as aluminum, titanium, and chromium. The vacuum induction process is also effective in removing oxygen and metallic oxides from the melt through interaction with carbon to form carbon monoxide, which is removed by the vacuum system. This is an important step in the production of high-quality VIM ingot for subsequent processing.

Most nickel–iron alloys for wrought application are then vacuum consumable arc remelted (VAR) or electroslag remelted (ESR) to improve homogeneity and ingot structure. Contemporary consumable remelted ingots of nickel–iron-base superalloys range from approximately 12 to 28 in. in diameter and weigh up to 15,000 lb. The ESR process is gaining greater acceptance in recent years due to improved ingot surface for better product yield and to the beneficial refining reaction with the slag, which allow removal of detrimental species such as sulfur, nitrides, and oxides.[47,48] The major disadvantage with ESR melting is the potential for loss of reactive elements, particularly titanium, requiring careful control of slag chemistry.

Ingot breakdown for nickel–iron-base superalloys often includes thermal homogenization steps to dissolve undesirable phases such as Laves and G phases and to reduce local chemistry gradients, particularly titanium and columbium. Homogenization and hot-working temperatures vary from alloy to alloy; however, furnace temperatures between 2000 °F (1100 °C) and 2200 °F (1200 °C) are typical. A thorough treatment of the melting of nickel and nickel–iron alloys is given in Chapter 14.

Structure–Property Control

Manufacture of finished forgings of nickel–iron superalloys requires manipulation of the metal for shape, structure, and properties. Chapter 16 describes details of the forging process for superalloys generally. This section will describe some of the metallurgical aspects of microstructure and property control of nickel–iron alloys during hot working and heat treatment.

The most powerful tool available to the metallurgist for controlling the properties of superalloys lies in the control of alloy grain size during forging and heat treatment. By proper control of processing parameters, fine-grain structures can be achieved that maximize tensile and fatigue properties at the expense of elevated-temperature creep rupture properties. Conversely, processes yielding coarser grain structures will maximize creep rupture properties at the sacrifice of tensile and fatigue resistance. Both the forging and heat treatment processes can be effective in controlling structure–property relationships.

The nickel–iron-base alloys offer a distinct advantage over nickel-base superalloys with regard to control of microstructure during processing.[20] This advantage can be directly attributed to the availability of η or δ phases for grain size control. In order to induce recrystallization either during forging or heat treatment, temperatures must exceed the solvus temperatures of either the γ′ or γ″ phases. The solvus temperatures for these phases for selected commercial alloys are shown in Table 2. If, then, recrystallization is allowed to occur at temperatures below the η- or δ-phase solvus, these phases will be effective in controlling grain growth. The η and δ solvus temperatures are shown in Table 2 for selected commercial alloys.

Table 2. Summary of Structure-Controlling Phases in Typical Nickel–Iron-Base Superalloys[a]

Alloy	Phase	Solvus Temperature (Limit of Stability) °F (°C)
A-286	γ′	1575 (855)
	η (Ni$_3$Ti)	1675 (915)
718	γ″	1675 (915)
	δ (Ni$_3$Cb)	1825 (995)
706	γ″ or γ′	1625 (885)
	η (Ni$_3$Ti) and/or δ (Ni$_3$Cb)	1750 (955)
901	γ′	1725 (940)
	η (Ni$_3$Ti)	1825 (995)

[a] From ref. 20.

Table 3. Effect of Grain Size on the High-Cycle Fatigue Properties
of Incoloy 901 and Inconel 718 (850 °F)

Alloy	Grain Size	Fatigue Strength ksi (10^7 cycles)	Fatigue Ratio (FS/UTS)[a]
Incoloy 901	ASTM 2	46	0.32
	ASTM 5	64	0.42
	ASTM 12	91	0.55
Inconel 718	ASTM 2	55	0.33
	ASTM 5	80	0.45
	ASTM 12	115	0.59

[a] FS/UTS = fatigue strength to ultimate tensile strength.

Figure 9 shows a variety of microstructures that can be developed for 718 following warm working and annealing as indicated. Heating below the γ'' solvus produces no change in grain size, grain boundaries are pinned by fine globular δ, and a background of overaged matrix γ'' (Fig. 9b) is formed. Figure 9c shows a duplex microstructure produced by heating the material above the γ'' but below the δ solvus. Here, recrystallization occurred due to the loss of the γ'' phase, but the grain boundaries of the recrystallized grains are inhibited in movement by the globular-to-platelet δ that formed during the 1750 °F (955 °C) solution treatment. Heating to 1900 °F (1040 °C) (Fig. 9d) above the δ solvus permits grain growth since the structure-controlling effect of the δ is lost.

A method for utilizing structure control concepts to achieve extremely fine-grain structures (ASTM 10 or finer) in 901 and 718 is described elsewhere.[45] The processing sequence, termed MINIGRAIN processing, utilizes forging processes below the η- or δ-phase solvus temperature with high forging reduction to develop a uniform

Table 4. Properties of 718 as Function of Heat Treatment[a]

Solution Treatment[b]	Tensile 70 °F (21 °C)				Stress Rupture 1200 °F (650 °C), 100 ksi		
	0.2% YS (ksi)	UTS (ksi)	EL (%)	RA (%)	Life (hr)	EL (%)	RA (%)
None (direct aged)	193	221	19	34	95	24	31
1725 °F (940 °C), 1 h AC	180	212	18	34	194	11	16
1750 °F (955 °C), 1 h AC	717	206	20	38	122	14	19
1775 °F (970 °C), 1 h AC	166	204	23	41	218	13	15
1800 °F (980 °C), 1 h AC	170	204	24	43	200	6	10
1850 °F (1010 °C), 1 h AC	172	202	22	46	270	6	12
1900 °F (1040 °C), 1 h AC	169	198	25	48	225[c]	2	8

[a] See Table 1 for definition of abbreviations.
[b] All aged 1325 °F (720 °C), 8 h cool 100 °F (55 °C)/h to 1150 °F (620 °C), 8 h AC.
[c] Notch sensitive at Kt = 3.8.

dispersion of the phase. This reference also presents the effect of grain size variations from ASTM 2 to ASTM 12 on tensile, creep, and fatigue properties. Table 3 shows the effect of grain refinement on the high-cycle fatigue strengths of 901 and 718,[45] showing that fatigue strength is doubled as a result of refinement in grain size from ASTM 2 to ASTM 12. Fatigue ratio (fatigue strength to ultimate tensile strength) also increased significantly.

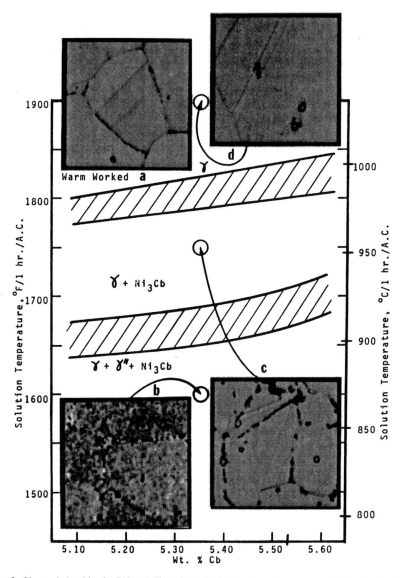

Fig. 9. Phase relationships for 718 and effect of solution heat treatments on warm-worked microstructure. All 1000×. Adapted from ref. 20.

A recent innovation in the processing of 901 and 718 involves the use of controlled forging practice to introduce hot deformation at or below the η or δ solvus followed by quenching from the forging press. The material is then directly aged, bypassing the solution heat treatment operation associated with conventional processes. Exceptionally high strength levels are achieved as a result of a retention of forging deformation from the forging process. This approach, termed direct-age processing, has extended the strength of nickel–iron alloys to levels approaching those of the high-volume-fraction γ' nickel-base alloys produced by powder metallurgy techniques. Table 4 illustrates the tensile and stress rupture properties of 718 processed using various solution heat treatment temperatures as well as direct-age processing. Interrelationships between heat treatment, properties, and microstructure can be seen by comparing Table 4 and Fig. 9.

RECENT AND FUTURE DEVELOPMENTS

It is doubtful whether further alloy development efforts will be successful in extending the temperature capability of the nickel–iron alloys much beyond those of today's commercially available materials. Higher temperature applications, above 1200 °F (650 °C), will more than likely be served by the nickel-base superalloys. For applications below 1200 °F (650 °C), the nickel–iron alloys have a firm foothold, and future alloy and process development efforts will likely be successful in achieving further improvements in strength for the higher stressed environments of advanced gas turbine engines. Direct-age processing approaches will undoubtedly play a role in further extending the properties of nickel–iron alloys. Control of defects will become increasingly important as strength capability increases. This will necessitate the use of advanced melting processes, such as electron beam hearth melting, which has been recently shown to be effective in eliminating oxide and nitride defects.[48]

In summary, the future looks bright for the nickel–iron superalloys where high mechanical properties up to 1200 °F (650 °C) are combined with good processability and low cost relative to nickel-base superalloy alloys.

REFERENCES

1. C. P. Sullivan and M. J. Donachie, *J. Met. Eng. Q*, **11**, 1 (1971).
2. D. L. Klarstrom, H. M. Tawancy, and M. F. Rothman, *J. Met. Sol. AIME*, 553 (1984).
3. D. F. Smith, D. J. Tillack, and J. P. McGrath, *ASME*, 85-1GT-140, 1985.
4. D. Raynor and J. M. Silcock, *Met. Sci. J.*, **14**, 121 (1970).
5. D. J. Dyson and B. Holmes, *JISI*, **208**, 469 (1970).
6. R. F. Decker, in *Symposium: Steel Strengthening Mechanisms, Zurich 1969*, Climax Molybdenum, Greenwich, CT, 1970, pp. 147–170.
7. K. J. Irvine, D. T. Llewellyn, and F. B. Pickering, *JISI*, **199**, 153 (1961).
8. M. C. Chaturvedi and Y. Han, *Met. Sci.*, **17**, 145 (1983).
9. D. F. Paulonis, J. M. Oblak, and D. S. Duvall, *Trans. ASM*, **62**, 611 (1969).
10. Y. Han, P. Deb, and M. C. Chaturvedi, *Met. Sci.*, **16**, 555 (1982).
11. R. Cozar and A. Pineau, *Met. Trans.*, **4**, 47 (1973).

12. E. L. Raymond and D. A. Wells, *Superalloys-Processing, Battelle Columbus Laboratories, MCIC 72-10*, Columbus, Ohio 1972, pp. N1–N21.
13. H. J. Wagner and A. M. Hall, DMIC-217, 1965.
14. V. Ramaswamy, P. R. Swann, and D. R. West, *J. Less-Common Met.*, **27**, 17 (1972).
15. B. S. Natapov, V. E. Ol'shanetskii, and E. P. Ponomarenko, *Met. Sci. Heat Treat*, **1–2**, 11 (1966).
16. J. R. Mihalisin and R. F. Decker, *Trans. TMS-AIME*, **218**, 507 (1960).
17. J. G. Hoag, DMIC-84, Battelle Memorial Institute, Columbus, OH, February 6, 1961.
18. G. Chen et al., in *Superalloys 1984*, M. Gell, C. S. Kortovich, R. H. Bricknell, W. B. Kent, and J. F. Radavich (eds.), TMS-AIME, Warrendale, PA, 61 (1984).
19. J. M. Moyer in *Superalloys 1984*, M. Gell, C. S. Kortovich, R. H. Bricknell, W. B. Kent, and J. F. Radavich (eds.), TMS-AIME, Warrendale, PA, 443 (1984).
20. D. R. Muzyka, *Met. Eng. Q.*, **11**, 12 (1971).
21. R. F. Decker and R. R. DeWitt, *J. Met.*, **17(2)**, 139 (1965).
22. H. J. Beattie, Jr. and W. C. Hagel, *Precipitation Processes in Steels*, Iron and Steel Institute, London, 1959, pp. 108–117.
23. H. J. Beattie, Jr. and W. C. Hagel, *Trans. TMS-AIME*, **233**, 277 (1965).
24. H. L. Eiselstein, "Advances in the Technology of Stainless Steels and Related Alloys," STP 369, ASTM, Philadelphia, PA, 1965, pp. 62–67.
25. C. C. Clark and J. S. Iwanski, *Trans. AIME*, **215**, 649 (1959).
26. J. F. Radavich, in *Advances in X-ray Analysis*, Vol. 3, W. M. Mueller (ed.), Plenum, New York, 1960, pp. 365–375.
27. J. M. Silcock and N. J. Williams, *JISI*, **204**, 1100 (1966).
28. G. B. Heydt, *Trans. ASM*, **54**, 220 (1961).
29. G. N. Manier and H. M. James, *Trans. ASM*, **57**, 368 (1964).
30. R. F. Decker and J. R. Mihalisin, *Trans. ASM*, **62**, 481 (1969).
31. J. M. Oblak, D. F. Paulonis, and D. S. Duvall, *Met. Trans.*, **5**, 143 (1974).
32. I. Kirkman and D. H. Warrington, *Met. Trans.*, **1**, 2667 (1970).
33. Carpenter Technology Corporation, unpublished research, Reading, PA.
34. B. R. Clark and F. B. Pickering, *JISI*, **205**, 70 (1967).
35. F. G. Wilson and F. B. Pickering, *JISI*, **204**, 628 (1966).
36. L. K. Singhal and J. W. Martin, *JISI*, **205**, 947 (1967).
37. D. Dulieu and B. Aronsson, *Jerkont. Ann.*, **150**, 787 (1966).
38. R. F. Decker and S. Floreen, in *Precipitation From Iron-Base Alloys*, G. R. Speich and J. B. Clark (eds.), Gordon & Breach, New York, 1965, p. 69.
39. F. G. Wilson and F. B. Pickering, *JISI*, **207**, 490 (1969).
40. H. J. Beattie, Jr. and W. C. Hagel, *Trans. TMS-AIME*, **209**, 911 (1957).
41. I. Kirkman, *JISI*, **207**, 1612 (1969).
42. J. F. Barker, E. W. Ross, and J. F. Radavich, *J. Met.*, **22(1)**, 31 (1970).
43. J. P. Stroup and R. A. Heacox, *J. Met.*, **21(11)**, 46 (1969).
44. R. T. Weiner and J. J. Irani, *Trans. ASM*, **59**, 340 (1966).
45. E. E. Brown, R. C. Boettner, and D. L. Ruckle, *Superalloys-Processing*, Battelle Columbus Laboratories, Columbus, Ohio MCIC-72-10, 1972.
46. J. H. Moll, G. W. Maniar, and D. R. Muzyka, *Met. Trans.*, **2**, 2153 (1971).
47. G. L. R. Durber, C. L. Jones, and A. J. Dykes, 433 (1984).
48. E. E. Brown, J. E. Stulga, L. Jennings, and R. W. Salkeld, 159 (1980).
49. G. K. Bouse and M. F. Collins, Nov. 1985 unpublished work.
50. W. J. Boesch and H. B. Canada, *J. Met.*, **21(10)**, 34 (1969).
51. D. S. Keiser and H. L. Brown, *Idaho National Engineering Laboratory Report #ANCR 1292-UC-25*, 1976.
52. C. H. Lund, *Physical Metallurgy of Nickel-Base Superalloys*, DMIC-153, Columbus, OH, 1961.
53. C. P. Sullivan and M. J. Donachie, Jr., *Met. Eng. Q.*, **7**, 36 (1967).
54. G. P. Sabol and R. Stickler, *Phys. Stat. Sol.*, **35**, 11 (1969).
55. C. T. Sims, *J. Met.*, **18(10)**, 1119 (1966).

Chapter 7

Directionally Solidified
Superalloys

DAVID N. DUHL

Pratt & Whitney, East Hartford, Connecticut

Directionally solidified (DS) columnar-grained (CG) and single-crystal (SC) superalloys have the highest elevated-temperature capability of any superalloy. For this reason, they are finding widespread use as turbine airfoils, an application that demands the most in alloy properties at elevated temperatures. There are two primary reasons that explain why DS superalloys are superior to conventionally cast (CC) superalloys. Alignment, or elimination in the case of SC superalloys, of the grain boundaries normal to the stress axis enhances elevated-temperature ductility by eliminating the grain boundary as the failure initiation site. This permits the γ' microstructure to be refined with a solution heat treatment that increases alloy strength. The second reason is that the DS process provides a preferred low-modulus $\langle 001 \rangle$ texture or orientation parallel to the solidification direction. This results in a significant enhancement in thermal fatigue resistance, so important in elevated-temperature components.

VerSnyder and coworkers[1,2] were the first to demonstrate for superalloys that by aligning the grain boundaries parallel to the principal stress axis, the stresses acting at elevated temperatures on the weak grain boundaries could be minimized, thus delaying failure initiation and enhancing creep rupture life. The DS process normally is employed to align the grain boundaries parallel to the solidification direction. The resultant structure consists of an array of columnar grains all aligned parallel to the solidification direction, similar to a bundle of sticks. Each grain has the low-modulus $\langle 001 \rangle$ orientation aligned parallel to the grain axis, but the direction about the $\langle 001 \rangle$ zone is random. A slight modification of the DS process is made

Conventional casting   **Directional solidification**

Equiaxed grain   **Columnar grain**   **Single crystal**

Fig. 1. Turbine airfoil castings made by conventional and directional solidification and the resultant grain structures.

to provide a SC with no grain boundaries.[3-5] Single-crystal superalloys also have the ⟨001⟩ low-modulus orientation parallel to the solidification direction and a random secondary orientation in the plane normal to the direction of solidification. Other specific primary or secondary orientations are possible with the use of seed crystals. The three solidification modes, CC, CG, and SC, are shown in Fig. 1 as macroetched turbine blade castings.

Directionally solidified, CG, or SC superalloys have excellent creep strength resulting from the alignment or elimination of grain boundaries normal to the major stress axis and superior thermal fatigue resistance resulting from the low-modulus ⟨001⟩ texture provided by the DS process. The advantage DS superalloys exhibit over CC alloys lies in the improvement afforded in these two critical areas by the DS process.

Aligned CG and SC superalloys can also be obtained with a solid-state process based on secondary grain growth.[6] However, it has been found that the higher energy liquid-to-solid phase transformation is easier to control than the solid-state process. In addition, the preferred texture produced by solid-state processing is a function of alloy composition, while the DS process provides the desirable thermal fatigue-resistant, low-modulus texture independent of alloy composition. For these reasons the DS process is the only one used commercially; all subsequent remarks in this chapter will focus on it.

Basic superalloy strengthening mechanisms that depend on dislocation interactions with the γ' strengthening phase and the γ matrix are similar in all superalloys independent of solidification mode and have been covered in Chapter 3. This chapter will concentrate on those characteristics of DS, CG, and SC superalloys that are different from the behavior of CC superalloys.

DIRECTIONAL SOLIDIFICATION PROCESS

Both CG and SC castings are made by the same basic process.[7] Elimination or alignment of the grain boundaries parallel to the casting or airfoil axis can best be done by utilizing the high-energy, liquid-to-solid phase transition in superalloys, referred to as solidification. Allowing solidification to occur in a controlled thermal gradient produces elongated grains aligned parallel to the gradient, which by definition aligns the grain boundaries parallel to the thermal gradient.

Directional solidification is accomplished in vacuum, as shown in Fig. 2, by pouring molten alloy into a ceramic shell mold that has been preheated to a temperature above the liquidus of the superalloy. (See Chapter 15 for a discussion of investment casting.) The preheated shell mold is open at the bottom and sits on a water-cooled copper chill plate. The molten superalloy solidifies upon contact with the copper chill to form a thin layer of equiaxed grains. Subsequent growth favors those grains most closely aligned with the ⟨001⟩ direction parallel to the thermal gradient and produces a columnar array of grains with a common vertical ⟨001⟩ orientation. Solidification proceeds in a directional manner as a consequence of the thermal gradient created between the upper portion of the ceramic shell mold being in a heated furnace (which maintains the alloy in the upper portion of the mold molten) and the chill plate, which is extracting heat from the bottom of the mold. Heat extraction by conduction through the solidified superalloy is difficult, so after solidification has started, the water-cooled chill plate is lowered gradually, withdrawing the ceramic shell mold filled with molten superalloy from the heated furnace. Heat loss is now controlled by radiation from the shell mold to the cold vacuum chamber walls. A baffle can be placed at the bottom of the furnace, as shown in Fig. 2, to increase the thermal gradient. The initial CG structure, which began in the starter

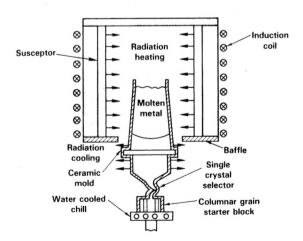

Fig. 2. DS process [3].

block above the chill plate, proceeds to fill the entire mold cavity, resulting in a CG airfoil casting as shown in the middle of Fig. 1.

A SC casting is obtained by inclusion above the starter block of a helical constriction that acts as a filter and only permits a single grain to pass through. This is because superalloys solidify by dendritic growth. Each dendrite can grow only in the three mutually orthogonal $\langle 001 \rangle$ directions. The continually changing direction of the helix combined with the orthogonal nature of dendritic growth gradually constricts all but one favored grain, resulting in a single crystal emitting from the top of the helix, as shown in Fig. 3. The selected grain then fills the shell cavity as for the case with the CG casting. A $\langle 001 \rangle$-oriented SC airfoil casting results, as shown on the right side of the Fig. 1. The DS process as described above is currently employed to produce hollow SC and CG turbine airfoil castings in production quantities.

Crystal Growth from Seeds

The DS process permits SC castings with any orientation to be grown. Single-crystal turbine airfoils usually are cast with the $\langle 001 \rangle$ orientation aligned parallel to the airfoil axis, as described above, but with the use of seed crystals any orientation can be achieved using the process illustrated in Fig. 2, with the helix and starter block replaced by a seed crystal. The seed crystal should be of the desired alloy or one with an equivalent or higher melting temperature. It is positioned so that its orientation will be repeated in the alloy that fills the mold cavity. The seed sits on the chill, and the temperature at the top of the seed is controlled so that the seed crystal does not melt completely, thereby allowing the molten alloy in the mold cavity to solidify with the same orientation as the seed.

Whether a seed crystal or a helical constriction is used, solidification proceeds by growth of dendrites in the three orthogonal $\langle 001 \rangle$ orientations, with the $\langle 001 \rangle$ orientation most closely aligned to the thermal gradient being the primary growth

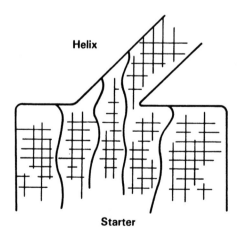

Fig. 3. Helix in SC mold acts as a grain filter for alloys solidifying by dendritic growth [3].

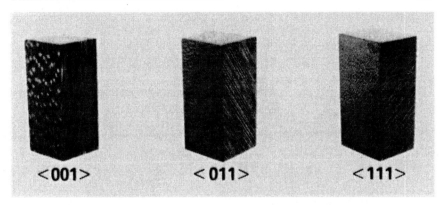

Fig. 4. SC superalloys grown in any orientation solidify by dendritic growth in the $\langle 001 \rangle$ direction [5].

direction. This can be seen in Fig. 4, where crystals grown in the three major orientations have been etched to show the dendrites. For the $\langle 001 \rangle$-oriented crystal the $\langle 001 \rangle$ dendrites are parallel to the growth axis in both $\{001\}$ faces. In the $\langle 011 \rangle$-oriented crystal, with $\{011\}$ and $\{001\}$ faces, the $\langle 001 \rangle$-oriented dendrites are parallel to the crystal axis in the $\{011\}$ face and at 45° in the $\{001\}$ face. For a $\langle 111 \rangle$ crystal the $\langle 001 \rangle$-oriented dendrites can be seen in the $\{112\}$ and $\{011\}$ faces.

Casting Defects

Grain Boundary Cracking in CG Castings.   When hollow air-cooled turbine airfoil castings are required, a ceramic core, which is subsequently leached out, is used to provide the hollow cavity of required dimensions in the casting. The ceramic core has a lower coefficient of expansion than the DS superalloy, so hoop stresses are set up in the alloy during cooling as the superalloy shrinks around the ceramic core. In the case of CG castings these stresses can result in cracks forming at the grain boundaries. To avoid this, 0.75–2% hafnium is added to CG superalloys.[8]

Equiaxed Grains and Freckles.   The DS process requires control of the solidification parameters to produce castings free of defects. To prevent equiaxed grains from forming, sufficient heat must be removed at the solidus interface by conduction into the solid so that the heat of solidification, ΔH, does not accumulate, altering the sign or direction of the thermal gradient, G_s, at the solidus into the two-phase mushy zone above it.[9] This can be expressed in terms of solidification rate R and thermal conductivity K_T of the solid as

$$R = \frac{K_T G_s}{\Delta H} \tag{1}$$

At solidification rates higher than given in equation (1) equiaxed grains will be formed. This bounds the area on a plot of growth rate versus thermal gradient where

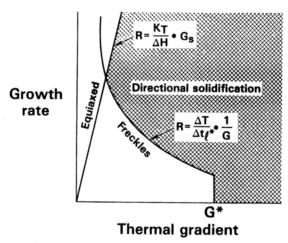

Fig. 5. DS process requires control of casting parameters such as thermal gradient G and growth rate R to avoid defects [9].

the DS process can be carried out successfully, as shown in Fig. 5. Another consideration in the control of the DS process involves the formation of freckles or chains of equiaxed grains.

With the liquid phase being above the solid phase in the DS process, solute elements such as aluminum and titanium that are rejected from the solid during solidification are enriched in the liquid near the bottom of the two-phase mushy zone adjacent to the solidified alloy. The aluminum- and titanium-enriched liquid has a lower density, on average, than the alloy and the liquid above it, which is closer to the liquidus and not as enriched in aluminum and titanium. This density difference can result in jets of liquid rising to the top of the mushy zone, breaking off dendrite tips in the process. These dendrite tips act as nuclei for chains of equiaxed grains, referred to as freckles. The presence of equiaxed grains defeats the reason for having a DS component and can result in premature failure. Freckles and other equiaxed grains are not acceptable, and solidification conditions must be controlled to prohibit their formation.

For steady-state solidification there is a critical thermal gradient, G^*, at which the height of the mushy zone is too small to permit a sufficiently large difference in liquid density to support the formation of the inversion jets. Therefore, at thermal gradients above G^* freckles will not form.[9] The value of G^* is a function of alloy chemistry. At lower thermal gradients the formation of the inversion jets is controlled by the difference between the liquidus and solidus temperatures, ΔT; the time required for the mushy zone to pass a point in the casting, referred to as the local solidification time, Δt_l; and the gradient in the mushy zone, G. If there is a critical time Δt_l^* required to produce a detectable freckle trail, the solidification rate above which freckle formation will not occur is

$$R = \frac{\Delta T}{\Delta t_1^*} \frac{1}{G} \tag{2}$$

Equation (2) has been drawn in Fig. 5 to show with equation (1) the cross-hatched area on the plot of growth rate versus thermal gradient over which the controlled DS process occurs.

Recrystallized Grains. Directionally solidified alloys are subject to other defects not normally encountered in CC superalloys. Among these are recrystallized grains that can result from cold work imparted to the casting during postsolidification processing and handling, which is followed by a high-temperature exposure. If the exposure temperature is sufficiently high, insufficient γ' is present to inhibit grain boundary motion, and normal recrystallization occurs. If the exposure is at lower temperatures, where the dislocations must cut through the γ' precipitates, recrystallization proceeds more slowly and is less extensive. This type of recrystallization, referred to as cellular recrystallization, involves dissolution of the γ' particles ahead of the moving grain boundary, which then reprecipitates behind it, after the boundary has moved. This results in a slower moving boundary. Normally recrystallized grains can contain twins while cellular recrystallized grains do not. Recrystallized boundaries are relatively free of precipitates or the beneficial elements hafnium, zirconium, carbon, or boron that are found in as-cast grain boundaries as a result of the segregation associated with solidification. With only an occasional $M_{23}C_6$ carbide these recrystallized boundaries are weak and act as crack initiation sites upon loading, especially those normal to the stress axis.

Other Casting Defects. As columnar grains grow, they align themselves parallel to the thermal gradient and normal to the solidification front, which is not perfectly flat but develops a curvature. Grains tend to diverge as the solidification front proceeds up the casting with the number of grains becoming less as the CG casting length increases. This can result in boundaries intersecting the critical leading or trailing edges of an airfoil, which is not desirable. The angle that the grain boundaries make with these critical regions of an airfoil casting can be controlled by improving solidification front planarity and are usually restricted to about 10° or less.

In general, any defect that involves grain boundaries with a component normal to the stress or airfoil axis, such as freckles or recrystallized grains, is not acceptable. Directionally solidified superalloys are subject to most of the quality requirements for CC alloys and are inspected by using radiographic and fluorescent inspection techniques. Porosity is not as severe in DS components due to the presence of a liquid head that feeds the shrinkage throughout the entire solidification process. Pore size and incidence are both reduced from that in CC alloys.

Single-crystal castings are subject to the same solidification defects as CG castings except for the concern about boundary alignment with CG castings. However, SC castings are subject to the formation of low angle boundaries. These are boundaries with a small lattice misorientation (usually less than 15°) but that, nonetheless, can act as crack initiation sites. Although the permissible low angle boundary misorientation is a function of alloy and application, boundaries above 10° are usually not permitted. In addition to the absence of boundaries, the primary orientation of a SC is required to be within 10°–15° from a particular crystallographic axis such as the $\langle 001 \rangle$ axis.

MICROSTRUCTURE

The microstructure of DS superalloys is similar to that of CC superalloys consisting primarily of γ' precipitates in a γ matrix with a few carbides and borides as described in Chapter 4.

Gamma Prime

Hafnium in CG alloys promotes the formation of a eutectic γ' in the interdendritic regions, as shown in Fig. 6. The slower cooling rate in the DS process results in a coarser solid-state cuboidal γ' in the as-cast condition, but the volume fraction of γ' is unaltered. In the fully heat-treated condition the γ' in a DS alloy is finer and more uniform than in a CC alloy, as discussed more fully in the section on solution heat treatment.

Carbides

Hafnium usually is added to CG alloys to prevent grain boundary cracking during solidification of cored castings. The presence of hafnium alters carbide chemistry

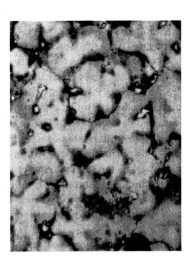
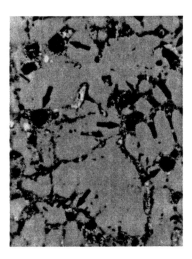

Fig. 6. Hafnium additions to DS alloys promote formation of $\gamma-\gamma'$ eutectic; (a) CG MAR-M 200 with few eutectic nodules; (b) CG MAR- M 200 + Hf with increased eutectic (arrows).

and morphology. When greater than about 1% hafnium is present, HfC forms in addition to the mixed MC carbide present in most high-strength cast superalloys. The high-melting-point HfC probably forms in the liquid, unlike the usual mixed MC carbide that forms in the liquid-plus-solid, or mushy, zone. Consequently, HfC is equiaxed and basically devoid of other metal atoms. The mixed MC carbide contains hafnium together with titanium, tantatum, columbium (Nb), or tungsten depending on the MC-carbide-forming elements present in the alloy. Having formed in the mushy zone, the mixed MC carbide tends to be more dendritic and less equiaxed than the HfC since carbide morphology is primarily a function of the thermal gradient in the mushy zone, becoming more dendritic and larger as the gradient decreases. Fatigue life is inversely related to defect size, with the smaller equiaxed carbides being the preferred morphology and solidification under a high thermal gradient the preferred process.

Another impact of hafnium on carbide microstructure is to tie up the carbon in the more stable hafnium-containing MC carbides that form earlier or at higher temperatures in the solidification process. This prevents the formation of $M_{23}C_6$ carbides in the as-cast alloy. Upon heat treatment at temperatures below 2000 °F (1090 °C) the MC carbides start to break down and the chromium-rich $M_{23}C_6$ carbide precipitates in grain boundaries as in a CC alloy. The M_6C formation also will be retarded by the presence of hafnium, which stabilizes the MC at the expense of the other carbides. The general rules that govern M_6C formation, wherein a high molybdenum or tungsten content is required, also hold true for DS alloys.

With SC the carbon content often is kept intentionally low (below 50 ppm), as there are no grain boundaries that require strengthening by carbides, so that the only carbides are a few very small $M_{23}C_6$ carbides. This is because carbon contents of about 100 ppm are required before MC carbides are formed; the minimum carbon content needed to form MC can be lower if there is a preponderance of strong carbide formers such as tantalum, titanium, or hafnium in the alloy.

Borides

Boron additions of about 0.015% are common in CG alloys with borides present in the grain boundaries in a morphology similar to that in CC alloys. Although most CG superalloys contain hafnium for transverse grain boundary strength, boron has been found to be almost as effective. When present as the primary grain boundary strengthener in CG alloys, it is at higher levels of about 0.2% and usually with very low carbon contents, similar to the boron and carbon levels of the CC BC alloys.[10] In high-boron alloys M_5B_3 is the principal boride similar to that found in CC alloys.

TCP Phases

Topologically close-packed (TCP) phases such as σ, μ, or Laves can form in DS alloys as they do in CC alloys. In fact, their formation is somewhat more likely in the as-cast condition due to the greater segregation present in DS superalloys than with the more rapidly solidified CC alloys. However, as discussed in the section on heat treatment, DS superalloys are given a solution heat treatment that homogenizes the alloy, reducing the likelihood of forming any TCP phases.

Microstructural Stability

Directionally solidified superalloys are required to be microstructurally stable. Phase control procedures, such as the use of electron hole (vacancy) numbers, are effective with DS alloys. When present, instabilities such as γ or μ phase are less detrimental in DS superalloys because the alloy matrix is inherently more ductile. Embrittlement due to phase instability, usually a major concern with CC alloys, is of less concern for DS alloys. A more relevant concern is the reduction of strength that can occur as a consequence of undesirable phase precipitation, which in time depletes the alloy of important strengthening elements. Unstable DS alloys have been observed to exhibit premature plastic deformation.

HEAT TREATMENT

Superalloys in general are given three types of heat treatments: (1) a solution heat treatment to dissolve the γ' so that it can subsequently be reprecipitated on a finer, more homogeneous scale to increase alloy strength; (2) a coating heat treatment to bond the coating onto the alloy substrate; and (3) an aging heat treatment to precipitate additional γ' and/or other phases such as carbides and borides at the grain boundaries.

Solution Heat Treatment

Directionally solidified superalloys are given a solution heat treatment to strengthen the alloy by refining the γ' precipitate size. Solution heat treatment results in reduced ductility and creep rupture life in high-strength CC alloys with high volume fractions (>0.5) of γ'. Strengthening the grains by heat treatment makes deformation in the vicinity of the grain boundaries, which is needed to accommodate changes in grain shape during deformation of a polycrystalline body, more difficult. As a result, grain boundary cracking is more likely to occur, reducing creep ductility and life. For DS superalloys, which are not limited by the ability to transfer strain across a grain boundary without initiating cracks and which in the as-cast condition have a coarser, less homogeneous precipitation of γ', a solution heat treatment is normally employed to develop optimum mechanical properties.[3,11]

Ideally, all the γ' in the as-cast (eutectic and coarse cuboidal) alloy should be taken into solution and the alloy composition homogenized so that upon cooling from the solution heat treatment temperature to below the γ' solvus (temperature below which γ' exists), a relatively fine uniform precipitation of γ' occurs throughout the microstructure. This is the optimum superalloy microstructure for maximum mechanical properties. It is a consequence of (1) segregated portions of the alloy, with a eutectic or coarse γ' microstructure, being weak areas that do not fully contribute to alloy strength and (2) alloy strength in high-volume-fraction γ' superalloys being inversely related to γ' size, as discussed in the section on mechanical properties. To achieve this microstructure, the alloy must be heated above the γ' solvus to dissolve the as-cast γ' and below the incipient melting temperature to prevent melting, which can result in solidification segregation, eutectic γ' formation, and shrinkage porosity. Both solvus and incipient melting temperatures are a function of alloy composition. It is best, therefore, initially to homogenize the alloy to make it easier to stay between these two critical temperatures. The ease of carrying out the solution heat treatment depends on the difference between the incipient melting and γ' solvus temperatures. With current furnace controls a difference of at least 15 °F (10 °C) is necessary.

With most CG alloys, especially those with hafnium additions, the difference between the incipient melting and solvus temperatures is negative, and these alloys cannot be completely solutioned. A solution heat treatment temperature just below the incipient melting temperature usually is selected. The inability of most CG alloys to be completely solutioned, a result in large measure of the need for hafnium to prevent cracking during the solidification cool-down cycle in cored CG castings, is one reason why SC superalloys, which have little if any hafnium, have greater mechanical strength.

In many CG and SC alloys a small volume percent (~5%) of the alloy consists of the interdendritic γ–γ' eutectic constituent. While it would be desirable to completely dissolve this eutectic γ', as well as the coarse solid-state γ', it is the latter, present in a much greater amount, that is the focus of the effort. The eutectic, being last to form, is therefore the most difficult γ' to dissolve and often is not. Due to its relatively low volume fraction in most alloys, the strength lost by leaving the eutectic γ' undissolved is negligible.

Once the γ' has been solutioned, it should be precipitated on a fine uniform scale. If the alloy is homogeneous, the γ' solvus will be the same throughout the alloy. Precipitation will initiate at the same temperature, and more importantly, coarsening or growth of the fine γ' particles will be uniform.

To control the size of the γ' precipitates, which affects mechanical properties, it is necessary to control the cooling rate from the solvus to that temperature below which γ' will not coarsen in short times, that is, about 2000 °F (1090 °C). The effect of cooling rate from the solution temperature on γ' size of PWA 1480, a representative SC alloy, is shown in Fig. 7. It is the cooling rate from the solution heat treatment temperature (actually the cooling rate from the solvus) that controls mechanical properties and not the cooling rate during the DS process. Solution heat treatment temperatures in excess of 2200 °F (1205 °C) usually are employed, so an inert atmosphere at reduced pressure is used to prevent oxidation. Full vacuum is not used to minimize loss of chromium. Cooling rates in excess of 100 °F (55 °C) per minute can be achieved by backfilling with cold argon and rapidly circulating the argon past the components being cooled.

The solution heat treatment affects other phases besides γ'. MC carbides start to go into solution. Insufficient time is available for significant beneficial $M_{23}C_6$ carbide precipitation during cooling, but greater quantities of carbon are available

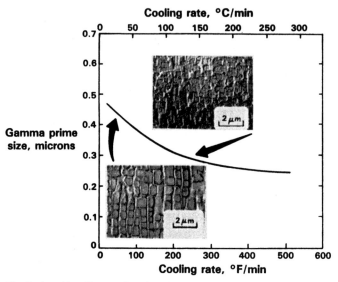

Fig. 7. A rapid cooling rate from the solution heat treatment temperature prevents coarsening, maintaining a fine γ' size.

for carbide precipitation during subsequent lower temperature heat treatments. In some hafnium-containing CG alloys, such as MAR-M 200 + Hf, this is important as there are no $M_{23}C_6$ carbides, which are beneficial to alloy properties, present in the as-cast condition. The increase in alloy homogeneity obtained with the solution heat treatment is helpful in preventing undesirable phase precipitation, such as σ or μ, in the more highly segregated interdendritic areas.

Coating Heat Treatment

The second heat treatment given many DS alloys is the coating heat treatment that bonds the coating to the alloy. Temperatures of 1800–2050 °F (980–1120 °C) for times up to 8 h are commonly used. These temperature–time combinations can result in growth of the γ' precipitates, which will alter alloy mechanical properties, as discussed in the section on mechanical properties. The short time spent in cooling from the coating heat treatment temperature is insufficient to result in significant growth of the γ' particles, so alloy properties are not sensitive to the rate of cooling from the lower coating heat treatment temperatures.

M_6C and some $M_{23}C_6$ precipitation occur during the coating heat treatment. The type of carbide depends on alloy composition, with $M_{23}C_6$ being present in almost all DS superalloys.

Aging Heat Treatment

The third type of heat treatment occurs at lower temperatures [1300–1650 °F (700–900 °C)] and for longer times of as much as 32 h. The primary purpose is to precipitate $M_{23}C_6$ carbides at grain boundaries to provide some resistance to grain boundary sliding. Additional very fine γ' also is precipitated at these lower temperatures, resulting in a bimodal size distribution. Yield strength below 1400 °F (760 °C) is a sensitive inverse function of γ' size and can be enhanced with a low-temperature heat treatment. For low-carbon SC alloys, where higher temperature creep strength is the primary objective, the aging heat treatment can be omitted with little, if any, effect on alloy creep properties.

ALLOY DESIGN

The history of metallurgical technology is such that whenever an improvement in processing such as DS is introduced, alloys that capitalize on the new process are developed. Such has been the case for CG and SC superalloys. The basic differences in composition between DS and CC superalloys are (1) the hafnium added to CG superalloys to prevent solidification cracking, (2) the balancing of alloying elements in SC superalloys to provide an incipient melting temperature above the γ' solvus, (3) the balancing of refractory alloying elements to minimize solidification defects such as freckles, and (4) the absence (or low level) of the grain-boundary-strengthening elements boron, carbon, hafnium, and zirconium in SC alloys. A list of some CG and SC superalloy compositions is given in Table 1.

Hafnium is added primarily to prevent grain boundary cracking during cooling following solidification of cored CG castings. The hafnium probably prevents O_2

Table 1. Directionally Solidified Superalloy Compositions

Alloy	Cr	Co	W	Mo	Ta	Cb	Ti	Al	Hf	B	Zr	C
				(a) Columnar-Grained Alloys								
MAR-M[a]												
200 + Hf	9	10	12	—	—	1.0	2.0	5.0	2.0	0.015	0.08	0.14
MAR-M												
246 + Hf[a]	9	10	10	2.5	1.5	—	1.5	5.5	1.5	0.015	0.05	0.15
MAR-M												
247[a]	8.4	10	10	0.6	3.0	—	1.0	5.5	1.4	0.015	0.05	0.15
René 80H[b]	14	9.5	4.0	4.0	—	—	4.8	3.0	0.75	0.015	0.02	0.08
				(b) Single-Crystal Alloys								
PWA 1480[c]	10	5.0	4.0	—	12	—	1.5	5.0	—			
CMSX-2[d]	8.0	5.0	8.0	0.6	6.0	—	1.0	5.5	—			
CMSX-3[d]	8.0	5.0	8.0	0.6	6.0	—	1.0	5.5	0.15			
SRR-99[e]	8.5	5.0	9.5	—	2.8	—	2.2	5.5	—			

[a] Martin Marietta Corp.
[b] General Electric Co.
[c] Pratt & Whitney
[d] Cannon-Muskegon Co.
[e] Rolls-Royce Ltd.

embrittlement of the boundary[12] under the hoop stresses set up during cooling as the metal with its higher thermal expansion shrinks around the ceramic core. Grain boundary cracking has also been observed in complex solid (uncored) parts where transverse stresses are imposed by the complex casting geometry. Hafnium enhances oxidation resistance, behaving in a manner similar to the active elements yttrium, cerium, lanthanum, and so on. In addition, some strengthening of the γ' phase probably results from the addition of hafnium. However, hafnium decreases the incipient melting temperature and increases the γ' solvus, making complete solutioning of the alloy difficult. For this reason, it is added in only limited amounts, if at all, to SC superalloys.

As γ' is the principal strengthening phase in superalloys, its solvus is directly related to the temperature at which an alloy weakens. Increasing the solvus increases the temperature capability of a superalloy. High-melting-point refractory elements with high solubilities in γ' such as tantalum, hafnium, columbium, and titanium raise the γ' solvus. However, these elements are added in controlled amounts since the incipient melting temperature must be kept above the γ' solvus in order to fully solution the alloy to achieve maximum strength. The addition of the refractory elements tungsten and rhenium and the absence of zirconium and boron increase the incipient melting temperature. Cobalt can be added to decrease the γ' solvus with little effect on the incipient melting temperature, thereby enhancing the heat treatment response. Care must be taken, however, as too much cobalt will decrease alloy stability, resulting in undesirable phase precipitation. Iron and chromium behave in a similar manner.

Freckle defects, as discussed in the section on solidification, result from local density inversions in the liquid-plus-solid, or mushy zone during solidification. Elements that segregate to the liquid (and are concentrated in the interdendritic areas of the casting) and have higher densities than the average liquid alloy density, such as tantalum, molybdenum, and hafnium, are beneficial in that they make freckle formation more difficult. However, low-density elements that segregate to the liquid (such as Al or Ti) together with high-density elements that segregate to the solid and are concentrated in the dendrite core (such as W and Re) promote the formation of freckles. To avoid freckle formation in a DS alloy, the detrimental elements must be maintained below that level that promotes freckles during solidification of a representative configuration under a typical thermal gradient.

For SC alloys the grain-boundary-strengthening elements boron and carbon are not necessary. This effectively eliminates borides or carbides, which can act as failure initiation sites in fatigue loading and in some creep-limited applications. Zirconium is another grain-boundary-strengthening element that is normally not added to SC superalloys as it reduces the incipient melting temperature. Other elements beneficial to achieving a balance in creep/fatigue strength and oxidation resistance can be added in place of these grain boundary strengtheners.

The strength improvements obtained with CC superalloys have been achieved by increasing the volume percent of γ' to about 60%. Further increases in γ' have proven to be less effective, as above 70% it becomes difficult to prevent the γ' from becoming the matrix phase, thereby reducing alloy properties. Further improvements in creep strength have come about with DS superalloys by more effective use of the available γ' in the form of a more uniform precipitate and by increasing the γ' solvus. The strength of the γ' also has been increased by adding refractory elements, which probably increases the antiphase boundary energy. Tantalum, tungsten, and rhenium, with their higher melting points, have been found to be more effective strengthening additions than columbium, vanadium, and molybdenum. The precious metals, though not usually added to superalloys for economic reasons, are effective strengtheners. Platinum is beneficial as both a strong γ' former with the anticipated increase in creep strength and as an enhancer of both alloy oxidation and hot-corrosion resistance.

Chromium is added to all superalloys to enhance hot-corrosion resistance. Higher titanium and titanium–aluminum ratios, as well as lowered molybdenum contents, have also been found to enhance hot-corrosion resistance, but chromium is the primary alloying addition used to enhance hot-corrosion resistance. For a fixed-strength level DS superalloys with higher chromium contents, and thereby greater hot-corrosion resistance, can be obtained than with CC superalloys. The directionally solidified alloys are beginning to be employed in hot-corrosion limited applications.

Oxidation resistance in superalloys can best be achieved by forming a thin (submicron) adherent film of Al_2O_3 which acts as a barrier to the diffusion of oxygen. When more than 4% aluminum is present, chromium enhances the activity of aluminum, enabling Al_2O_3 to form upon oxidation with its superior oxidation resistance.[13] Addition of chromium beyond that required to form Al_2O_3 does little to enhance oxidation resistance. The amount of chromium required to form Al_2O_3 decreases with increasing temperature as the available aluminum can more readily diffuse to the surface at the higher temperatures. Addition of active elements further enhances alloy oxidation resistance by making spallation during thermal cycling of the protective Al_2O_3 layer that forms during oxidation more difficult.[14] Elements such as titanium, columbium, and boron have proven to be detrimental to alloy oxidation resistance, while tantalum and platinum have been found to be beneficial.

The increased temperature capability of DS alloys has resulted in their use with oxidation-resistant coatings. Oxidation resistance now is measured in terms of how long a coating will survive on an alloy. For aluminide diffusion coatings the elements in the alloy become incorporated into the much higher aluminum content coating on the surface. The behavior of the various alloying elements in the coating is similar to their behavior in the uncoated DS superalloy. Failure in aluminide or overlay coated superalloys is a result of (1) the loss of aluminum needed to form a protective Al_2O_3 layer on the surface due to spallation during thermal cycling and (2) interdiffusion of nickel from the substrate, diluting the aluminum in the coating, thereby decreasing its ability to reform Al_2O_3 after the previous oxide layers have spalled off during thermal cycling.[15] Elements such as tantalum and platinum in the alloy have an affinity for nickel, which reduces the rate at which the nickel diffuses into the coating, diluting the aluminum and thereby promoting increased coating durability.

Several elements (i.e., Pb, Bi, Te, and Tl) with low melting points and low solubilities segregate to grain boundaries when present in cast nickel-base superalloys.

These low-melting-point elements, sometimes referred to as "tramp elements," embrittle CC superalloys at elevated temperatures by promoting premature grain boundary decohesion as a consequence of their (1) relatively high concentration at grain boundaries, (2) low melting points, and (3) low solubilities. The DS alloys, being less sensitive to grain boundary properties, are less susceptible to the presence of these low-melting-point tramp elements. However, controls on the tramp elements in DS alloys have generally been maintained at the same levels as in CC alloys.

MECHANICAL PROPERTIES

Elastic Behavior

All DS superalloys have a preferred orientation in at least one direction. The elastic, or Young's, modulus, E, is anisotropic and will vary depending on the direction in which the DS superalloy is loaded. For SC superalloys the modulus can be expressed as a function of orientation[16] over the standard stereographic triangle by

$$E^{-1} = S_{11} - [2(S_{11} - S_{12}) - S_{44}] [\cos^2\phi(\sin^2\phi - \sin^2\theta \cos^2\phi \cos^2\theta) \quad (3)$$

where θ is the angle made with $\langle 001 \rangle$ and ϕ is that made with the $\langle 001 \rangle$–$\langle 110 \rangle$ boundary of the triangle, as shown in Fig. 8. The terms S_{11}, S_{12}, and S_{44} are the elastic compliances. The elastic properties are insensitive to composition, and the values for nickel are a good approximation of the values for even the most highly alloyed nickel-base superalloys. The room temperature modulus of PWA 1480, a representative SC superalloy, is shown as a function of orientation in Fig. 9.

CG alloys in the longitudinal or growth direction have the lowest possible modulus as the grains are aligned with the $\langle 001 \rangle$ orientation parallel to the growth direction.

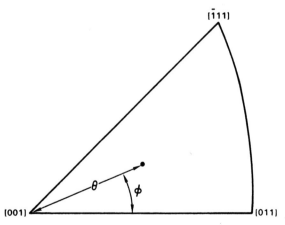

Fig. 8. Definition of the orientation parameters θ and ϕ in equation (3) describing the location of an arbitrary direction relative to the [001] orientation [17].

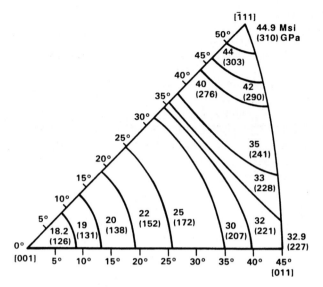

Fig. 9. Elastic modulus of SC superalloy PWA 1480 as a function of orientation at room temperature.

As the ⟨001⟩ orientation has the lowest possible modulus, any variation or deviation of the grain structure will increase the modulus. For most CG superalloys the longitudinal room temperature modulus is 19×10^6 psi (131 GPa) while for a well-aligned ⟨001⟩-oriented SC alloy it is 18×10^6 psi (124 GPa).

The modulus in the ⟨111⟩ orientation is the highest, with a value of about 45×10^6 psi (310 GPa), while the ⟨110⟩ direction has an intermediate elastic modulus of 33×10^6 psi (228 GPa). In torsion the reverse is true with the ⟨001⟩ orientation having the highest shear modulus, G, of 18×10^6 psi (124 GPa) and the ⟨111⟩ orientation having the lowest, about 8×10^6 psi (55 GPa). The DS superalloys stressed along the ⟨001⟩ growth axis, therefore, have low longitudinal or bending stiffness and high shear stiffness.

In the transverse plane normal to the growth direction of a ⟨001⟩-oriented SC alloy the elastic modulus is a function of the angle, ϕ, between the [100] and [010] directions and can be expressed as

$$E^{-1} = S_{11} - [2(S_{11} - S_{12}) - S_{44}] [\sin^2\psi \cos^2\psi] \qquad (4)$$

If this expression is averaged over all transverse directions, the transverse modulus for CG superalloys is found to be[18]

$$E_T = 2[S_{11}(2S_{11} + 2S_{12} + S_{44})]^{-1/2} \sim 24 \times 10^6 \text{ psi (165 GPa)} \qquad (5)$$

The longitudinal, or ⟨001⟩, modulus is independent of secondary orientation, ϕ. As a result of this, resonant frequency in bending also should be independent of secondary orientation. This has also been found to be the case in torsion.

Tensile Properties

Tensile properties of superalloys are primarily controlled by alloy composition and γ' size. Columnar-grained and SC superalloys with orientations near $\langle 001 \rangle$ deform by octahedral slip on the close-packed $\{111\}$ planes and exhibit yield strengths similar to, but somewhat higher than, those of CC superalloys.[19] Tensile ductilities are usually in excess of 10%.

For superalloys with a high volume fraction of γ' the $\langle 001 \rangle$ yield strength below about 1400 °F (760 °C), above which thermal activation become controlling, is inversely related to γ' size, as shown in Fig. 10 for PWA-1480, a representative SC alloy. The requirement imposed on DS alloys to maintain a minimum cooling rate from the solution heat treatment temperature is based in part on this inverse relationship. The inverse relationship between yield strength and γ' size is consistent with the rate-controlling step determining alloy yield strength being the cutting of the γ' by the leading dislocation of the superlattice pair.[19] The inverse relationship between γ' size and yield strength does not hold for high-modulus orientations.[20]

High-modulus SC alloys can exhibit lower yield strengths as a result of their deforming on $\{100\}$ cube planes that have a lower critical resolved shear stress. SC alloys in the $\langle 111 \rangle$ orientation exhibit lower yield strengths but very high ultimate strengths, in excess of 200 ksi (1380 MPa) at room temperature, indicative of the many active cube slip systems participating in the deformation process. By way of contrast, the $\langle 110 \rangle$ orientation has a yield strength intermediate between $\langle 100 \rangle$ and $\langle 111 \rangle$ with very little work hardening (low ultimate) as the $\langle 110 \rangle$ orientation is close to that for single slip under uniaxial loading.

The CG alloys in the longitudinal direction behave similarly to $\langle 001 \rangle$-oriented SC alloys. The presence of grains in the CG structure provide some constraint during deformation, which promotes multiple slip. This results in more work hardening

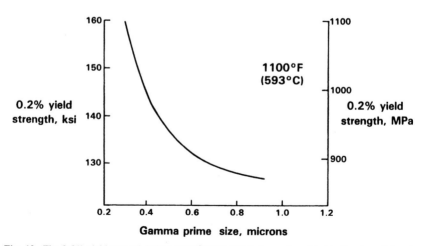

Fig. 10. The 0.2% yield strength below 1400 °F (760 °C) is inversely related to the size of the γ'.

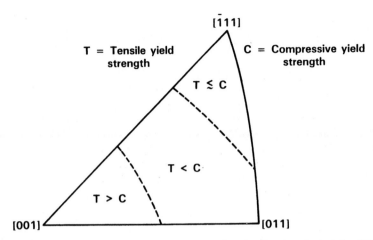

Fig. 11. Tension compression asymmetry in SC superalloys is a function of orientation [21].

and a somewhat higher ultimate strength than for ⟨001⟩-oriented SC alloys but with little effect on yield strength or ductility. Care should be exercised when testing CG alloys in the transverse direction to be sure that there are a sufficient number of grains in the cross section. Large variations in transverse properties are usually indicative of too few grains. The orientation of these grains sin the transverse direction is random, and properties representative of SC alloys with orientations anywhere between ⟨001⟩ and ⟨110⟩ can result. Transverse tensile ductility is not a sensitive indication of grain boundary strength, which is best measured by creep ductility.

Tensile failure usually initiates in planar bands of concentrated slip that are characteristic of γ′-strengthened alloys. Defects, unless they are fairly large and can act as slip initiation sites, play a minor role as failure initiation sites. The planar inhomogeneous nature of slip results in concentrated strains and ultimately slip plane failure with the formation of macroscopic crystallographic facets on the fracture surface of uniaxially loaded specimens. Under certain circumstances of temperature and orientation these facets, especially in SC alloys with no grain boundary constraints, can be quite large. Under higher magnification these facets can be seen to have numerous small steps where the slip deformation has translated to an adjacent parallel slip plane. The presence of these facets are not indicative of brittle behavior, as they occur in SC superalloys with very high fracture elongations. At temperatures above about 1700 °F (925 °C) deformation becomes more homogeneous and the facets become less pronounced.

An interesting characteristic of SC superalloys is that the yield strengths in compression and tension are not equal, with the difference being a function of orientation. In the ⟨001⟩ direction tensile yield strength is greater, while the reverse is true for the ⟨011⟩ orientation, as shown in Fig. 11. The asymmetry in the yield strength for SC alloys can be explained by the detailed dislocation γ′ precipitate interactions that define the yield strength.[21]

Creep Rupture

At temperatures above about 1500 °F (815 °C), where thermal activation is important, the creep rupture properties of superalloys with a high volume fraction of γ' are affected by the size of the γ' particles. The relationship is shown in Fig. 12 at 1800 °F for ⟨001⟩-oriented PWA 1480, a representative SC alloy. When the γ' particles are very small, it is easy for dislocations to climb around them, resulting in reduced creep strength. The greatest creep strength is achieved when the dislocations are forced to cut the γ'. As the γ' particles become larger, it is easier for dislocations to bow between them, reducing creep strength. This relationship between creep strength and γ' size is why it is necessary to control the final size of the γ' after heat treatment, as discussed in the section on heat treatment.

Directionally solidified superalloys exhibit the same three stages of creep as do CC alloys, but their primary creep behavior at higher uniaxial stresses and lower temperatures is different.[22] In the initial stage of creep deformation proceeds for extensive distances on single-slip systems, even in CG alloys whose grain boundaries are not effective barriers to slip. This is especially true at lower temperatures [1400–1500 °F (760–815 °C)] where thermally activated cross-slip is of little consequence and where high stresses are required for creep to occur in reasonable periods of time. The extensive slip distances afforded by DS superalloys results in high primary creep strains, with the extent of primary creep increasing with stress. At temperatures above 1600 °F (870 °C) the magnitude of this effect becomes much

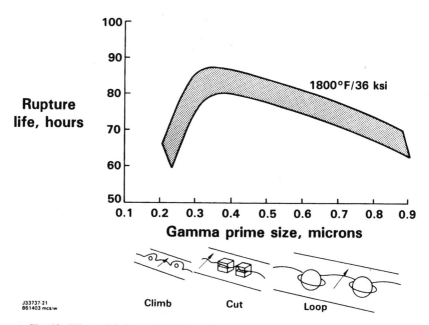

Fig. 12. Effect of γ' size on 1800 °F (982 °C) creep rupture life of a SC superalloy [5].

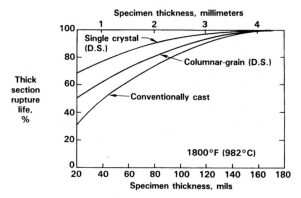

Fig. 13. DS superalloys (CG and SC) retain a higher fraction of their thick-section rupture life at thin sections below 150 mils (3.8 mm) than CC superalloys.

less as slip distances are reduced by thermally activated cross-slip. While extensive primary slip can occur in uniaxial creep tests, the more complex loading found in DS components under actual service conditions limits the extent of primary creep to values typical of CC alloys.

The secondary creep rate of DS superalloys is usually lower than that for CC alloys. This is a result of the finer and more uniform γ' achieved by solution heat treating DS superalloys.[11] Solution heat treatment of CC alloys with a high volume fraction of γ' also will increase creep strength (lower the creep rate) but will aggravate the difficulty in transmitting strain across grain boundaries, resulting in decreased ductility and a shorter rupture life. The alignment or elimination of grain boundaries in DS superalloys removes this limitation on alloy creep strength.

Tertiary creep is quite extensive in DS alloys as grain boundary failure initiation sites have been largely reduced or eliminated. In CG alloys the failure or crack initiation sites are either grain boundary segments normal to the applied stress or interdendritic areas that have not been thoroughly homogenized and are weaker, that is, have a coarser γ' microstructure than the surrounding regions. With SC alloys failure initiates at interdendritic areas having coarser γ' microstructures. The extended creep ductility of DS alloys is a result of these changes in failure initiation sites relative to the grain boundary initiation prevalent in CC alloys. Cracking usually initiates in third-stage creep at strains of 1–2% for CG alloys and 3–4% for SC alloys. Rupture elongations of 20–30% are not unusual for DS alloys.

At low temperatures the fracture surfaces of DS alloys are faceted, similar to what is seen with the higher strain rate tensile loading. At higher temperatures the facets become less pronounced.

As section size is reduced, creep rupture strength decreases, as shown in Fig. 13. The reduction is less for CG alloys than for CC alloys and smallest for SC alloys. The creep resistance of CC alloys increases as the number of grains per unit area increases.[23] When grain boundary decohesion occurs due to cavitation in CC alloys the effect on load carrying capability is greater in material with fewer grains per unit area where each grain represents a greater fraction of the cross section.[24] Creep failure in SC alloys is different initiating at interdendritic areas. Only at very thin sections would the interdendritic areas occupy a significant fraction of the cross section and affect creep-rupture properties. CG alloys with failure initiation sites at grain boundary segments perpendicular to the loading direction are better than CC alloys but not as good as SC alloys. This effect can be important as many cooled turbine airfoils involve localized wall thicknesses as thin as 20 mils (0.5 mm).

In the transverse direction, normal to the growth direction, CG superalloys exhibit lower creep properties than in the longitudinal direction parallel to the direction of grain alignment. This is because in the transverse direction there is a considerable extent of grain boundary area normal to the stress direction. Transverse CG superalloys behave similarly to CC alloys. The room temperature modulus in the transverse direction [24×10^6 psi (165 GPa)] is higher than the low $\langle 001 \rangle$ modulus in the longitudinal direction [19×10^6 psi (131 GPa)] but lower than the CC modulus, which is an average of all orientations [32×10^6 psi (221 GPa)].

The hafnium present in most CG superalloys enhances grain boundary ductility, so while transverse creep ductility is significantly lower than longitudinal creep ductility, it is still in the upper range of creep ductilities observed for CC superalloys. Transverse creep ductility is an effective way to measure grain boundary strength in CG alloys.

The creep properties of SC superalloys are anisotropic, depending on orientation. As an example of the large variations in creep anisotropy that can occur, creep curves for the three major orientations are plotted in Fig. 14 for the SC superalloy PWA 1480 at 1400 and 1800 °F (760 and 982 °C). At 1400 °F (760 °C) the $\langle 001 \rangle$ orientation is the strongest, with $\langle 111 \rangle$ being the weakest, while at 1800 °F (982 °C) the $\langle 111 \rangle$ orientation is the strongest. Not all SC superalloys exhibit the same type of anisotropy. Many alloys exhibit the highest creep strength in the $\langle 111 \rangle$

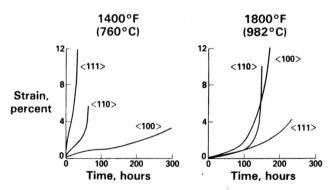

Fig. 14. Creep anisotropy of SC PWA 1480 is a function of temperature. At 1400 °F (760 °C) the $\langle 001 \rangle$ orientation is the most creep resistant, but at 1800 °F (982 °C) the $\langle 111 \rangle$ orientation is the most creep resistant [5].

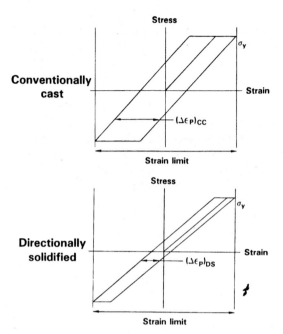

Fig. 15. Low-modulus DS superalloys have smaller plastic strain ranges in strain-controlled low-cycle fatigue than CC superalloys [5].

orientation at all temperatures. Stress, as well as temperature, can affect creep anisotropy. At the present time creep anisotropy in nickel-base superalloys is not well understood.

Thermal Fatigue

Thermal fatigue is a result of the strain induced due to the constraint of free thermal expansion in a component exposed to temperature gradients, and can result in cracking. The thermal fatigue strain is the product of the thermal expansion coefficient and the change in temperature. Strain-controlled low-cycle fatigue is a good way to model thermal fatigue. The hysteresis loop in a strain-controlled fatigue test is shown in Fig. 15. A CC superalloy exhibits a hysteresis loop as shown in the upper figure. The lower modulus DS superalloy has a narrower hysteresis loop, as shown in the lower portion of Fig. 15. This is because (1) the yield strength of the lower modulus DS alloy is the same as for the alloy in the CC form with the higher modulus and (2) the lower modulus requires less plastic strain to attain the same total strain. The plastic strain range for the higher modulus CC alloy $(\Delta\epsilon_p)_{CC}$ is higher than for the lower modulus DS alloy $(\Delta\epsilon_p)_{DS}$.

Low-cycle fatigue life (N_f) varies inversely with the plastic strain range,

$$N_f = K(\Delta\epsilon_p)^{-C} \tag{6}$$

where K and C are constants. Thus, the lower modulus DS alloy will have a greater fatigue life. In fluidized-bed tests, where specimens are alternately immersed in two fluidized beds at different temperatures and the number of cycles to first cracking is taken as the thermal fatigue life, the lower modulus DS superalloys exhibit significantly enhanced thermal fatigue lives compared to higher modulus CC superalloys.[25] The enhanced thermal fatigue capability of DS superalloys has been confirmed by engine testing air-cooled first-stage turbine blades.

Thermal fatigue commonly is measured using a thin-walled tube specimen driven through an independently controlled combined thermal and mechanical or thermomechanical fatigue (TMF) cycle.[26] A wide variety of cycles can be imposed, but that with the independently controlled temperature and strain 180° out of phase (maximum temperature at minimum strain), referred to as cycle I, is considered to represent realistic conditions for the leading edge of an air-cooled gas turbine airfoil. The strain range (tension plus compression) through which the specimen is cycled is plotted versus the number of cycles to 50% load drop, which is taken as the failure point or thermal fatigue life. The lower modulus DS alloys exhibit a significant improvement in thermal fatigue life compared to CC superalloys, as shown in Fig. 16. These tests generally are conducted with coated specimens, as DS superalloys usually are utilized in the coated condition. Crack initiation occurs at the coating surface and propagates into the superalloy substrate, so alloy defects do not play a significant role in determining thermal fatigue life. Creep during the higher temperature compression portion of the cycle I TMF test results in a higher mean stress and reduced fatigue life. Thermal fatigue life is related to alloy creep strength at the maximum cycle temperature, with stronger alloys exhibiting longer thermal fatigue lives. Fatigue crack propagation rate is also an important alloy property, as

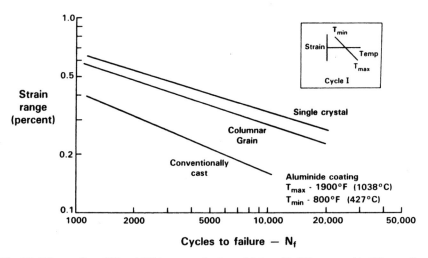

Fig. 16. DS superalloys (CG and SC) have superior thermal fatigue life (N_f) compared to CC superalloys.

a large fraction of total life is spent in propagation, and initiation life is relatively independent of alloy properties.

Load-Controlled Fatigue

The DS superalloys exhibit load-controlled fatigue characteristics similar to those of CC alloys. Although the low modulus would at first be expected to result in large strains and poor fatigue lives this is not usually observed as the fatigue life in most load controlled applications are controlled by stress concentrations which locally behave in a strain controlled manner. The higher ductility and yield strength of DS alloys also are responsible, in part, for their good fatigue properties.

Fatigue crack initiation in CG alloys is similar to that in CC superalloys with carbides or small pores acting as initiation sites, whereas microporosity is the major initiation site in carbide-free SC alloys. Elimination of the porosity by hot isostatic pressing can enhance significantly the fatigue life of SC superalloys. If present, defects such as freckles, recrystallized grains, and boundaries in SC alloys act as failure initiation sites. The fatigue capability of an alloy is a more sensitive function of such defects than are creep or tensile strength.

Fracture usually initiates normal to the stress direction; as the crack becomes larger, it converts to a crystallographic mode. The extent of crystallographic cracking increases as temperature is reduced and frequency increased.

OXIDATION–HOT CORROSION

Oxidation–hot-corrosion resistance is primarily a function of alloy composition and not solidification mode. Grain boundary oxidation is reduced or eliminated with DS superalloys, but this is not normally a problem at the high temperatures where DS superalloys are used. Improvements in alloy oxidation resistance are independent of how the alloy has been solidified. Little, if any, effect of crystallographic orientation has been observed on oxidation–hot-corrosion rate.

FUTURE DIRECTIONS

Single crystal superalloys with strength and temperature capabilities beyond those of the alloys currently available will be developed. Improvements in CG alloys also will be made, but SC superalloys always will have superior temperature capability. These improvements will be achieved by altering and optimizing alloy compositions and microstructures. It should be possible to further increase the refractory alloy content of SC superalloys for enhanced creep strength. Active element additions will be made for improvements in oxidation–hot-corrosion resistance. Solidification process modifications will enable DS alloys to be made less expensively and with higher quality. Processing modifications in such areas as heat treatment and hot isostatic processing also will extend the capability of these alloys.

The melting temperature of nickel, however, remains fixed. Therefore, there is a limit to the improvements in metal temperature capability that can be expected with nickel-base superalloys. It is possible that the application of the directional technology developed with nickel-base alloys to other high-temperature systems may extend the temperature capability of high-temperature alloys well beyond what is currently available.

REFERENCES

1. F. L. VerSynder and R. W. Guard, *Trans. ASM*, **52**, 485 (1960).
2. F. L. VerSnyder and M. E. Shank, *Mat. Sci. Eng.*, **6**, 213 (1970).
3. M. Gell, D. N. Duhl, and A. F. Giamei, in *Superalloys 1980, Proceedings of the Fourth International Symposium on Superalloys*, J. K. Tien et al. (eds.), ASM, Metals Park, OH, 1980, p. 205.
4. J. E. Northwood, *Metallurgia*, **46**, 437 (1979).
5. M. Gell and D. N. Duhl, in *Processing and Properties of Advanced High-Temperature Alloys*, S. Allen et al. (eds.), ASM, Metals Park, OH, 1986, p. 41.
6. S. E. Hughes and R. E. Anderson, Technical Report AFML-TR-79-4146, on USAF Contract F33615-76-C-5136, 1978.
7. J. S. Erickson, W. A. Owczarski, and P. M. Curran, *Met. Prog.*, **99**, 58 (1971).
8. D. N. Duhl and C. P. Sullivan, *J. Met.*, **23**, 38, (1971).
9. S. M. Copley, A. F. Giamei, S. M. Johnson, and M. F. Hornbecker, *Met. Trans.*, **1**, 2193 (1970).
10. D. H. Maxwell, F. J. Baldwin, and J. F. Radavich, *Metallurg. Met. Form.*, **42**, 332 (1975).
11. J. J. Jackson, M. J. Donachie, R. J. Henricks, and M. Gell, *Met. Trans. A*, **8A**, 1615 (1977).
12. D. A. Woodford, *Met. Trans. A*, **12A**, 299 (1981).
13. C. S. Giggins and F. S. Pettit, *J. Electr. Soc.*, **118**, 1782 (1971).
14. C. S. Wukusick and J. F. Collins, *Mater. Res. Stand.*, **4**, 637 (1964).
15. G. W. Goward, in *Proceedings of the Symposium on Properties of High Temperature Alloys with Emphasis on Environmental Effects*, Z. A. Foroulis and F. S. Pettit (eds.), Electrochemical Society, Princeton, NJ, 1976, p. 806.
16. J. F. Nye, *Physical Properties of Crystals*, Oxford University Press, London, 1957.
17. M. McLean, *Directionally Solidified Materials for High Temperature Service*, The Metals Society, London, 1983.
18. C. H. Wells, *Trans. ASM*, **60**, 270 (1967).
19. B. H. Kear and B. J. Piearcey, *Trans. Met. Soc. AIME*, **239**, 1209 (1967).
20. S. M. Copley and B. H. Kear, *Trans. Met. Soc. AIME*, **239**, 984 (1967).
21. D. M. Shah and D. N. Duhl, in *Superalloys 1984, Proceedings of the Fifth International Symposium on Superalloys*, M. Gell et al. (eds.), TMS-AIME, Warrendale, PA, 1984, p. 105.
22. G. R. Leverant and D. N. Duhl, *Met. Trans.*, **2**, 907 (1971).
23. E. G. Richards, *J. Inst. Met.*, **96**, 365 (1968).
24. T. B. Gibbons, *Met. Technol.*, **8**, 472 (1981).
25. D. A. Spera, M. A. Howes, and P. T. Bizon, "Thermal Fatigue Resistance of 15 High Temperature Alloys Determined by the Fluidized-Bed Technique," NASA TMX-52975, Cleveland, OH 1971.
26. S. W. Hopkins, in *Thermal Fatigue of Materials and Components*, D. A. Spera and D. F. Mowbray (eds.), ASTM STP 612, American Society for Testing and Materials, Philadelphia, PA, 1976, p. 157.

PART THREE

BEHAVIOR OF ALLOY
SYSTEMS

Chapter 8

Prediction of Phase Composition

CHESTER T. SIMS

Rensselaer Polytechnic Institute, Troy, New York

In the early 1960s a potentially dangerous phenomenon was discovered in the nickel alloy IN-100 by Wlodek[1] and by Ross.[2] A hard compound called σ phase was found to precipitate, which caused significant degradation in creep rupture properties. Subsequently, similar effects were found in a number of other superalloys. An intense effort was started to understand this effect and to attempt to control it, usually through alloy composition modification, as was done successfully for IN-100. Then, by application of solid-state physics in the form of electron concentration theory, methods evolved to predict whether alloys would be "safe" or not.

This chapter first discusses some of the phase relationships in the occurrence of hard compounds such as σ, the appearance and morphology of σ, and some of its effects on mechanical properties. Subsequently, the development of the major commercially utilized calculations related to controlling the occurrence of σ is covered, including a review of the underlying reasons. Some of the problems concomitant with phase computation and calculation are discussed, and new concepts in phase computation systems are reviewed.

PHASE RELATIONS

Illustrations in earlier chapters and in Appendix A show the quaternary alloy systems from which all austenitic high alloys of these types form. They amply illustrate that in quaternary phase space superalloys have their continuous matrix composition in the face-centered-cubic (FCC) austenitic (γ') field. This field is widely separated from the other major single-phase terminal volumes of the quaternaries, the body-centered cubic (BCC) field. Between these two lies a band of numerous small single-phase volumes identified as σ, μ, R, and the like. These are hard intermetallic compounds unfit as ductile alloy bases and not yet generally found useful as strengthening phases. In superalloys their formation is avoided at all costs.

Also, related to phase space location and the presence of impurity elements such as silicon, the well-known Laves phase may precipitate. The chemical and mechanical effects of Laves are assumed here to be similar with that of the quaternary-related hard phases previously described, although its morphology and precipitation locale vary somewhat. It is occasionally present in superalloys, but it is usually considered undesirable.

The intermediate hard phases in the quaternary systems of interest are electron-bonded compounds. Characteristically, one or more of the elements demonstrate an electropositive character, (i.e., -chromium, molybdenum, tungsten).

In the iron–chromium system σ phase forms around the composition CrFe. In superalloys a typical composition for a σ might be $(Cr,Mo)_x(Ni,Co)_y$ where x and y can vary from 1 to 7 but most often are approximately equal. The μ phase occurs at somewhat similar compositions, but molybdenum and cobalt dominate; iron also stabilizes μ. Sigma forms from elements of nearly equal atom size, while μ forms from elements of greater size differences. The Laves phase, with an A_2B formula, contains atoms bonded by "size" factors; examples are Co_2Mo or Co_2Ta.

The lattice structures of all of these phases are complex, lack multiple slip systems, and thus are nondeformable. For instance, σ contains 30 atoms per body-centered tetragonal (BCT) unit cell, with $c/a \approx 0.52$ (c/a = unit cell–size ratio); μ contains 13 atoms per rhombohedral unit cell. Laves possesses a Zr_2Mg structure. The cell structures of σ, μ, and Laves are characterized by close-packed layers of atoms with the layers separated by relatively large interatomic distances. If one observes a model of σ, the layers of close-packed atoms are displaced from one another by sandwiched larger atoms, developing a characteristic "topology." Beattie and Hagel[3] have characterized all of these compounds as possessing a *topologically close-packed* structure; thus, these phases are termed TCP phases. Conversely, A_3B-type γ' compounds, close packed in all directions, are characterized as *geometrically close-packed* (GCP) phases.

Interestingly, σ is structurally related closely to the common $M_{23}C_6$ carbide formed in many nickel-base alloys. If one were to remove all carbon atoms from an $M_{23}C_6$ lattice, only a slight shift in atom-to-atom dimensions would yield the σ structure. Since $M_{23}C_6$ (often with the composition $Cr_{21}Mo_2C_6$) contains many of the chromium and molybdenum atoms needed for σ formation, this relationship becomes important. Considerable lattice coherency exists between σ and $M_{23}C_6$, so that σ often nucleates on $M_{23}C_6$ (Figs. 1a, 1b). Further, it has been observed that the decarburization of a σ-prone alloys containing $M_{23}C_6$ can lead to the formation of σ on the $M_{23}C_6$ sites, and carburization to its elimination.

The same reasoning can be applied to the μ phase. Mu has a similar close structural relationship with the carbide M_6C. In alloys that develop M_6C, excessive

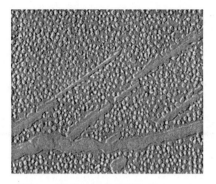

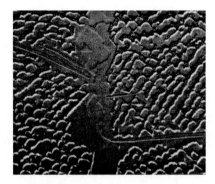

a. Sigma Plates in U-500; Note Coherency with $M_{23}C_6$ [4] 5,625X

b. Sigma Plates in N-115; Note Coherency with $M_{23}C_6$ [5] 5,625X

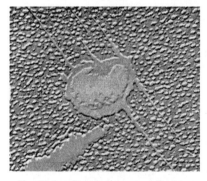

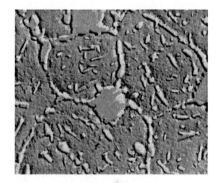

c. Mu Plates Developing near Degenerating MC Particle in AF-1753 4225X

d. Laves Plates in S-590, with M_6C and CbC [6] 5,625X

Fig. 1. Microstructures of TCP phases formed in austenitic alloys.

concentrations of molybdenum and chromium can be expected to lead to formation of μ rather than σ, although the rule is not fast. Thus, alloys that form $M_{23}C_6$ can form σ (such as U-700, René-80 and IN-100), while alloys that form M_6C (such as M-252 and René 41) tend to form the μ phase. Conversely, no similar relationship has yet been noted between the Laves phase and a carbide of commonality in nickel- or cobalt-base alloys.

MICROSTRUCTURES OF TCP PHASES

Identification of TCP phases in austenitic alloys was originally made only by metal-lographic techniques. Visual identification is still a vital tool (Figs. 1 and 2). In almost all cases in nickel alloys TCP phases occur in a platelike morphology, which, of course, appears as needles on a single-plane microstructure. This platelike structure

a. Blocky sigma in U-500 grain boundary[6] 5,300X

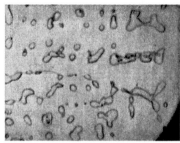

b. Blocky sigma in Disco-3[7] 265X

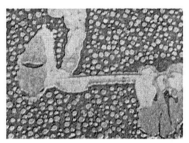

c. Sigma plate with $M_{23}C_6$ at U-500 grain boundary. 5,300X

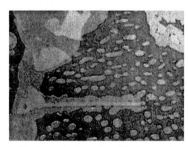

d. Blocky grain boundary $M_{23}C_6$ growing to plate in U500[4] 5,300X

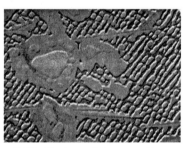

e. N-115 with sigma plates or needles (above) and $M_{23}C_6$ plates (lower center) 3710X

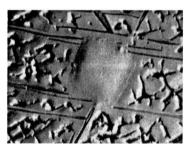

f. Face and edge view of sigma plates IN-100. 1000 hrs. at 1500 F (815C) 2650X[8]

Fig. 2. Microstructure details of the occurrence of σ phase in austenitic superalloys.

is a critical property since it usually affects mechanical properties negatively. Several examples of platelike phases are illustrated in Fig. 1, including σ, μ, and Laves.

TCP phases usually generate initially near grain boundaries where concentrations of elements critical to their formation, such as chromium and molybdenum, often concentrate. Layers of σ phase may also be found as a substrate under coatings of superalloys. However, the concentration of highly reactive elements is restricted to the coating alloy interlayer, and the σ phase does not spread and permeate the structure.

Nucleation of σ phase often occurs on carbide particles. Figure 1 shows that $M_{23}C_6$ particles at the grain boundary have clearly provided a nucleus. Several microstructures from Fig. 1 also illustrate another point, which is that σ grows principally from the alloy matrix. The indigenous γ' present is more stable and is not contributing to the chemistry of σ. In growing through the alloy, the σ plates have displaced the residual γ' but have not created it.

TCP and Laves phases are not the only phases that crystallize in platelike morphologies in nickel austenitic alloys. It is also common to see precipitation of η (eta), nitrides, and $M_{23}C_6$ in plates or semiplatelike fashion. However, the TCP phase will appear as very straight-sided plates (Fig. 2b). When polished and observed under light, it possesses no characteristic color; nitrides appear pinkish.

$M_{23}C_6$ often precipitates along slip lines, stacking faults, or twin boundaries in a linear fashion and has been mistaken for σ. However, under high magnification the "plates" will be shown composed of short individually aligned particles or rods. Positive identification of σ can be made by phase separation and identification techniques.

Some of the mechanisms by which the σ phase may disrupt or destroy alloy properties are discussed subsequently. Certainly its hardness and platelike morphology in nickel alloys has a significant effect on fracture initiation and propagation; however, it has been conjectured that massive σ with a blocky structure might not affect properties negatively.

Some massive blocky TCP phases have been observed. Figure 2b shows blocky σ in an experimental cobalt alloy. Figures 2c and 2d illustrate additional grain boundary phenomena in U-500 of obvious interest; phases of MC, $M_{23}C_6$, γ, γ', and σ were identified in the heat illustrated. Figure 2c shows a σ plate imbedded in $M_{23}C_6$, with which it forms a coherent interface, and in grain boundary γ'. Figure 2d is an enigma; it probably shows a grain boundary $M_{23}C_6$ particle growing a rare plate. However, if the plate is σ, the blocky grain boundary particles are as well. It was not possible to confirm σ by stain etching in this sample.

Figure 2e shows *both* σ plates and $M_{23}C_6$ plates together in N-115 but etching in a completely different fashion. The σ is above, and the $M_{23}C_6$ is below. Further, a few blocky particles may also be σ, based on the etching response. Attempts to create blocky σ in nickel-base alloys for strengthening have not yet been successful, although creation of platelike σ for strengthening in austenitic alloys is a possibility.

EFFECT ON MECHANICAL PROPERTIES

The effect of TCP σ plates in moderate to large quantities on the mechanical properties of superalloys has been almost uniformly disastrous.[9] In fact, it was creep-weakening effects that focused attention on them.

Rupture Properties

The most feared effect is that on creep rupture at elevated temperature. Examples, as originally reported, are shown in Fig. 3. Such weakening by σ is usually accompanied

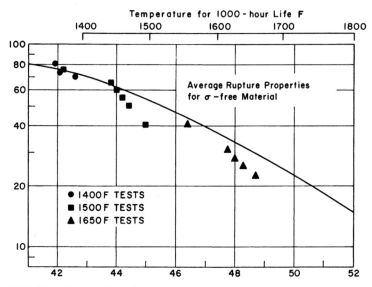

Fig. 3. Original data showing effect of excessive σ formed during rupture testing on properties of IN-100 (after ref. 2.)

by a reduction in ductility as well, a sign of rather severe degradation of alloy properties, not merely the loss of a significant strengthening mechanism. Rupture strength degradation has been observed from σ in IN-100, U-700, N-115, and Astroloy when composition was not carefully controlled.

The μ phase, however, has not been observed to generate such a totally severe effect. It has degraded properties in AF-1753 (Fig. 1c), but long-time serviced M-252, which developed the μ phase, did not show rupture weakening.[10] René-41 also has been utilized with μ present, but IN-625 clearly shows strength retention properties when μ (and Laves) are present in the range 1000–1200 °F.

Dreshfield and Ashbrook[11] have conducted a detailed study of the effects of σ on the mechanical properties of IN-100 by increasing aluminum and titanium contents to create several levels of σ-prone alloy. Microstructures illustrating the increasing amounts of σ (Fig. 4) show the observed reduction in rupture properties of IN-100 after unstressed exposure for 2500 h. These authors confirmed a general reduction in ductility with increasing σ content, evidence that a significant strength-weakening mechanism was operative.

Tensile Properties

The effect of copious amounts of the σ phase on low-temperature tensile properties, particularly ductility, for *ferritic* materials such as 446SS are well known. Although the tensile strength properties may be lowered significantly, ductility is often lowered to zero.

Similar effects can be expected for austenitic nickel- and cobalt-base alloys if large amounts of σ are formed at high temperatures.[1] A study of σ in IN-100 (Table 1) showed that tensile strength and yield strength reductions were moderate at room temperature and negligible at high temperatures. However, the sharp reductions of ductility found are certainly reminiscent of the properties found for highly sigmatized ferritic materials.

Temperature, Structure, and Stress

The occurrence of σ as a function of temperature for cast or wrought structure is shown in Fig. 5. Sigma forms earlier in cast IN-100 due to greater inhomogeneity. Figure 5 also confirms the most critical σ-forming temperatures to be approximately 1550–1700 °F (840–925 °C).

A number of observations have been made of the effect of stress on σ formation, as would be expected for precipitation of a phase during thermal stabilization of an alloy structure. Wlodek[1] stated that studies of σ generation in IN-100 showed that stress did induce precipitation. Mihalisin et al.[12] found that prone INCO-713C stress rupture bars showed more σ in the reduced section than the unstressed threaded section. Observations in the author's laboratory have shown similar effects.

However, other studies are less clear. Collins[13] reported a complex effect in IN-100 and U-700; stress promoted σ in IN-100 at 1500 °F (815 °C) but had no effect

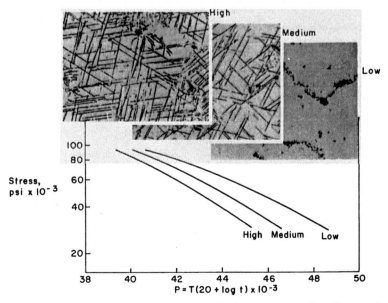

Fig. 4. Data showing effect of preformed σ in IN-100 rupture properties with respective structures illustrated. Properties determined at 1360–1675 °F (740–910 °C) following 2500 h exposure at 1550 °F (840 °C). (Adapted from ref. 11.)

at 1600 °F (870 °C), suggesting that an optimum stress was necessary for formation. Support for this view has been provided by Dreshfield and Ashbrook.[11]

Weakening Mechanisms

There are several explanations for how TCP phases cause rupture weakening:

1. *Fracture site and path.* Brittle σ platelets originating at grain boundary carbides offer an ideal morphology for initiation and propagation of fracture, zigzagging from one plate to another. Ross[9] has termed this "intersigmatic" fracture. Figure 6 shows fractures of sigmatized and nonsigmatized superalloy bars. This is certainly the major mechanism for σ, but μ, when deleterious, appears more to relate to the next issue.

2. *Compositional weakening.* TCP phases contain large quantities of refractory elements. The average amount of chromium, molybdenum, or tungsten in σ or μ is approximately 50 a/o. These elements, scavenged from the alloy matrix, would thus be expected to lower solution strengthening effects. The μ phase has caused some alloys to become weaker but more ductile, which is observed in certain SX alloys. Blocky σ in UMCo-50 is considered a strengthener.

   Also, σ or another TCP phase would be expected to change the chemical balance in the host alloy, so that factors such as γ–γ′ lattice mismatch would change, perhaps causing a reduction in strength.[11]

Table 1. Effect of Unstressed Thermal Exposure on Tensile Properties of IN-100 with Varying σ-Forming Tendencies[a]

Condition	σ-Forming Tendency	UTS (psi)	0.2% YS (psi)	Elongation (%)	RA (%)
		Room Temperature			
As cast	Low	141.5	101.7	12.0	11.0
Exposed 2500 h	Low	131.0	106.5	6.5	5.5
at 1500 °F	Medium	118.0	101.3	1.5	2.0
	High	117.1	96.4	1.0	2.0
		1400 °F			
As cast	Low	144.2	118.3	8.0	8.0
Exposed 2500 h	Low	137.0	112.0	9.7	9.3
at 1500 °F	Medium	141.8	119.4	2.0	2.0
	High	140.1	121.0	2.0	2.0

[a] From ref. 11. Abbreviations: UTS, ultimate tensile strength; YS, yield strength; RA, reduction of area.

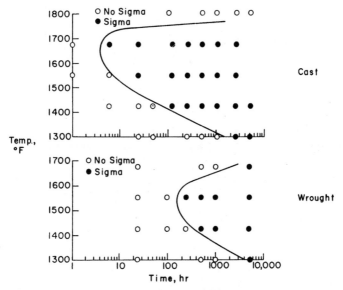

Fig. 5. Occurrence of σ in fine-grained unstressed IN-100. (Adapted from ref. 11.)

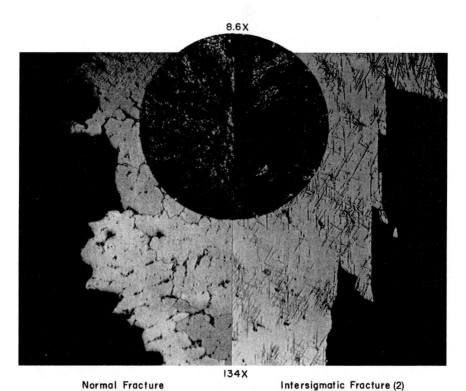

Fig. 6. Effect of σ on rupture fracture of Ni-base superalloys. Shiny areas at 1:00 o'clock of inset are light reflections from sigmatized fracture facets.

3. *Structural effect*. It has been suggested that σ divides the alloy grains into finer sections with a resulting reduction in creep strength à la Petch.[11] However, weakening also occurs where the plates do not completely traverse the original grains, casting some doubt on this idea.

ELECTRON HOLE NUMBERS

Computerized calculations are now used to predict and control the precipitation of TCP compounds in austenitic alloys. This primary technique is called PHACOMP, an acronym for phase computation.[14] The calculations are based on electron hole (previously called electron vacancy) theory and, of course, also can be viewed as an effort to calculate the phase boundary for the TCP phase as it actually occurs in a complex alloy system.

In 1938 Linus Pauling[15] studied the magnetic properties of chromium, manganese, iron, cobalt, and nickel, the first long period of transition elements (see inside cover). These elements have similar atomic diameter; their $3d$ electron shell is only partially filled. Pauling reasoned that the five d orbitals of each spin divided into 2.56 bonding orbitals and 2.44 nonbonding orbitals. The 2.56 bonding orbitals then hybridized with the p and s orbitals to create metallic bonding.

Pauling also assumed that chromium utilized 5.78 hybrid (spd) electrons in bonding. Six can be obtained from the $3d$ and $3s$ orbitals, leaving a remainder $(6.00 - 5.78)$ of 0.22 electrons; these add to the $3d$ nonbonding orbitals. Conversely, there are $2 \times 2.44 = 4.88$ holes (vacancies) in the d shell, or $4.88 - 0.22 = 4.66$ net holes for chromium, the electron "hole" number.

Similar assumptions were made for nickel, cobalt, iron, and manganese, where Pauling assumed that the number of spd bonding electrons remained at approximately 5.78. This would cause an increase in nonbonding electrons to 1.22 for manganese, 2.22 for iron, 3.22 for cobalt, and 4.22 for nickel. Pauling[15] then suggested the following density-of-states diagrams for the $3d$ orbitals, where $N(E)$ is the electron energy state density:

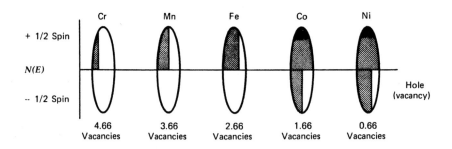

As can be seen for nickel and cobalt, the electron holes are matched by available unpaired electrons, but for chromium, manganese, and iron all the nonbonding electrons are unpaired.

Since the magnetic moment at absolute zero should be equal to the average number of unpaired electrons, Pauling's predictions suggest that the moment for nickel should be 0.66, for cobalt 1.66, and for iron 2.22. Measurements have given 0.61, 1.71, and 2.22, respectively.

In the past it has generally been the practice to extend the elemental electron hole numbers N_v to transition elements other than those in the first long period by assuming that the numbers are constant within a periodic group; for instance, Mo and W = Cr at N_v = 4.66. If so, the elements that interest us for superalloy calculations would possess electron hole numbers as shown in Table 2.

Some doubt on this reasoning has been case by Hume-Rothery,[16] who has shown that the compressibility of chromium is 0.6, that of molybdenum is 0.36, and that of tungsten is 0.3. This means stronger bonding in tungsten than in molybdenum than in chromium, indicating that the effective valence of chromium should be less than molybdenum or tungsten, and thus N_v should be greater for molybdenum and tungsten than for chromium. Insufficient data are available in the literature to establish whether this is true; however, the empirical calculations of the PHACOMP system, which usually assumes that all numbers within a group *are* constant, lead to occasional problems, further suggesting that N_v may, indeed, vary within a periodic group (discussed subsequently).

The effect of temperature on N_v was considered by Kittel,[17] who showed that although above the Curie temperature thermal effects redistributed the holes between positive or negative spins to yield a zero net magnetic moment (loss of magnetism), the number of $3d$ electron holes was unchanged. Thus, it should be possible to use N_v unchanged over a range of temperatures.

CORRELATION OF ELECTRON HOLE THEORY AND ALLOY PHASES

In the early 1950s Beck et al.[18-20] were attempting to correlate the occurrence of σ phase in binary and ternary alloy systems. Since it had been suggested previously[21] that σ was an electron compound, it was realized that it might be possible to characterize the σ range by calculating an average electron hole number $\overline{N}_v$ for the

Table 2. Possible Electron Hole Numbers for Transition Elements[a]

	Group VIA	Group VIIA	Group VIIIA		
First long period	Cr (4.66)	Mn (3.66)	Fe, 2.22	Co, 1.71	Ni, 0.61
Second long period	Mo (4.66)	Tc (3.66)	Ru (2.66)	Rh (1.66)	Pd (0.66)
Third long period	W (4.66)	Re (3.66)	Os (2.66)	Ir (1.66)	Pt (0.66)

[a] Numbers in parentheses are extrapolations and estimates only from Pauling's original predictions (from ref. 15). Fe, Co, and Ni are determined by magnetic mismatch measurements.

solid-solution matrix from its components. The systems Cr–Co–Fe, Cr–Co–Mo, and Cr–Ni–Mo were studied by use of the equation

$$\bar{N}_v = 4.66(\text{Cr} + \text{Mo}) + 3.66(\text{Mn}) + 2.66(\text{Fe}) + 1.71(\text{Co}) + 0.61(\text{Ni})$$
(1)

where each coefficient is the N_v characteristic of each element and the parenthesised element symbolizes the atom percent of each present.

It was found that the range of a composition could be characterized by N_v values of 3.16–3.66. In a further empirical change an N_v of 5.6 was used for molybdenum, which narrowed the range of N_v values to 3.35–3.68. In this work an important first step had been made; the composition range of a σ phase was being calculated using N_v. Other important early contributors were Bloom and Grant[22] and Shoemaker,[23] who suggested that electron density determines the Brillouin zones for the σ structure.

A second significant step was taken later; a large series of alloys with widely varying known σ content were evaluated for $\bar{N}_v$.[5] Despite considerable scatter, a linear relation was observed. The σ boundary was established at approximately $\bar{N}_v = 2.50$; "pure" σ was at ~3.4.

Subsequently, Beck et al.[18–20] studied a number of binary and ternary systems in attempting to establish more exact values for the elemental N_v numbers. Receiving the most attention were manganese, iron, vanadium, and molybdenum since they gave the most difficulty. For instance, study of the V–Ni–Co ternary showed that vanadium could be assigned $N_v = 5.66$, as expected from its periodic position. However, in combination with other electropositive elements, $N_v = 4.88$ for vanadium. It seemed that N_v for vanadium was undeniably composition dependent.

From study of the systems Mo–Fe–Co, Mo–Fe–Ni, and Mo–Ni–Co, it appeared to Das et al.[19] that $N_v = 4.66$ for molybdenum as originally hypothesized. However, molybdenum appeared to contribute at least 5.6 electron holes and more if more chromium than molybdenum were present. In fact, use of values to $N_v \approx 10.00$ have been suggested with some justification based on the appearance of phase diagrams. Thus, N_v for molybdenum is also dependent on composition; one would expect that tungsten should follow suit.

From this study of relatively simple systems it seems clear that N_v for elements in the second and third long periods can be sharply compositional dependent, and N_v may not be the same as for the first period. This has inhibited the use of electron hole calculations to estimate phase diagrams in alloy systems not determined experimentally.

ELECTRON HOLE CALCULATIONS IN SUPERALLOYS (PHACOMP)

The electron hole work described previously was principally of fundamental interest; it is not known by this writer that any attempt was made to apply these ideas to industrial alloys until the mid-1960s. At this time it was realized by workers at Special Metals Corporation and General Electric that it might be possible to utilize electron hole theories to calculate the occurrence of unwanted TCP phases such as σ, μ, or Laves in nickel-base superalloys. Boesch and Slaney[24] published first and are credited for this discovery. Woodyatt et al.[5] independently developed the idea and published some months later, leading to the first industrial specification and the title PHACOMP.

However, prior to discussing the specific calculations accomplished, several physical factors related to phase diagrams should be clarified. First, the work of Beck et al.[18-20] had as objective the prediction of an N_v for pure σ phase. However, the interest of superalloy metallurgists is to predict the very first precipitation of σ, that is, the $\gamma/\gamma + \sigma$ boundary.

Further, in order to apply calculations to the $\gamma/\gamma + \sigma$ boundary, it is helpful for that boundary to lie parallel, or approximately parallel, to the $\gamma + \sigma/\sigma$ boundary, or at least to the centerline of the σ field. Fortunately, this is usually the case. Also, if the $\gamma/\gamma + \sigma$ phase boundary line does not follow a constant electron hole value identifying the occurrence of σ with a simple function $\overline{N}_v$ phase boundaries are termed iso–electron hole lines or iso-PHACOMP lines.

Another consideration is that of homogeneity in the alloy system. Nickel-base alloys are extremely complex. The effects of chemical inhomogeneity on phase presence and reactions on calculations such as those attempted by PHACOMP could be extremely complex and deceptive.

Finally, a practical point. The original "PHACOMP" calculation, found to be practical and applicable to control commercial superalloys, was very quickly incorporated into practice. It spread from melter to melter quickly and has been used now for nearly 20 years. It works well, tending to discourage change. Meanwhile, of course, many thinking superalloys scientists have developed improved systems based on more accurate compositional analysis and advances in application of basic concepts. Their approaches are reviewed later in this chapter, but the system described here is the only one detailed since it is the system utilized in industrial practice.

Source of TCP Phases

Because Beck et al.[18-20] studied essentially "pure" systems without intentional carbides, γ', or borides present, they were nicely restricted to precipitation into the alloy matrix of an electron compound. In nickel- and cobalt-base commercial alloys phases such as carbides, borides, γ', and even other TCP phases exist and must be accounted for in calculation.

In the first approaches to PHACOMP calculations, the important assumption was made that the σ phase or any other TCP phase precipitated directly and only from the austenitic matrix and that it did not, at least during formation, enter into extensive chemical reaction with other phases present such as carbides, borides, or γ'.[24] Metallographic evidence indicates that this assumption is probably valid.

In order to calculate whether σ will occur in a given superalloy, it is essential to determine the composition of the austenite from which the σ will form. First, the composition of all second phases in the alloy is estimated. Next, one must calculate the amounts of elements consumed in these phases, assuming the alloy is fully aged and in service. Then the composition of the residual austenite can be calculated by subtraction of the second phases from the whole alloy chemistry. The residual austenite is a ternary or quaternary system, much like that with which Beck et al. worked.

Phase Relations

A physical fact has materially aided this work. To date, no observation of a "quaternary" phase between metals has been reported. (A quaternary phase is one that would appear in quaternary phase space but uniquely would not extend to any ternary face.) Thus, regardless of the number of elements involved, all phase relations applying to any nickel- or cobalt-base superalloy can be expressed in terms no more complicated than quaternary diagrams. The composition of a superalloy austenite (γ) from which a TCP phase may precipitate can always be expressed in terms of one or another quaternary phase diagram even if such a diagram may have many additional elements in solid solution. Further, it becomes practical and reasonable to relate alloy compositions of interest to the "closest" ternary phase diagrams as well; this is subsequently discussed.

A typical example of a quaternary is the Ni–Co–Cr–Mo quaternary (Chapters 3 and 6). A simple hand calculation using logical compositions for borides, carbides, and γ' present will show that the residual alloy austenitic composition often very loosely approximates a value of 30Ni–30Co–30Cr–10Mo for nickel-base alloys such as the U-700/N-115/R-77/U-500 type, foremost when PHACOMP was developed. This group occurs at or near the center of the ternary Ni–Cr–Co field, elevated toward molybdenum a few percent. Further, a visual observation of the quaternary will show that the objective of any calculation should be to establish if the alloy matrix composition had pierced through the critical $\gamma/\gamma + \sigma$ boundary that is near this area. If not, the alloy is "safe" and not precipitate σ; if the composition exceeds the boundary, the alloy is "unsafe," and σ would be present at equilibrium.

Calculation Assumptions

Using reasoning obvious from the previous discussion, electron hole theory has been applied in phase calculations to predict the occurrence of σ or other TCP phases. The assumptions are:

1. All phases expected in the alloy have formed and the alloy is close to equilibrium.
2. The TCP phase that might precipitate will form entirely from the alloy matrix.
3. The $\overline{N}_v$ will be a linear function of the matrix composition.

PHACOMP Calculations

The calculations are made in two major steps:

1. calculation of the composition of all secondary phases followed by their discard from the whole alloy to find the composition of the residual austenite, and

2. calculation of $\overline{N}_v$ from the austenite composition.

A number of variations first evolved to calculate both the composition of austenite and the resultant $\overline{N}_v$. The systems described here are those devised by Woodyatt et al.,[5] differing from other similar systems only in details such as the type of elemental N_v used (i.e., 1.61 vs. 1.67 for Co) and whether carbides should be included in the calculation or ignored.† The method chosen uses the following rules to calculate the composition of the various phases present.

1. Convert the alloy to atomic percent composition.

2. Calculate the composition of the second phases present as follows:

   (a) One half of the carbon is assumed to form monocarbide in the order preferred, which is TaC, CbC, ZrC, TiC, and VC. The remaining carbon forms $Cr_{23}C_6$ or $C_{21}(Mo,W)_2C_6$ when molybdenum or tungsten are present. Above Mo + ½W = 6.0 wt. %, $(NiCo_2Mo_3)C$ (i.e., M_6C) forms in place of the $M_{23}C_6$.

   (b) Nickel, chromium, titanium, and molybdenum are calculated to form a boride as follows:

$$(Mo_{0.5}Ti_{0.15}Cr_{0.25}Ni_{0.10})_3B_2$$

   (c) The total aluminum titanium, and columbium left following carbide and boride precipitation plus chromium equal to 3% of its original atomic percent present combine with three times as much nickel to form $Ni_3(Al,Ti,Cb,Cr)$, γ'. If there is insufficient nickel, the product will be a mixture of $Ni_3(Al,Ti,Cb,Cr)$ and $Ni(Al,Cb,Ti,Cr)$, β.

3. The phase amounts calculated in item 2 are then summed and subtracted from the whole alloy. The residual composition (all austenite) is then scaled to 100.

4. The new matrix composition is then used to calculate the mean electron hole number by summation:

$$\overline{N} = \sum_{i=1}^{n} m_i(N_v)_i \tag{2}$$

---

† The Boesch and Slaney[24] method first reported ignored carbide precipitation, which is a conservative step and creates a system quite applicable to alloys such as U-500 and U-700. No diligent application of the Boesch–Slaney system has failed to provide σ-free U-500 or René 77.

where $\overline{N}_v$ is the average electron hole number for the alloy, m_i is the atomic fraction of particular element, N_v is the individual electron hole number of particular element, and n is the number of elements in the matrix.

5. The value of $\overline{N}_v$ is then examined critically. The most widely used early technique was to assume simply that if $\overline{N}_v \geq 2.45$–2.50, the alloy is prone to form σ phase or another TCP phase; if $\overline{N}_v < 2.45$–2.50, the alloy was considered safe.

This calculation has been applied to many compositions of nickel-base alloys in both analytical and synthetic study. However, the first General Electric calculations on Udimet 700 generated a set of results that perhaps remain the most classic example of the feasibility of PHACOMP calculations. The calculation model was applied to 12 compositions that were known to form σ in some cases and not in others. The results achieved were those shown in Table 3.

It is obvious from Table 3 that alloys with $N_v \geq 2.50$ formed σ, while those below did not. Subsequent study of the calculation system, revision of elemental N_v numbers, and application of new knowledge, particularly about alloy chemistry, has produced a continuous series of changes, but none revolutionary. In general, in contemporary applications for alloys of the U-700/N-115/R-77 type, which can form σ if not carefully controlled, the breakpoint is considered to be in the range 2.45–2.49.

However, if the alloy forms μ or Laves instead, the TCP phase will form at a lower $\overline{N}_v$ than for σ. Alloys where (Mo + W/2) $\geq 6\%$ (i.e., René 62, AF-1753, René 41, and M-252), containing a nominal amount of iron or cobalt, are inclined

Table 3. Chemistries, Phase Occurrences, and Average Electron Hole Numbers of Individual Heats of Udimet 700

Heat	Composition (wt. %)										$\overline{N}_v$
	C	Al	Ti	Cr	Mo	Co	Fe	B	Zr	Ni	
1	0.05	4.70	3.60	14.45	5.00	19.00	0.00	0.020	0.00	53.17	2.67[a]
2	0.13	4.43	3.49	15.30	5.00	18.60	0.13	0.015	0.02	52.92	2.62[a]
3	0.06	4.49	3.44	15.10	4.95	18.70	0.15	0.014	0.05	53.05	2.62[a]
4	0.06	4.45	3.45	15.20	4.95	18.70	0.00	0.031	0.00	53.16	2.61[a]
5	0.07	4.40	3.43	14.60	5.10	18.00	0.30	0.030	0.05	54.00	2.56[a]
6	0.05	4.30	3.31	15.20	5.00	18.40	0.10	0.030	0.05	53.60	2.55[a]
7	0.06	4.37	3.40	14.60	4.45	17.60	0.00	0.028	0.00	55.49	2.48
8	0.08	4.50	3.47	15.28	4.28	19.17	0.00	0.030	0.00	57.47	2.45
9	0.07	4.20	3.23	14.70	4.70	18.00	0.00	0.030	0.00	55.07	2.44
10	0.05	4.20	3.19	14.50	4.50	17.50	0.10	0.030	0.05	55.90	2.41
11	0.05	4.20	3.19	14.50	4.50	17.50	0.00	0.028	0.00	56.03	2.40
12	0.05	3.91	2.98	14.80	4.45	17.50	0.00	0.030	0.00	56.27	2.32

[a] Sigma-prone heats.

toward Laves or μ, while those with lesser molybdenum or tungsten (U-700, U-500, and N-115) will be σ prone. This rule of thumb for μ or Laves fails if the iron or cobalt becomes very high, that is, over 20 or 30%. Inspection of appropriate phase diagrams will show why these rules can be used.

As more complicated alloys have developed, refinement of the calculations has continued. The N_v numbers have been further revised, and it is practice to study each alloy carefully, establish its number, and specify that number to the melter. For instance, the critical $\overline{N}_v$ value for René 80 is 2.32, and for IN-738 it is 2.38. The PHACOMP number is customized for each alloy in practice.

INDUSTRIAL APPLICATIONS OF PHACOMP

A PHACOMP calculation is complex. It has been computerized. Computer programs built on the principles given previously are now used in engineering specifications and instructions for superalloys to which these calculations have been applied in the free world. Thousands of tons of alloys for application in aircraft engines and industrial gas turbines have been ordered to these specifications. As of 1985 this author knows of no heat purchased to which PHACOMP has been properly applied where a TCP phase was generated in service. However, the newer alloy compositions containing large amounts of tantalum, columbium, and other elements have each required individual attention.

An advantage is that use of PHACOMP in specifications actually loosens the chemical requirements. Previously, alloy melters found it necessary to carefully control the chemistry level of elements that they felt were individually critical to formation of the σ phase, such as chromium or molybdenum. With PHACOMP the tendency of all elements to promote or negate the formation of a TCP phase is considered, depressants as well as promoters. This allows wider ranges of individual elements to be called out in specifications since the $\overline{N}_v$ number is the *net* chemistry control.

A provocative use of these calculations is in the evaluation or development of new alloy compositions. The system tends to eliminate the possibility of creating an alloy that would precipitate a TCP phase unexpectedly. This allows more concentration on other important factors such as rupture strength, tensile properties, and corrosion behavior as a function of the alloy composition. In this regard, PHACOMP has been a significant aid to the advance of nickel alloy development. For instance, Shaw[25] utilized PHACOMP extensively in the development of the alloy IN-939, which is as strong as IN-738 even though it contains 22% chromium.

PROBLEMS IN PHASE CALCULATIONS

Initially, PHACOMP calculations appeared applicable to virtually all types of nickel-base alloys. Subsequently, certain alloys were found not to respond. Some of the problems were basic and should be discussed.

Misplaced Cobalt

INCO-713C is a well-known investment-cast aircraft engine alloy. In 713C σ phase was reported at $\overline{N}_v$ values below 2.20, a considerable deviation from the behavior expected. An interesting analysis of the 713C problem was published by Slaney.[26] He noted that several underlying factors were involved in the development of early phase calculations. These were:

1. The phase analysis for γ' used in all early systems for PHACOMP calculation was the one reported by Decker and Bieber[27] for INCO-713C; it was the only γ' analysis available. INCO-713 is one of the few nickel superalloys that contains no cobalt. Thus, the phase analysis of 713C showed no cobalt in the γ'.
2. The first calculations of PHACOMP by all parties involved were rationalized on alloys of the type U-700. However, U-700 *does* contain considerable (18%) cobalt, as do a majority of nickel alloys.
3. Thus, original calculations concerned with σ in U-700[5,24] were made utilizing a composition of γ that contains no cobalt. Subsequently, γ' in U-700 was found to contain considerable cobalt substituted for nickel, as expected.[28]

When these calculation systems were applied to 713C, the intrinsic error remained; the composition of γ' used in the calculation was now clearly correct for 713C (no cobalt), but the calculations had been developed based on alloys containing cobalt. Accordingly, from a calculative view, 713C was handled improperly, resulting in anomalous σ predictions.

Another way of correcting 713C is to use a new composition for γ'[13], $Ni_{2.95}$ $(Mo + W)_{0.05}(Ti + Al + Ta + Cb + Cr + 0.5V + 0.3Cr)$, as found for the alloy TAZ8B.

Lund et al.[29] studied the effect of cobalt in MAR-M 421. The $\overline{N}_v$ for first σ formation increased with cobalt content, probably related to the stabilizing effect of cobalt in regularizing the $\gamma/\gamma + \sigma$ boundary to parallelism with the location of the σ phase itself (Fig. 7).

Use of Phase Diagrams

Murphy et al.[30] evaluated 713C using a phase diagram approach. In the development of PHACOMP, the alloys studied have usually been those with residual matrix compositions on or close to the Ni–Cr–Co ternary diagram (Fig. 7) where the iso-PHACOMP line (N_v = const) is essentially parallel to the $\gamma/\gamma + \sigma$ boundary. In Fig. 7 it has been drawn directly through a barely σ-free matrix composition of U-700 plotted on the ternary; this corrects for temperature effects, minor element effects, and so on. The iso-PHACOMP line is actually a "real-life" phase boundary and will predict σ for alloys that relate to the Ni–Cr–Co ternary.

However, 713C does not intentionally contain cobalt; its residual matrix does *not* lie on the Ni–Cr–Co ternary but on the Ni–Cr–Mo ternary, on which two

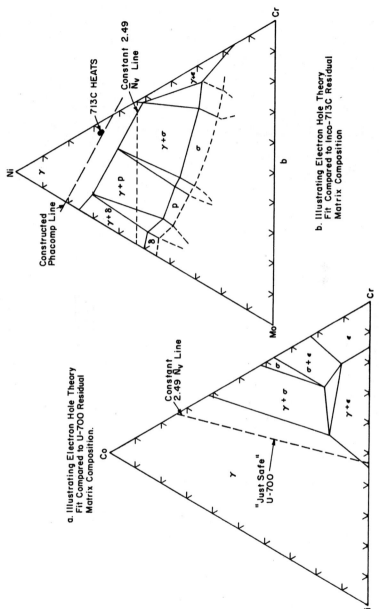

Fig. 7. Ni–Cr ternary phase diagrams with Co and Mo at 2200 °F (1200 °C).[29]

235

Table 4. Developments in Phase Calculations for Superalloys

Topic or Name	Reference	System Basis	Results	Method	Remarks
New PHACOMP	31	Uses both atom size and electronegativity to predict boundaries	Tracks/predicts phase boundaries in austenitic alloys more accurately than PHACOMP	Utilizes DV-Xα method to calculate parameter $\overline{M}^d$ (average energy level of d orbitals of transition metals in alloyed state); correlates both with electronegativity and atomic size)	Appears to replicate phase boundaries well; could be very useful for alloy development; does not account for lesser precipitate phases
Conjugate phase estimation	32	Geometric (analytical geometry) method to interpret phase diagrams	Modeled the 2-phase (γ and γ') region of the Ni–Al–Cr–Ti–W–Mo systems; can describe two phases in commercial alloys	Utilizes an analytic geometry approach to interpret applicable phase diagrams; deduces tie line behavior and then (by computer) calculates γ and γ' by locating the intersection of a tie line and the solvus hypersurface	A technique in development; estimates elementary composition of γ and γ' but not accurately yet
Critical electron vacancy number	33	Calculates separate electron vacancy numbers for each alloy	Replicates applicable phase diagram γ solubility limits with greater accuracy; appears to	Estimate N_v^c from solid solubilities in binary phase diagrams; features temperature correction to consti-	A customized PHACOMP; needs more refinement; cannot separate the IN-713C

Method	Ref.	Basis	Results	Method description	Comments
			predict phase behavior with greater accuracy than PHACOMP	tutional limits; compares a critical N_v^c to $\bar{N}_v$ for the residual matrix	maverick, but recommended for alloy development
Sigma safe	34	Phase diagram basis, with like elements grouped	Predicted 24 safe and σ-prone superalloys correctly; missed μ in M-252; predicted phase boundaries between σ, γ, and γ' correctly	Utilizes phase diagrams to determine degrees supersaturation of σ; the saturation number, Δ, correlates to wt. % σ; Constructs a Ni–Mo–Al–Cr-type quaternary	Appears to improve prediction of σ; μ not understood; Needs exposure through practice and more data
Modification of N_v^c	35	Residual aluminum assumed σ-former	Applied to 19 commercial alloys with mixed results; correctly predicted σ in six varying heats of IN-100	Based on Barrows and Newkirk's N_v^c; residual Al in the γ' is entered as an A-type (σ former) rather than B-type (γ stabilizer)	A relatively small corrective effect, but appears to be an effective improvement for N_v^c when rigorously applied
Direct measurement of N_v^c	36	Measurement of alloy electron energy with electron spectrum chemical analyzer	Correctly predicted σ when present in bulk or formed by argon bombardment.	Samples of alloy are examined in the ESCA by irradiation of Al Kα X-rays; plotted spectra is the base for calculated N_v^c	A straightforward analytical tool; very sensitive

237

matrix compositions for INCO-713C have been plotted that formed σ at very low $\bar{N}_v$ values.

The iso-PHACOMP line on Fig. 7 is *not* parallel to the 2200 °F (1200 °C) γ/γ + σ boundary but intersects it at approximately 30°, and the 2.49 iso-PHACOMP "just-safe" line is not just-safe. Conventional PHACOMP does not predict the Ni–Cr–Mo diagram γ limit with even passable accuracy. Describing the problem graphically results in the solution

$$\bar{N}_v = 0.61\text{Ni} + 4.66\text{Cr} + 1.71\text{Co} + 10\text{Mo} \qquad (3)$$

This has also been proposed by Mihalisin et al.,[12] but the solution is not general for other alloys. We have attempted to define an iso-PHACOMP plane in quaternary space, but the four intercepts on binary legs do not all lie on a plane surface, and all coefficients must be adjusted to obtain a best fit.

An observation to be gained from such studies relates to the method of problem solution. When applying phase calculations to a balky alloy system, the composition should be assessed for its location in ternary or quaternary phase space and then related to rationalized alloys.[30]

Developments in Phase Calculations

There has been considerable subsequent activity in the fascinating field of superalloy phase calculations. Following initial publications and practice, new systems were proposed that were usually improved in some way over PHACOMP. While PHACOMP is built on the basis only of electron hole relationships (i.e., electronegativity), subsequent investigators[31–36] have called into play considerations such as atom size effects, geometric interpretations of phase diagrams, temperature corrections, and determination of supersaturation effects. A summary of the leading works is given in Table 4. Published results show some significant advances, such as greater ability to predict phase boundaries accurately over an apparent wide range of alloys, while PHACOMP must be specified individually for each alloy. Accordingly, it is recommended that metallurgists studying subtle alloy effects or involved in alloy development investigate these methods as potentially more sensitive for phase control.

However, as stated, the rapid industrialization of the original system which led to broad use by alloy developers, melters, and users is so established that no new system, however improved, appears likely to replace PHACOMP for commercial alloy production control.

PHASE CALCULATIONS IN AUSTENITIC IRON AND COBALT ALLOYS

Iron-Base Alloys

Sigma was a problem during World War II in aircraft engines utilizing iron-base austenitics with 15–25% chromium. To eliminate σ, a number of empirical relationships between alloying elements were formulated such as the following by Gow and Harder[37]:

$$\text{Ratio factor} = \frac{\text{percent Cr} - 16(\text{percent C})}{\text{percent Ni}} \qquad (4)$$

A ratio of 1.7 approximated the boundary between σ-free and σ-prone alloys; an alloy with a ratio less than 1.7 would be σ free. Such relationships applied only to a single alloy and are too qualitative to be useful in alloy development. Nonetheless, σ in iron-base alloys can be controlled readily by applying simple procedures such as equation (4).

Cobalt-Base Alloys

Cobalt-base alloys may be different. The residual matrix composition of many cobalt-base alloys is startingly similar to nickel-base alloys. Cobalt alloys also precipitate σ and μ; Laves has been found in L-605.

Attempts to apply PHACOMP to cobalt alloys[38,39] use steps identical to the nickel alloy program discussed earlier except that the γ'-forming step is eliminated. Limited data indicate a critical $\overline{N}_v$ of about 2.70, above which TCP phases can be expected. Correlations of $\overline{N}_v$ are readily made when the alloy matrix exists near the ternary system Ni–Co–Cr and refractory metal additions are moderate. However, refractory metal elements can exert a powerful effect to reduce critical $\overline{N}_v$, and L-605 (15% W) forms Laves at an $\overline{N}_v \simeq 2.48$, considerably below the critical $\overline{N}_v$ general for other cobalt alloys; this may be due to silicon.

Herchenroeder[40] used PHACOMP to develop HS-188 from L-605. In experimental heats aged 200 h at 1600 °F (870 °C) with a variety of $\overline{N}_v$ (calculated by the Boesch–Slaney method; carbides excluded), Laves formed above $\overline{N}_v \simeq 2.66$. HS-188 was then assigned to a chemistry range yielding $\overline{N}_v \simeq 2.42$ by controlling nickel, chromium, and tungsten. While HS-188 can still form a small amount of Laves after thousands of hours at temperature (Chapter 5), use of phase calculations sharply reduced the severe ductility problem suffered by L-605 from massive Laves.

Application of PHACOMP made to 10 cobalt-base alloys containing tantalum[7] showed σ in two of the alloys, both with $\overline{N}_v > 2.75$ (Fig. 2b); all others were less than 2.75. Thus, while PHACOMP is not used in the melt chemistry control of commercial cobalt alloys, it is an important alloy development tool.

REFERENCES

1. S. T. Wlodek, *Trans. ASM*, **57**, 110 (1964).
2. E. W. Ross, "Recent Research on IN-100" AIME Annual Meeting, Dallas, TX, February 1963.
3. H. J. Beattie, Jr. and W. C. Hagel, *Trans. AIME*, **233**(2), 277 (1965).
4. H. J. Murphy, C. T. Sims, and G. R. Heckman, *TMS-AIME*, **239**, 1961 (1967).
5. L. R. Woodyatt, C. T. Sims, and H. J. Beattie, Jr., *TMS-AIME*, **236**, 519 (1966).
6. H. J. Beattie, Jr. and W. C. Hagel, *TMS-AIME*, **221**, 28 (1961); **215**, 973 (1959).
7. N. T. Wagenheim, *Cobalt*, **48**, 129 (1970).

8. R. F. Decker, "Strengthening Mechanisms in Nickel-Base Superalloys," Climax Molybdenum Company Symposium, Zurich, May 5–6, 1969.

9. E. W. Ross, *J. Met*, **19**, 12 (December 1967).

10. H. M. Fox, "The Structure of M-252 After Exposure for 1,000,000 Hours," AIME Annual Meeting, Cleveland, OH, October 1970.

11. R. L. Dreshfield and R. L. Ashbrook, NASA TN D-5185, April 1969; NASA TN D-6015, September 1970.

12. J. R. Mihalisin, C. G. Bieber, and R. T. Grant, *TMS-AIME*, **242**, 2399 (1968).

13. H. E. Collins and C. S. Kortovich, "Research on Microstructural Instability of Nickel-Base Superalloys," Interim Engineering Progress Report 5, U.S.A.F. Contract AF33(615)-5126, TRW, Inc., October 1967.

14. C. T. Sims, *J. Met.*, **18**, 1119 (October 1966).

15. L. Pauling, *Phys. Rev.*, **54**, 899 (1938).

16. W. Hume-Rothery, *The Structure of Metals and Alloys*, The Institute of Metals, London, 1954.

17. C. Kittel, *Introduction to Solid State Physics*, Wiley, New York, 1956, p. 329.

18. P. Greenfield and P. A. Beck, *TMS-AIME*, **200**, 253 (1954).

19. D. K. Das, S. P. Rideout, and P. A. Beck, *TMS-AIME*, **194**, 1071 (1952).

20. P. Greenfield and P. A. Beck, *TMS-AIME*, **206**, 265 (1956).

21. A. H. Sully and T. J. Heal, *Research*, **1**, 228 (1948).

22. D. S. Bloom and N. J. Grant, *TMS-AIME*, **197**, 88 (1953).

23. D. P. Shoemaker, C. B. Shoemaker, and F. C. Wilson, *Acta Crystallogr.*, **10**, 1 (1957).

24. W. J. Boesch and J. S. Slaney, *Met Prog.*, **86**(1), 109 (1964).

25. S. W. K. Shaw, *Met. Prog.*, 47 (March 1979).

26. J. Slaney, "A General Method for the Prediction of Precipitate Compositions," International Symposium on Structural Stability in Superalloys, Seven Springs, PA, September 4–6, 1968.

27. R. F. Decker and C. G. Bieber, "Microstructure of a Cast Age-Hardenable Nickel Chromium Alloy," Symposium on Electron Metallography ASTM Special Technical Publication 262, American Society for Testing and Materials, Philadelphia, PA.

28. R. W. Guard and J. H. Westbrook, *Trans. AIME*, **215**, 807 (1959).

29. C. H. Lund, M. J. Woulds, and J. Hockin, "Cobalt and Sigma: Participant, Spectator, or Referee?" International Symposium on Structural Stability in Superalloys, Seven Springs, PA, September 4–6, 1968.

30. H. J. Murphy, C. T. Sims, and A. M. Beltran, *J. Met.*, **20**, 46 (November 1968).

31. M. Morinaga, N. Yukawa, H. Adachi, and H. Ezaki, TMS-AIME, Warrendale, PA, 1984, p. 525.

32. R. Dreshfield, "Estimation of Conjugate γ and γ' Compositions in Ni-Base Superalloys," NB5 SP-496, January 10–12, 1977.

33. R. Barrows and J. Newkirk, *Met. Trans.*, **3**, 2889 (1972).

34. E. Machlin and J. Shao, "Sigma-Safe; A Phase Diagram Approach to the Sigma Phase Problems," personal communication, 1977.

35. W. Wallace, *Met. Sci.*, **9**, 547 (1975).

36. P. Cowley, *Direct Measurement of The Stability of Alloys*, National Gas Turbine Establishment, Hants, England, personal communication.

37. J. T. Gow and J. E. Harder, *TMS-AIME*, **30**, 855 (1942).

38. C. T. Sims, *J. Met.*, **21**, 27 (December 1969).

39. A. M. Beltran, General Electric Co., private communication.

40. R. B. Herchenroeder, "Haynes Developmental Alloy No. 188 Aging Characteristics," International Symposium on Structural Stability in Superalloys, Seven Springs, PA, September 4–6, 1968.

Chapter 9

Mechanical Behavior

STEPHEN FLOREEN

Knolls Atomic Power Laboratory, Schenectady, New York

Superalloys continue to be active areas for research and development. However, there are notable changes in the recent subject matter. Newer research topics, such as creep crack growth and stress corrosion cracking (SCC) behavior, are becoming popular. Material developments are focusing less on the traditional intermediate temperature range. Instead, there has been considerable activity to develop new materials for both lower temperature and higher temperature applications. Work in these different areas is somewhat fragmented because different groups and organizations have specialized in specific subjects. The main objective of this overview is to compare the mechanical properties within these diverse developments.

This discussion is divided into three different temperature ranges: from ambient to about 1000 °F (538 °C), an intermediate range from 1000 to 1400 °F (538–760 °C), and above 1400 °F (760 °C). This division is arbitrary but useful for identifying temperature regimes where major changes in material behavior and material applications are encountered. Below ~1000 °F (538 °C) creep deformation does not play a major role, and critical properties are usually yield strength plus resistance to corrosion and SCC. It is often assumed that time-dependent changes in properties could be neglected at these temperatures, but newer research indicates that this is not always correct.

The intermediate range has been the classical operating temperature for turbine disks. In this range creep and microstructural stability are important variables, and intergranular crack growth is an endemic concern. Special processing steps produce extremely high strength materials, but resistance to cracking suffers. Consequently, considerable attention is being paid to fracture problems in this temperature range.

Above 1400 °F (760 °C) the major product of interest is turbine blades, and creep strength is the major parameter of interest. Considerable work is being done to develop new and sometimes unique processing steps to achieve significant increases in creep resistance.

One common theme in much of the recent literature is the pervasive effect of the environment on the properties of superalloys. These effects include SCC and hydrogen embrittlement in aqueous or sour-gas environments and creep and fatigue crack growth in oxidizing, sulfidizing, or other gaseous environments at higher temperatures. The degradation caused by the environment will be discussed extensively because of its importance to alloy design and usage and an understanding of the fracture behavior of superalloys.

PROPERTIES OF SUPERALLOYS AT LOWER TEMPERATURES

Superalloys have been used for years in various lower temperature applications where creep strength is not a concern. In the past little was said about these applications, and this temperature range generally suffered from benign neglect as far as studies of superalloy behavior were concerned. Recently, however, considerable activity has arisen in two subject areas.

The first area of interest is the stress corrosion of certain nickel-base superalloys in nuclear reactor applications. The second, somewhat related area is the desire to find high-strength materials for use in deep sour-gas wells. The high concentrations of H_2S in these gas deposits make it impossible to use conventional high-strength steels and prompted extensive studies of current nickel-base alloys and development of a new alloys for these applications. These studies indicate that the cracking susceptibility of superalloys can be considerably affected by the alloy composition and microstructure.

Stress Corrosion in Nuclear Environments

The nuclear environments are either pressurized water reactor (PWR) or boiling water reactor (BWR) waters at temperatures ~600 °F (316 °C). PWR water is typically high-purity water with oxygen contents of several parts per billion and with minor chemical additions to adjust the pH. BWR water is similar except that the oxygen is generally about 10 ppm. Various impurities are sometimes added to the test solutions to simulate upset or contaminated conditions.

The alloys of primary concern are X-750 and, to a lesser extent, 718 and A-286. These alloys were originally developed for elevated-temperature service, and their compositions and heat treatments were optimized primarily on the basis of creep testing. Various structural components such as bolts, springs, pins, and so on, made from these alloys have experienced long-time use in PWR and BWR water in nuclear plants. After varying lengths of service failures were detected in some of these components. Most fractures were detected by inspections during routine plant shutdowns, and plant safety has not been jeopardized by these failures. Nonetheless, the failures caused concern, and extensive efforts are being made to determine the cause of cracking and find remedies for the problem.

Garner and Smith[1] reviewed the service experience of these components and the metallurgical factors that are pertinent. Most failures have been in X-750, but this may simply be the consequence of the much greater use of this alloy. Failures have occurred in both PWR and BWR environments. Failed components were always highly stressed, on the order of the yield strength of the material. Because of the uncertainties in the actual service stresses, it has not been possible to specify a threshold stress below which cracking would not occur. The heat treatment used on the X-750 also has a profound effect. The heat treatments originally developed for creep applications do not produce the best resistance to SCC. A new heat treatment incorporating a high-temperature solution anneal before aging provides much better properties. Tests have also shown that conventional creep-based heat treatments with alloys 718 and A-286 produce fairly poor stress corrosion resistance.[2] Thus, one lesson that has emerged is that heat treatments optimized for creep resistance may not be best for lower temperature properties. This result is not surprising, but it is worth emphasizing. Other types of heat treatments can also be useful. There is evidence that overaging treatments that promote more homogeneous slip are helpful in PWR water.[3]

Since SCC of X-750 in nuclear environments is commonly intergranular, variations in grain boundary chemistry are significant. Grove and Petzold[4] have shown that the susceptibility to cracking in PWR water is related to the phosphorous content of the grain boundaries. Figure 1 shows the incidence of a failure in 680 °F (360

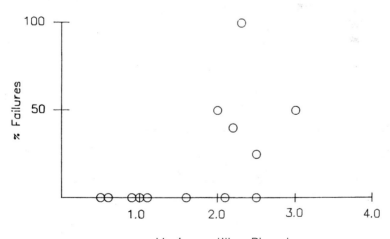

Fig. 1. Effects of grain boundary phosphorous content on the SCC susceptibility of alloy X-750 in primary water.[4] Reprinted with permission from C. A. Grove and L. D. Petzold, *J. Mat. Ener. Sys.*, **7**, no. 2, 147–162 (1985), American Society for Metals, Metals Park, OH.

°C) water versus the maximum concentrations of phosphorus in the boundaries. The data scatters at higher phosphorous levels but shows that cracking becomes significant only at concentrations above about 2%. They also found that in addition to cracking at 680 °F, there was a second regime of even more rapid crack growth centered at temperatures ~200 °F (93 °C), which also could be related to the phosphorous content. Crack growth in both cases was hypothesized to be caused by hydrogen embrittlement resulting from the general corrosion of the X-750 in the water environment. The susceptibility to cracking was believed to be enhanced by the presence of phosphorus in the grain boundaries.

A number of studies using high-purity nickel as a starting material have demonstrated that phosphorus and/or sulfur tend to segregate to the grain boundaries and that such segregation can promote intergranular cracking. Refining additions that either prevent segregation or minimize its consequences should be helpful.

Grain boundary carbides are also important. Considerable work has been done to explore the role of carbides on the stress corrosion susceptibility of alloy 600 in various aqueous environments. This alloy is essentially X-750 with much lower aluminum and titanium levels and does not precipitation harden. The stress corrosion resistance of Inconel 600 in some environments can be improved by heat treating at about 1300 °F (704 °C) to precipitate grain boundary carbides. The reason that carbide precipitation improves the stress corrosion resistance of alloy 600 is unclear. Suggested mechanisms include gettering of deleterious impurities, introduction of compressive residual stresses around the grain boundary carbides to provide innocuous trap sites for hydrogen, and promoting more uniform slip around the grain boundaries. Grove and Petzold[4] found that precipitation of $M_{23}C_6$ particles and/or the absence of MC-type carbides improved the cracking resistance of X-750 at 680 °F (360 °C) but had no effect in the low-temperature regime.

Precipitation of chromium-rich carbides at grain boundaries can cause sensitization because the depletion of chromium in the matrix surrounding the carbides makes these areas susceptible to corrosion attack. It would be advisable to minimize sensitization if superalloys are to be used in aggressive environments. The corrosion data suggests that sensitization is possible after the precipitation hardening treatments commonly used to harden most superalloys. Several remedies are available to minimize sensitization. Age hardening for longer times permits chromium to diffuse back into the chromium-depleted regions surrounding the carbides. Precipitation of $M_{23}C_6$ during aging also can be minimized by alternate heat treatments, tying up the carbon with stronger carbide formers, and/or keeping the carbon level of the alloy sufficiently low. It should be noted that the results with alloys 600 and X-750 in PWR and BWR environments suggest that $M_{23}C_6$ precipitation and sensitization can be beneficial in some cases, and eliminating carbides may not always be desirable.

Sour-Gas Well Alloys

Deep sour-gas wells are intended to tap large deposits of methane at various locations throughout the world. Typically, these deposits are several miles or more below the earth's surface and contain high concentrations of H_2S, CO_2, and brine. Superalloys and cold-worked nickel- and cobalt-base alloys have been evaluated for these applications. While they generally displayed much better H_2S resistance than high-strength steels, they are susceptible to hydrogen embrittlement.

An interesting observation in these tests was the degradation in resistance to hydrogen embrittlement produced by long-time exposures at relatively low temperatures. Portions of the tubing in these wells could be exposed for long periods to temperatures of ~600 °F (316 °C). To evaluate the effects of such exposures, speciments were thermally treated in air, either stressed or unstressed, and their cracking resistance at room temperature in NACE H_2S test solutions was then evaluated.

Figure 2 shows an example of the results from such tests.[5] The alloy is Hastelloy C-276, a solid-solution alloy that was cold rolled 37, 48, or 59% to produce three

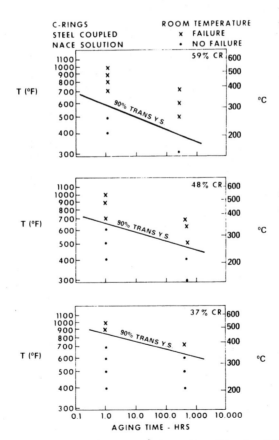

Fig. 2. Effects of thermal exposures on the stress corrosion cracking of Hastelloy C-276 in H_2S solutions. Samples tested at 90% of the transverse yield strength.[5]

strength levels. The specimens were galvanically coupled to steel in the H_2S test because such coupling is likely in service. Exposures of several hundred hours in the 400–700 °F (204–371 °C) range markedly reduced the resistance of the alloy to hydrogen embrittlement in all three cold-worked conditions.

The behavior illustrated in Fig. 2 has been found in many other nickel- and cobalt-base alloys, including cold-worked and precipitation-hardened alloys, as long as the yield strengths were in excess of 100 ksi (689 MPa). The reasons for the degrading effects of the low-temperature exposures are complex and can involve several phenomena depending on the alloy. One common cause is the segregation of phosphorus to crack nucleation sites such as grain boundaries or particle surfaces. Such segregation could impair the resistance to hydrogen embrittlement, and in the case of alloy C-276, heats having extremely low phosphorus contents showed better properties.[6]

Exposures of C-276 at temperatures of 932 °F (500 °C) also produced an A_2B-type long-range ordered structure in the matrix. Ordering promoted planar slip, which was considered harmful, because it promoted local stress concentrations, because sharper slip steps might nucleate cracks, or because hydrogen transport into the lattice by moving dislocations could be facilitated. As shown in Fig. 3, the shortening of failure times in the NACE H_2S environment for C-276 was a result of phosphorous segregation after short aging times and the ordering reaction after aging for about 100 h.[7]

Ordering reactions such as those seen in C-276 appear confined to certain alloys, and other materials do not appear to suffer from this type of problem. However,

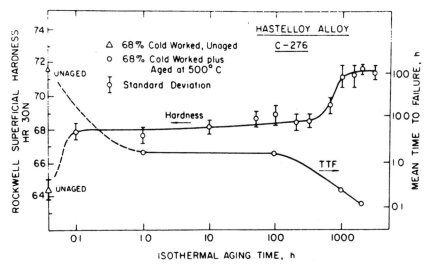

Fig. 3. Effects of aging time at 932 °F (500 °C) on the hardness and stress corrosion failure times in H_2S solutions of Hastelloy C-276.[7] Reprinted with the permission from *Met. Trans. A*, **12A**, 653–658, a publication of the Metallurgical Society, Warrendale, PA.

segregation of phosphorus or other impure elements to interfaces is possible in any alloy because the solubility for such elements decays rather sharply with temperature. Mulford[8] has shown, for example, that the solubility of sulfur in various nickel alloys decreased an order of magnitude at ~900 °F (500 °C) compared with 1300 °F (704 °C). In many alloy systems a 10-fold drop in solubility can produce a comparable enrichment of impurity elements at the grain boundaries.

The question, then, is whether the kinetics permit significant enrichment over the time–temperature conditions of service. With cold-worked alloys dislocation, pipe diffusion, or possibly transport by moving dislocations, could provide rapid enrichment. With volume diffusion Mulford estimated that about 2000 h would be required at 932 °F (500 °C) for the grain boundary concentration of sulfur to reach 40% of the equilibrium value. At lower temperatures volume diffusion would be slower, but sufficient enrichment to adversely affect the properties could still occur if there were a long enough exposure time.

Variations in the γ' lattice mismatch and composition can influence the environmental susceptibility. Thompson and Brooks[9] found that minor changes in alloy composition, which increased the mismatch between the γ' particles and the matrix, improved the resistance of A-286 to hydrogen embrittlement. They proposed that the greater mismatch helped trap hydrogen in innocuous sites around the γ' particles and, thus, minimized hydrogen buildup at crack nucleation sites.

The pitting susceptibility of nickel-base alloys could be related to the preferential dissolution of coarser γ' particles.[10] Bulk single-crystal specimens of various γ' compositions were examined by anodic polarization tests. The corrosion behavior of the alloys containing coarse γ' particles correlated with the behavior of the bulk γ' samples.

Carbides and γ' will often be thermodynamically unstable in corrosive environments. For example, the oxidizing reactions given by

$$Cr_{23}C_6 + 3O_2 \rightarrow 23Cr + 6CO$$

and

$$2Ni_3Al + 3/2O_2 \rightarrow Al_2O_3 + 6Ni$$

are both driven to the right by low oxygen activities. In most instances chromia, alumina, or other protective surface films will prevent such reactions from taking place. However, should the films be chemically or mechanically disrupted, the local oxygen activity can be sufficient to oxidize carbide and γ' particles. Other aggressive environments can produce analogous effects.

As discussed below, various gaseous environments markedly accelerate creep and fatigue crack growth at intermediate temperatures. Although the temperatures and environments are different, the cracking observations made at higher temperatures are often consistent with lower temperature aqueous behavior. Economy et al.[11] showed that stress corrosion of X-750 in steam–hydrogen gas mixtures at 750 °F (399 °C) was consistent with results in PWR water at lower temperatures. More work bridging lower temperature aqueous and high-temperature gaseous cracking behavior would considerably broaden our understanding of environmental damage in these alloys.

PROPERTIES OF SUPERALLOYS AT INTERMEDIATE TEMPERATURES

During the 1970s and early 1980s considerable effort was directed toward increasing the design strengths of turbine disks by raising first the creep and later the fatigue strengths. New processing techniques, such as gas atomizing and hipping, allowed development of alloys with higher γ' contents. Strength was further boosted by warm-working practices that left considerable work hardening in the structure or that promoted recrystallization into extremely fine grain sizes. With such processing, room temperature yield strengths well over 200 ksi (1379 MPa) could be achieved after precipitation hardening.

Special grain shapes, such as serrated grain boundaries or necklace structures consisting of coarser elongated grains surrounded by much finer equiaxed grains, also were developed. Very high creep and fatigue strengths could be achieved in these high-strength materials, but the ductility and crack growth resistance often were impaired. In response to these latter problems, there has been extensive theoretical and practical work to better understand the creep fracture behavior of superalloys at intermediate temperatures.

Creep fracture at intermediate temperatures often begins with the nucleation and growth of cavities on grain boundaries oriented perpendicular to the load axis. Nucleation frequently is seen where slip bands intersect grain boundary particles.

Nucleation of cavities requires the local tensile stress σ_n to be $\sigma_n \geq 2\gamma_s/r$, where γ_s is the interfacial energy and r is the cavity radius. This sensitivity to the interfacial energy creates a dependence of the cavitation rate on variations in the local chemistry at the cavity sites. Yoo and Trinkhaus[12] have calculated, for example, that if a mobile grain boundary impurity in a nickel-base alloy reduced γ_s by 50%, the threshold stress for cavitation would be decreased by a factor of 3 and the cavity nucleation rate would increase by ~20 orders of magnitude. Further complicating the estimates of cavitation rates are observations that cavity nucleation often takes place continuously with creep deformation.

Cavity growth has been somewhat easier to treat. The diffusion-controlled cavity growth is viewed as progressing by grain boundary diffusion of vacancies to boundaries perpendicular to the stress axis. These models usually predict a threshold stress for cavity growth, a growth rate proportional to the stress, and growth kinetics limited by grain boundary diffusion or surface diffusion.

Stress-controlled cavitation growth has also been modeled. Grain boundary shear is important in forming wedge-shaped cracks at grain boundary intersections. Stress-controlled models generally predict a stress dependency for cavity growth similar to the stress dependency for creep and a temperature dependence for growth comparable to that for second-stage creep.

Comparisons of diffusion and stress-controlled models show that diffusion control should dominate at small size cavities and stress control at larger sizes. The transition cavity size is typically several micrometers. Cavities this size are often seen on the fracture surfaces of creep specimens. The overlap in growth mechanisms has led to the consideration of various sequential or coupled mechanisms. One such is illustrated in Fig. 4, where the cavity is surrounded by a zone of diffusion control, which in turn is coupled to a cage of material undergoing stress-controlled creep deformation.[13]

Experiments have shown that under controlled conditions the models can predict experimentally determined cavity sizes accurately. Generally, these experiments employed materials having a reasonably well-defined cavity distribution developed by some sort of processing so that additional cavity nucleation during creep testing could be neglected. Also, it was sometimes necessary to make assumptions about the growth mechanism controlling the behavior. Thus, while some of the models appear correct in detail, it is not yet possible to use them in any general way to predict the cavitation behavior of alloys.

Interest in applying fracture mechanics techniques to elevated-temperature problems arose because of the need to better characterize the loading conditions responsible for creep or fatigue crack growth. Fracture mechanics was originally developed for low-temperature conditions in which time-dependent creep flow in the crack tip region could be neglected. At elevated temperatures this plasticity must be considered in order to identify the proper stress parameter. As shown schematically in Fig. 5, with larger sized specimens of brittle materials, linear elastic fracture mechanics can be used, and the stress intensity factor K is the most suitable parameter.[14] With more ductile materials in which the amount of creep flow around the crack tip is larger, K can no longer be used, and the most suitable parameter appears to be the time-dependent J integral, or c^*, defined as

$$c^* = \frac{-1}{B} \frac{du}{da}$$

CAGE OF CREEPING MATRIX

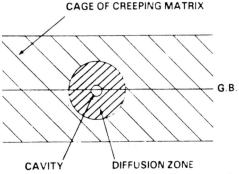

G.B.

CAVITY   DIFFUSION ZONE

Fig. 4. Schematic model of cavity growth involving coupled diffusion-controlled and creep-controlled zones.[13]

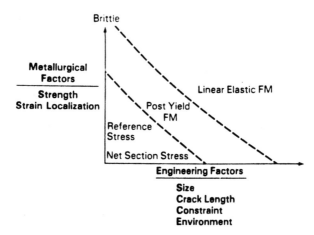

Fig. 5. Effects of metallurgical and engineering factors on the stress parameters describing fracture.[14]

where B is the specimen thickness and du/da is the time-dependent area under the load displacement curve per unit increase in crack length. With very ductile materials in which creep takes place over the entire cross section before crack growth occurs, parameters such as the reference stress or the net section stress should be used.

At present there is no standardized method that prescribes how to select the correct stress parameter to characterize elevated-temperature fractures. Riedel and Rice[15] have suggested a method to distinguish between K and $c*$ by estimating the time to relax the stress in the vicinity of a stationary crack tip. If rapid relaxation takes place, the loading condition is best described by $c*$. Conversely, if relaxation is slow, the creep zone around the crack tip will be small and elastically confined, and K is the correct parameter. Using displacement gauges to estimate the relaxation, one can experimentally determine which parameter to employ. Another experimental method sometimes employed is to test two different types of specimens, for example, edge-notched and center-notched samples, and compare the results in terms of $c*$ and K. The parameter that gives the closest agreement between the two types of specimens is then considered the correct one. Note that the choice of parameters becomes crucial only when trying to predict service behavior or for comparison with other results. If the tests are only for internal comparison of materials, heat treatments, and so on, the choice of parameters is not vital as long as identical specimens are used. The reason for this is because K, $c*$, and the net section stress are proportional to each other for a fixed geometry and crack length.

A number of fracture-mechanics-based studies have been made of creep and fatigue crack growth in superalloys. The results generally suggest K can be used to characterize the creep behavior of the high-strength alloys. For stainless steels and heat-resistant alloys that are not age hardened, $c*$ appears to be a more appropriate parameter for creep loading. For fatigue loading K or ΔK appear valid for both types of materials.

Creep crack growth at intermediate temperatures seems to be completely inter-granular in age-hardened alloys. Threshold K values below which no crack growth occurs are frequently observed. These thresholds can be quite low. Figure 6 compares K values for alloy 718 for fracture in conventional rising load fracture toughness tests with the threshold K values observed under static load tests.[16] Note the order of magnitude drop and the K values under 20 ksi $\sqrt{\text{in}}$. (22 MPa $\sqrt{\text{m}}$) in the threshold load tests.

At K values above the threshold the creep crack growth rates a frequently can be expressed:

$$a = A K^m$$

where A and m are constants. Values for m are generally from 2 to 7 but do not correlate well with stress exponents determined from creep tests. Crack growth rates increase with increasing temperature, and sometimes Arrhenius plots yield well-defined apparent activation energies. However, the measured values vary widely with different alloys and test environments, and no consistent rate-controlling mech-anism can be deduced from these results.

Microstructural parameters significantly influence the crack growth behavior. Fine grain size is helpful for retarding fatigue crack initiation. Coarser grain sizes,

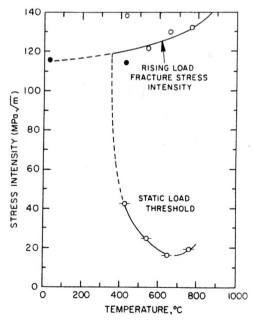

Fig. 6. Stress intensity values for rising load and static load fractures of alloy 718 at various temperatures.[16] Reprinted with permission from the American Society of Mech. Engineers.

however, are better for retarding creep or fatigue crack growth. Thus, there is frequently an optimum grain size that provides the best compromise between these conflicting requirements. Grain shape is also important; serrated grain boundaries or the more irregularly shaped boundary of cast structures are superior to planar boundaries. The age-hardening heat treatments are also important. Overaging is observed to be beneficial, but whether this is a result of slip homogenization or perhaps of local alteration of the grain boundary microstructure is not certain. The effects of grain boundary carbides are not well explored. Some testing of X-750 showed that the best carbide morphology for retarding creep crack growth varied with the test environment and suggested that environmental factors rather than inherent plastic flow variables were dominating the carbide effects.[17]

The environment strongly influences the creep and fatigue behavior. The earlier work using smooth samples is difficult to summarize because a variety of different, sometimes contradictory, results were obtained. For example, creep rupture lives may be shorter in air versus vacuum because of oxygen embrittlement of the grain boundaries, or they may be longer in air because internal oxidation retards creep.

Crack growth studies in superalloys have shown that air, specifically oxygen, can markedly increase the rates of creep or fatigue cracking compared with tests in vacuum or inert environments. Figure 7 illustrates this change for alloy 718 at 1200 °F (649 °C), showing that the creep crack growth rate in air was about 100 times greater than in helium.[18] Sulfidizing environments are still more ferocious; even minor amounts of H_2S or SO_2 added to helium can produce very rapid crack growth. Adding salt to sulfidizing environments, in the manner of hot-corrosion mixtures, further promotes sulfidation damage. Although sulfur forms a low-melting-point eutectic with nickel, there is no evidence to suggest that liquid film formation is necessary for rapid cracking. Carburizing environments are less aggressive, producing about threefold increases in crack growth rates over that in helium. Attempts to relate the cracking rates to the thermodynamic activities of the environments have met with very limited success, probably because of the many complicating, inhibiting, or catalytic effects, stress effects, and so on, that can alter the chemical reactions.

A number of mechanisms have been suggested to account for environmental effects. These can be broken down into three general categories.[19]

Type A. Cracking due to adsorption of an active element at the crack tip surface

Type B. Cracking due to the formation of a corrosion product at the crack tip

Type C. Cracking as a result of diffusion of the active element into grain boundaries ahead of the crack tip

While type A mechanisms have been proposed several times, there is no direct evidence that they play a major role in high-temperature crack growth. However, type B mechanisms have been widely observed. Numerous investigators have found oxide wedges or spikes at surface grain boundaries after creep exposures in air. Cracks will nucleate at these particles, and machining off the particles can significantly improve the subsequent life. Mechanical loading can considerably accelerate the rate of oxidation. Measurements by Skelton and Bucklow[20] of the effects of cyclic

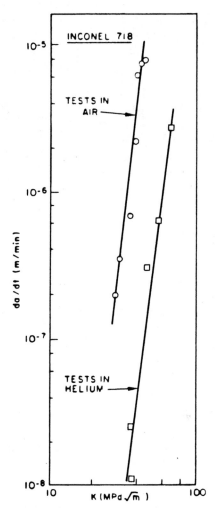

Fig. 7. Creep crack growth rates of alloy 718 versus stress intensity at 1202 °F (650 °C).[18]

loading on the rate of oxidation of a chromium–molybdenum–vanadium steel showed that at higher strain ranges the oxide growth rate per cycle is increased by several orders of magnitude. These investigators suggested that the fatigue crack growth in air [$(da/dN)_A$] could be expressed as

$$\left(\frac{da}{dN}\right)_A = \left(\frac{da}{dN}\right)_v + k\left(\frac{da}{dN}\right)_0$$

where $(da/dN)_v$ is the crack growth rate in vacuum, k is a constant, and $(da/dn)_o$ is the oxidation rate per cycle. Subsequent investigators have noted similar additive fatigue crack growth rates in other alloys. It should be remembered that the rate of oxidation during fatigue may be orders of magnitude greater than observed in conventional static oxidation tests. It does not appear possible to use static oxidation data to predict da/dN in air.[19] This lack of correlation is also brought out in creep crack growth studies where some of the very aggressive sulfidizing environments produced negligible attack on unstressed samples. Static high-temperature corrosion data may, therefore, be very nonconservative in terms of predicting environmental effects during creep or fatigue.

Type C damage is demonstrated by the studies of Bricknell and Woodford.[21] These investigators conducted various tests in which samples were exposed to environments at high temperatures and then tested at lower temperatures. Even after visible surface damage was machined off, the specimens frequently displayed a marked loss in properties as a result of the diffusion of aggressive species into the grain boundaries during the initial exposure.

Bricknell and Woodford found that several different types of damage could be produced by these exposures. In nickel of various purities oxygen interacted with grain boundary carbides to form CO bubbles.[21] In nickel–manganese–sulfur alloys they found that oxidation of MnS particles could liberate sulfur in the grain boundaries.[22] They also suggested, in other cases, that oxygen or oxides at the grain boundaries could suppress sliding, thus inhibiting the relief of local stresses during creep deformation.[23] Subsurface oxidation of carbides and γ' particles has also been observed a number of times in superalloys, which could clearly weaken the structure. It is not necessary to completely oxidize such particles to produce significant damage since destroying the bonds between the particles and the matrix would create cavities.

The relationships between metallurgical parameters and environmental degradation are not clear. Strength appears to be a significant factor. Severe environmental effects are generally confined to high-strength alloys. It must be noted, however, that not all high-strength superalloys display the marked sensitivity to the environment exemplified by the 718 results shown in Fig. 7. Generally, materials that are more notch sensitive tend to be more environment sensitive.

The roles of structural variables, for example, carbide and γ' morphologies, grain size, and shape, on the crack growth behavior in air are becoming clearer. It is still uncertain whether changes in these parameters influence the properties because of internal reasons such as dislocation motion or external reasons related to the air environment.

Grain boundary chemistry, particularly minor additions of elements such as boron, zirconium, magnesium, and hafnium, is quite important in preventing crack growth. Many experiments have demonstrated that boron in particular can be effective in either minimizing the degrading effects of the environment and/or increasing the inherent resistance of the grain boundaries to fracture. A number of mechanisms have been proposed to account for the benefits of elements such as boron to the boundaries. These include

1. increasing the bonding strength of nickel atoms at the boundaries,
2. reducing the rate of grain boundary diffusion,

3. gettering or neutralizing of harmful impurities from either internal or external sources,

4. modifying the composition or morphology of grain boundary carbide or γ' particles, and

5. stabilizing more protective oxides at intersections of the grain boundaries with exterior surfaces.

Quantum-mechanical calculations made by Briant and Messmer indicate that boron additions can promote better bonding of nickel atoms across the grain boundaries.[24] Various experimental results support the other mechanisms. This list is not all-inclusive, and it is also clear that some of the observations could have common origins in more basic mechanisms. At present it seems unrealistic to attribute the benefits of elements such as boron to any single cause. The converse, in fact, seems much more likely—that these elements can behave differently in different alloys. Additional experiments are necessary to establish their roles and to learn how to optimize the grain boundary chemistry.

PROPERTIES OF SUPERALLOYS AT HIGH TEMPERATURES

With increasing temperatures creep resistance becomes the dominating parameter dictating superalloy design and usage. The minimum second-stage creep rate, $\dot{\epsilon}$, in many alloys can be expressed as

$$\dot{\epsilon} = A(\sigma - \sigma_0)^n \exp\left(\frac{-Q}{RT}\right)$$

where σ is the applied stress; Q is the activation energy; R is the gas constant; T is the temperature; and A, σ_0, and n are constants. The σ_0 term is often a function of the initial stress, temperature, and microstructure. With suitable values of σ_0 the apparent activation energy for creep, Q, will approximately equal the Q for self-diffusion of nickel, and the stress exponent n will be on the order of 4.

Creep in superalloys is very dependent on a number of microstructural parameters. Of primary importance are the γ' precipitate volume fraction, lattice mismatch, and morphology. As illustrated in Fig. 8, creep life commonly increases roughly linearly with the γ' volume fraction.[25] Older wrought nickel- or iron-base superalloys typically contain 20–30 vol. % γ' and usually have inferior creep strength to cast alloys and newer nickel-base alloys that contain about double this amount of γ'.

The lattice mismatch between the matrix and γ' will also influence the creep strength. As shown in Fig. 9, the creep strength can be increased by optimizing the lattice mismatch because of the coherency strains developed at the γ' matrix interfaces.[26] This strengthening contribution tends to be most effective for short-time, low-temperature conditions because increasing lattice mismatch also promotes instability of the γ'. Instability of the γ' is usually undesirable because coarsening

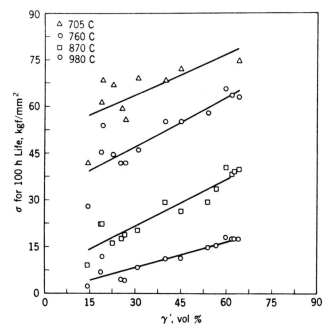

Fig. 8. Creep strength versus γ' content of superalloys at various temperatures.[25]

or transformation of the γ' degrades the creep strength. In some instances, however, such as in the development of the raft-type γ' morphology discussed below, the lattice mismatch and γ' instability are used as a means to promote the formation of a more desirable microstructure during creep loading.

For the best creep strength the γ' particles should be very small, but this often causes undesirable losses in ductility or notch rupture strength. In most alloys an optimum γ' size of ~0.1–0.5 μm is typically used to achieve a good combination of strength and ductility. Bimodal or even trimodal γ' size distributions incorporating coarser γ' particles are common in some alloys. Coarse γ' particles often can be found in cast alloys because microsegregation of alloying elements during solidification produces primary γ' particles that are very difficult to dissolve by heat treatment. Coarser particles also are developed during multistep aging treatments of either cast or wrought alloys. The coarser γ' particles contribute little to the creep strength but help to disperse slip and reduce the notch sensitivity.

For good creep resistance it is essential to strengthen the grain boundaries compared with the interior of the grains. Carbides play an essential role of pinning the grain boundaries and preventing sliding or migration. Carbide particles, however, are commonly sites of cavity formation during creep. Thus, there is usually an optimum carbon content of ~0.5–0.15 wt. % in wrought nickel-base alloys and somewhat higher in cast alloys that represents the best compromise between strengthening the grain boundaries and introducing fracture sites.

Various carbides can be precipitated in the grain boundaries depending on the alloy composition and heat treatment. Some alloys are heat treated at temperatures of 1900–2000 °F (1038–1093 °C) to produce Cr_3C or M_6C. Other alloys are heat treated in the 1350–1550 °F (732–843 °C) range to develop primarily $M_{23}C_6$, which appears to be the stable carbide at lower temperatures. The general goal for creep strength is a regular arrangement of small (0.1–0.3 μm) stable and discrete carbide particles at the grain boundaries. Cellular carbides or continuous carbide films are often detrimental to the rupture life and ductility. Also to be avoided are carbide precipitation reactions that rob the adjoining matrix of γ'-forming elements, leaving a zone free of γ' precipitates adjacent to the grain boundaries. Such precipitate-free zones, if wide enough, can be very weak and provide an easy path for deformation and cracking.

Phase instabilities during creep can cause progressive deterioration in properties. One type of instability is particle coarsening, that is, dissolution of finer particles and growth of larger particles. Coarsening of grain boundary carbides can weaken the boundaries and lead to intergranular cracking. Coarsening of γ' will reduce the creep strength. Decomposition of metastable phases and/or precipitation of new phases may also take place during creep. Less stable carbides can transform to more stable forms. Decomposition of γ' and the formation of deleterious phases

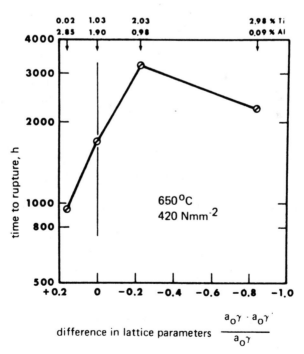

Fig. 9. Effects of lattice mismatch on creep rupture strength of several superalloys.[26]

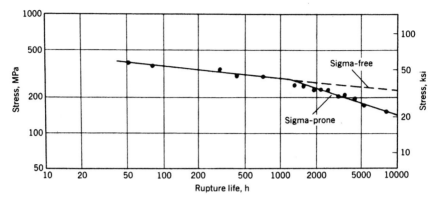

Fig. 10. Effects of sigma formation on the stress rupture life of Udimet 700 at 1562 °F (850 °C).[27]

such as δ, η, σ and so on, can take place. As illustrated in Fig. 10, such changes noticeably impair the creep strength.[27] The loss in properties associated with the formation of phases such as σ usually is either the result of the depletion of alloying elements and consequent loss in strength associated with the formation of the phase or is due to cracking in the new phase or at the new phase–matrix interface.

Grain size and shape are also important variables. Fine grain size will provide higher strengths under lower temperature–higher stress conditions. At high temperatures coarser grain sizes will help minimize grain boundary sliding at lower stresses and is beneficial to the creep strength. The grain size must be limited in relation to the section size, however, in order to ensure a minimum of about 10 grains per cross section. This proviso applies to equiaxed grain structures. As discussed below, with special grain shapes the number of grain boundaries per cross section is not limited.

A number of steps have been taken to improve the creep properties of superalloys. Improved production methods have helped minimize segregation of alloying and impurity elements during solidification. The use of alloying guidelines such as PHACOMP have further helped to minimize the formation of deleterious phases such as σ in newer alloys.

Another major step was to minimize the weakest link in conventional superalloys, the grain boundaries lying perpendicular to the tensile axis. This has been achieved by directional solidification or directional heat-treating techniques that produce grain structures having either extremely high aspect ratios or the complete absence of transverse grain boundaries. As shown in Fig. 11, the creep rupture strength of an advanced alloy increased significantly with increasing grain aspect ratio because of the minimization of cavitation sites on transverse grain boundaries.[28]

Further improvement in properties can be achieved by controlling the crystallographic texture of the directionally solidified material. As illustrated in Fig. 12, the creep strength is highly sensitive to the crystallographic orientation.[29] However, the optimum orientation in service applications is not always dictated primarily by creep strength considerations. In some of the directionally solidified materials the ⟨100⟩ lattice planes were oriented parallel to the loading axis. This was done to minimize the stresses introduced by thermal fatigue cycles in service.

Producing alloys as single crystals permits further improvements in properties. The absence of grain boundaries makes possible the removal of alloying elements such as carbon, boron, zirconium, and hafnium that are used to optimize the grain boundaries in polycrystalline materials. Removal of these alloying elements causes a significant increase in the melting point in most alloys. This in turn permits the alloys to be solution treated at higher temperatures, which brings more γ' into solid solution, thus increasing the amount of fine γ' precipitation produced by age hardening. Progressing from equiaxed to directionally solidified to single-crystal structures can produce significant increments in creep strength,[30] as exemplified in Fig. 13.

Further strengthening in single crystals has been achieved by development of the γ' raft structure morphology. With suitable control of the lattice mismatch between the γ' and the matrix, the precipitation of γ' under stress will produce γ' platelets oriented perpendicular to the tensile stress axis. These platelets or rafts can reach the dimensions of the crystal. Strengthening depends on the amount of interfacial area between the γ' and the matrix, and better creep strength is achieved by finer scale γ' rafts. By careful tailoring of the lattice mismatch, very effective γ' strengthening can be developed in single crystals.

The strength of the γ' phase itself and the consequent strength of γ'-strengthened alloys are sensitive to the temperature. Depending on the compositions, the yield strength of γ' achieves a peak value at temperatures of ~1300–1400°F (704–760 °C). At higher temperatures the strength of γ' decays, and alloys strengthened by γ' tend to lose strength rapidly as the temperature approaches 1800 °F (982 °C). For service at these temperatures other strengthening mechanisms have been developed

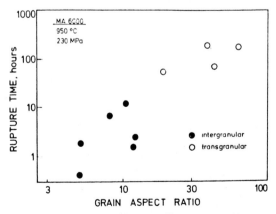

Fig. 11. Rupture life of MA-6000 versus grain aspect ratio.[28] Reprinted with permission from *Superalloys 1984*, by F. Arzt and R. F. Singer, The Metallurgical Society, 420 Commonwealth Dr., Warrendale, PA.

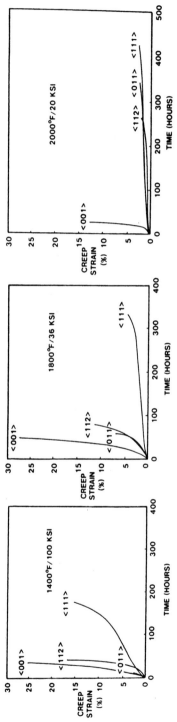

Fig. 12. Influence of crystallographic orientation on the creep properties of superalloy single-crystal specimens.[29] Reprinted with the permission from *Superalloys 1984*, by R. P. Delal, et al., the Metallurgical Society, 420 Commonwealth Dr., Warrendale, PA.

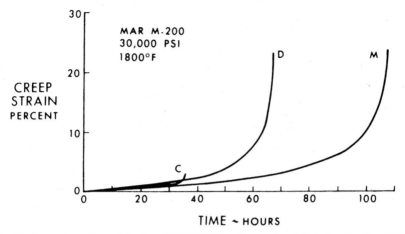

Fig. 13. Comparison of creep behavior of MAR-M 200 as C = equiaxed, D = directionally solidified and M = single crystal.[30] Reprinted with permission from *Mat. Sci. Eng.* **6**, 235 (1970).

to replace γ'. Directionally solidified eutectics incorporating phases such as Ni_3Al, Ni_3Cb, TaC, and Cr_3C_2 have been investigated. After directional solidification these structures ideally consist of aligned uniformly spaced intermetallic or carbide fibers in an alloy matrix. Further alloying has been done in some systems to strengthen the matrix by γ' precipitation. These materials have shown good creep strengths at high temperatures, but their commercial usage has been inhibited by the low solidification rates needed to produce optimum fiber morphologies.

High-temperature creep strength also has been achieved by oxide dispersion strengthening. These materials are produced by powder metallurgy techniques and usually contain Y_2O_3 particles for strengthening purposes and control of the grain size and shape. At lower temperatures these alloys are usually weaker than γ'-hardened alloys because of the difficulty in achieving the large, uniformly distributed volume fraction of oxide particles needed to produce very high strengths. The much greater thermal stability of the oxide particles makes them more effective than γ' at higher temperatures, and at temperatures above ~1800 °F they have superior creep strength to γ'-hardened alloys. Recent work has focused on combining oxide dispersion hardening with γ' hardening in order to achieve high creep strengths over a broader range of temperatures.

Many of these new materials are anisotropic and can display marked deterioration in properties when stressed in off-axis directions. This has led to problems in certain areas, such as where turbine blades are attached to the disk. Thus far, these limitations have not precluded their use as blades. Little has been done thus far to extend the use of these interesting materials beyond jet engine applications. There may be lower temperature uses, such as fasteners or tubing, in which anisotropic properties could be used to advantage, and it might prove fruitful to look for broader applications of these specialized materials.

REFERENCES

1. G. L. Garner and J. L. Smith, EPRI Contract Report, Palo Alto, CA, RP-2058-4, 1984.
2. I. L. W. Wilson and T. R. Mager, *Corrosion* **42**, 352 (1986).
3. S. Floreen and J. L. Nelson, *Met. Trans. A*, **14A**, 133 (1983).
4. C. A. Grove and L. D. Petzold, *ASM J. Mat. Ener. Sys.*, **7**, 147 (1985).
5. R. D. Kane, M. Watkins, D. F. Jacobs, and G. L. Hancock, *Corrosion*, **33**, 309 (1977).
6. B. J. Berkowitz and R. D. Kane, *Corrosion*, **36**, 24 (1980).
7. R. J. Coyle, J. A. Kargol, and N. F. Fiore, *Met. Trans. A*, **12A**, 655 (1981).
8. R. A. Mulford, *Met. Trans. A*, **14A**, 865 (1983).
9. A. W. Thompson and J. A. Brooks, *Met. Trans. A*, **6A**, 1431 (1975).
10. S. Floreen, J. M. Davidson, and P. E. Morris, Int. Nickel Co., Sterling Forest, NY, unpublished.
11. G. Economy, R. J. Jacko, A. W. Klein, and F. W. Pement, Presented at NACE Corrosion 85, Boston, MA, 1985.
12. M. H. Yoo and H. Trinkhaus, *Met. Trans. A*, **14A**, 547 (1983).
13. G. H. Edwards and M. F. Ashby, *Acta Met.*, **27**, 1505 (1979).
14. D. B. Gooch, J. R. Haigh, and B. C. King, *Met. Sci.*, **11**, 545–550 (1977).
15. H. Riedel and J. R. Rice, *Fracture Mechanics*, ASTM STP 700, American Society for Testing and Materials, Philadelphia, PA, 1980, p. 112.
16. K. Sadananda and P. Shahinian, *J. Eng. Mater. Tech.*, **100**, 381 (1978).
17. S. Floreen, *Elastic-Plastic Fracture Second Symposium*, American Society for Testing and Materials, Philadelphia, PA, 1983, p. 708.
18. S. Floreen and R. H. Kane, *Fat. Eng. Mat. Struct.*, **2**, 401 (1980).
19. S. Floreen and R. Raj, *Flow and Fracture at Elevated Temperatures*, ASM, Metals Park, OH, 1985, p. 383.
20. R. P. Skelton and J. I. Bucklow, *Met. Sci.*, **12**, 64 (1978).
21. R. H. Bricknell and D. A. Woodford, *Acta Met.*, **30**, 257 (1983).
22. R. H. Bricknell, R. A. Mulford, and D. A. Woodford, *Met. Trans. A*, **13A**, 1223 (1982).
23. R. H. Bricknell and D. A. Woodford, *Met. Trans. A*, **12A**, 425 (1981).
24. R. P. Messmer and C. L. Briant, *Acta Met.*, **30**, 457 (1982).
25. R. F. Decker, in *Steel Strengthening Mechanisms*, Climax Mo Co., Greenwich, CT, 1970, p. 147.
26. F. Schubert, *Superalloys Source Book*, ASM, Metals Park, OH, 1984.
27. F. R. Morral (ed), *Superalloys Source Book*, ASM, Metals Park, OH, 1984.
28. E. Arzt and R. F. Singer, *Superalloys, 1984*, TMS-AIME, Warrendale, PA, 1984, p. 367.
29. R. P. Delal, C. R. Thomas, and L. E. Dordi, *Superalloys 1984*, TMS-AIME, Warrendale, PA, 1984, p. 185.
30. F. L. VerSnyder and M. E. Shank, *Mat. Sci. Eng.*, **6**, 235 (1970).

Chapter 10

Fatigue

ROBERT V. MINER

NASA Lewis Research Center, Cleveland, Ohio

The fatigue behavior of superalloys is not a narrow subject. The name superalloy is applied to alloys ranging from solid-solution-strengthened nickel- and cobalt-base alloys to nickel-base alloys containing about 65v/o of the γ' phase and to forms ranging from single-crystal castings to fine-grained wrought powder metallurgy products. Discussion of fatigue behavior must also include the interaction of creep and environmental damage mechanisms since superalloys are used at high temperatures in aggressive environments. Also, the stages of cyclic deformation, crack initiation, and crack propagation all must be considered for the most efficient design of complex machines such as gas turbine engines.

CYCLIC DEFORMATION

To predict stresses in structures that experience cyclic plasticity, either on a large scale or locally at stress concentrations, it is necessary to know the cyclic stress–strain behavior of the material. Knowledge of cyclic deformation mechanisms also provides insight into fatigue crack initiation and growth behavior. These processes involve intense local cyclic deformation at slip bands, the region around material defects, or at crack tips. Understanding of cyclic stress–strain behavior requires knowledge of how microstructure changes with the accumulation of plastic strain, recovery, and precipitate coarsening and how alloy composition, temperature, time, and strain rate affect these processes.

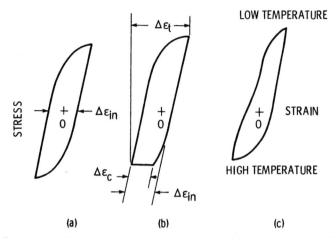

Fig. 1. Cyclic stress–strain hysteresis loops: (a) continuous cycling; (b) creep-fatigue cycle with compressive stress dwell; (c) out-of-phase thermomechanical fatigue cycle.

To define some terms for future reference, schematic hysteresis loops are shown in Figs. 1a–c for an isothermal fatigue test, an isothermal creep fatigue test with a compressive stress dwell, and a thermomechanical fatigue test with the highest and lowest temperatures coinciding with the maximum compressive and tensile strains, respectively. The meanings of the total, inelastic, and creep strain ranges ($\Delta\epsilon_t$, $\Delta\epsilon_{in}$, and $\Delta\epsilon_c$) are indicated for the creep fatigue hysterisis loop. For each cycle the ratio of the minimum strain to the maximum strain, R_ϵ, is -1.

γ'-Strengthened Superalloys

Udimet 700 is a prototype superalloy, studied extensively by Wells and Sullivan[1,2] that shows many aspects of fatigue behavior general to the γ'-strengthened nickel-base superalloys. The cyclic stress response of Udimet 700 at temperatures of 70–1700 °F (21–927 °C) is presented in Fig. 2. It may be seen that for the lowest $\Delta\epsilon_{in}$, the material does not harden or soften with cycling. For $\Delta\epsilon_{in}$ greater than about 10^{-3}, Udimet 700 initially hardens, except at high temperatures, where it softens continually from the first cycle. Hardening is brief and followed by continual softening, except at temperatures around 800 °F (427 °C), where hardening continues about 10 times longer than at either higher or lower temperatures, and then fracture occurs before softening.

While the cyclic stress response of γ'-strengthened superalloys is similar in several respects to alloys strengthened with other shearable precipitates, it is unique in other respects due to the special attributes of the γ' precipitate. A tendency for planar slip results from the characteristics of the ordered compound, its coherency with the matrix, and relatively difficult cross-slip of dislocations around the particles. Stacking fault energy is sufficiently low in the Ni–Cr–X superalloy matrix that cross-slip is difficult, particularly at low temperatures. The concentration of strain in relatively few, coarse planar slip bands is associated with small γ' size, large grain size, low temperature, high strain rates, and low $\Delta\epsilon_{in}$.

The hardening followed by softening exhibited by γ'-strengthened superalloys under some conditions (Fig. 1) is not unique to alloys with ordered precipitates. This is illustrated by the behavior of Cu–2 a/o cobalt single crystals[3] shown in Fig. 3. The precipitate in this alloy is essentially cobalt. Aging to produce various precipitate sizes shows the range of behaviors possible. When the precipitates are small, cyclic hardening is little different from that of the supersaturated solid solution. Hardening and subsequent softening are at a maximum for the peak-aged condition in which dislocations first pile up and then cut the precipitates. For still larger precipitate sizes it becomes possible for the dislocations to loop around the particles. For the largest precipitate size the cyclic stress amplitude initially increases but then

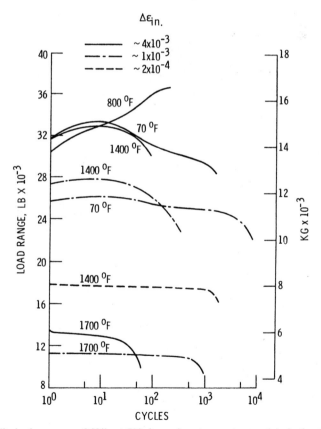

Fig. 2. Cyclic load response of Udimet 700 for various temperatures and inelastic strain ranges, 0.03 Hz, $R_\epsilon = -1$.[1,2] Reprinted with permission from C. H. Wells and C. P. Sullivan, *ASM Trans.*, **60**. American Society for Metals, Metals Park, OH, 1967, p. 217; C. H. Wells and C. P. Sullivan, *ASM Trans.*, **61**, American Society for Metals, Metals Park, OH, 1968, p. 149.

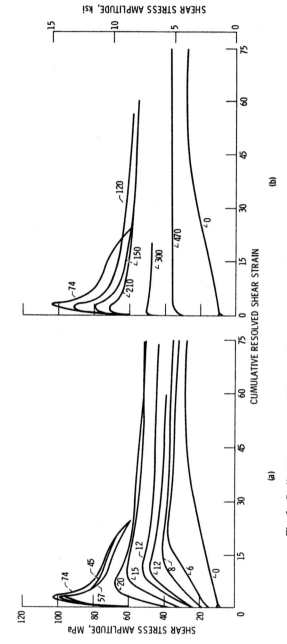

Fig. 3. Cyclic stress response of Cu-2 a/o Co single crystals aged to various precipitate sizes, resolved plastic shear strain amplitude 10^{-3}, $R_\epsilon = -1.$[3] Particle radii are shown in angstrom.

approaches a characteristic very stable condition. Similar effects of precipitate size are observed in superalloys.[4,5]

Once cutting of the γ' begins, work hardening probably also contributes to the initial cyclic hardening. The rapid work hardening of γ' and its complex dependencies on temperature and crystal orientation have been discussed recently by Staton-Bevan.[6] Cyclic hardening followed by softening is observed in other single-phase ordered intermetallic compounds.[7] One contribution to the eventual softening is disordering in the slip bands. Stoltz and Pineau[5] demonstrated that if Waspaloy is briefly aged after cycling into the softening regime, the initial hardening behavior returns upon subsequent cycling. Cyclic softening has also been tied to the decrease in cross-sectional area of the precipitates in the sheared plane.

Many authors have shown that slip planarity in superalloys during fatigue is associated with the γ' cutting that occurs for underaged or peak-aged conditions.[5,8,9] Among commercial superalloys having multiple γ' size distributions, those not aged to produce a large volume fraction of large particles, ~ 0.2 μm, exhibit the most planar slip and the least cyclic hardening. In overaged alloys little particle cutting occurs, and cyclic stress response is typically stable after some initial hardening. Dislocations loop around the particles and pack into the matrix. Slip bands may be observed, but they are less planar and more homogeneously distributed. Large grain size also increases slip planarity. This effect is very noticeable in cyclic crack growth, as will be discussed later.

Also associated with the peak-aged, or somewhat underaged, condition is a very coarse, inhomogeneous planar slip distribution.[8,9] Figure 3 presages what will be stressed later, that the inhomogeneous slip distribution associated with peak aging leads to early crack initiation in the planar slip bands in γ'-strengthened superalloys. For the peak-aged condition the maximum shear measured in any slip band can be three to four times greater than for either the underaged or overaged conditions. The extreme hardening–softening associated with peak aging (Fig. 2) reflects an unstable flow situation. The peak-aged particles are very strong, but once slip is initiated on a few relatively weak planes, they soften drastically, concentrating further strain in these few slip bands. Fournier and Pineau[10] show that coarse slip bands appear only after the onset of cyclic softening.

As shown previously (Fig. 2) cyclic hardening also depends on $\Delta\epsilon_{in}$. This relationship is expressed by the cyclic stress–strain curve. The value of $\Delta\sigma$ at half the number of cycles to failure is commonly taken as the "stabilized" stress range, and $\Delta\sigma$ as a function of $\Delta\epsilon_{in}$ for various tests defines a cyclic stress–strain curve. In the range of $\Delta\epsilon_{in}$ of common interest, the cyclic stress–strain curve may be approximated by the Holloman relation

$$\Delta\sigma = k \, \Delta\epsilon_{in}^{n'}$$

where k and n' are constants. For superalloys $n' \simeq 0.1$ at low and intermediate temperatures. Landgraf[11] shows this is intermediate behavior between classes of alloys. Precipitation-hardened aluminum alloys, maraging steels, and titanium alloys have $n' \leq 0.1$, while nickel, copper, low-alloy steels, and solid-solution-strengthened alloys have $n' > 0.2$. The low value of n' for the precipitation-hardened, planar slip materials reflects the easily reversed deformation in the softened slip bands. More homogeneous deformation at high temperatures is reflected in increased values of n' for superalloys, to as high as about 0.3.

The formation of persistent slip bands in the γ'-strengthened superalloys is different from that in alloys with wavy slip character in that little, if any, general homogeneous deformation precedes their formation. Transmission electron microscopic (TEM) studies of superalloys cycled to failure at low $\Delta\epsilon_{in}$ show few dislocations outside of a few slip bands. For low $\Delta\epsilon_{in}$ intense damage occurs only very locally at defects or coarse slip bands. Deformation is relatively reversible in the persistent slip bands and does not spread to regions between the slip bands. Deformation is also inhomogeneously distributed among the grains in polycrystals. Grains in which strain is accommodated elastically are seen to bear orientations with respect to the applied stress giving them low elastic moduli and/or resolved shear stress.

With increasing $\Delta\epsilon_{in}$ the density of slip bands increases, and finally cyclic hardening increases as other slip systems are forced to operate and deformation becomes more homogeneous. The complete cyclic stress–strain curve for a single-crystal superalloy at room temperature is similar to those of pure FCC metals.[12] At very low and very high $\Delta\epsilon_{in}$, $\Delta\sigma$ increases with increasing $\Delta\epsilon_{in}$, but in between, $\Delta\sigma$ is relatively constant. This plateau is thought to represent the shear stress necessary to create a persistent slip band, each of which allows a minute amount of strain. Increasing $\Delta\epsilon_{in}$ can be accommodated at about the same $\Delta\sigma$ until some critical density of persistent slip bands is reached. In polycrystals the plateau region has a positive slope because the critical resolved shear stress for some grain orientations is higher than for others.

Slip planarity in the γ'-strengthened superalloys is reduced with increasing temperature. Various effects contribute to this change. Stacking fault energy in the matrix increases with increasing temperature[13] allowing easier cross slip. The primary cube slip system may be activated in grains suitably oriented. The strength of the γ' increases with temperature up to intermediate temperatures, and the tendency for dislocations to loop around the particles is increased. Dislocation climb is facilitated by increased diffusion rates. Similarly, climb is facilitated by lower strain rates and creep dwell periods in the cycle. Because of these effects, cyclic softening due to cutting of the γ' particles is reduced. Microstructures after high-temperature cyclic deformation show rapid γ' coarsening as in aging under static stress, only a lamellar morphology does not develop.[14] Dislocations are seen stored at the interfaces of the coarsened γ' particles, like misfit networks. Some of these interface dislocations must provide plastic strain by moving back and forth across the γ channels.

Creep processes introduce a time dependence to cyclic stress–strain behavior. Low frequencies reduce cyclic stresses both because of the lower strain rate and because of dynamic recovery processes. Increased time per cycle allows static recovery of the dislocation substructure as well as coarsening of the γ' precipitates. In polycrystalline alloys grain boundary cavitation becomes an important damage process.

Cyclic deformation of the γ'-strengthened superalloys mirrors deformation of γ' in additional ways. Octahedral or cube slip may be observed depending on orientation both for single crystals and for individual grains in polycrystals. The tension–compression anisotropy of yielding for orientations near $\langle 001 \rangle$ and $\langle 011 \rangle$ and continued anisotropic hardening in the highest strength direction has been observed both for single-phase γ'[15] and for superalloys.[16,17]

Strain aging effects at intermediate temperatures, such as increasing flow stress with decreasing strain rate and/or serrated yielding, have been reported in several nickel-base alloys.[1,17–19] For instance, in Udimet 500, $\Delta\sigma$ for tests at a given $\Delta\epsilon_{in}$ increases with decreasing test frequency for temperatures below about 1400 °F (760 °C).[18] Slip bursts in some alloys may persist for many cycles[1,17] and are most noticeable for single-slip orientations in single crystals.[17] The greater cyclic hardening at intermediate temperatures shown in Fig. 1 may be associated with dynamic strain aging as well as the intrinsic behavior of the γ' phase.

Other Superalloys

Inconel 718 deserves separate discussion since it is widely studied and exhibits unique behavior due to its strengthening by γ''. At room temperature and high $\Delta\epsilon_{in}$ Inconel 718 exhibits cyclic hardening and then softening, like a γ'-strengthened alloy, but at higher temperatures within its use range it only softens.[4,10] Increased softening and inhomogeneity of the planar deformation was observed with increasing temperature.[10] Shearing of the γ'' particles converted them to the β structure of Ni_3Nb.

The solid-solution carbide-strengthened superalloys are weaker than γ'-strengthened superalloys but can be used in applications where greater plasticity is enforced. Consider Hastelloy X as an example. Fatigue cycling at large $\Delta\epsilon_{in}$ leads to a very dynamic and complicated cyclic stress–strain behavior, though Hastelloy X is a relatively simple alloy. At low temperatures it is capable of considerable cyclic hardening but does not subsequently soften like γ'-strengthened alloys.[19] It creeps readily at the top of its temperature use range; however, at temperatures around 800–1000 °F (427–538 °C) it exhibits an inverse dependence on strain rate because of strain aging. Strain aging leads to a cyclic stress response that is very history dependent, requiring additional complication in any nonisothermal model of cyclic stress–strain behavior.[20]

Modeling

Of great current interest is the quantitative prediction of cyclic stress–strain, or constitutive, behavior. This is a large field in itself, but it should at least be mentioned here. Constitutive models predict the relation between stress and strain rate considering the rate of strain hardening, competing recovery processes, and their effects on the cyclically evolving state of the material. These processes are given temperature dependencies, and the material also is given a memory of its prior deformation history. Attempts are also being made to account for additional complexities such as multiaxial stress states, material anisotropy as in single crystals, and other orientation effects specific to superalloys such as the operation of octahedral or cube slip and tension–compression anisotropy. Much of the model development has been done for problems in the nuclear industry.[21] NASA has supported the development of models particularly suited to the needs of gas turbine engine manufacturers.[22,23]

Even in the advanced constitutive models, simplified and sometimes empirical relationships are often employed because of the number of variables to be considered. Still, these models may contain 10 or more experimentally determined constants. The challenge in predicting the complex shape of a hysterisis loop for a thermo-mechanical fatigue cycle such as that shown in Fig. 1c is readily apparent. Further complexity will be required to predict effects of thermomechanical history such as strain aging.[20]

Available computing power allows the use of such complex systems of equations. Still, the cost lies not only in computer time but also in the experimental determination of all the constants. More mechanistic understanding is required to refine the existing semiempirical models and to hopefully reduce the number of empirical constants.

CRACK INITIATION

Philosophically, fatigue life should be considered as divided into crack initiation and propagation phases since their kinetic laws are known to be different. Exper-imentally, however, it is difficult to obtain separate and complete information on these two phases. Smooth specimen fatigue tests unavoidably measure some portion of the crack propagation phase, and it is difficult to obtain information about the very important initial phase of crack growth in tests of precracked specimens. Thus, we can only examine the available information from both types of tests and try to join it into a complete picture.

Fatigue

Pure fatigue damage is produced by time-independent inelastic deformation. However, superalloys are used at high temperatures in aggressive environments, and damage is often due to the combined effects of fatigue, creep, and environment. Still, other factors being equal, fatigue life is governed by $\Delta \epsilon_{in}$. The Manson–Coffin relation that expresses this relationship is often written in the transposed form

$$\Delta \epsilon_{in} = a N_f^b$$

where N_f is the number of cycles to failure, and a and b are constants. The initial constant varies widely among materials, generally increasing with ductility.[24] For many metals and alloys $b \simeq -0.6$, though for superalloys more negative values are often reported. It is useful to consider the Manson–Coffin behavior as the norm, a basis of comparison in assessing the importance of other damage mechanisms.

However, in many practical situations we are interested in fatigue life for a given total strain range, or stress range. Figure 4 shows the life of PM IN-100 at 1200 °F (649 °C) on the basis of the total strain range ($\Delta\epsilon_t$) and the elastic ($\Delta\epsilon_{el}$) and inelastic components.[25] It is seen that over the range of the data, $\Delta\epsilon_{el}$ as well as $\Delta\epsilon_{in}$ bears a power law relationships to N_f. The Basquin relationship

$$\Delta\epsilon_{el} = cN_f^d$$

follows if both N_f and $\Delta\sigma$ bear power law relations to $\Delta\epsilon_{in}$ as in the Manson–Coffin and Holloman relations, and d will equal $n'b$. The constant c is roughly proportional to ultimate tensile strength divided by elastic modulus.[24] The effect of elastic modulus is readily apparent in the fatigue lives of single crystals of various orientations, since modulus varies by a factor of greater than 2.[17,26]

The generalizations relating fatigue behavior to monotonic tensile strength and ductility may be observed in the comparison of several superalloys[25] presented in Fig. 5. These alloys all were developed for gas turbine disk application. In the regime where $\Delta\epsilon_t$ is mostly elastic, life among the alloys increases roughly with increasing tensile strength. In the regime of larger $\Delta\epsilon_{in}$ life increases roughly with increasing tensile ductility. But these correlations are only rough. Fatigue behavior does not simply reflect tensile behavior for a number of reasons. Tensile behavior is much less sensitive to material defects and does not reflect the cycle and time-dependent damage mechanisms present in high-temperature fatigue.

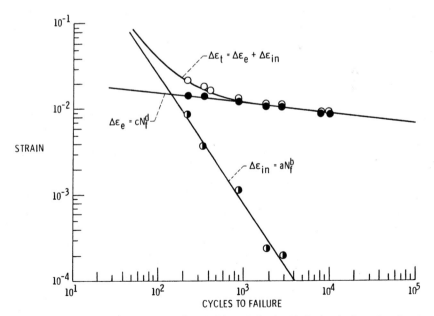

Fig. 4. Fatigue life of PM IN-100 at 1200 °F (649 °C) on the basis of inelastic, elastic, and total strain, 0.33 Hz, $R_\epsilon = -1$.[25]

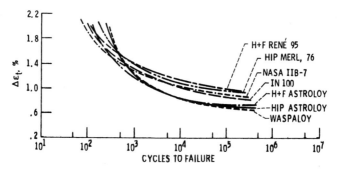

Fig. 5. Total strain-fatigue life behavior of several superalloys at 1200 °F (649 °C), 0.33 Hz, $R_\epsilon = -1$.[25]

Figure 5 shows that material selection depends on the strain regime in which the component will operate. Applications such as high-temperature pressure vessels or gas turbine engine combustors where large thermally induced $\Delta\epsilon_{in}$ are unavoidable dictate materials such as the lower strength, solid-solution-strengthened superalloys that have a much greater tolerance for large $\Delta\epsilon_{in}$. The fatigue life of an alloy like Hastelloy X for a given $\Delta\epsilon_{in}$ is at least 100 times greater than that for any of the γ'-strengthened superalloys shown in Fig. 5. Still, the goal of alloy development for any application is to raise both the inelastic and elastic life lines as much as the generally contrary requirements will allow.

Figure 6 shows the fatigue life of Nimonic 90 as a function of $\Delta\epsilon_{in}$ on a log-log basis for several temperatures.[27] The slope of the life lines, b, decreases with increasing temperature from about -0.5 at 1202 °F (650 °C) to about -0.8 at 1652 °F (900 °C). Values of b more negative than -0.6 are not uncommon for superalloys at high temperatures. Such steep slopes probably indicate the shortening of life by time-dependent damage mechanisms in the low-strain, long-life regime.

Commonly b is constant over several orders of $\Delta\epsilon_{in}$. However, approximately bilinear life curves have been found for Inconel 718[28,29] and Waspaloy.[29] For low $\Delta\epsilon_{in}$ life appears to be reduced (b is more negative). The bilinear life line appears to be associated with coarse slip. It does not occur for Waspaloy at intermediate temperatures where slip becomes finer. However, deformation in 718 becomes increasingly coarse with increasing temperature,[10] and the bilinear life line persists.[28] If the change in b corresponded to a change in deformation mechanism, it might be expected that it would be reflected in the cyclic stress-hardening exponent, n'. An inverse relationship between the cyclic strain-hardening exponent, n', and $-b$ is predicted in several theories. Morrow[30] originally suggested the form

$$-b = 1/(1 + n')$$

However, no change in n' is apparent for either alloy. The shortened lives in the long-life regime may be an environmental effect in the coarse slip bands.

The effect of increasing temperature on fatigue life viewed on the basis of $\Delta\epsilon_{in}$ reflects not only other accelerated damage processes but also the change in ductility with temperature. The life of cast René 80, shown in Fig. 7, actually increases with temperature above about 1400 °F.[31] The low life at 1400 °F reflects, at least in part, the ductility minimum generally observed for γ'-strengthened superalloys at intermediate temperatures. Increased life, on a $\Delta\epsilon_{in}$ basis, observed for some materials at high temperatures may also be a result of an oxidation crack blunting process.[32]

Material defects, pores or inclusions, become more significant in advanced high-strength alloys because they are used at higher stresses, and as will be shown, they do not necessarily have improved crack growth resistance. Defects largely control the fatigue life of advanced superalloys at low $\Delta\epsilon_t$ and particularly at low to intermediate temperatures. At high temperatures environmental or creep damage may become most important. Defects in advanced PM superalloys have recently prompted large programs to improve processing methods (see Chapter 17). In polycrystalline or directionally solidified castings carbides often govern fatigue life.[33] Carbon is not intentionally added to most single-crystal alloys, and small micropores have become life limiting. To extract the last bit of performance, planar front solidification methods and hot isostatic pressing are being examined to minimize porosity.

It was discussed previously that peak aging leads to early crack initiation for a given $\Delta\epsilon_{in}$ (Fig. 3) because of the concentration of strain in a few intense slip bands. More homogeneous distribution of slip associated with under- or overaging

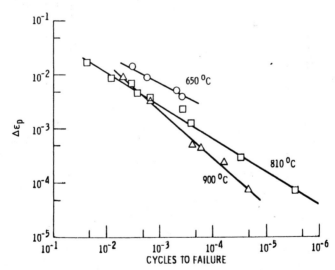

Fig. 6. Inelastic strain-fatigue life behavior of Nimonic 90 at various temperatures, 0.41 Hz, $R_\epsilon = -1$.[27] The original version of this paper was first published in AGARD Conference Proceedings 243, by the Advisory Group of Aerospace Research and Development, North Atlantic Treaty Organization (AGARD/NATO) 1978.

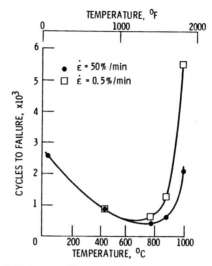

Fig. 7. Fatigue life of René 80 in tests with constant $\Delta\epsilon_{in} = 0.05\%$ for various temperatures and two strain rates.[31]

provides greater resistance to crack initiation in slip bands. There is also some evidence that crack initiation at defects is retarded by overaging. Still, the possible benefit of overaging heat treatments must be weighed against a decrease in strength. Increased homogeneity of deformation may be one source of the improvement in fatigue strength resulting from thermomechanical processing. The remnant dislocation substructure should act to homogenize subsequent cyclic deformation.

It is well known that fatigue crack initiation behavior is improved by fine grain size at low temperatures where creep and environmental damage are minimal. In the conventional cast and wrought superalloys, even with their large carbides, an intense slip band or twin boundary crossing a large grain may represent the largest defect.[1] The apparent defect sensitivity of the fine-grain PM alloys occurs, in part, because the grains are smaller in relation to the defect size. Still, crack initiation occasionally occurs in accidental large grains in powder metallurgy superalloys[34,35] rather than at inclusions or pores. These rare grains may be 10 times larger than average and result from powder particles that are off composition. Grain boundary morphology affects fatigue behavior at low temperatures less than grain size, since cracking is not usually intergranular.

Mean stress, like $\Delta\epsilon_{in}$, affects fatigue life. High mean stresses lead to early crack initiation at material defects. A cycle between unbalanced stress limits has a mean stress, of course. However, a mean stress may also be developed in some strain-controlled cycles. As shown in Fig. 8a, a cycle between unbalanced strain limits develops a mean stress on the first cycle that is largest for cycles with small $\Delta\epsilon_{in}$. The tests shown have $R_\epsilon = 0$. For cycles with large $\Delta\epsilon_{in}$ not only is the mean stress small for the first cycle but also it continues to decrease with cycling because of dynamic recovery processes driven by inelastic strain. Tests with very small $\Delta\epsilon_{in}$ may maintain a stable mean stress for the life of the test.

Figure 8b compares the fatigue lives of PM IN-100 and Waspaloy on the basis of $\Delta\epsilon_t$ in tests with $R_\epsilon = 0$ and $R_\epsilon = -1$.[25] The $R = -1$ fully reversed strain tests have no mean stress. In these tests the stronger PM IN-100 is greatly superior to Waspaloy at low $\Delta\epsilon_t$. These are the same data shown in Fig. 5. However, in the

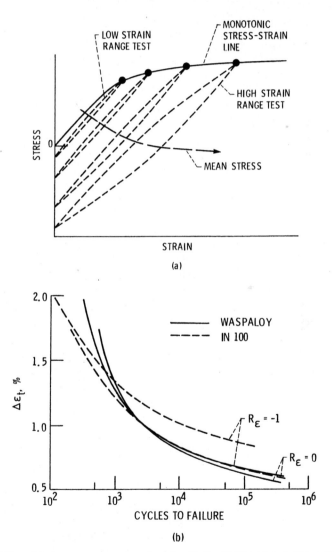

Fig. 8. (a) Effect of strain range on mean stress for tests with $R_\epsilon = 0$; (b) total strain-fatigue life relations for PM IN-100 and Waspaloy at 1200 °F (649 °C), 0.33 Hz, $R_\epsilon = 0$ or -1.[25]

$R = 0$ tests the weaker Waspaloy yields and quickly sheds its mean stress, while IN-100 cannot in the tests at low $\Delta\epsilon_t$. Thus, the life of Waspaloy is little different between $R = 0$ and $R = -1$, but that of IN-100 is greatly reduced for $R = 0$ at low $\Delta\epsilon_t$, becoming nearly the same as that of Waspaloy. The same large effect of R ratio was observed in another high-strength alloy studied, H + F René 95. These R ratio effects illustrate that the choice of the best material depends on the specific cycle.

Creep and Environmental Effects

At elevated temperatures fatigue life becomes dependent not only on the number of cycles but also on the time per cycle. Generally, fatigue life is reduced either by low cyclic frequencies or by dwell periods in the cycle. In polycrystalline alloys at intermediate temperatures time-dependent damage is largely a stress–corrosion effect of the environment. True creep damage such as grain boundary cavitation becomes the dominant time-dependent contribution at higher temperatures.

Though air probably affects crack growth most, it also affects crack initiation. Cracks initiate in oxidized grain boundaries.[36,37] Coffin[37] studied the low-cycle fatigue behavior of the iron–nickel-base superalloy A-286 at 70 and 1100 °F (21 and 593 °C) at various frequencies in air and vacuum. Compared on a $\Delta\epsilon_{in}$ basis, lives in vacuum at both temperatures were nearly the same at all frequencies. For tests at 1100 °F (593 °C) in air, life for a given $\Delta\epsilon_{in}$ decreased roughly by half for each decade increase in the time per cycle. Increased oxidation at local surface sites was observed with increased time per cycle, and these were taken to be sites of early crack initiation.

At very high temperatures, however, oxidation may increase fatigue life by crack blunting. Duquette and Gell[32] demonstrated that fatigue life of MAR-M 200 single crystals is greater in air than in vacuum at 1700 °F. The increase in fatigue life with increasing temperature from 1400 to 1800 °F (760 to 982 °C) for René 80 illustrated in Fig. 7 may be partly due to oxidation crack blunting.

There is an especially strong stress–corrosion effect on fine-grained alloys for tests in air at intermediate temperatures. Figure 9 shows the fatigue life of PM IN-100, which has a grain size of about 5 μm, for tests at 1200 °F (650 °C) with various dwells at the maximum tensile stress.[25] It may be seen that a 15-min dwell reduced life by about one decade for $\Delta\epsilon_t = 1\%$ even though the amount of creep strain per cycle was very small, less than 0.01%. The dwell tests produced intergranular crack initiation; however, there was no evidence of cavitation in grain boundaries near the fracture surfaces or secondary surface cracks indicating a purely creep damage process. Since the greatest decrease in life is in the long-life regime where crack initiation should represent the major portion of life, the dwell does appear to have reduced the crack initiation portion of life.

The dwell effect is greatest in fine-grained alloys. In addition to PM IN-100, all the alloys shown in Fig. 5 were also tested in a 15-min tensile dwell test at 1200 °F (649 °C)[25] Whereas the life difference between the fine-grained, high-strength alloys and the larger grained, lower strength alloys may be seen to be roughly 10

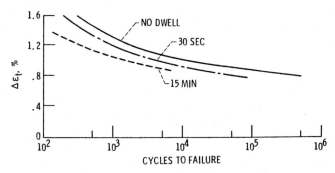

Fig. 9. Total strain-fatigue life relation for PM IN-100 at 1200 °F (649 °C) for 0.33 Hz tests and tests with 30-s or 15-min tensile dwells at the maximum tensile strain, $R_\epsilon = -1$.[25]

times for the 0.33-Hz tests, it was less than 3 times for the 15-min tensile dwell tests.

However, compressive dwell cycles have generally been found to be even more damaging to the γ'-strengthened superalloys than cycles with tensile or tensile and compressive dwells.[1,38,39] Similarly, tests having strain rates that are fast in tension and slow compression are very damaging. All of several studies on γ'-strengthened superalloys in a 1978 conference on this subject found the compressive dwell or slow compressive strain rate cycle the most damaging (AGARD-CR-243, 1978).

One way of comparing the different cycle types on the basis of $\Delta\epsilon_{in}$ is the Strain Range Partitioning Method.[40] In this view there are four basic cycle types, pp, pc, cp, and cc, where p indicates high strain rate plastic deformation, c indicates creep deformation, and the first and second letters represent the tension and compression halves of the cycle, respectively. Recognizing there is some pp strain in any creep fatigue cycle (see Fig. 1b), a method is given for calculating the life of a "pure" creep fatigue cycle type as a function of $\Delta\epsilon_{in}$. Figure 10 shows the life lines for the four "pure" cycle types for two alloys.[38] The behavior of cast IN-100 is typical of γ'-strengthened superalloys. The behavior of A-286 is more typical of alloys that cavitate more readily than the high-strength superalloys. A tensile bias is necessary to produce grain boundary cavitation.

For the γ'-strengthened superalloys differences in life among the various cycle types for the same $\Delta\epsilon_{in}$ appear, at least in part, to be related to the stresses developed. Low life for the pc cycle may be related to the tensile mean stress illustrated in Fig. 1b. The other cycle types have zero or compressive mean stresses. Also, highest maximum tensile stress,[39] or stress range,[41] have been correlated with the finding of lowest life for the pc cycle. Thus, some aspect of the cyclic stresses does appear to be important in determining life, probably through its role in driving crack growth. Differences in the grain boundary damage mechanisms are probably also important. Wells and Sullivan[1] observed that their cp cycle produced round grain cavities in the grain boundaries, while the more damaging pc cycle produced sharp cracks. Supporting such a view is the finding that in the single-crystal superalloy

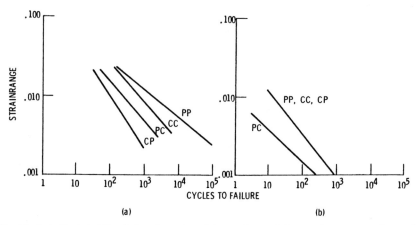

Fig. 10. Inelastic strain-life relations for various fatigue and creep fatigue cycle types according to the method of Strain Range Partitioning[38]: (a) A-286 at 1000 °F (593 °C); (b) cast IN-100 at 1700 °F (927 °C).

PWA-1480 the pc cycle was found to be only slightly worse than the other cycle types.[41]

A similar effect occurs in thermomechanical fatigue cycling. The most damaging cycles are those where the strains are compressive at high temperatures and reversed in tension at low temperature.[42,43] Such a cycle occurs upon start-up at the thin leading and trailing edges of a turbine engine vane. Thermal expansion, which is constrained by the still cool bulk of the vane, produces compressive stresses. The opposite condition occurs upon cooling. Not only is the tensile strain imposed in the temperature range of least ductility but also high tensile stresses are developed in reversing the inelastic compressive strain that develops readily at the high temperature end of the cycle (Fig. 1c).

Purely thermal fatigue tests are not used as much as they once were. In Glenny-type tests thermal strains are produced as they are in actual components by differences in heating rates in the wedge-shaped specimen. However, the stresses and strains must be calculated. The trend now is to adapt low-cycle fatigue testing for rapid heating and cooling and apply this well-characterized life data to the component. This way the mechanical and thermal analyses need only be performed on the component, not the specimen as well. Still, thermal fatigue tests offer a simple comparison of performance among materials, and they do capture effects missing in thermomechanical fatigue tests. The microstructure of cast to size wedge specimens approximates that at the leading edge of a turbine engine blade in terms of grain orientation and size. Also, burner-heated tests may simulate oxidation or hot-corrosion damage.

Figure 11 gives comparison of the thermal fatigue resistance of several superalloys, conventionally cast, directionally solidified, and some coated.[44] The specimen shape is also shown. The test cycle was 3 min each in fluidized beds at 600 and 1990 °F

(316 and 1088 °C). Although the latest single-crystal alloys are not represented, this figure shows the progress that has been achieved in superalloy and coating development. The largest single source of improvement apparent in these data is directional solidification. Directional solidification both removes the grain boundaries transverse to the principal stress axis and produces an elastic modulus in that direction that is about 40% lower than for a polycrystalline structure.[45] The lower modulus reduces the stresses developed by the enforced thermal strains.

Life Prediction

This is a large subject that must be summarily treated. Still, some references can be provided. The most used high-temperature creep fatigue life prediction method is the time and cycle fraction method specified in the ASME Boiler and Pressure Vessel Piping Code. However, this method does not recognize differences among all the cycle types previously shown. Advanced life prediction models account for these differences in various ways. The strain range partitioning method introduced previously allows different life$-\Delta\epsilon_{in}$ relations for the pp, pc, cp, and cc cycle types and provides rules for separating $\Delta\epsilon_{in}$ for complex cycles into components belonging to each of the cycle types.[40] Other models also relate life to inelastic strain but

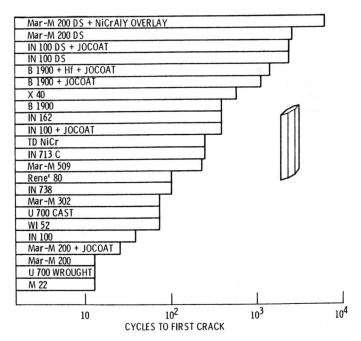

Fig. 11. Thermal fatigue resistance of several superalloys cycled between fluidized beds at 1990 and 600 °F (1088 and 315 °C), immersion time 3 min in each bed.[44]

include various modifications to account for time-dependent damage. The frequency separation method includes the time under tension per cycle and the ratio of time in tension to that in compression.[46] The frequency-modified damage function includes time per cycle and makes the maximum tensile stress the difference among the cycle types.[39] The damage rate model considers the maximum tensile and compressive inelastic strains and strain rates.[47]

CRACK PROPAGATION

Cyclic crack growth rates may be correlated with crack length, applied load, and geometry of the body through the linear elastic stress intensity factory ΔK as long as the plastic zone ahead of the crack is small with respect to the size of the body and the crack length. The linear elastic crack growth regime is probably of most practical interest for the high-strength superalloys since they cannot tolerate large cyclic plasticity. However, crack growth in the plastic fields due to notches, rapid surface heating and cooling, and the more large scale yielding possible in structures made of the solid-solution-strengthened superalloys will be briefly touched upon later.

In the intermediate regime cyclic crack growth rate, da/dN, is usually described by the Paris–Erdogan equation,

$$\frac{da}{dN} = C \, \Delta K^n$$

where C and n are constants. For low ΔK, da/dN decreases more rapidly with decreasing ΔK, eventually reaching some threshold value, ΔK_{th}, below which a crack will not propagate. For ΔK above the intermediate regime da/dN increases rapidly and at some ΔK becomes catastrophic. We will first discuss behavior at low temperatures where creep and environmental effects are negligible and crack growth is only dependent on the number of cycles and not on the time per cycle.

Fatigue

At room temperature the gross differences among materials in da/dN for a given ΔK is largely an effect of elastic modulus. Hoffelner[48] has shown that the equation $da/dN = 1.7 \times 10^6 \, (\Delta K/E)^{3.5}$ meters per cycle, proposed earlier by Speidel, predicts growth rates within about ± 2 times for an extensive list of superalloys. Still, this range of cyclic crack growth performance among alloys is not inconsequential, and as will be discussed later, differences among alloys become even more significant in the important threshold regime of ΔK and at higher temperatures. It has become increasingly clear that for any given alloy cyclic crack growth behavior is influenced by microstructure.

The planar slip character of the γ'-strengthened superalloys imparts a strong tendency for crystallographic cracking along slip bands, or stage I cracking. This tendency is strongest at low temperatures, high strain rates, and low ΔK. The size of the region ahead of the crack undergoing plastic deformation increases with increasing ΔK. When this plastic zone is within a single grain, crack extension may occur on one of its shear planes. As a crack grows, ΔK increases. Mutual accommodation of deformation among several grains within the larger plastic zone leads to noncrystallographic transgranular cracking normal to the applied stress axis, or stage II cracking. The effect of increasing ΔK has also been observed in a transition from stage I to stage II crack initiation among low cycle fatigue tests with increasing stress range.

Increasing grain size is the most cited microstructural route to improved cyclic crack growth resistance. This improvement occurs both for low-temperature transgranular crack growth and, as will be discussed later, for higher temperature intergranular crack growth. Large grain size has been shown to decrease the rate of transgranular cyclic crack growth in the intermediate growth rate regime for high-strength aluminum and titanium alloys as well as for several superalloys.[49-52] In this regime $da/dN \propto d^k$, where d is the grain size and k is -1 to $-\frac{1}{2}$.[50,51] The improvement in cyclic crack growth resistance has been explained as a result of more planar, reversible slip character. Of course, decreasing yield strength must be weighed against the benefits of increasing grain size.

Larger grain size also increases the plastic zone size, or ΔK, where the transition to stage II occurs. King[53] has demonstrated this for Nimonic AP1 (HIP Astroloy). In very fine grain PM superalloys cracks initiate in a stage II mode, skipping stage I entirely. or in single-crystal superalloys at low and intermediate temperatures, stage I cracking may continue across the entire diameter of LCF specimens. (This tendency produces great complications in crack growth testing of single crystals at these temperatures, and even at high temperatures, where stage II crack growth occurs, the rate is found to vary with crystallographic direction.[26])

Another benefit of large grain size is that it raises the ΔK at which stage II growth begins, and more cycles are spent in stage I. King[53] shows a two- to threefold increase in da/dN upon the transition to stage II growth. This may be interpreted as an effect of reduced slip reversibility. The persistent slip bands operating in the stage I regime allow deformation at the crack tip with low cyclic hardening. In the stage II regime the plastic zone of the crack encompasses more grains, and mutual accommodation of deformation activates intersecting slip systems, producing more rapid hardening.

Lower cyclic crack growth rates for larger grain sizes extend into the threshold regime for tests using long crack specimens at low loads[53,54]; however, this appears to be largely an effect of crack closure and may not occur for short cracks such as at the initial stages of propagation from defects. Short crack behavior is very complex, and as pointed out by Brown et al.[54] and others, closure effects are not a complete explanation; however, it is clear that grain size effects are much lessened relative to long crack behavior. Their results on Astroloy (Fig. 12a), and Waspaloy show both higher cyclic crack growth rate for small cracks and a small grain size effect relative to that for large cracks. Similar results have been obtained in titanium alloys (see several papers in ref. 56).

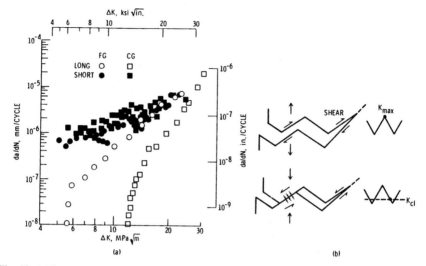

Fig. 12. (a) Room temperature cyclic crack growth rate in Nimonic AP1 (HIP Astroloy) with fine and coarse grain sizes (FG and CG) in both long and short crack tests, 40 Hz, $R = 0.1$.[54] Fine and coarse grain sizes were about 12 and 50 μm, respectively. (b) Illustration of crack closure due to crystallographic shear.

Figure 12b shows schematically that for a "long" sharply angled crystallographic crack the faces may be offset by shear upon opening. When the load is decreased, the crack faces close before zero load, and thus the effective ΔK is lower than indicated by the load range. Closure by this mechanism is expected for low ΔK, where cracking is crystallographic, and for large grain sizes because both increase the roughness of the crack faces. Grain size should not affect short crack growth, since when the crack is still within one grain, there is no length of rough crack face behind the crack tip. Grain size effects near threshold are also greatly reduced in long crack tests when closure is prevented by use of high R ratios.

It is felt that these observations on short crack behavior make it easier to connect what is known about the effects of grain size on both cyclic crack growth rate and fatigue crack initiation. If increasing grain size decreased cyclic crack growth rate for all ΔK, even for short cracks, it would be difficult to rationalize the general rule that decreasing grain size increases cyclic endurance stresses even if a somewhat longer crack initiation period exists for the fine-grained material. Also indicated is the necessity of understanding short crack behavior in order to best design alloys for improved defect tolerance.

The size of the γ' precipitates affects cyclic crack growth behavior for low temperatures and cycle types where crack growth is transgranular. Otherwise grain boundary morphology and composition are more important. For transgranular crack growth the decreased cyclic yield stress produced by overaging increases cyclic crack growth rate. Further, it appears that for the same yield strength underaging is preferable to overaging.[52,55] This was attributed to a more planar, reversible slip character. Refined γ' and/or more planar slip also have been related to increased ΔK_{th}.[53,56]

Creep and Environmental Effects

Cyclic crack growth rates increase with increasing temperature and become dependent on the time per cycle. As temperature increases, elastic modulus decreases, deformation becomes less planar, the rate of environmental attack increases, and eventually creep processes are activated. Crack initiation may remain transgranular, but subsequent propagation in polycrystals becomes increasingly intergranular with increasing temperature, lower frequencies, longer dwell times, and finer grain sizes.[51,57–61]

Figure 13 shows the effect of cyclic frequency, or time per cycle, and temperature on the cyclic crack growth rate and the mode of cracking in the cobalt-base superalloy Haynes 188.[59] At high frequencies cracking is transgranular and da/dN is independent of frequency, but with decreasing frequency and increasing temperature the cracking mode becomes intergranular and da/dN increases with decreasing frequency. When the slope reaches -1, da/dN is directly proportional to the time per cycle. The extent of the intergranular cracking regime with respect to temperature and frequency depends on alloy environmental and creep resistance.

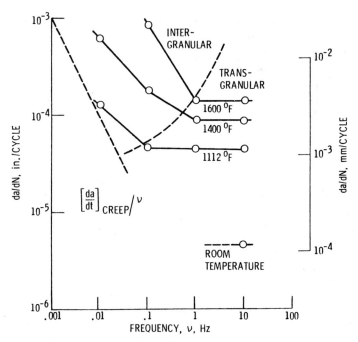

Fig. 13. Cyclic crack growth rate of Haynes 188 at $\Delta K = 50$ ksi as a function of frequency for several temperatures, $R = 0.08$.[59]

At intermediate temperatures the time-dependent, intergranular cyclic crack growth is by a stress–corrosion effect of the air environment at grain boundaries.[49,51,58,62] Sufficient oxygen is present even in laboratory-grade argon to produce significant effects.[63] Fine grain size greatly exacerbates the effect of environments,[51] and conversely, cyclic crack growth of single-crystal superalloys exhibits little time dependence.[64] Very inhomogeneous planar slip, as in Inconel 718 at intermediate temperatures, may also interact with the environmental effect to promote intergranular cracking.[10,50] Overaging improves the environmental resistance of 718, whether through the finer slip distribution or a change in grain boundary chemistry.[50]

The alloys represented in Fig. 14 demonstrate the effect of grain size on the rate and mode of cyclic crack growth rates at 1200 °F (649 °C).[25] For 0.33-Hz tests cracking was transgranular for the alloys with grain sizes larger than about 15 μm, Waspaloy, HIP Astroloy, and HIP MERL-76. In NASA IIB-7 and IN-100, both having 5-μm grain sizes, cracking was intergranular, while in H+F Astroloy and René 95 cracking was intergranular in the fine necklace grains and transgranular in the large unrecrystallized grains. For the 15-min dwell tests cracking was intergranular in all the alloys; yet the difference in cyclic crack growth rate among the alloys was magnified.

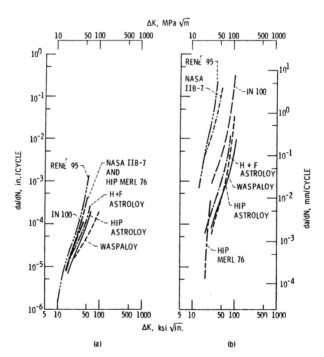

Fig. 14. Cyclic crack growth rates of several superalloys at 1200 °F (649 °C) in tests at 0.33 Hz and tests with a 15-min dwell at maximum load, $R = 0.05$.[25]

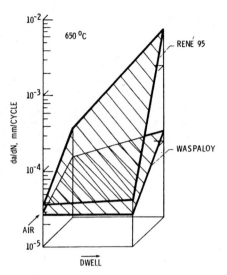

Fig. 15. Cyclic crack growth rate of HIP René 95 and Waspaloy at 1200 °F (649 °C) in 0.1 Hz tests and tests with 2-min dwells at the maximum load both in vacuum and in air, $R -0.05$.[65]

The separate contributions of environmental and creep damage are illustrated in Fig. 15 for Waspaloy and HIP René 95.[65] Their respective grain sizes were about 40 and 8 μm. Cyclic crack growth rates at 1200 °F (649 °C) for $\Delta K = 30$ MPa $\sqrt{m}$ are shown for 0.1 Hz and 2-min dwell tests in air and vacuum. It is clear that in vacuum the creep contribution alone to da/dN is very small even in Waspaloy. Crack growth rates in air were considerably higher for the fine-grain HIP René 95, particularly in the 2-min dwell tests even though the mode of growth was intergranular in both alloys. In contrast, a conventional René 95 forging with a 150-μm grain size studied earlier[35] had about the same crack growth behavior as the Waspaloy represented in Fig. 15. Relatively, the increase in da/dN for the 2-min hold test in air over that for the "pure" fatigue test in vacuum was more 100 times greater for the fine-grain René 95 than that for the large-grain René 95.

Grain size accounts for much of the differences among alloys in resistance to the environmentally controlled, intergranular mode of cyclic crack growth just discussed. Additional effects due to alloy composition appear to be relatively small. This is suggested by the comparison of large-grain forms of René 95 and Waspaloy above. Additionally, it was found that for PM forms of Astroloy, René 95, and IN-100, all having grain sizes of 3–10 μm, cyclic crack growth at 1200 °F and 0.33 Hz was intergranular, and the rate was very nearly the same among the alloys.[51] Minor element or impurity concentrations at grain boundaries may be more significant than those of the major elements. This appears to be the case for pure nickel; however work to date on nickel-base superalloys appears inconclusive.

Strain range and environment have an interesting interaction in determining the location of failure initiation and the size of defects observed at these sites.[66] This is illustrated in Fig. 16 by observations on failure initiation sites in three forms of René 95 tested at 1200 °F (649 °C).[35] In high $\Delta\epsilon_t$ tests cracks initiate readily, even at the many near average size defects, and one connected to the surface grows quickly to failure because of the access of air. In low $\Delta\epsilon_t$ tests cracks initiate only at the rare large defects. These are most likely to be subsurface since they are randomly distributed in the volume of the specimen. The variation in the fraction of crack propagation life during which the environment is excluded contributes to the large variance in life observed for the PM superalloys. Of course, the largest contribution to this variance is the sizes and types of defects discussed in Chapter 17.

Effects of air environment on superalloys without grain boundaries, on the other hand, appear small and possibly beneficial in some respects. Fatigue life tests of single crystals have shown improvement in air environment relative to vacuum,[32] and crack growth rate tests have shown little time dependence in the air environment.[64]

While the environmental influence is strong at intermediate temperatures, creep becomes the dominant time-dependent influence at higher temperatures. Internal damage, cavitation, occurs at grain boundaries ahead of the crack, and crack growth occurs by linkage of the internal cavities. Thus, environment has somewhat less of an effect. The transition to creep-dominated crack growth in HIP Astroloy was studied by Pelloux and Huang.[67] For test at 1 Hz, da/dN increased less than 2-fold between 1200 and 1400 °F (649 and 760 °C). For 15-min dwell tests, a 20-fold increase in da/dN was observed at 1400 °F, but there was little difference between tests in air and vacuum. Fracture surfaces for the 1400 °F dwell tests showed that grain boundary cavitation had become the dominate damage mechanism. Hofflener

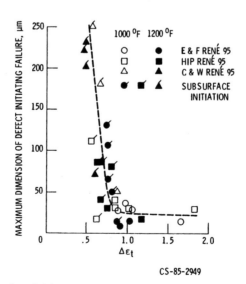

CS-85-2949

Fig. 16. Size and location of defects at failure origin for fatigue tests of three forms of René 95 at 1200 °F (649 °C), 0.33 Hz, $R_\epsilon = -1$.[35]

and Speidel[62] showed for cast IN-738 LC and IN-939 at 1562 °F (850 °C) that decreasing frequency increased da/dN as expected for both air and vacuum tests, but the difference in da/dN between air and vacuum diminished to near nothing at low frequencies. Scarlin[61] had attributed similar findings for IN-738 LC to crack branching that occurred in the air tests, effectively lowering ΔK.

Creep-dominated crack growth behavior is complicated for other reasons. Though ΔK often provides adequate correlation, the question of the best parameter for problems with large-scale plasticity, ΔJ, C^*, and so on, remains unsettled. Also, creep produces different effects at low and high stress intensities. It is typically observed that ΔK_{th} is higher for creep fatigue than for fatigue cycles, but at higher stress intensities creep fatigue crack growth becomes faster. When it is attempted to extend a sharp crack initiated in fatigue in a creep dwell cycle at low ΔK, crack growth may slow and even arrest completely. If K for the dwell period is below K_{th} for creep, the crack cannot propagate during the dwell and only blunts, slowing the crack advance for each cycle.

WORK FOR THE FUTURE

The following is a short list of areas where increased understanding may be expected in the near future because of programs underway in industrial, university, and government laboratories: (1) constitutive behavior and life modeling for thermo-mechanical cycling including multiaxial loading and material anistropy effects; (2) life modeling for components with environmental coatings; and (3) short crack growth including general plasticity, environmental, and material microstructure and anisotropy effects.

REFERENCES

1. C. H. Wells and C. P. Sullivan, *Trans. ASM*, **60**, 217 (1967).
2. C. H. Wells and C. P. Sullivan, *Trans. ASM*, **61**, 149 (1968).
3. V. Gerold and D. Steiner, *Scripta Met.*, **16**, 405 (1982).
4. H. F. Merrick, *Met. Trans.*, **5**, 891 (1974).
5. R. E. Stoltz and A. G. Pineau, *Mat. Sci. Eng.*, **34**, 275 (1978).
6. A. E. Staton-Bevan, *Phil. Mag.*, **47**(6), 939 (1983).
7. C. E. Feltner and P. Beardmore, *Achievement of High Fatigue Resistance in Metals and Alloys*, ASTM-STP 467, American Society for Testing and Materials, Philadelphia, PA, 1970, p. 77.
8. H.-P. Klein, *Z. Metalk.*, **61**, 573 (1970).
9. M. Graf and E. Hornbogen, *Scripta Met.*, **12**, 147 (1978).
10. D. Fournier and A. Pineau, *Met. Trans. A*, **8A**, 1095 (1977).
11. R. W. Landgraf, *Work Hardening in Tension and Fatigue*, TMS-AIME, Warrendale, PA, 1977, p. 240.
12. L. G. Fritzmeier, "The Cyclic Stress-Strain Behavior of Nickel-Base Superalloy Poly- and Single Crystals," Eng. Sc. D. Thesis, Columbia University, 1984.
13. L. Remy, A. Pineau, and B. Thomas, *Mat. Sci. Eng.*, **36**, 47 (1978).
14. S. D. Antolovich, E. Rosa, and A. Pineau, *Mat. Sci. Eng.*, **47**, 47 (1981).
15. S. S. Ezz, D. P. Pope, and V. Paidar, *Acta Met.*, **30**, 921 (1982).
16. D. A. Jablonski and S. Sargent, *Scripta Met.*, **15**, 1003, (1981).

17. T. P. Gabb, J. Gayda, and R. V. Miner, *Met. Trans. A*, **17A**, 497 (1986).
18. L. F. Coffin, Jr., *Met. Trans.*, **2**, 3105 (1971).
19. C. E. Jaske, R. C. Rice, R. D. Buchheit, D. B. Roach, and T. L. Porfilio, NASA CR-135022, May 1976.
20. D. N. Robinson and P. A. Bartolotta, NASA CR-174836, March 1985.
21. C. E. Pugh and D. N. Robinson, *Pressure Vessels and Piping: Design Technology—1982—A Decade of Progress*, ASME, New York, 1982, p. 171.
22. G. A. Swanson and R. C. Bill, AIAA Paper 85-1421, 1985.
23. A. Kaufman, J. H. Laflen, and U. S. Lindholm, AIAA Paper 85-1418, American Institute of Aeronautics and Astronautics, New York, 1985.
24. S. S. Manson, *Exp. Mech.*, **5**(7), 193 (1965).
25. B. A. Cowles, J. R. Warren, and F. K. Haake, NASA CR-165123, July 1980.
26. D. L. Anton, *Acta Met.*, **32**, 1669 (1984).
27. G. F. Harrison and M. J. Weaver, AGARD-CP-243, Advisory Group for Aerospace Research and Development, NATO, Neuilly-Sur-Seine, France, August 1978, p. 6-1.
28. C. R. Brinkman and G. E. Korth, *J. Test. Eval.*, **2**(4), 249 (1974).
29. M. Clavel, C. Levaillant, and A. Pineau, *Creep-Fatigue-Environment Interactions*, TMS-AIME, Warrendale, PA, 1980, p. 24.
30. J. D. Morrow, *Internal Friction, Damping, and Cyclic Plasticity*, ASTM STP-378, American Society for Testing and Materials, Philadelphia, PA, 1964, p. 45.
31. B. Boursier, Evaluation of Damage Mechanisms in the Ni-Base Superalloy René 80 under Low Cycle Fatigue in the Temperature Range 75–1400 °F, M.S. Thesis, University of Cincinnati, Cincinnati, OH, 1981.
32. D. J. Duquette and M. Gell, *Met. Trans.*, **3**, 1899 (1972).
33. M. Gell, G. R. Leverant, and C. H. Wells, *Achievement of High Fatigue Resistance in Metals and Alloys*, ASTM STP-467, American Society for Testing and Materials, Philadelphia, PA, 1969, p. 113.
34. D. R. Chang, D. D. Kruger, and R. A. Sprague, *Superalloys 1984*, TMS-AIME, Warrendale, PA, 1984, p. 245.
35. R. V. Miner and J. Gayda, *Int. J. Fat.*, **6**(3), 189 (1984).
36. L. F. Coffin, Jr., *Met. Trans.*, **3**, 177 (1972).
37. C. J. McMahon and L. F. Coffin, Jr., *Met. Trans.*, **1**, 3443 (1970).
38. M. H. Hirschberg and G. R. Halford, NASA TN D-8072, January 1976.
39. W. J. Ostergren, *1976 ASME-MPC Symposium on Creep-Fatigue Interaction*, ASME, New York, 1976, p. 179.
40. S. S. Manson, G. R. Halford, and M. H. Hirschberg, *Symposium on Design for Elevated Temperature Environment*, ASME, New York, 1971, p. 12.
41. R. V. Miner, J. Gayda, and M. G. Hebsur, NASA TM-87110, September 1985.
42. K. D. Sheffler, NASA CR-134626, June 1974.
43. G. R. Leverant, T. E. Strangman, and B. S. Langer, *Superalloys: Metallurgy and Manufacture*, Claitors, Baton Rouge, LA, 1976, p. 285.
44. P. T. Bizon and D. A. Spera, *Thermal Fatigue of Materials and Components*, ASTM STP-612, American Society for Testing and Materials, Philadelphia, PA, 1976, p. 106.
45. A. E. Gemma, B. S. Langer, and G. R. Leverant, *Thermal Fatigue of Materials and Components*, ASTM STP-612, American Society for Testing and Materials, Philadelphia, PA, 1976, p. 199.
46. L. F. Coffin, Jr., *1976 ASME-MPC Symposium on Creep-Fatigue Interaction*, ASME, New York, 1976, p. 349.
47. S. Majumdar and P. S. Maiya, *1976 ASME-MPC Symposium on Creep-Fatigue Interaction*, ASME, New York, 1976, p. 323.
48. W. Hoffelner, *Superalloys 1984*, TMS-AIME, Warrendale, PA, 1984, p. 771.
49. H. F. Merrick and S. Floreen, *Met. Trans. A*, **9A**, 231 (1978).
50. S. Floreen, *Creep-Fatigue-Environment Interactions*, TMS-AIME, Warrendale, PA, 1980, p. 112.
51. J. Gayda, R. V. Miner, and T. P. Gabb, *Superalloys 1984*, TMS-AIME, Warrendale, PA, 1984, p. 733.

52. B. Lawless, S. D. Antolovich, C. Bathias, and B. Boursier, *Fracture: Interaction of Microstructure, Mechanisms, Mechanics*, TMS-AIME, Warrendale, PA, 1984, p. 285.
53. J. E. King, *Met. Sci.*, **16**, 345 (1982).
54. C. W. Brown, J. E. King, and M. A. Hicks, *Met. Sci.*, **18**, 374 (1984).
55. E. Hornbogen and K.-H. Zum Gahr, *Acta Met.*, **24**, 581 (1976).
56. R. A. Venables, M. A. Hicks, and J. E. King, *Fatigue Crack Growth Threshold Concepts*, TMS-AIME, Warrendale, PA, 1984, p. 341.
57. H. G. Popp and A. Coles, AFFDL TR 70-144, Air Force Flight Dynamics Laboratory, Wright-Patterson Air Force Base, Ohio, 1970, p. 71.
58. H. D. Solomon and L. F. Coffin, *Fatigue at Elevated Temperatures*, ASTM STP-520, American Society for Testing and Materials, Philadelphia, PA, 1973, p. 112.
59. T. Ohmura, R. M. Pelloux, and N. J. Grant, *Eng. Fract. Mech.*, **5**, 909 (1973).
60. P. Shahinian and K. Sadananda, *1976 ASME-MPC Symposium on Creep-Fatigue Interaction*, ASME, New York, 1976, p. 365.
61. R. B. Scarlin, *Advances in Research on the Strength and Fracture of Materials* (Fracture 1977), vol. 2B, Pergamon, Oxford, 849.
62. W. Hoffelner and M. O. Speidel, *Advances in Fracture Research* (Fracture 1981), Pergamon, Oxford, 1981, p. 2431.
63. S. Golwalkar, N. S. Stoloff, and D. J. Duquette, *Proc. 7th Int. Conf. on Strength of Metals and Alloys*, vol. 2, Pergamon, Oxford, 1983, p. 879.
64. D. P. DeLuca, and B. A. Cowles, AFWAL-TR-84-4167, Wright-Patterson Air Force Base, Ohio, February 1985.
65. J. Gayda, T. P. Gabb, and R. V. Miner, NASA TM-87150, September 1985.
66. J. M. Hyzak and I. M. Bernstein, *Met. Trans. A*, **13**A, 45 (1982).
67. R. M. Pelloux and J. S. Huang, *Creep-Fatigue-Environment Interactions*, TMS-AIME, Warrendale, PA, 1980, p. 151.

PART FOUR

SURFACE STABILITY

Chapter 11

High-Temperature Oxidation

JAMES L. SMIALEK and GERALD H. MEIER

NASA Lewis Research Center, Cleveland, Ohio,
and University of Pittsburgh, Pittsburgh, Pennsylvania

Resistance to oxidizing environments at high temperatures is a requisite for superalloys whether used coated or uncoated. Therefore, an understanding of superalloy oxidation and how it is influenced by alloy characteristics and exposure conditions is essential for effective design and application of superalloys. In this chapter a brief review of the fundamentals of the oxidation of metals and alloys is presented, followed by a discussion of simple Cr_2O_3- and Al_2O_3-forming alloys. The effects of common alloying elements on the oxidation behavior of these base systems are then discussed, which provides the basis for a broad treatment of the oxidation behavior of complex superalloys.

FUNDAMENTAL PRINCIPLES OF ALLOY OXIDATION

The major aspects of the oxidation of metals and alloys will be briefly reviewed in this section.

Pure Metals

The diffusion-controlled growth rate of an oxide layer of thickness x is observed usually to follow the equation

$$dx/dt = \frac{k'}{x} \tag{1}$$

where k' is called the "parabolic rate constant" with units cm^2/sec. The integrated form of this equation is

$$x_t^2 - x_{t_0}^2 = 2k'(t - t_0) \tag{2}$$

where t_0 is the time at which diffusion control begins. The extent of reaction may also be expressed in terms of mass change per unit area, $\Delta m/A$, according to

$$\left(\frac{\Delta m}{A}\right)_t^2 - \left(\frac{\Delta m}{A}\right)_{t_0}^2 = 2k''(t - t_0) \tag{3}$$

where

$$k'' = \left(\frac{8}{V}\right)^2 k' \tag{4}$$

and V is the equivalent volume of the oxide. The units of k'' are g^2/cm^{-4}. (See Chapter 3 of Birks and Meier[1] for a discussion of units and conversions.)

Wagner[2] provided a theoretical treatment for k' based on the following assumptions:

1. The oxide layer is a compact, perfectly adherent scale.
2. Migration of ions or electrons across the scale is the rate-controlling process.
3. Thermodynamic equilibrium is established at both the metal–oxide and oxide–gas interface.
4. The oxide shows only small deviations from stoichiometry.
5. Thermodynamic equilibrium is established locally throughout the scale.
6. The scale is thick compared with the distances over which space charge effects occur.
7. Oxygen solubility in the metal may be neglected.

The results of this treatment may be expressed by the following expression:

$$k' = \frac{1}{RT} \int_{\mu_0'}^{\mu_0''} \left(D_O + \frac{Z_M D_M}{2}\right) d\mu_O \tag{5}$$

where D_M and D_O are the diffusivities of M and O, respectively, in the oxide, and Z_M is the metal volume, and μ_O' and μ_O'' are the chemical potentials of O at the metal–oxide and oxide–gas interfaces, respectively. Good agreement has been found between experimentally determined rate constants and those calculated from the Wagner theory for a number of metals reacting with oxygen, sulfur, and halogens.[3]

The majority of data, however, deviate from the Wagner model because one or more of Wagner's assumptions are not satisifed in the real systems. This is particularly true when assumption 2 is not satisfied, in which case the rate of scale growth will not follow a simple parabolic expression such as equations (2) or (3). Nevertheless, the use of parabolic rate constants is a convenient way to compare the relative rates of oxidation for the many cases that exhibit approximately parabolic behavior, as illustrated in Fig. 1.

Alloys

The oxidation of alloys is more complex than that of pure metals. Following Wagner[4] alloys may be grouped as (1) noble parent with alloying elements of a more base (less noble) nature and (2) base parent with base alloying elements. Typical oxidation morphologies within these two groups are presented in Figs. 2 and 3. A comprehensive treatment of the diffusion processes leading to these and other morphologies has been given by Whittle.[5] In Fig. 2 element A represents a noble parent (i.e., it will not react with oxygen at the ambient pO_2) and element B represents a base solute (i.e., the dissociation pressure of the oxide BO is less than the ambient pO_2). If B is present in dilute concentration and oxygen is soluble in A, a morphology such as Fig. 2a will result, in which BO precipitates internally. In the simplest case for which BO is very stable and $D_B << D_O$ in the alloy the depth of internal oxidation X may be written as

$$X(t) = \left(\frac{2N_O^{(s)} D_O t}{N_B^{(0)}} \right)^{1/2} \tag{6}$$

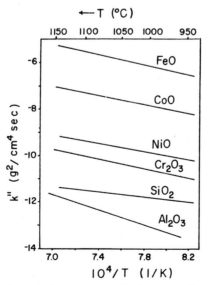

Fig. 1. Order-of-magnitude rate constants for the growth of selected oxides.

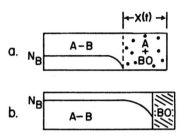

Fig. 2. Schematic cross sections and concentration profiles for oxidation of a noble metal, A, alloyed with a reactive metal, B. (a) Alloy dilute in B showing internal oxidation of B. (b) Alloy concentrated in B showing formation of an external layer of BO.

where $N_O^{(s)}$ is the oxygen solubility and $N_B^{(0)}$ is the bulk alloy concentration of B both expressed as atom fraction. Equation (6) indicates that $X(t)$ decreases as $N_B^{(0)}$ increases, and when sufficient B is present, the outward flux of B results in BO being formed as a continuous surface layer (Fig. 2b). This transition from internal to external oxidation occurs when $N_B^{(0)} > N_B^{(\text{critical})}$, where

$$N_B^{(\text{critical})} = \left(\frac{\pi g^*}{3} N_O^{(s)} \frac{D_O V_M}{D_B V_{ox}} \right)^{1/2} \tag{7}$$

where V_M and V_{ox} are the molar volumes of alloy and oxide, respectively, and g^* is the critical volume fraction of oxide required for the transition, which is often around 0.3.

Figure 3 presents the more general case where the oxides of both A and B are stable in the gas but BO is more stable than AO (in superalloys A generally represents Ni or Co and B represents Cr, Al, Ti, etc.). For low concentrations of B an external layer of AO will form, and internal oxides of BO will precipitate in the alloy (Fig. 3a). If the concentration of B is increased to exceed the critical concentration for transition to external oxidation, a morphology such as that in Fig. 3 develops. The

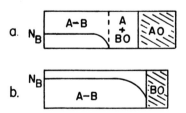

Fig. 3. Schematic cross sections of an A–B alloy where both components form stable oxides but BO is more stable than AO. (a) Alloy dilute in B showing internal oxidation of B under an external layer of AO. (b) Alloy concentrated in B showing continuous external BO.

formation of a continuous layer of BO precludes any further formation of the less stable AO, although some AO will generally form before the BO layer becomes continuous, that is, "transient oxidation." The morphology depicted in Figure 3b is the objective of alloying for oxidation resistance. That is, an alloying element B, which forms an oxide that is both very stable and slow growing, is added in sufficient quantity to form a protective surface layer by "selective oxidation."[6]

The oxidation rate of an alloy such as that in Fig. 3b will be essentially parabolic with a rate constant characteristic of BO. However, as seen from the concentration profile of B, selective oxidation depletes B from the alloy under the scale. This depletion will eventually result in an enrichment of oxides of A in the scale with the rate increasing toward that characteristic of AO. The length of time required for the transition to a more rapid rate will depend on a number of factors such as temperature, specimen size, diffusivities in the alloy and scale, and initial concentration of B in the alloy. The transition is hastened by any process that lessens the protectiveness of the BO layer. This can occur by evaporation of a volatile oxide species (e.g., CrO_3 from Cr_2O_3) that thins the protective layer or by mechanical damage to the scale by the action of erosive particles or cracking and spalling due to applied or generated stresses. The latter situation is particularly important since most superalloys undergo thermal cycling during service that is accompanied by stress generation due to thermal expansion mismatch between the oxide and alloy. The results of thermal cycling are shown schematically in Fig. 4, which compares the isothermal and cyclic oxidation weight changes of a given alloy. Initially the rate of weight gain in cyclic exposure is similar to the isothermal rate but eventually goes through a maximum and then decreases. This decrease, which is due to oxide spalling,

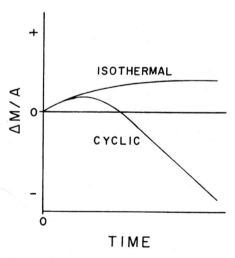

Fig. 4. Schematic plot of weight change vs. time comparing isothermal and cyclic oxidation behavior of the same alloy.

actually corresponds to a more rapid rate of consumption of the alloy. In many situations cyclic oxidation rates such as that in Fig. 4 may be approximated by

$$\frac{\Delta m}{A} = k^{1/2}_1 t^{1/2} - k_2 t \tag{8}$$

where k_1 is a growth constant and k_2 a spalling constant.[7]

OXIDATION OF M–CR ALLOYS (CR$_2$O$_3$ FORMERS)

A large number of nickel-base superalloys and virtually all Co- and Fe-base superalloys rely on the formation of Cr_2O_3 scales for oxidation resistance. Therefore, the oxidation of binary alloys with chromium will be discussed here. The oxidation behavior of nickel–chromium alloys may be separated into three groups.[8] In Group I, which corresponds to dilute alloys ($< 10\%$ Cr), the oxidation morphologies are similar to Fig. 3a, that is, external NiO scales and internal Cr_2O_3 precipitates. For this group the rate constants are somewhat larger than those for pure nickel due to "doping" of the NiO scale with chromium, which results in an increased diffusivity of nickel ions. In Group III ($\geqslant 30\%$ Cr) the oxidation morphologies are similar to Fig. 3b, that is, external Cr_2O_3 scales. For this group the rate constants are several orders of magnitude smaller than those for pure nickel. In Group II external scales of Cr_2O_3 were observed over alloy grain boundaries, while external NiO scales formed over alloy grains with corresponding formation of internal Cr_2O_3. Concentrations of chromium of 30 wt. % or more are required for purely external Cr_2O_3 scales. The oxidation of cobalt–chromium and iron–chromium alloys are qualitatively similar to nickel–chromium. However, the rate constants for Cr_2O_3 growth on nickel-, cobalt-, and iron-base alloys vary by two orders of magnitude with no apparent correlation with substrate alloy.[9]

Growth of Cr$_2$O$_3$

The apparent variability of the rate constant for Cr_2O_3 growth on various substrates requires discussion of the defect structure and growth mechanism of Cr_2O_3. Kofstad and Lillerud[10] reviewed the existing literature on nonstoichiometry and concluded that the predominant defects in Cr_2O_3 at low pO_2 are chromium interstitials compensated by electrons and that at high pO_2, Cr_2O_3 is an intrinsic semiconductor. Seebeck coefficient measurements on sintered Cr_2O_3 also indicate chromium interstitials to be the predominant ionic defect at low pO_2.[11] However, the pO_2 dependence of the chromium diffusivity in single-crystal Cr_2O_3 (see ref. 12) and the deviation from stoichiometry in polycrystalline Cr_2O_3 (see ref. 13) are consistent with chromium vacancies being the predominant defect. However, these measurements are not necessarily contradictory since the highest pO_2 used indicating chromium interstitials in the Seebeck effect measurements was about four orders of magnitude lower than the lowest pO_2 used in the diffusion measurements. Therefore, it is possible that the predominant ionic defects in Cr_2O_3 are chromium vacancies at high pO_2 and chromium intersitials at low pO_2, perhaps with a zone of intrinsic semiconduction separating the two regimes.

The above discussion of lattice defects in Cr_2O_3 cannot be applied, in most cases, to the growth of Cr_2O_3 on pure chromium or chromium-containing alloys since it has become clear that short-circuit diffusion predominates during this process. The rate constant for Cr_2O_3 growth calculated from equation (5) using the lattice diffusion coefficient for chromium in Cr_2O_3 is always many orders of magnitude smaller than the experimental value for the growth of polycrystalline Cr_2O_3.[12,14] Also in rare instances when single-crystal Cr_2O_3 scales form, the growth rate is orders of magnitude slower than for polycrystalline Cr_2O_3.[15] The single-crystal scales are uniform and planar, whereas the polycrystalline scales become buckled and detached from the substrate. These observations have led to the proposal that grain boundaries in the oxide provide the short-circuit paths for diffusion of cations and, perhaps, anions.[15] Indeed, grain-size-dependent growth rates for Cr_2O_3 have been observed.[16] Therefore, it appears likely that much of the variability of Cr_2O_3 growth rates is associated with the scale microstructure and the availability of short-circuit diffusion paths. However, a quantitative correlation between growth rate and detailed scale microstructure is still needed.

Effect of Oxygen-Active Elements and Oxide Dispersions

Small alloy additions of rare-earth elements and other oxygen-active elements have been found to alter the oxidation resistance of Cr_2O_3-forming alloys.[17–20] The effects of these additions usually include (1) formation of continuous Cr_2O_3 scales at lower alloy chromium concentrations, (2) reduction in the rate of Cr_2O_3 growth, (3) improved scale adhesion, (4) change in the primary growth mechanism of the oxide from outward cation migration to inward anion migration, and (5) a reduction of the grain size in the Cr_2O_3 scale. Similar effects have also been observed when the oxygen-active elements are present as a fine oxide dispersion in the alloy prior to oxidation.[21–25] The effects of cerium on reducing growth rates and increasing adherence of Cr_2O_3 scales are illustrated in Fig. 5.

These observations have been explained in terms of (1) dispersoid accumulating at the metal–oxide interface and eventually blocking diffusive transport,[21] (2) dispersoid particles acting as heterogeneous nucleation sites for oxide grains to reduce the internuclear distance and allowing more rapid formation of a continuous Cr_2O_3 film with finer grain size and fewer short-circuit paths for cations (probably dislocations) so that anion diffusion becomes rate controlling,[24] or (3) oxide particles of the oxygen-active element serving as nucleation sites to produce a fine-grained scale and the ions of these elements blocking grain boundary transport through the Cr_2O_3 scale.[20] Similar effects have been observed when CeO_2 powder was applied to the surface of a nickel–chromium alloy and ion implantation of elements such as yttrium and cerium have been observed to decrease Cr_2O_3 growth on nickel–chromium[26] and iron–nickel–chromium[27] alloys.

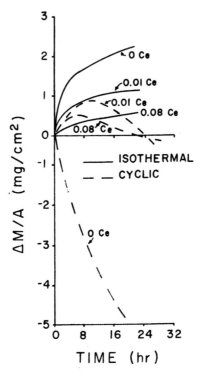

Fig. 5. Effect of Ce additions on the isothermal and cyclic oxidation behavior of Ni–50Cr (at. %) in O_2 at 1100 °C (2010 °F).[20]

Oxide Volatility Effects

Alloys that form protective Cr_2O_3 scales are susceptible to accelerated degradation at very high temperatures in high pO_2 gases due to the reaction

$$Cr_2O_3(s) + \frac{3}{2} O_2(g) = 2CrO_3(g) \tag{9}$$

The evaporation of CrO_3 results in the continuous thinning of the Cr_2O_3 scale so the diffusive transport through it is rapid. The effect of the volatilization on the oxidation kinetics have been analyzed.[28] The results for the oxidation of chromium are shown in Fig. 6. Initially, when the diffusion through a thin scale is rapid, the effect of CrO_3 volatilization is not significant, but as the scale thickens, the rate of volatilization becomes comparable and then equal to the rate of diffusive growth. This situation, paralinear oxidation, results in a limiting scale thickness, x_0, for which $dx/dt = 0$, as shown schematically in Fig. 6a. The occurrence of a limiting scale thickness implies protective behavior, but in fact, the amount of metal consumed increases until a constant rate is achieved. This can be seen more clearly by considering the metal recession (Fig. 6b). This problem is one of the major limitations on the

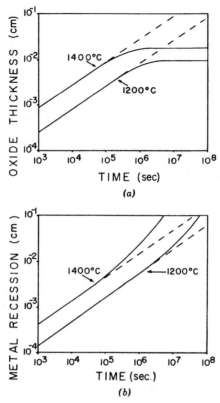

Fig. 6. Effect of CrO$_3$ evaporation on the oxidation behavior of Cr. (a) Oxide thickness reaches steady state with time. (b) Metal recession rate increases with time.[28]

very high-temperature use of Cr$_2$O$_3$-forming alloys and coatings. It becomes significant at temperatures around 1000 °C and even lower in the presence of high-velocity gases. In these situations Al$_2$O$_3$-forming alloys are more resistant since there are no vapor species in the aluminum–oxygen system with significant vapor pressures.

AL₂O₃-FORMING SYSTEMS; COATING ALLOYS

In addition to chromium, aluminum plays a critical role in the oxidation of γ'-strengthened superalloys. This element also provides the basis of oxidation-resistant NiAl diffusion aluminide and MCrAlY overlay coatings described in Chapter 13. This section is therefore concerned with the mechanism of exclusive Al$_2$O$_3$ growth and adhesion.

Effect of Aluminum Content

On the basis of thermodynamics alone, exclusive Al$_2$O$_3$ scales should be formed on nickel–aluminum alloys at levels as low as 1 ppm aluminum.[29] In actuality, the process is kinetically limited by the opposing diffusional fluxes in the alloy of oxygen inward and aluminum outward. These promote nonprotective internal oxidation or protective external scales, respectively. Analogous to nickel–chromium oxidation, exclusive Al_2O_3 scales will be formed as opposed to discontinuous internal Al_2O_3 precipitates when the aluminum content is sufficiently high to produce the critical mole fraction of oxide particles required for particle link-up. Growth of external Al_2O_3 scales can be sustained when the flux of aluminum in the alloy exceeds that consumed by scale thickening:

$$J_{alloy}^{Al} > J_{oxide}^{Al} \qquad (10)$$

The nickel–aluminum system is characterized by three regions shown in Fig. 7a: (I) 0–6% aluminum, internal $Al_2O_3(+NiAl_2O_4)$ oxides + NiO external scales result; (II) 6–17% aluminum, external Al_2O_3 forms initially, cannot be sustained due to an inadequate supply of aluminum and is overtaken by a fast growing $NiO+NiAl_2O_4+Al_2O_3$ oxide mixture; (III) $\geqslant 17\%$ aluminum, the external Al_2O_3 scale is maintained due to a sufficient supply of aluminum. Higher temperature extends this region to lower aluminum contents.

Changes in oxidation kinetics go hand in hand with the scales formed (Fig. 7b). At low aluminum contents, $k_p(NiO)$ is increased by approximately one order of magnitude due to Al^{3+}-doped NiO scales plus some internal oxidation. As more Al_2O_3 is formed (region II) k_p drops by one to two orders of magnitude. Lower growth rates are observed for higher temperatures because of the increased tendency for an exclusive Al_2O_3 scale. At 25% (region III), only Al_2O_3 is formed, and k_p is reduced by another one to three orders of magnitude.

Effects of Chromium Additions

Ni(Co,Fe) base alloys with *both* chromium and aluminum additions benefit from a remarkable synergistic effect of these elements amounting to great technological importance. For example, chromium additions of about 10 w/o can enable external Al_2O_3 scale formation on alloys having aluminum levels as low as 5 w/o. (This is in contrast to $\geqslant 17\%$ Al for binary alloys.) This phenomenon has allowed for the design of more ductile and diffusionally stable MCrAl coatings as well as the matrix compositions for ODS alloys.

The compositional control of oxidation is most easily summarized in oxide maps of scale components and weight change as a function of chromium and aluminum content. In Fig. 8a three primary regions of oxidation can be seen corresponding to (I) NiO external scales + Al_2O_3/Cr_2O_3 subscales; (II) Cr_2O_3 external scale + Al_2O_3 subscales; and (III) external scales of only Al_2O_3. Long-duration cyclic oxidation has the effect of substantially contracting the region of optimum performance to higher aluminum and chromium contents (Fig. 8b). A wide transition zone forms in its place due to oxide spalling, accelerated aluminum removal, and the formation of $NiO+NiCr_2O_4+NiAl_2O_4+Al_2O_3$ composite scales.

The role of chromium in producing Al_2O_3 scales at much lower aluminum contents than in binary nickel–aluminum alloys can be described by the general

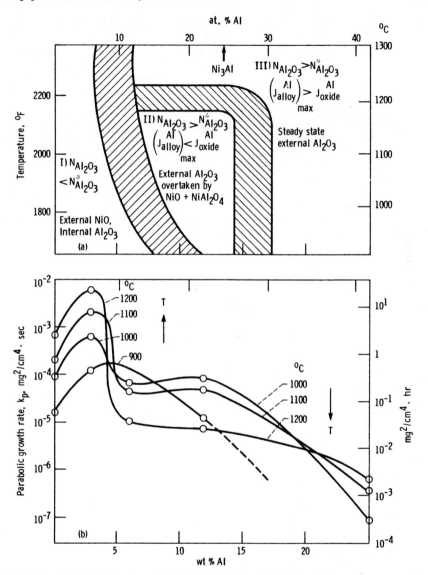

Fig. 7. Composional effects on the oxidation behavior of binary Ni–Al alloys. (a) Temperature-composition oxide phase map. (b) Scale growth rates corresponding to regimes in (a).[29]

oxidation phenomenon known as "gettering."[34] The mechanistic details of NiCrAl transient oxidation best illustrates this phenomenon[30,35] and is described for Ni–15Cr–6Al specifically in the schematic of Fig. 9.[35] The initial oxide (a) contains all the cations of the immediate alloy surface resulting in 15% NiO–85% Ni(Cr,Al)₂O₄ coverage (fast cation transport also causes some overgrowth of NiO); (b) subscale formation of Cr₂O₃ occurs because it is stable at the low oxygen activity defined

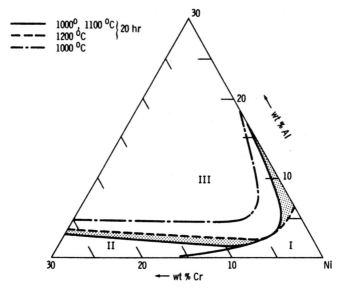

I   External NiO, Internal $Cr_2O_3/Al_2O_3/Ni(Al, Cr)_2O_4$
II   External Cr_2O_3, Internal Al_2O_3
III   External Al_2O_3

(a) **Isothermal map.**

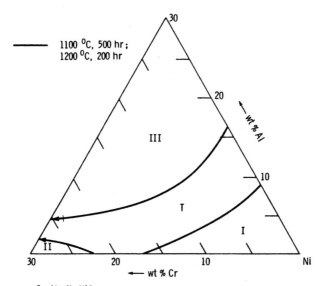

I   Mostly NiO
II   Mostly Cr_2O_3
III $Al_2O_3(+NiAl_2O_4)$
T   NiO, $NiCr_2O_4$, $NiAl_2O_4$, Al_2O_3 Transition

(b) **Cyclic map.**

Fig. 8. Compositional effects on the oxidation behavior of Ni–Cr–Al ternary alloys.[30,31,32,33]

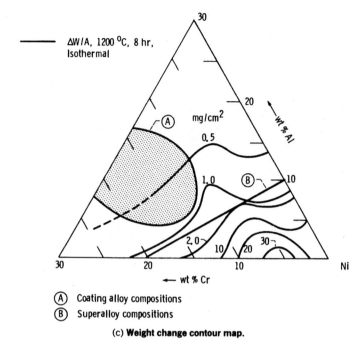

(c) **Weight change contour map.**

Fig. 8. (*Continued*)

by the NiO-alloy equilibrium, and internal oxidation of Al_2O_3 occurs ahead of this front since it is stable at the even lower activities here. The high chromium content results in a continuous Cr_2O_3 subscale (c) that defines a lower scale-alloy oxygen activity, reduces oxygen diffusion, and curtails internal Al_2O_3 formation. Further $NiO/Ni(Cr,Al)_2O_4$ growth is also blocked. Eventually, the Al_2O_3 subscale becomes continuous and rate controlling. Steady state is generally achieved in less than 1 h at 1000 °C. This cooperative action of chromium is also seen to have a very subtle nature: intimate mixtures of crystallographically coherent 0.1-μm subgrains of $Ni(Cr,Al)_2O_4$- and $(Cr,Al)_2O_3$-alloyed scales precede pure α-Al_2O_3 nucleation at the oxide–metal interface.[38]

Oxide growth rates can generally be grouped by the regions, and $k_p(I) > k_p(II) > k_p(III)$ by more than an order of magnitude for each grouping, as shown in Fig. 10.[30,36,37,39] This figure provides an interesting perspective on the comparison of pure nickel to Group I, Ni–30Cr to Group II, and Ni–25Al to Group III alloys. The three differentials caused by a ternary addition are the results of (1) aluminum- and chromium-doped NiO, (2) aluminum gettering assistance for Cr_2O_3 control, and (3) greater chromium doping of Al_2O_3 (for groups I, II, and III, respectively).

Somewhat more weight change detail can be seen from the contour map of Fig. 8c. Maximum oxidation resistance is suggested for compositions greater than 10%

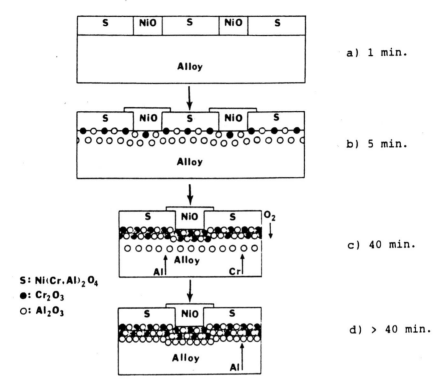

Fig. 9. Synergistic effect of Cr on the production of Al_2O_3 scales; gettering during transient oxidation of Ni–15Cr–6Al at 1000 °C (1830 °F): (a) 1 min; (b) 5 min; (c) 40 min; (d) >40 min (see text for discussion).[35]

chromium and 10% aluminum, that is, the region used for actual NiCrAl coatings. There is some indication that chromium additions under 10% are detrimental for all nickel–aluminum compositions, especially in long-term (100–500-h) or cyclic conditions.[32,33] This is in apparent disagreement with some studies[30,40] and would not be predicted on the basis of oxide phase maps alone (Figs. 8a,b). On the other hand, aluminum increases beyond 3–5 are always found to be beneficial to oxidation resistance (cf. Fig. 7).[29,32,33,39,40]

Effects of Oxygen-active Dopants: Mechanism of Al_2O_3 Scale Growth and Adhesion

Rare-earth or oxygen-active elements produce dramatic modifications to MCrAl oxidation behavior that are equal in importance to those arising from aluminum or chromium additions. Small amounts (<1%) of these dopant elements prevent the Al_2O_3 scales from otherwise spalling at the oxide–metal interface. Discussion of the adherence effect is usually entwined with diffusion, growth, and morphological phenomena, as presented below.

Diffusion in Al₂O₃. Oxygen diffusion in α-Al₂O₃ occurs predominantly by grain boundaries.[41-43] The grain boundary contribution for a typical 1-μm grain size is about four orders of magnitude greater than the lattice contribution. Data derived from an Al₂O₃ scale growth model for FeCrAl–Y₂O₃ oxidation also predicts grain boundary dominance over lattice diffusion.[44]

Aluminum transport in the lattice is considerably faster than oxygen.[45] No direct measurements of aluminum boundary diffusion are available. For typical Al₂O₃ scale grain sizes of 1μm, aluminum lattice diffusion can be on the same order as the oxygen boundary contribution, with increasing importance as the grain size increases (and vice versa). This is especially true when compared against the oxygen boundary coefficients calculated from actual Al₂O₃ scale growth models.[44] Changes in grain size with time, temperature, or position in the scale can be significant (Fig. 11) and need to be addressed in any diffusional growth model.[41]

The question of whether Al₂O₃ scales grow inward by oxygen diffusion or outward by aluminum diffusion can only be answered unequivocally by tracer oxidation experiments.[46] The few studies available indicate that oxygen boundary diffusion does indeed predominate at high temperatures (1100 °C, 2010 °F) for

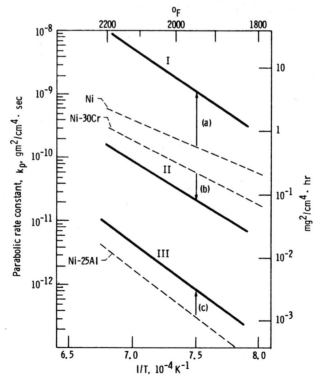

Fig. 10. Comparison of the oxidation kinetics of NiCrAl alloys (groups I, II, and III) with pure Ni, Ni–Cr, and Ni–Al alloys; Arrhenius plots.[30]

(a) Oxide - gas surface

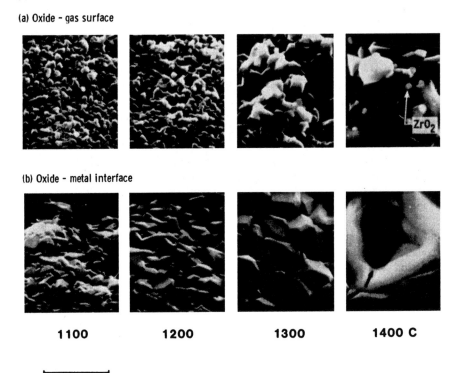

(b) Oxide - metal interface

1100   1200   1300   1400 C

├───────────────┤
5 μm

Fig. 11. Development of Al_2O_3 grains upon oxidation of FeCrAl+Zr at high temperatures for 100 h. (a) Oxide–gas surface and appearance of ZrO_2 precipitates. (b) Oxide–metal surface and dimples in centers of large Al_2O_3 grains.

NiCrAl+Y (or Zr).[47] However, some aluminum outward growth is observed, which gains in significance for low dopant values or low temperatures (900 °C, 1650 °F) where metastable transition aluminas and fast-cation transport are common.[38,48-50] Aluminum diffusion appears to be significant for undoped β-NiAl in 1100 °C, 3 h oxidation.[46]

Al_2O_3 scales show very low bulk solubilities for the much larger cations used as dopants. This results in segregation to oxide grain boundaries and precipitation of yttrium, zirconium, and thorium grain boundary oxides here (Fig. 11).[44,51] The precise effect of segregation on oxygen or aluminum boundary diffusion has not been measured but is likely to be considerable.

Dopant Effects on Kinetics, Morphology, and Adhesion. For alloys with no dopants, oxidation results in nonadherent Al_2O_3 films which spall cleanly at the oxide–metal interface. The addition of about 0.01–0.1 wt. % oxygen-active or-rare earth elements (including but not limited to Sc, Y, Zr, La, Hf, Ce, Yb, and

Th) results in strongly adherent scales. The effects of dopant level on isothermal and cyclic oxidation are summarized schematically in Fig. 12 for CoCrAl+Y, Hf and NiCrAl+Zr.

Region I.   Small additions are found to reduce somewhat the rate of scale growth (Fig. 12) and the amount of compressive growth stress within the scale.[52-55] The reduction in scale weight is attributed to a decrease in the amount of compressive buckling and total amount of oxide area as opposed to oxide thickness, as shown in Figs. 13a,b. This observation is often related to a growth stress argument based on production of oxide within the scale at its grain boundaries.[56-58] It is supported by tracer studies that found ¹⁸O buildups within preexisting scales on undoped alloys of NiAl.[46] This suggests that new oxide can be formed within the scale and that aluminum outward diffusion is responsible for growth stress and buckling. At very low dopant levels interfacial porosity can be effectively eliminated (Fig. 13b).[52,56,59] Finally, oxide adherence is conferred rather precipitously with dopant level as indicated by the amount of total oxidation in cyclic tests (Fig. 12).

Region II.   These alloys exhibit the optimum combination of isothermal and cyclic performance having both excellent adherence and minimal growth rates. Buckled scales and convoluted metal surfaces give way to flat interfaces and no voidage (Figs. 13b,c). Fingerlike Al₂O₃ protrusions (pegs) initiated by dopant-rich oxides grow from the oxide–metal interface into the alloy.[52-54]

Region III.   High dopant levels cause excessive weight gains due to subscale formation (Figs. 12 and 13d). Exceeding the solubility limit of the dopant element X often results in nonprotective oxidation of coarse M–X precipitates to form voluminous Al–X mixed oxides.[60]

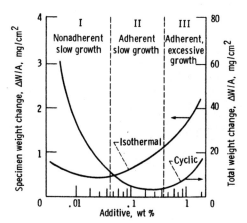

Fig. 12. Schematic comparison of oxygen-active dopant effects on isothermal and cyclic oxidation [NiCrAl+Zr and CoCrAl+Y, Hf alloys, 1200 °C (2200 °F) for 100 h].[52-54]

The boundaries of the three regions do not occur at a fixed or consistent value of the dopant level for different base alloys or oxidation temperatures. Consequently, the minimum dopant levels required to impart adhesion vary considerably,[52-57,61-63] and no universal optimum level exists. A simple generalization is that 0.1 wt. % is usually sufficient for adhesion and 1.0 wt. % usually results in excessive subscales.

Adherence Models. The models most often quoted to explain adhesion are listed as 1–4 in Table 1, as discussed elsewhere.[64] They have basically evolved from explanations of regions I, II, and III behavior and remain controversial. Much of the controversy arises from three factors. First, each possesses a certain degree of elegance because it is consistent with most of the kinetic, morphological, and adherence effects. Second, despite this agreement, there is at least one inconsistency between observed adherence and the mechanism by which adherence is proposed. For example, adherence without pegs, spalling without voids, spalling without stressed or buckled scales, and adherence in highly deformed scales have all been observed.[56,59,61] Third, partial contributions to adherence may indeed occur from all four models.

The most recent model (5 in Table 1) is based on the observation that sulfur segregation to the oxide–metal interface can be prevented by the oxygen-active

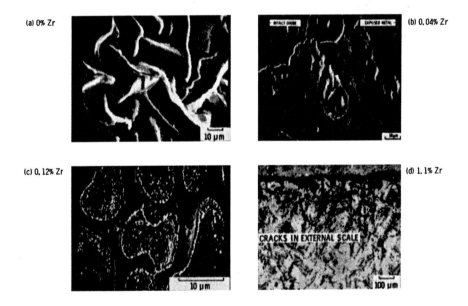

Fig. 13. Changes in scale morphology with Zr dopant level in NiCrAl alloys. (a,b,c) 8 h at 1100 °C (2010 °F); (d) 200 h at 1200 °C (2200 °F). (a) 0% Zr, buckled and cracked (arrow) scale. (b) 0.04% Zr, partial spalling and rumpling, no voids. (c) 0.12% Zr, no spalling or buckling. (d) 1.10% Zr, excessive subscale formation.[47,53]

Table 1. Models of Al₂O₃ Scale Adherence

Designation	Dopant effect
1) Pegging:	Oxide intrusions mechanically key scale
2) Vacancy sink:	Interfacial voids prevented by vacancy accommodation
3) Growth stress:	Al^{3+} counterdiffusion and growth within oxide prevented
4) Scale plasticity:	Finer Al$_2$O$_3$ grains, higher D$_{gb}$ promote stress relief by Coble creep
5) Chemical bond:	Sulfur "bond poison" gettered by dopants

dopants.[65-68] The role of sulfur here is believed to be analogous to sulfur grain boundary embrittlement of metals caused by capturing metal atom electrons and "poisoning" the bond.[69] Additional strong support for this proposal comes from experiments where adherent scales are produced for NiCrAl alloys having *no* oxygen-active yttrium, hafnium, or zirconium dopants by simply reducing or avoiding the initial sulfur impurity.[70,71]

The observed segregation of the dopant elements themselves has spawned some speculation that this further increases the strength of the interfacial bond.[68] The chemical factors tied to successful adherence additives, as proposed in this model, are a highly negative ΔG_f (oxide) and ΔG_f (sulfide), an electron configuration with empty d orbitals, and a low solubility in Ni(Co,Fe).[68]

EFFECTS OF OTHER COMMON ALLOYING ELEMENTS

Superalloys contain a number of significant alloying elements in addition to chromium and aluminum, including manganese, titanium, silicon, and the refractory metals. In many instances these elements exert significant influence on the oxidation resistance of Cr_2O_3- or Al_2O_3-forming alloys. In this section the effects on model alloys will be reviewed.

Manganese, Titanium, and Silicon

Manganese may be viewed as a potential, but much less effective, replacement for chromium in establishing healing layers of Cr_2O_3 scales. Manganese promotes Cr_2O_3 formation in Ni–20Cr,[72,73] but up to 30% manganese was not effective in Co–19Cr.[74] Additions of manganese have been shown to maintain Al_2O_3 scales on Fe–Al alloys by preventing iron-rich nodular eruptions.[75,76]

Additions of titanium promote Cr_2O_3 formation on both Ni–20Cr and Co–20Cr but do not significantly affect the Cr_2O_3 growth rate.[77,78] Adherence has been reported to degrade on Ni–20Cr, improve on Ni–50Cr, and remain unchanged on Co–20Cr when titanium is added.[78-80]

Titanium slightly increases the growth rate of Al_2O_3 scales on β-NiAl and does little to promote exclusive Al_2O_3 formation on γ'-Ni_3Al or γ-Ni(Al) alloys.[40,81] Titanium has a deleterious effect on adherence on a nickel-base superalloy[82] but improves it on Fe–18Cr–6Al.[83]

Of all the elements with oxygen reactivity intermediate between that of aluminum and nickel, silicon causes the most dramatic beneficial effects. The isothermal and cyclic behavior of some Ni–Cr, Fe–Cr, and Ni–Al alloys can be improved by silicon so as to equal the performance of exclusive Cr_2O_3- or Al_2O_3-forming alloys.[40,81,84] These benefits stem from the formation of SiO_2 subscales and the prevention of Ni(Fe) oxides. Additions of 0.5–1.3% silicon are very beneficial to B-1900, improving its cyclic oxidation resistance to the level of an aluminide coating on B-1900.[85] Similar additions to MAR-M 200 and IN-713 also caused improvements but not to the same extent as B-1900.[86] However, these benefits could not be made practical in that the silicon additions severely degraded mechanical properties even at the 0.5% level. Silicon has also found increasing usage in protective coatings based on NiCr or NiCrAl systems.[87] High-silicon coatings actually rely on the formation of a protective SiO_2 scale that can be as slow growing as Al_2O_3 films for certain Ni–Cr–Si compositions.[88] Smaller silicon additions to conventional PVD Ni–Cr–Al coatings are also used, presumably to stabilize Al_2O_3 scale formation for a longer period of time.

Effects of Refractory Elements

Refractory elements such as molybdenum, tungsten, and tantalum are used rather extensively in nickel- and cobalt-base superalloys as strengtheners, participating in γ' formation, carbide formation, and through solution effects. Other refractory elements such as columbium, hafnium, and zirconium are also utilized for strengthening purposes including the formation of Ni_3Cb.

The literature, which has recently been reviewed,[89] indicates that the refractory elements can produce three effects on the oxidation of nickel- and cobalt-base alloys. One effect is beneficial and arises since these elements can be considered to be oxygen getters and assist in the formation of Al_2O_3 and Cr_2O_3 healing layers. The other two effects are deleterious. First, refractory elements decrease the diffusion of aluminum, chromium, and silicon, which opposes healing layer formation. Second, the oxides of the refractory metals are generally nonprotective (i.e., low melting points, high vapor pressure, high diffusivities, etc.) and are consequently undesirable as components of external scales. Thus, the deleterious effects produced by the refractory metals outweigh the beneficial effects and, therefore, they are not usually added to superalloys to improve oxidation behavior. On the other hand, some of these elements do appear to be preferable to others. For example, tantalum does not appear to produce deleterious effects as severe as do tungsten or molybdenum, and hence is probably one of the preferred refractory elements. Tungsten, molybdenum, and vanadium are similar, but tungsten does evidently decrease alloy interdiffusion rates more than the other elements and therefore may have more of an adverse effect on selective oxidation. Columbium oxides are not protective and the presence of this element in oxide scales is not desired. Rhenium has been used in superalloys to a limited extent and appears to exert similar effects. Hafnium and zirconium are often present in superalloys at low concentrations and significantly improve the adherence of oxide scales.

COMPLEX NICKEL-BASE SUPERALLOYS

The many alloying elements in superalloys over and above the NiCrAl-base compositional variations, provide such a diversity in oxidation behavior that classification into a few simple categories is at present impossible. Therefore, the specific behavior of individual alloys and an attempt at some broad generalization is presented.

General Mechanisms

Kinetics. The description of the oxidation kinetics of superalloys by gravimetric data is complicated in that substantial fractions of the total scale are produced during a linear rate transient period (≤ 2 h). This is often followed by more than one parabolic regime requiring two or three rate constants, which precludes a simple comparison of alloys.[90–92] The temperature dependence of the first parabolic constant, k_{pI} (Fig. 14) indicates the rates for most superalloys fall between that for Cr_2O_3 growth (Ni–30Cr),[8] and Al_2O_3 growth (Ni–14Cr–12Al).[30] Most of the superalloy rates are within a factor of 2 of each other at 1800 °F, and the activation energies are closer to that for Cr_2O_3 growth (≈ 60 kcal/mol) than for Al_2O_3 (≈ 120 kcal/mol).[90–99] No consistent behavioral trend is apparent as the composition varies from high Cr, low Al to low Cr, high Al as would be predicted from oxidation maps for ternary Ni–Cr–Al alloys, nor do any of these alloys approach the oxidation resistance of exclusive α-Al_2O_3 formers, thus indicating the influence of elements other than chromium and aluminum.

Scale Components. The changes in rate parameters are linked to changes in the oxide phases comprising the scale, as shown in the time–temperature–oxide map for Udimet 700 (Fig. 15). The initial linear region (1) is characterized by a thin film of Al_2O_3 which is being overgrown by colonies of Cr_2O_3 originating at alloy grain boundaries.[90] Upon completion of this process, parabolic kinetics ensue and internal oxidation is observed, region 2. During the entire process depletion zones are formed ahead of the oxidation front by selective removal of the active elements (usually Al, Ti, and Si). Depletion zones and internal oxidation fronts also adhere to parabolic growth rates, which increase with temperature.

A secondary parabolic rate, k_{pII}, may occur for some alloys[92] and is usually lower than k_{pI}. Extended oxidation of Udimet 700 is characterized by $NiCr_2O_4$, NiO, and TiO_2 overgrowths at the gas–$(Cr,Al)_2O_3$ surface, region 3. A peculiar feature of Udimet 700 is the formation of a protective $Ni(Al,Cr)_2O_4$ spinel scale at ~ 1900 °F (1030 °C). This scale prevented overgrowths and resulted in a growth rate *less* than that at 1800 °F (982 °C).

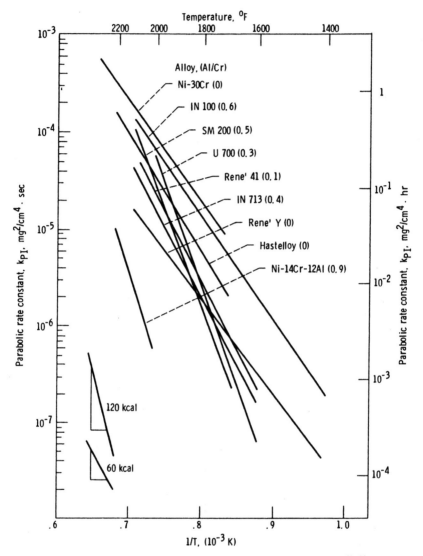

Fig. 14. Temperature dependence of $k_{p,I}$ for commercial superalloys.[90-96]

The three oxidation regimes for a number of alloys are summarized in Table 2. Oxides Al_2O_3, Cr_2O_3, or SiO_2 are often the thin initial scales in region 1. These oxides, coupled with TiO_2 or TiN, also make up most of the internal oxide phases that appear throughout stages 2 and 3. Advanced oxidation (3) is often characterized by the appearance of (Ni or Mn) (Al or $Cr)_2O_4$ spinels, NiO, and TiO_2 and $NiTiO_3$. This chart emphasizes the complexity of the scales and scale evolution schemes for common superalloys. The widespread occurrence of nickel-containing oxides possessing high diffusivities accounts for the high oxidation rates with respect to NiCrAl in Fig. 14.

Performance–Mechanism Correlations

The complex nature of superalloy oxidation and the relative performance of four superalloys as a function of time, temperature, aluminum and chromium content, refractory metal content, and type of test (static vs. high velocity, isothermal vs. cyclic) have been identified in a series of comparative studies.[97–101] The isothermal and cyclic oxidation performance of B-1900 and NASA-TRW VIA at 900–1100 °C (1650–2010 °F) was generally excellent; IN-713C was intermediate, and IN-738X was poor by comparison. For example, the total surface degradation, defined as alloy consumption plus depletion zone, was only ~1 mil for B-1900 and VIA after 100 h cyclic oxidation at 1100 °C. This is in comparison to 3.5 mils for IN-713C and 12.7 mils for IN-738X. Parallel trends in the isothermal and cyclic weight change were also observed.

The overall correlations of alloy composition, oxide phase content, and performance can be seen in Fig. 16. Here the type (a) oxides (Al_2O_3, $NiAl_2O_4$, and MR_2O_6) are associated with better performance compared to type (b) oxides (Cr_2O_3, $NiCr_2O_4$, and $NiTiO_3$). These phases are ultimately tied to the alloy compositions designated as types (a) or (b), that is, (Al+R) or (Cr+Ti), where R = Cb, Ta, W, Mo.

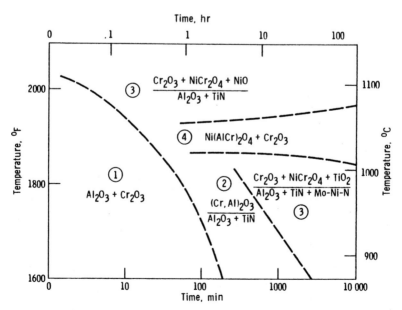

Fig. 15. Map of scale constituents for isothermal oxidation of Udimet 700; (external scale)/[(internal oxides) + subscale].[90]

Table 2. Oxidation Stages of Superalloys[a]

Low temp, short time ——————————————→ High temp, long time

	(1)	(2)	(3)
RENE' Y (22Cr,1Si,1Mn,9Mo)	Cr_2O_3	$\dfrac{Cr_2O_3 + MnCr_2O_4}{SiO_2 + Mn_xO_y}$	$\dfrac{MnCr_2O_4 + Cr_2O_3 + NiO}{SiO_2 + Mn_xO_y}$
HASTELLOY X (22Cr,1Si,0.5Mn,9Mo)	$Cr_2O_3 + NiCr_2O_4 + SiO_2$		$\dfrac{Cr_2O_3 + NiCr_2O_4}{SiO_2 + NiCr_2O_4}$
RENE' 41 (20Cr,1Al,3Ti,10Mo)	$Al_2O_3 + Cr_2O_3$	$\dfrac{(Cr,Al)_2O_3}{Al_2O_3}$	$\dfrac{Cr_2O_3 + NiCr_2O_4 + NiO + (TiO_2)}{Al_2O_3 + TiN}$
UDIMET 700 (15Cr,4Al,4Ti,4Mo)	$Al_2O_3 + Cr_2O_3$	$\dfrac{(Cr,Al)_2O_3}{Al_2O_3 + TiN}$	$\dfrac{Cr_2O_3 + NiCr_2O_4 + NiO + (TiO_2)}{Al_2O_3 + TiN}$
IN 713 C (14Cr,6Al,1Ti,6Mo)	$NiCr_2O_4 + NiO + Cr_2O_3$	$\dfrac{NiCr_2O_4 + (Cr_2O_3)}{Al_2O_3}$	$\dfrac{Al_2O_3 + NiAl_2O_4 + (NiCr_2O_4)}{TiO_2 + Cr_2O_3 + (TiN)}$
SM 200 (9Cr,4Al,2Ti,12W)	Cr_2O_3	$\dfrac{Ni(Cr,Al)_2O_4 + TiO_2 + NiO}{Al_2O_3}$	$\dfrac{NiCr_2O_4 + NiAl_2O_4 + NiWO_4 + NiTiO_3}{Al_2O_3 + TiO_2 + TiN}$
IN 100 (10Cr,6Al,4Ti,3Mo)	$NiO + TiO_2$	$\dfrac{(Ni,Co)O + (Ni,Co)Cr_2O_4 + NiTiO_3}{Al_2O_3 + TiO_2 + (TiN)}$	$\dfrac{NiCr_2O_4 + NiTiO_3 + NiAl_2O_4 + Al_2O_3}{TiO_2 + TiN}$

EXTERNAL SCALE
INTERNAL OXIDES OR SUBSCALE
( ) = present in part of region (3)

[a] 1400–2200 °F (760–1200 °C), $1–10^5$ min. isothermal (from refs. 90–92).

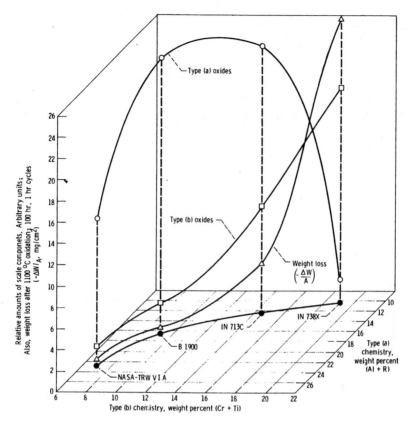

Fig. 16. Dependence of oxidation behavior and scale constituents on the composition of four superalloys.[97,98]

The tri-rutile or tapiolite MR_2O_6 oxides may be beneficial in assisting Al_2O_3 healing layers and preventing NiO overgrowths. The role of $NiTiO_3$ and TiO_2 is not known. However, substantial titanium enrichment is usually associated with the outer surface of the scales, especially for IN-713C and IN-738X, as shown in Fig. 17. Evident are a titanium-containing outer layer of Cr_2O_3 and $NiCr_2O_4$, internal oxide fingers of $NiTa_2O_6$ and Al_2O_3, and the γ alloy depletion zone denuded of γ' particles. Variations on this microstructure exist for the other alloys; in the extreme B-1900 exhibits only a thin $Al_2O_3 + NiAl_2O_4$ external scale.

The changes in oxidation behavior with time, temperature, and cycling are difficult to generalize for all alloys and conditions. Low-temperature (900–1000 °C, 1650–1830 °F) or isothermal conditions often favor the formation of Al_2O_3 scales, while higher temperature (1100 °C, 2010 °F) or cycling promotes NiO, $NiCr_2O_4$, and Cr_2O_3 in comparison. The major scale constituents increase monotonically with time, except for IN-738X, where Cr_2O_3 reaches a maximum and then is overtaken by substantial NiO and $NiCr_2O_4$ scales that ultimately cause breakaway cyclic behavior.[97] In these four cases high aluminum and refractory

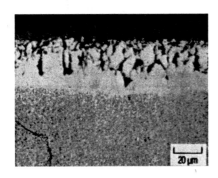 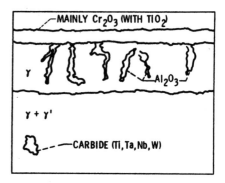

Fig. 17. Microstructure of IN-738 after 24 h oxidation at 975 °C (1790 °F) showing outer Ti-rich Cr₂O₃ scale, internal oxidation, and depletion zones.[100]

metal content appear preferable to high chromium or titanium. This basic theme repeats itself in the following discussion of compositional effects. However, it is emphasized that these rules must be applied with caution to avoid overgeneralization. It is not only the elements taken individually but the combined effects of many complex interactions *between* the elements that fully determines oxidative behavior of commercial superalloys.

Effects of Multielement Variations

Model Alloys (Statistical Studies). The precise compositional effects on the oxidation mechanisms of complex superalloys involve multielement interactions that are beyond present-day understanding. However, cyclic oxidation performance (weight change) was successfully correlated with systematic multielement variations in two extensive, statistically designed studies.[102,103] In the first study seven elements were varied at two levels; in the second study five elements were varied at five levels, amounting to nearly 100 alloys in all. The metal consumption of each alloy was described by an attack parameter, K_A, which takes into account both the growth and spallation of scales. Multiple linear regression analysis produced equations relating K_A to composition for each study.

Insights are also gained by ranking the alloys numerically from best to worst and identifying compositional trends.[104] A summary of one such ranking is shown in Table 3. In (a) the alloys are grouped by performance, the average values of performance and composition shown, and consistent trends within a group (if any) identified in boldface. This technique illustrates that all the best groups are consistently high in aluminum. Also the performance decreases by orders of magnitude if the aluminum content is decreased. Part (b) presents some specific alloys that were selected to represent the extremes of either performance or composition.

Taken together, these charts from both studies allowed the following conclusions:

1. High aluminum (6%) is the most important factor for good performance.
2. High chromium (15%) is not always needed for good performance.

Table 3. Cyclic Oxidation Ranking and Compositional Trends for Model Superalloys[a]

(a) Group averages[a]

Rank (group), number	$\overline{\Delta W/A}_{100}$ hr, mg/cm²	$\overline{K}_A$	$\overline{Al}$ 3.2/6.2	$\overline{Cr}$ 6.0/18.0	$\overline{Ti}$ 1.0	$\overline{Co}$ 0/20	$\overline{Cb}$ 1.0	$\overline{Ta}$ 0/8.0	$\overline{Mo}$ 0/4.0	$\overline{W}$ 2.0	Generalization
(a) 1-7	-1.7	.46	**5.5**	11.6	1.0	9.3	1.0	**5.4**	1.6	2.0	High Al, high Ta
(b) 8-13	-3.1	.76	**5.5**	13.0	→	8.3	→	**4.7**	**3.0**	→	High Al, high Ta, high Mo
(c) 14-19	-7.6	1.8	**5.3**	13.0	→	8.3	→	3.3	2.3	→	High Al
(d) 20	-12.	2.8	4.8	12.0	→	**10.0**	→	**4.0**	2.0	→	Midpoint composition, 8 replicates
(e) 21-29	-28.	7.5	4.8	13.0	→	**10.6**	→	3.8	1.7	→	
(f) 30-38	-150.	42.	4.0	**13.7**	→	**10.6**	→	3.3	2.1	→	Low Al, high Cr
(g) 39-45	-476.	108.	4.0	**9.0**	→	9.3	→	3.7	2.1	→	Low Al, low Cr

(b) Specific endpoint compositions[b]

Rank (group), number	$\overline{\Delta W/A}_{100}$ hr, mg/cm²	$\overline{K}_A$	$\overline{Al}$ 3.2/6.2	$\overline{Cr}$ 6.0/18.0	$\overline{Ti}$ 1.0	$\overline{Co}$ 0/20	$\overline{Cb}$ 1.0	$\overline{Ta}$ 0/8.0	$\overline{Mo}$ 0/4.0	$\overline{W}$ 2.0	Generalization
(a) 1	-1.4	.30	**5.5**	**15.0**	1.0	**15.0**	1.0	**6.0**	1.0	2.0	High Al, low Mo, high others
(b) 2	-1.7	.32	**5.5**	9.0	→	**15.0**	→	**6.0**	**3.0**	→	High Al, low Cr, high others
(c) 9	-2.9	.73	**5.5**	**15.0**	→	**15.0**	→	**6.0**	**3.0**	→	High Al, high others
(d) 20	-12.	2.8	4.8	12.0	→	10.0	→	**4.0**	2.0	→	Midpoint composition
(e) 33	-90.	29.	4.0	**15.0**	→	5.0	→	2.0	1.0	→	Low Al, high Cr
(f) 43	-496.	114.	4.0	9.0	→	5.0	→	2.0	1.0	→	Low Al
(g) 45	-748.	158.	4.0	9.0	→	5.0	→	2.0	**3.0**	→	Low Al, high Mo

[a]Boldface represents consistent trend within group.
[b]Boldface represents high levels.

[a]1100 °C (2010 °F), 100 h, 1-h cycles. (a) Group averages. (b) Specific endpoint compositions. (From refs. 103 and 104.)

319

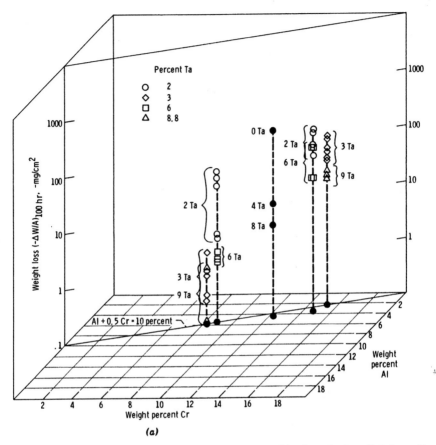

Fig. 18. Multielement oxidation maps illustrating some compositional trends that affect the cyclic oxidation of superalloys [1100 °C (2010 °F), 100 h, 1-h cycles]. (a) Beneficial trends with high Al and Ta for model superalloys.[102-104]

3. High chromium (15%) cannot compensate for low aluminum (2–4%).

4. Tantalum (3–9%) appears in best groupings and alloys.

5. Low titanium (<2%) appears in most of the better alloys.

The apparent beneficial effect of tantalum is shown graphically in the multielement oxidation map in Fig. 18a, where the higher tantalum alloys consistently exhibited less weight loss compared to other alloys of the same Al–Cr levels. However, as is the case for chromium, tantalum cannot fully compensate for lower aluminum contents (<5%). To allow direct comparison, the alloys shown here have Cr–Al contents in the range of cast superalloys where Al+½Cr = 10±2. Further generalization regarding, cobalt, columbium, molybdenum, and tungsten becomes dangerous because of the complex multielement interactions. These charts also identified

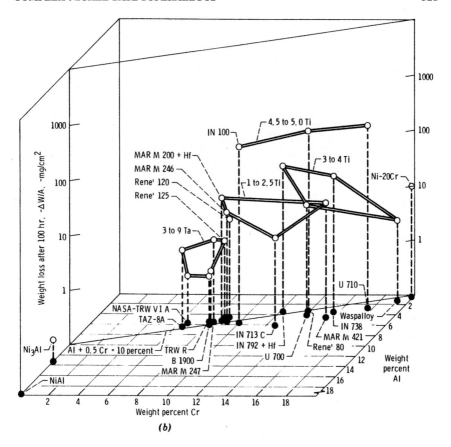

Fig. 18. *(Continued)* (b) Beneficial trends with high Al and Ta and low Ti for commercial superalloys.[104,105]

optimum alloys with weight losses <2 mg/cm² after 100 h, representing extremely good oxidation resistance, that is, surface recession < 0.1 mils per side.

The oxide phases produced by these alloys were broadly grouped into Al_2O_3, $NiAl_2O_4$, and $Ni(Ta,Cb,W,Mo)_2O_6$ tri-rutiles for the high-aluminum alloys and NiO, $NiCr_2O_4$, and $Ni(W,Mo)O_4$ for the low-aluminum alloys. This again illustrates the desirability of aluminum-containing oxides as well as the acceptability of the tri-rutile refractory metal oxide.

Commerical Superalloys. A similar ranking analysis for ~20 commercial superalloys is presented in Table 4. The alloys are arbitrarily grouped into classes whose average weight loss after 100 h, 1100 °C (2010 °F) cyclic oxidation increases about threefold from group to group. Compositional effects are much more difficult

Table 4. Cyclic Oxidation Groupings and Compositional Trends for Commercial Superalloys.[a]

Group	Alloy	$(-\Delta W/A)_{100hr}$ −mg/cm²	Al	Cr	Cb	Ta	Mo	W	Ti	Co	Zr	Hf	Al (trend)	Ta (trend)	Ti (trend)
A	NASA TRW VI A	.6	5.4	6.1	0.5	9.0	2.0	5.8	1.0	7.5	0.13	0.4	High 5–6	High 4–9	Low 0–3
	TRW R	.6	5.3	8.1	.3	6.3	2.8	4.0	.8	8.0	0.12	1.1			
	B 1900	.9	6.0	8.0	—	4.0	6.0	—	1.0	10.0	0.1	—			
	B 1900 + Hf	1.3	6.0	8.0	—	4.0	6.0	—	1.0	10.0	0.1	1.5			
B	TAZ 8A	2.7	6.0	6.0	2.5	8.0	4.0	4.0	—	—	1.0	1.5			
	RENE' 125	2.7	4.9	8.7	—	3.8	1.7	7.1	2.4	10.1	.05	—			
	WASPALLOY	2.9	1.3	19.5	—	—	4.3	—	3.0	13.5	.06	—			
	MAR M 247	3.0	5.5	8.2	—	3.0	.6	10.0	1.0	10.0	.09	—			
	IN 713 C	5.5	6.1	12.5	2.0	—	4.2	—	.8	—	.10	—			
	RENE' 120	6.4	4.3	9.0	—	3.8	2.0	7.0	4.0	10.0	—	0.1			
C	UDIMET 700	10.0	4.0	14.2	—	—	4.9	—	3.9	16.4	.02	—	Medium 4–5	Low 0–2	Medium 2–4
	MAR M 246	10.0	5.5	9.0	—	1.5	2.5	10.0	1.5	10.0	.05	—			
	MAR M 421	12.4	4.3	15.8	2.0	—	2.0	3.8	1.8	9.5	.05	—			
	MAR M 200 + Hf	16.4	5.0	8.6	.9	—	—	11.7	1.8	9.2	.03	2.0			
	MAR M 200	17.4	5.0	9.0	1.0	—	—	12.0	2.0	10.0	.05	—			
D	IN 738	31.2	3.4	16.0	0.9	1.7	1.7	2.6	3.4	8.5	.10	—	Low 2–4		High 3–5
	IN 792 + Hf	45.2	3.3	12.3	.0	3.9	1.9	4.1	4.0	8.9	.07	—			
	IN 939	79.4	1.9	22.4	1.0	1.4	—	2.1	3.7	19.0	.11	—			
E	IN 100	164.	5.5	10.0	—	—	3.0	—	4.7	15.0	.06	.06			
	RENE' 80	187.	3.0	14.0	—	—	4.0	4.0	5.0	9.5	.03	.03			
	UDIMET 710	202.	2.5	18.0	—	—	3.0	1.5	5.0	15.0	—	—			
	MAR M 211	450.	5.0	9.0	2.7	—	2.5	5.5	2.0	10.0	—	.05			

[a] 1100 °C (2010 °F), 100 h, 1-h cycles (from refs. 104 and 105).

to prove because there is no systematic variation as in the statistically designed studies. Nevertheless broad trends appear (indicated as composition blocks) which parallel those already put forth: 5–6 Al, 3–9 Ta, and 0–2 Ti are characteristically present in alloys with optimum oxidation resistance.

These trends are presented graphically in the multielement oxidation map (Fig. 18b) where significant differentials in cyclic weight change are associated with aluminum, tantalum, and titanium contents. Data for binary Ni–Al and Ni–Cr alloys are shown for comparison purposes.[105] It is also useful to note that the superalloy line, Al+½Cr = 10, lies for the most part in an optimum regime of the NiCrAl map shown in Fig. 8c. This suggests the overriding effects of the many alloying elements apart from just chromium and aluminum. Furthermore, commercial alloys with high aluminum also have low chromium, and vice versa. Therefore the Cr/Al ratio, often used to describe superalloy oxidation, actually overspecifies composition. Indeed, high Al model alloys are the most oxidation resistant, irrespective of the chromium content or the Cr/Al ratio.[102–104]

The effect of substantial amounts of cobalt (20%) in many nickel-base superalloys provides another important consideration. Reducing the cobalt level to 0–5% benefited the cyclic oxidation performance of a number of alloys at or above 1100 °C (2010 °F).[106] This is consistent with the generally inferior oxidation resistance of cobalt-base alloys compared to nickel-base alloys. These phenomena arise from the fast growth rates of highly defective cobalt oxides as well as the high refractory metal and low aluminum contents typical of these alloys.

Futher optimization of the oxidation resistance of superalloys could be achieved by in-depth studies of the key elements alluded to throughout this section. However, the most demanding applications employ current alloys coated with oxidation-resistant MCrAlY or NiAl materials. These Al_2O_3-forming alloys have therefore received a greater degree of attention, which ultimately led to a more fundamental understanding of scale growth, adhesion, and degradation mechanisms than currently exists for the superalloys.

REFERENCES

1. N. Birks and G. H. Meier, *Introduction to High Temperature Oxidation of Metals*, Arnold, Baltimore, 1983.
2. C. Wagner, *Z. Phys. Chem.*, **B21**, 25 (1933).
3. R. A. Rapp, *Met. Trans.*, **15A**, 765 (1984).
4. C. Wagner, *Ber. Bunsenges Phys. Chem.*, **63**, 772 (1959).
5. D. P. Whittle, in *High Temperature Corrosion*, R. A. Rapp (ed.), NACE, Houston, TX, 1983, p. 171.
6. C. Wagner, *Z. Elektrochem.*, **63**, 772 (1959).
7. C. E. Lowell, J. L. Smialek, and C. A. Barrett, in *High Temperature Corrosion*, R. A. Rapp (ed.), NACE, Houston, TX, 1983 p. 219.
8. C. S. Giggins and F. S. Pettit, *Trans. TMS-AIME*, **245**, 2495 (1969).
9. D. P. Whittle and H. Hindam, in *Corrosion-Erosion-Wear of Materials in Emerging Fossil Energy Systems*, A. V. Levy (ed.), NACE, Houston, TX, 1983, p. 54.
10. P. Kofstad and K. P. Lillerud, *J. Electrochem. Soc.*, **127**, 2410 (1980).
11. E. W. A. Young, P. C. M. Stiphout, and J. H. W deWit, *J. Electrochem. Soc.*, **132**, 887 (1985).

12. K. Hoshino and N. L. Peterson, *J. Am. Ceram. Soc.*, **66**, C202 (1983).
13. C. Greskovich, *J. Am. Ceram. Soc.*, **67**, C111 (1984).
14. H. V. Atkinson, *Oxid. Met.*, **24**, 177 (1985).
15. D. Caplan and G. I. Sproule, *Oxid. Met.*, **9**, 459 (1975).
16. G. M. Ecer and G. H. Meier, *Oxid. Met.*, **13**, 119 (1979).
17. S. D. Sehgal and D. Swarup, *Trans. Ind. Inst. Met.*, **15**, 177 (1962).
18. T. Nakayama and Y. Watanabe, *Trans. Iron Steel Inst. Jpn*, **8**, 259 (1968).
19. E. J. Felten, *J. Electrochem. Soc.*, **108**, 490 (1961).
20. G. M. Ecer and G. H. Meier, *Oxid. Met.*, **13**, 159 (1979).
21. C. S. Giggins and F. S. Pettit, *Metall. Trans.*, **2**, 1071 (1971).
22. H. H. Davis, H. C. Graham, and I. A. Kvernes, *Oxid. Met.*, **3**, 431 (1971).
23. G. R. Wallwork and A. Z. Hed, *Oxid. Met.*, **3**, 229 (1971).
24. J. Stringer, B. A. Wilcox, and R. I. Jaffee, *Oxid. Met.*, **5**, 11 (1972).
25. O. T. Goncel, D. P. Whittle, and J. Stringer, *Oxid. Met.*, **15**, 287 (1981).
26. F. H. Stott, J. S. Punni, G. C. Wood, and G. Dearnaley, in *Proceedings of the Conference on Modifications of Surface Properties of Metals by Ion Implantation, Manchester, U.K.*, 1981, p. 245.
27. M. J. Bennett, in *High Temperature Corrosion*, R. A. Rapp (ed.), NACE, Houston, TX, 1983, p. 145.
28. C. S. Tedmon, *J. Electrochem. Soc.*, **113**, 766 (1966).
29. F. S. Pettit, *Trans TMS-AIME*, **239**, 1296 (1967).
30. C. S. Giggins and F. S. Pettit, *J. Electrochem. Soc.*, **118**, 1782 (1971).
31. G. R. Wallwork and A. Z. Hed, *Oxid. Met.*, **3**, 171 (1971).
32. C. A. Barrett and C. E. Lowell, *Oxid. Met.*, **11**, 199 (1977).
33. A. S. Tumarev and L. A. Panyushin, NASA TT F-13221, 1970; from *Izv. Vyssh. Uchebn. Zaved. Chern. Metall.*, **9**, 125 (1959).
34. C. Wagner, *Corr. Sci.*, **5**, 751 (1965).
35. B. H. Kear, F. S. Pettit, D. E. Fornwalt and L. P. Lemaire, *Oxid. Met.*, **3**, 557 (1971).
36. F. H. Stott and G. C. Wood, *Corr. Sci.*, **11**, 799 (1971).
37. R. Kosak and R. A. Rapp, in *Interamerican Conference on Materials Technology*, David Black (coord.), Centro Regional de Ayuda Technical, Mexico, 1972, pp. 813, 823.
38. J. L. Smialek and R. Gibala, *Met. Trans.*, **14A**, 2143 (1983).
39. I. A. Kvernes ad P. Kofstad, *Met. Trans.*, **3**, 1511 (1972).
40. G. J. Santoro, D. L. Deadmore, and C. E. Lowell, NASA TN D-6414, 1971.
41. J. L. Smialek and R. Gibala, in *High Temperature Corrosion*, R. Rapp (ed.), NACE, Houston, TX, 1983, p. 274.
42. K. P. Reddy and A. R. Cooper, *J. Am. Ceram. Soc.*, **65**, 634 (1982). Also K. P. Reddy, Ph.D. Thesis, Case Western Reserve University, May 1979.
43. J. D. Cawley, J. W. Halloran, and A. R. Cooper, NASA TM-83622, June 1984.
44. T. A. Ramanarayanan, M. Raghavan, and R. Petkovic-Luton, *J. Electrochem. Soc.*, **131**, 923 (1984).
45. A. E. Paladino and W. D. Kingery, *J. Chem. Phys.*, **37**, 957 (1962).
46. E. W. A. Young and J. H. W. deWit, *Sol. St. Ion.*, **16**, 39 (1985).
47. K. P. R. Reddy, J. L. Smialek, and A. R. Cooper, *Oxid. Met.*, **17**, 420 (1982).
48. G. C. Rybicki and J. L. Smialek, *Electrochem. Soc.*, Extended Abstracts, 1984 Spring Meeting (1984).
49. J. K. Doychak, T. E. Mitchell, and J. L. Smialek, *Mat. Res. Soc. Symp. Proc.*, **39**, 475 (1985).
50. J. K. Doychak, J. L. Smialek, and T. E. Mitchell, *Int. Cong. Met. Corr. Proc.*, **1**, 35 (1984).
51. J. L. Smialek and B. C. Buzek, *Bull. Am. Ceram. Soc.*, **58**, 144 (1979).
52. I. M. Allam, D. P. Whittle, and J. Stringer, *Oxid. Met.*, **12**, 35 (1978).
53. A. S. Khan, C. E. Lowell, and C. A. Barrett, *J. Electrochem. Soc.*, **127**, 670 (1980).
54. C. A. Barrett, A. S. Khan, and C. E. Lowell, *J. Electrochem. Soc.*, **128**, 25 (1981).
55. J. C. Pivin et. al., *Corr. Sci.*, **20**, 351 (1980).

56. F. A. Golightly, F. H. Stott, and G. C. Wood, *Oxid. Met.*, **10**, 163 (1976).
57. F. H. Stott, F. A. Golightly, and G. C. Wood, *Corr. Sci.*, **19**, 889 (1979).
58. F. H. Stott, G. C. Wood, and F. A. Golightly, *Corr. Sci.*, **19**, 869 (1979).
59. J. K. Tien and F. S. Pettit, *Met. Trans.*, **3**, 1587 (1972).
60. J. K. Kuenzly and D. L. Douglass, *Oxid. Met.*, **8**, 139 (1974).
61. C. S. Giggins and F. S. Pettit, ARL 75-0234, Aeronautical Research Laboratory, Dayton, OH, Final Report, June 1975.
62. E. Bullock, C. Lea, and M. McLean, *Met. Sci.*, **13**, 373 (June 1979).
63. J. H. Davidson et. al., in *Petten International Conference Proceedings*, The Metals Society, London, 1980, p. 209.
64. D. P. Whittle and J. Stringer, *Phil. Trans. Roy. Soc. Lond.*, **A295**, 305 (1980).
65. A. W. Funkenbusch, J. G. Smeggil, and N. S. Bornstein, *Met. Trans.*, **16A** 1164 (1985).
66. J. G. Smeggil, A. W. Funkenbusch, and N. S. Bornstein, *Met. Trans.* **17A**, 923 (1986).
67. K. L. Luthra and C. L. Briant, *Electrochem. Soc.*, Extended Abstracts, 1984 Spring Meeting, 1984, p. 26.
68. J. L. Smialek and R. L. Browning, NASA TM-87168, 1985. Also in *Electrochemistry Society, High Temperature Chemistry III Proceedings*, 1986, p. 258.
69. R. P. Messmer and C. L. Briant, *Acta Metall.*, **30**, 457 (1982).
70. J. G. Smeggil, Proc. 3rd. Corrosion/Erosion Materials Conf., A. Levy (ed.), NACE, Houston, TX, 1986.
71. J. L. Smialek, *Met. Trans.*, **18A**, 164 (1987).
72. C. E. Lowell, *Oxid. Met.*, **7**, 95 (1973).
73. D. L. Douglass and J. S. Armijo, *Oxid. Met.*, **2**, 207 (1970).
74. G. N. Irving, J. Stringer, and D. P. Whittle, *Oxid. Met.*, **8**, 393 (1974).
75. P. Tomaszewicz and G. R. Wallwork, in *High Temperature Corrosion*, R. A. Rapp)(ed.), NACE, Houston, TX, 1983, p. 258.
76. P. R. S. Jackson and G. R. Wallwork, *Oxid. Met.*, **21**, 135 (1984).
77. H. Nagai and M. Okabayashi, *Trans. Jap. Inst. Met.*, **22**, 691 (1981).
78. K. S. Chiang, Ph.D. Thesis, University of Pittsburgh, Pittsburgh, PA, 1980.
79. P. Elliot and A. F. Hampton, *Oxid. Met.*, **14**, 449 (1980).
80. G. M. Ecer and G. H. Meier, in *Properties of High Temperture Alloys*, Z. A. Foroulis and F. S. Pettit (eds.), The Electrochemical Society, Princeton, NJ, 1977, p. 279.
81. C. E. Lowell and G. J. Santoro, NASA TN D-6838, 1972.
82. S. W. Yang, *Oxid. Met.*, **15**, 375 (1981).
83. G. H. Meier and F. S. Pettit, unpublished research, 1986.
84. A. Kumar and D. L. Douglass, *Oxid. Met.*, **10**, 1 (1976).
85. R. V. Miner and C. E. Lowell, NASA TN D-7989, 1975.
86. R. V. Miner, *Met. Trans.*, **8A**, 1949 (1977).
87. H. W. Grunling and R. Bauer, *Thin Solid Films*, **95**, 3 (1982).
88. E. Fitzer and J. Schlichting, in *High Temperature Corrosion*, R. A. Rapp (ed.), NACE, Houston, TX, 1983, p. 604.
89. F. S. Pettit and G. H. Meier, in *Refractory Alloying Elements in Superalloys*, in J. K. Tien and S. Reichman (eds.), ASM, Metals Park, OH, 1984, p. 165.
90. S. T. Wlodek, *Trans. TMS-AIME*, **230**, 1078 (1964).
91. S. T. Wlodek, *Trans. TMS-AIME*, **230**, 177 (1964).
92. G. E. Wasielewski, AFML-TR-67-30, January 1967.
93. S. K. Rhee and A. R. Spencer, *Met. Trans.*, **1**, 2021 (1970).
94. S. K. Rhee and A. R. Spencer, *J. Electrochem. Soc.*, **119**, 396 (1972).
95. S. K. Rhee and A. R. Spencer, *Oxid. Met.*, **7**, 71 (1974).
96. S. K. Rhee and A. R. Spencer, *Oxid. Met.*, **8**, 11 (1974).
97. C. A. Barrett, G. J. Santoro, and C. E. Lowell, NASA TN-D-7484, 1973.
98. R. G. Garlick and C. E. Lowell, NASA TM X-2796, 1973.
99. C. E. Lowell and H. B. Probst, NASA TN D-7705, 1974.

100. G. C. Fryburg, F. J. Kohl, and C. A. Stearns, NASA TN-D-8388, 1977.
101. S. R. Smith, W. J. Carter, G. D. Mateescu, F. J. Kohl, C. Fryburg, and C. A. Stearns, *Oxid. Met.*, **14**, 415 (1980).
102. C. A. Barrett, R. V. Miner, and D. R. Hull, *Oxid. Met.*, **516**, 255 (1983).
103. C. A. Barrett, NASA TM-83784, 1984.
104. J. L. Smialek and C. A. Barrett, unpublished research, 1986.
105. C. A. Barrett, R. G. Garlick, and C. E. Lowell, NASA TM-83865, 1984.
106. C. A. Barrett, NASA TM 87297, 1986.

Chapter 12

Hot Corrosion

F. S. PETTIT and C. S. GIGGINS

University of Pittsburgh, Pittsburgh, Pennsylvania,
and Pratt & Whitney, East Hartford, Connecticut

Hot corrosion of metallic alloys is a form of degradation that has been important for at least the last 50 years.[1] It occurs whenever salt or ash deposits accumulate on the surfaces of alloys and alter the environment–alloy reactions that would have occurred had the deposit not been present. Deposit-modified corrosion, or hot corrosion, is observed in boilers, incinerators, diesel engines, mufflers of internal combustion engines, and gas turbines. The severity of hot corrosion in combustion processes can vary substantially and is significantly affected by the type of fuel used and its purity as well as the quality of the air required to support the combustion. For example, hot corrosion is observed to a greater extent in industrial and marine gas turbines as compared to aircraft gas turbines. The nature of the hot-corrosion attack almost always involves more severe degradation of the alloy compared to that occurring in the same gas environment in the absence of the deposit. The increase in attack often is initially minimal with the alloy showing little effects of the deposit. Eventually, however, the degradation mechanism is modified by the deposit in a way such that the rate of attack is increased by an order of magnitude or more. Finally, the deposit is usually liquid.

Since hot corrosion is an important form of degradation, considerable research and development has been performed to advance the theory of hot corrosion and to develop alloys and coatings with improved resistance to this form of degradation. A number of comprehensive reviews on hot corrosion have been prepared.[1-6] Significant advances have been achieved, but total agreement with regard to mechanisms and the effects of various elements is not currently available. In this chapter the hot-corrosion mechanisms of metals and alloys will be considered and then the behavior of some specific superalloys will be described.

EFFECT OF OPERATIONAL AND TEST CONDITIONS

The hot-corrosion process is markedly dependent on parameters such as alloy composition, gas composition, deposit composition, and temperature. This dependence manifests itself not only as changes in the rates of attack but also, more importantly, as changes in the mechanisms by which the hot-corrosion attack occurs. Field or service experience is of critical importance. It provides the initial data necessary to define the problem. In most instances, however, the operational parameters are not constant or well defined, and the results obtained are difficult to use for mechanism formulation. As a result of the service failures, laboratory hot-corrosion tests are developed to simulate the conditions encountered in practice. The microstructures of the degraded alloys are therefore important to compare the degradation incurred in the field and in the simulation test. In Fig. 1 the microstructural features resulting from the hot-corrosion attack in gas turbines used in aircraft and ship propulsion are presented. The specimen from aircraft service contains sulfides as part of its identifiable features, whereas the attack developed in marine service does not exhibit such phases.

The laboratory tests that are developed to study hot corrosion must produce degradation microstructures that are similar to those generated in the service application of interest. Furthermore, it is usually necessary to have the laboratory test be an accelerated test for economic reasons. Laboratory tests can be divided into two general types. The first type places emphasis on simulating the conditions of the service application of interest. An example of this type of approach in the gas turbine industry is the use of burner rigs.[7,8] The second type stresses the definition of the test conditions. In this latter test, conditions representative of the application are considered, but definition of conditions receives emphasis as opposed to simulation. Laboratory tube furnace tests are utilized in the second approach.[9,10]

Burner Rig Tests

A variety of tests have been used in attempts to simulate the service conditions for a gas turbine.[7,8] Generally, the apparatus consists of a burner for gas or liquid fuels, attendant air and fuel supplies, a combustion chamber, and a specimen chamber in which the specimens may be rotated or stationary. Specimen shapes are often cylindrical, but airfoils or other configurations have also been used. A contaminant, such as seawater, may be injected into the combustion chamber or it may be mixed with the fuel. The object is to have the contaminant cause a deposit to accumulate upon the specimens.

The extent of hot corrosion may be assessed by several different kind of measurements, but the two most frequently used are weight loss after descaling and

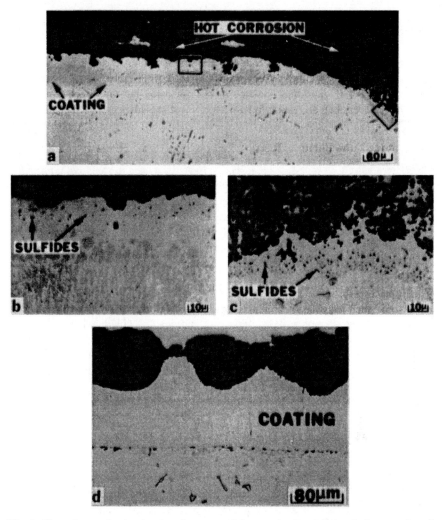

Fig. 1. Photomicrographs showing the microstructural features developed during the hot corrosion of gas turbine materials in aircraft and marine service. (a) High-temperature (aircraft) hot corrosion of an aluminide-coated nickel-base superalloy (B-1900), Type I. The characteristic features developed in the coating and superalloy substrate where the coating was penetrated are shown in (b) and (c), respectively. (d) Low-temperature (marine) hot corrosion of a CoCrAlY coating after 4200 h of service, Type II.

penetration depth of the reaction products.[11] Detailed metallographic analysis is also mandatory to thoroughly characterize the specific features common to the type of hot corrosion being induced.

For design purposes a correlation of burner rig data with performance in service must be established. Burner rig conditions are usually much more severe than service exposure, and extrapolations must be performed with care since different degradation mechanisms may be operative. The burner rig tests are adequate for ranking alloys, but their value for predicting lives in specific applications is limited to rather special cases.

Most burner rigs operate at ambient pressures, but there are a few that can establish greater pressures.[12] The results obtained from these high-pressure rigs show that pressure can affect the rate of hot corrosion but the effects are not so great as to necessitate all testing to be performed at pressures greater than ambient.

Laboratory Tube Furnace Tests

There are a variety of tube furnace tests that have been used to study hot corrosion. The simplest is to coat the specimen with the desired deposit and then to expose it to a controlled gas mixture at a specific temperature.[9] Deposits are applied by spraying warm [~100 °C (212 °F)] test coupons with solutions containing the compound of interest. Continuous weight change versus time measurements can be recorded to attempt to define the hot-corrosion kinetics. Depending on the test conditions, vaporization of the deposit may occur, which complicates the interpretation of such data, and detailed metallographic analyses of the exposed specimens are usually necessary.

In some such tests the specimens are removed periodically from the hot zone of the furnace in an attempt to induce cracking and spalling of the oxide scales. For resistant alloys the deposit must be removed to permit observations, and then a fresh deposit is applied for subsequent testing. At times it may be appropriate to immerse the specimens in crucibles containing the deposit.[9] However, substantially different results can be obtained for the hot corrosion of alloys even where the only variable is the thickness of the molten deposit covering their surfaces.

Utilization of Electrochemical Cell

Since the hot corrosion of alloys usually involves liquid deposits, a number of investigations have attempted to study this process by using the techniques employed for studying aqueous corrosion. In such tests the specimens are usually exposed to environments similar to that of a crucible test, since the experimental arrangement involves an electrochemical cell composed of a fused salt as the electrolyte, a reference electrode, a working electrode, and perhaps auxiliary electrodes. The purpose of the test may be either to investigate the properties of the salt mixture[13,14] or to determine the corrosion resistance of the specimen (working) electrode.[15,16]

Typical results are presented in Fig. 2 for some superalloys exposed to a $(Na,K)_2SO_4$ melt at 900 °C (1650 °F). These data compare the maximum depth of attack for 100 h exposure as a function of electrode potential. The electrode potential can be used to attempt to define the conditions in the melt. For example, at positive potentials the following reaction can be expected:

$$SO_4^{2-} \rightarrow SO_3 + 1/2O_2 + 2e \qquad (1)$$

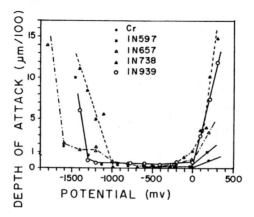

Fig. 2. Results obtained from an electrochemical test at 900 °C to compare the hot corrosion of some superalloys (ref. 15).

Hence, the melt can be considered to be acidic since SO_3 is produced.[1] In the case of highly negative potentials reactions of the following type prevail:

$$^1/2O_2 + 2e \rightarrow O^{2-} \tag{2}$$

$$SO_4^{2-} + 2e \rightarrow SO_3^{2-} + O^{2-} \tag{3}$$

$$SO_4^{2-} + 6e \rightarrow S + 4O^{2-} \tag{4}$$

and the melts are expected to be basic since oxide ions are produced. The results from such tests can be used to compare how various alloys may behave in melts with such compositions. For example, in the case of the results presented in Fig. 2 it was concluded that the potential range over which negligible attack was observed could be used as an indication of hot-corrosion resistance. Hence the hot-corrosion resistance of these alloys was concluded to increase in the order IN-738 < IN-657 < IN-939, IN-597.

The problem with using such tests is that the specimens generally must be exposed to thick melts. Furthermore, one must know which melt composition is the one that is important in the particular application of interest. Finally, at potentials where there is not substantial attack, the difference in behavior between alloys cannot be determined.

HOT-CORROSION DEGRADATION SEQUENCE

One of the problems in discussing the hot corrosion of alloys is that the degradation mechanisms change with time. Examination of hot-corrosion data obtained as a

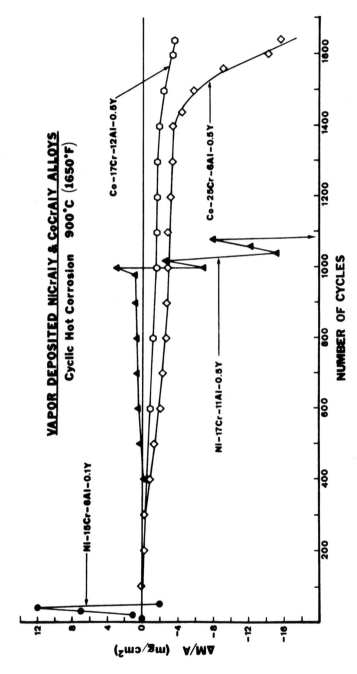

Fig. 3. Weight change versus time for the cyclic hot corrosion (1-h cycles) of Na_2SO_4-coated ($\sim$1 mg/cm^2 Na_2SO_4 applied every 20 h) alloys in air. Severe hot-corrosion attack was evident at times where abrupt weight increases or decreases occurred.

332

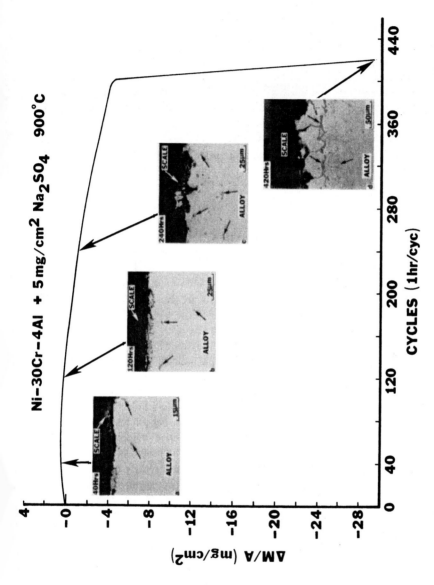

Fig. 4. Weight change versus time and corresponding microstructural features for the cyclic hot corrosion of Na_2SO_4-coated (applied every 20 h) Ni–30Cr–4Al in air. The amount of sulfide particles (small black arrows) increases until the oxidation of the sulfide phases significantly affects the rate of attack.

function of time (Fig. 3) shows that there are two distinct stages of attack, an initial stage during which the attack is not too severe and a later stage for which the attack is substantially increased. Examination of exposed specimens as a function of time (Fig. 4) shows that the microstructural features developed during the degradation undergo a marked change as the severity of the attack increases. Initially, the corrosion microstructures are not much different than those that would have developed in the absence of the salt deposit, but after the rate increases, they are substantially different. This tendency for the hot-corrosion process to consist of two stages, namely an initiation stage and a propagation stage, is not an uncommon phenonema. Corrosion-resistant alloys depend on selective reaction product barrier formation for their resistance, and hence a degradation sequence consisting of the eventual displacement of a more protective reaction product barrier by a less protective product is usually followed.

INITATION STAGE OF HOT CORROSION

During the initiation stage elements in the alloy are oxidized and electrons are transferred from metallic atoms to reducible substances in the deposit. In most hot-corrosion processes the reduced substances initially are the same as those that would have reacted with the alloy in the absence of the deposit. Consequently, the reaction product barrier that forms beneath the deposit on the alloy surface usually exhibits primarily those features resulting from the gas–alloy reaction. As the hot-corrosion process continues, however, features appear that indicate that the salt deposit is affecting the corrosion process. For example, in some forms of hot corrosion an increasing amount of sulfide particles becomes evident in the alloy beneath the protective reaction product barrier (Fig. 4). In others small holes become evident in the protective reaction product barrier where the molten deposit begins to penetrate it. Eventually the protective barrier formed via selective oxidation is rendered ineffective, and the hot-corrosion process enters into the propagation stage.

The hot-corrosion degradation sequence is not always clearly evident, and the time for which protective reaction products are stable beneath the salt layer is influenced by a number of factors. This sequence is usually evident whenever time is required before the deposit can cause nonprotective reaction products to be formed. The alloy must be depleted of certain elements before nonprotective products can be formed, or the composition of the deposit must change to prevent the formation of protective scales. There are also probably cases where the initiation stage does not exist at all and the degradation process is in the propagation stage as soon as the molten deposit comes into contact with the alloy at elevated temperatures. In Fig. 5 the factors that affect the initiation of hot-corrosion attack are presented. These factors have significance since they can be considered to precondition the alloy and determine the characteristics of the propagation mode that will be followed. Examples of the effects produced by some of these factors on the hot corrosion of alloys are presented in the following.

HOT CORROSION CHRONOLOGY

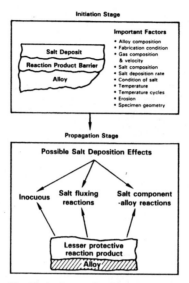

Fig. 5. Schematic diagram to identify the factors that determine the time at which the transition from the initiation stage to the propagation stage occurs and to illustrate the two general categories of protective scale breakdown.

Alloy Composition

No alloy is immune to hot-corrosion attack indefinitely, although there are some alloy compositions that require extremely long initiation times before the propagation stage is evident. In order to discuss the effects of specific elements on hot corrosion, the various types of propagation mechanisms must be examined. At this point it is sufficient to show that alloy composition can be a critical factor in the initiation of attack, as shown in Fig. 3, where the length of the initiation stage for hot corrosion induced by Na_2SO_4 in air increases as the aluminum content of nickel–chromium or cobalt–chromium alloys is increased from 6 to 11%. From these data it is also apparent, that for the test conditions used, longer times are required to initiate attack in cobalt–chromium–aluminum alloys than in nickel–chromium–aluminum alloys. In discussing the resistance of certain alloys to hot-corrosion attack, it is necessary to specify the conditions causing the attack since the alloy may behave much differently under other conditions. For example, whenever SO_3 is present in the gas and the temperature is in the range of 650–750 °C (1200–1380 °F), hot corrosion attack of cobalt–chromium–aluminum alloys can be in the propagation stage after short exposure times (Fig. 6).

Temperature

Hot-corrosion processes are dependent on temperature in a number of ways. The time to initiate attack can be decreased as the temperature is increased. Hot-corrosion

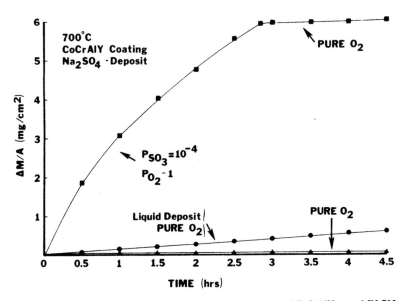

Fig. 6. Weight change versus time for the isothermal hot corrosion of CoCrAlY-coated IN-792. Hot corrosion was induced via Na_2SO_4 deposits (~ 1 mg/cm^2). In one experiment a Na_2SO_4–40 mol % $MgSO_4$ deposit was used to obtain a liquid deposit at the test temperature. The gas was flowing oxygen except for one experiment where a SO_2–O_2 mixture was passed over a Pt catalyst to develop a SO_3 pressure of 10^{-4} atm during the first 2.9 h exposure.

conditions also exist where the attack becomes less severe as the temperature is increased. In burner rig tests it is more or less common procedure to ingest a controlled amount of salt, which is then deposited on specimens during test. For a constant ingestion rate of salt, less is deposited on the specimens as the temperature is increased, and it is possible to observe less attack at the higher temperatures because of the small amount of salt on the specimens. Temperature is an extremely important factor in the hot corrosion of alloys since different mechanisms with different initiation times can be important at different temperatures. Hence, rate constants change not only because the kinetics are being influenced by temperature, but also because completely new reaction mechanisms can become operative.

Gas Composition

The composition of the gas phase can produce very substantial effects on the initiation of attack, the rate of attack, and the particular propagation mode that is followed. In Fig. 6 weight-change-versus-time data are compared for the oxidation of a Na_2SO_4-coated CoCrAlY coating in oxygen and in oxygen containing SO_3 at 10^{-4} atm. The hot-corrosion attack is initiated virtually from the beginning of weight increase measurements in the gas with SO_3, but no attack was observed after 20 h in pure oxygen. The influence of the SO_3 in this example is twofold. Sodium sulfate is not liquid at 700 °C. When oxidation of CoCrAl occurs at this temperature in SO_3, a liquid solution of Na_2SO_4–$CoSO_4$ is formed. Hot-corrosion attack is more easily induced when a liquid phase is present. Sulfur trioxide, however, also influences the rate at which the attack is propagated. For example, the attack in oxygen is not as severe as in oxygen with SO_3 even when a deposit of Na_2SO_4–$MgSO_4$ is used that is liquid at 700 °C (1290 °F) in oxygen (Fig. 6).

Salt Composition and Deposition Rate

The composition of the deposit and the rate at which it accumulates on the surfaces of alloys affects not only the time required to initiate hot-corrosion attack but also the type of propagation mode that is followed. In Fig. 7 photographs are presented to compare the degradation microstructures developed in coatings exposed to Na_2SO_4 containing different amounts of NaCl. The degradation becomes more severe as the NaCl concentration in the deposit is increased. As will be shown subsequently, the NaCl in the deposit causes the hot-corrosion degradation to be different than that

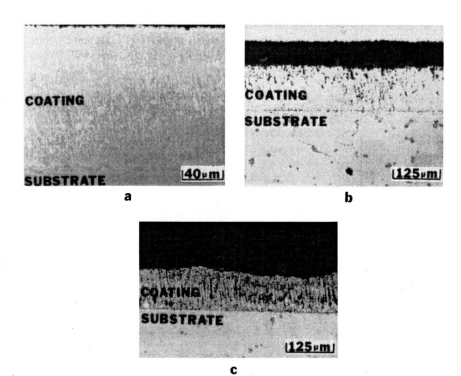

Fig. 7. Photomicrographs to compare the degradation of CoCrAlY coatings on IN-738 after exposure at 899 °C (1650 °F) in air to Na_2SO_4 deposits containing different amounts of NaCl: (a) 500 h with Na_2SO_4; (b) 500 h with Na_2SO_4–5 wt. % NaCl; (c) 40 h with Na_2SO_4–90 wt. % NaCl.

induced by pure Na_2SO_4. Numerous other examples are available to illustrate the importance of salt composition on hot-corrosion attack.[1]

Hot-corrosion attack occurs because the deposit modifies the type of reaction that takes place between the alloys and the gas environments. The condition of the deposit plays an important role in determining how the reaction is modified. Normally, a liquid deposit is most effective in causing hot-corrosion attack (Fig. 6), but it is not required to produce such attack. Very dense solid deposits can cause the chemical potential of the reactants in the gas to be much different at the alloy–deposit interface compared to the bulk gas[17], and therefore cause certain types of hot-corrosion attack.

The amount of salt that is present at the surface of alloys exerts very significant effects on the initiation, the rates, and the mechanisms of the hot-corrosion attack. The amount of salt present on the surfaces of alloys affects their hot corrosion in two ways. All degradation mechanisms are not self-sustaining. Salt is consumed in the corrosion processes that are not self-sustaining, and therefore, the more salt present, the more attack occurs (Fig. 8). Other mechanisms require the salt to have a particular composition for the initiation of attack. Such a composition is developed at the salt–alloy interface via reaction with the alloy. The thickness of the deposit influences the time required to obtain the composition necessary to initiate attack. When attack occurs because of the development of a gradient across the salt from the gas phase, thicker deposits cause the attack to be initiated sooner than thinner deposits. On the other hand, when attack occurs as a result of the accumulation of elements from the alloy in the deposit, attack can be observed sooner with thinner deposits.

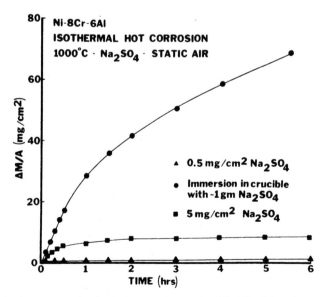

Fig. 8. Weight change versus time for the hot-corrosion attack of Ni–8Cr–6Al specimens with different amounts of Na_2SO_4. The amount of degradation increases as the amount of deposit is increased.

Other Significant Factors

Alloy composition, temperature, gas composition, and deposit composition probably exert the greatest influence on the initiation of hot-corrosion attack and the type of propagation mode that is followed. However, there are other factors that also significantly affect the hot-corrosion process. Factors that damage protective oxide scales such as thermal cycles or erosion are important. Similarly, specimen geometry also affects cracking and spalling of scales. The velocity of the gas is another significant parameter. Gas velocity effects are especially important in cases where volatile components play a role in the hot-corrosion process. For example, the accumulation of MoO_3 in Na_2SO_4 on a Ni-8Cr-6Al-6Mo alloy causes very severe hot-corrosion attack. The attack of this alloy is initiated in static air much sooner than in flowing oxygen because less MoO_3 is lost from the Na_2SO_4 to the gas in the static environment. Velocity-induced effects can be especially prevalent in burner rig experiments where gas velocities in excess of 300 m/s can be achieved.

The initiation of hot-corrosion attack frequently occurs at compositional inhomogeneities in alloys. Since cast alloys are often less homogeneous than wrought alloys or alloys made by consolidating powders, the fabrication condition can also affect the hot-corrosion behavior of alloys.[18]

PROPAGATION MODES OF HOT CORROSION

When a deposit develops on the surface of an alloy at elevated temperatures, reactants from the gas must diffuse through the deposit to the alloy, or elements from the alloy must dissolve into the deposit and diffuse to the deposit-gas interface. In many cases the diffusion of the reactants through the deposit is not rapid enough to prevent some change in its composition, and such changes can play significant roles in most hot-corrosion processes.

Thermodynamic stability diagrams are useful to help describe the effects of deposit compositional changes.[19] Sodium sulfate is frequently a major component of deposits inducing hot corrosion; it will be utilized here in discussing deposit effects. In Fig. 9 a sodium-oxygen-sulfur stability diagram is used to illustrate the various compositions that may exist or be developed in Na_2SO_4 when it is present as a liquid on the surface of an alloy. Three important effects are worth discussion. First, there are some compositional changes that will produce sulfur activities greater than those in the gas from which the Na_2SO_4 condensed. Second, the deposit may become more basic, which means the concentration of oxide ions, or the activity of Na_2O, is increased. Finally, the concentration of oxide ions may be decreased, which means that the deposit has become more acidic.

The thermodynamic stability diagrams can also be used to attempt to describe the stability of oxides on alloys beneath deposits of Na_2SO_4. A typical stability diagram is presented in Fig. 10, where the stability of phases containing aluminum, chromium, and nickel in Na_2SO_4 are compared. Such diagrams can also be used in conjunction with microstructural analyses for mechanism development (Fig. 11).

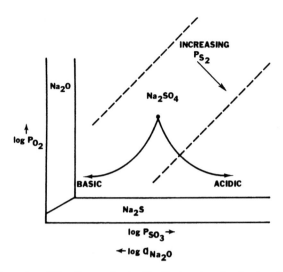

Fig. 9. Thermodynamic stability diagram for the Na–O–S system showing how the composition of Na$_2$SO$_4$ may change due to reaction of the alloy with the deposit.

Within the oxide stability ranges of these diagrams, a certain amount of oxide is soluble. Rapp[13,20-22] and Stern[23] have determined solubility curves for a number of oxides in Na$_2$SO$_4$ as a function of SO$_3$ pressures. Some typical curves are presented in Fig. 12. It should be noted that these curves usually have minimums bounded by regions of increasing solubilities.

The propagation modes for hot corrosion are intimately related to the reactions between the molten deposits and the alloys. In particular, the deposits cause nonprotective reaction products to be formed. Such reactions may be based on solubility changes within the oxide stability regions or the formation of phases outside these regions. In discussing the propagation modes, it is convenient to divide them into two groups. One group consists of modes having a common feature that the nonprotective reaction product is formed because of some "fluxing" action of the molten deposit. The other group has the common feature that a component of the deposit (e.g., S or Cl) plays a dominant role in the process, causing a nonprotective reaction product to be formed. In some instances the deposit may not exert a significant affect on the alloy–gas reaction. In cases where the deposit is innocuous it often exists on the surfaces of alloys as a porous, solid phase. The propagation mode for an alloy with an innocuous deposit will be determined by reaction between the alloy and the gas.

Salt-"Fluxing" Reactions

Those processes by which the reaction product barrier becomes nonprotective due to the formation of species that are soluble in the liquid deposit have been called "fluxing" reactions. There are a number of processes by which this can occur and the particular one that is important depends on the experimental conditions, in particular, alloy composition, gas composition, temperature, and deposit characteristics.

In a molten sulfate deposit the following equilibrium can be used to define the acidity or basicity:

$$SO_4^{2-} = SO_3 + O^{2-} \tag{5}$$

with the equilibrium condition

$$K = P_{SO_3} a_O^{2-} \tag{6}$$

Equation (6) is the basis for relating the SO_3 pressure and the oxide ion activity in Figs. 9 and 10. The magnitude of K is determined by the standard free energies of formation for Na_2SO_4, SO_3, and Na_2O, and a_O^{2-} is the activity of oxide ions in the melt, which equals the activity of Na_2O in such melts. Equivalent expressions can be formulated for other melts such as carbonates, hydroxides, nitrates, and so on. In the case of sulfates the acidity is determined by the SO_3 pressure, and the acidity increases as the SO_3 pressure is increased. The acidity need not only be controlled

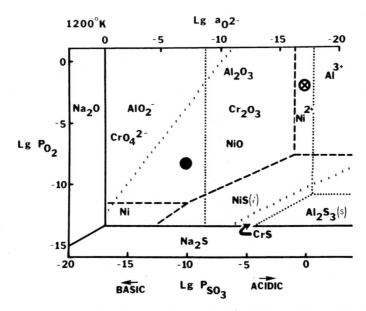

Fig. 10. Thermodynamic stability diagram to illustrate the phases of Ni (---), Al (···) and Cr (x) which can exist in a Na_2SO_4 layer on a NiCrAl alloy. The Na_2O and Na_2S boundaries are indicated by solid straight lines. In Na_2SO_4 of composition ⊗, NiO will dissolve making the Na_2SO_4 more basic, whereas Al_2O_3 will not react with the melt. In Na_2SO_4 of composition ●, Al_2O_3 will dissolve making the melt more acidic while Cr_2O_3 will not react.

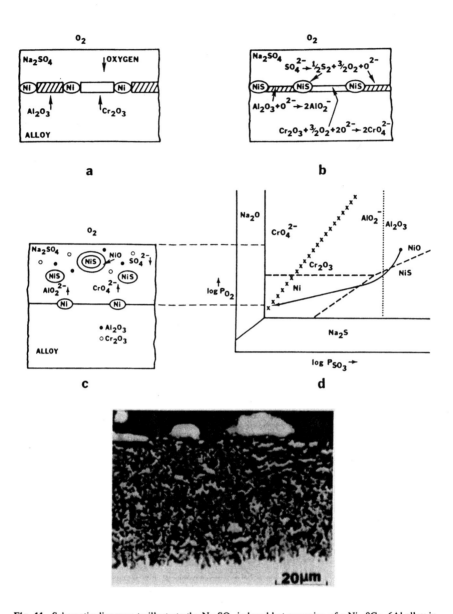

Fig. 11. Schematic diagrams to illustrate the Na$_2$SO$_4$-induced hot corrosion of a Ni–8Cr–6Al alloy in air. Oxygen depletion occurs, (a), and sulfide formation results in the production of oxide ions that react with Al$_2$O$_3$ and Cr$_2$O$_3$, (b). At higher oxygen pressures the Cr$_2$O$_3$ and Al$_2$O$_3$ precipitate from the melt, (c). A phase stability diagram, (d) is used to account for the stability of phases in (c). Typical microstructural features of the rapid attack that occurs via basic fluxing is shown, (e), for the alloy after exposure at 1000 °C (1830 °F) in air to 5 mg/cm^2 Na$_2$SO$_4$ for 2 min.

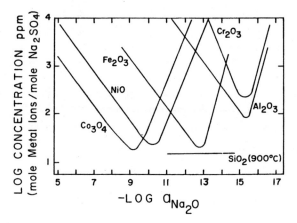

Fig. 12. Measured oxide solubilities in fused Na_2SO_4 at 927 °C (1700 °F) and 1 atm O_2 (ref. 22).

by the SO_3 pressure in the gas since there are other ingredients in some systems that may affect acidity. One example is vanadium, which may be a component in the alloy, or an impurity in fuels, whose oxide can react with Na_2SO_4 via a reaction such as

$$Na_2SO_4 + V_2O_5 \rightarrow 2NaVO_3 + SO_3 \tag{7}$$

to affect the acidity of the melt. Molybdenum and tungsten in alloys can cause similar effects when their oxides are formed as corrosion products.

Basic Fluxing. The initial concept of basic fluxing was first proposed by Bornstein and DeCrescente[24–26] and then described in thermodynamic terms for the hot corrosion of nickel by Goebel and Pettit.[9] An essential feature of this process is that oxide ions are produced in the Na_2SO_4 deposit due to removal of oxygen and sulfur from the deposit via reaction with the alloy or metal (Figs. 9 and 11). Furthermore, oxide scales (MO) that normally would form as protective barriers on the surfaces of alloys can react with these oxide ions via reactions such as

$$MO + O^{2-} \rightarrow MO_2^{2-} \tag{8}$$

and their protective properties are destroyed. Depending on conditions in the molten deposit, the oxide may reprecipitate in the melt.

Such basic fluxing has a number of distinct features. Sulfides are usually found in the alloy substrate as a result of sulfur removal from the Na_2SO_4 to produce oxide ions. Furthermore, the amount of attack depends on the production of oxide ions by the melt. Hence, a supply of Na_2SO_4 is necessary for the attack to continue. In other words, this form of basic fluxing hot corrosion is often not self-sustaining (Fig. 8). Finally, this form of hot corrosion is usually restricted to high temperatures (above ~900 °C, 1650 °F) since the processes that produce oxide ions proceed slowly at lower temperatures. Furthermore, it is more likely to be important in gases that do not contain an acidic component (e.g., SO_3); however, basic fluxing can occur in gases with acidic components when the concepts of Rapp and Goto are used.[27]

Rapp and Goto[27] have proposed that protective scales on alloys could be made nonprotective when the solubility gradients of the protective oxides in the molten deposit were negative since continuous dissolution and reprecipitation of oxide is then possible. The hot corrosion of nickel at 1000 °C (1800 °F) in air satisfies these requirements. The solubilities of NiO are such that dissolution of NiO as NiO_2^{2-} at the oxide–salt interface and reprecipitation of porous NiO at the salt–gas interface, where the solubility is lower, can occur. The Rapp–Goto mechanism was developed to permit fluxing, either basic or acidic, without the need for a source or sink for oxide ions. This is an important condition since it means that attack may continue with no additional supply of the deposit. Shores[28] has examined the Rapp–Goto precipitation criterion for a variety of conditions and has shown that the fluxing reactions are not always self-sustaining. For example, in the case of hot corrosion of nickel induced by Na_2SO_4 in air at 1000 °C (1830 °F), the molten deposit gradually becomes more basic at the gas interface and the attack eventually stops[9] unless more Na_2SO_4 is deposited. It should be noted, however, that hot-corroson effects need not be self-sustaining in order to exert very significant effects on the degradation of corrosion-resistant alloys. All systems degrade when used even in the absence of a molten deposit. The intermittent deposition of a deposit that produces hot-corrosion attack via a process that is not self-sustaining will cause the alloy to develop the less protective reaction product barriers sooner even when the deposition rate is small.

Acidic Fluxing.   Acidic fluxing[29,30] is the development of a nonprotective oxide scale via reactions whereby oxide ions are donated to the molten deposit by oxide that would have developed as a protective barrier had the deposit not been present, namely,

$$MO \rightarrow M^{2+} + O^{2-} \tag{9}$$

Acidic conditions can be developed in molten deposits on alloys by at least two different processes. In particular, a component can be present in the gas, which causes the deposit to be acidic, and an oxide of an element present in the alloy can cause the melt to become acidic. These two processes result in gas phase and alloy-induced acidic fluxing, respectively.

Gas Phase Acidic Fluxing.   In the case of gas phase acidic fluxing the acidic component is supplied to the liquid deposit by the gas. Two examples of acidic components in gases are SO_3 and V_2O_5 vapor. As these species are incorporated into the liquid deposits, the important reactions are

$$SO_3 + SO_4^{2-} = S_2O_7^{2-} \tag{10}$$

$$V_2O_5 + SO_4^{2-} = 2VO_3^- + SO_3 \tag{11}$$

Gas phase acidic fluxing has some very distinctive features. First, it is observed most often, at least in regards to SO_3-induced attack, at temperatures between about 650 and 800 °C (1200 °F to 1470 °F) and hence is often called "low-temperature" or type II hot corrosion. This low-temperature restriction results from the need to form sulfates such as $CoSO_4$, $NiSO_4$, and $Al_2(SO_4)_3$ (not necessarily at unit activity), which require higher SO_3 pressures as temperature is increased. Furthermore, the SO_3 pressures in many combustion environments frequently decrease as temperature is increased. Second, gas-phase-induced hot corrosion exhibits certain microstructural characteristics that depend on the alloy composition and the environment causing the attack. Typical degradational microstructures for a CoCrAlY alloy are presented in Figs. 1d and 13. Virtually no alloy-depleted zone is evident in advance of the corrosion front (Fig. 13), and when sulfur is detected in the corrosion product, it is found near to the alloy–corrosion product interface but always associated with oxygen. When nickel is added to the alloy composition, sulfur becomes more readily detected in the corrosion microstructure as discrete sulfides phases.

At least three models have been developed to account for the gas-phase-induced hot corrosion of alloys. One model[31-34] is unique from the others in that it proposes the formation of nonprotective Al_2O_3 and Cr_2O_3 scales on alloys as a result of the rapid removal of cobalt and nickel from the alloy surfaces via the deposit. The nonprotective Al_2O_3 and Cr_2O_3 form since the oxides cannot develop continuity due to the rapid removal of cobalt and nickel. Cobalt and nickel are removed from the alloy since their oxides are soluble in the acidic melt but can become stable at the surface of the deposit. Another model[35] proposes that the oxygen pressure over the alloy, beneath the deposit, is too low to form oxides of cobalt or nickel. Moreover, the oxygen pressure is sufficiently low whereby a significant concentration of SO_3^{2-} ions is formed:

$$SO_4^{2-} = SO_3^{2-} + 1/2O_2 \tag{12}$$

It is proposed that the Al_2O_3 is not stable in such melts since the following reaction is favored at the alloy–melt interface:

$$3SO_2 + Al_2O_3 \rightarrow 2Al^{3+} + 3SO_3^{2-} \tag{13}$$

Nonprotective oxide particles are formed in the melt away from the alloy surface where the oxygen pressure is higher. This model can be criticized since it does not appear that the deposit will be liquid at the low SO_3 pressures required to prevent oxides of cobalt and nickel from being formed. On the other hand, the degradational microstructures do indicate that oxides such as cobalt and nickel are not present in the alloy at the alloy–corrosion product interface (Fig. 13c).

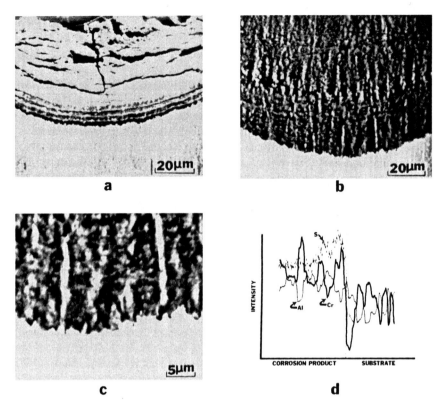

Fig. 13. Microstructural features developed in a CoCrAlY alloy during 17.3 h exposure to a Na_2SO_4 deposit (2.5 mg/cm^2) and oxygen containing SO_3 (7×10^{-4} atm) at 704 °C. Ghost images of the corrosion front, (a), and the α-cobalt phase in the alloy, (b) and (c), are evident. Results obtained from microprobe analyses of the corrosion product are presented in (d).

A third model[36-38] proposes that sulfides are formed at the corrosion–product alloy interface and that nonprotective oxide phases form as a result of their oxidation. Degradation microstructures similar to that shown in Fig. 4 are developed. There does not appear to be any question that this last mechanism occurs in alloys that contain significant amounts of nickel. However, it may not be valid for cobalt–chromium–aluminum alloys. An important point to be resolved is the interrelation of these models as a function of alloy composition and low-temperature hot-corrosion test conditions.

Alloy-Induced Acidic Fluxing.   In alloy-induced acidic fluxing elements such as molybdenum,[29,39] tungsten,[30] or vanadium[30] cause melts to become acidic as oxides of these elements are incorporated into the melts. Typical reactions would be

$$Mo + 3/2O_2 \rightarrow MoO_3 \qquad (14)$$

$$MoO_3 + SO_4^{2-} \rightarrow MoO_4^{2-} + SO_3 \qquad (15)$$

$$Al_2O_3 + 3MoO_3 \rightarrow 2Al^{3+} + 3MoO_4^{2-} \qquad (16)$$

where the Al_2O_3 dissolves into the melt at regions where MoO_3 activity is high and could precipitate out wherever the activity of MoO_3 is low, as could occur at the surface of the melt where MoO_3 is being lost to the gas phase. In acidic fluxing the deposit may become acidic, as, for example, Na_2SO_4 containing increased amounts of Na_2MoO_4, or the Na_2SO_4 can be totally converted to Na_2MoO_4 with increasing amounts of MoO_3 in the Na_2MoO_4.[39] In alloy-induced acidic fluxing the normally protective oxide can be destroyed by dissolving in the melt or by dissolving into the melt and reprecipitating as a nonprotective oxide. Misra[40] has studied the Na_2SO_4-induced corrosion of molybdenum containing nickel-base superalloys at 950 °C (1740 °F) and proposed that alloy-induced acidic fluxing occurs via a mechanism similar to that of Luthra[31-34] whereby Ni^{2+} ions are formed in the acidic melt and precipitated near the gas interface as NiO. The removal of nickel from the alloy prevents the development of continuous layers of Al_2O_3 and Cr_2O_3 over the alloy surface.

As in the case of the other propagation modes, conclusive remarks concerning the alloy-induced fluxing mechanism are not possible at present. The most significant feature of alloy-induced acidic fluxing is that it is self-sustaining and only one application of the liquid deposit can cause total destruction of the alloy. It is frequently observed in structural alloys since the solid-solution strengtheners are the elements whose oxides can make the melts acidic. Alloy-induced acidic fluxing is usually observed at higher temperatures due to the substantial amount of oxidation required to provide sufficient amounts of refractory metal oxides to make the melts acidic. It is not uncommon to have another propagation stage precede alloy-induced acidic fluxing.[41] This first propagation stage is responsible for introducing refractory metal oxides into the melt.

Deposit-Component-Induced Hot Corrosion

Another important characteristic of hot corrosion is the reduction of the oxygen activity over the alloys by the deposit. This occurs because elements are present in the alloys that have high affinities for oxygen, and access of oxygen to the alloy is restricted due to the fact that it must diffuse through the deposit (Fig. 9). Consequently, conditions become more favorable for other elements in the deposit to react with the elements in the alloys. In principal, a great variety of elements in deposits could affect the hot-corrosion process. Two elements that affect the hot-corrosion process in a very significant manner are sulfur and chlorine.

Sulfur Effects. Especially at elevated temperatures a substantial amount of sulfide phases can be found in alloys beneath the molten deposits for reasons discussed previously (Fig. 4). Eventually, as these sulfides begin to oxidize, the resulting phases are not protective. Such a condition was observed during some of the earlier investigations on the hot corrosion of alloys in gas turbines.[42,43] Hence, hot corrosion was believed to proceed as a result of the oxidation of sulfides and was called sulfidation. It is important to note that while some hot-corrosion attack can proceed via a mechanism called sulfidation, not all hot-corrosion processes occur in this manner. The mechanism by which the sulfidation form of hot corrosion proceeds unquestionably involves the oxidation of sulfides (Fig. 14). It appears that nonprotective oxides may form during the oxidation of sulfides due to SO_2 evolution as well as the oxides being in tension.[44]

Much of the earlier studies of hot corrosion used oxygen as the gaseous reactant. When low-temperature Type II hot corrosion became important, more studies were done using O_2 + SO_2 gas mixtures. Such investigations[37,38] have now shown that the sulfidation propagation mechanism is important throughout the entire temperature range for which hot corrosion is observed. At the lower temperatures SO_2 must be present in the gas phase in order for this form of degradation to be observed.

Chlorine Effects. Chlorine in deposits may affect the hot corrosion of alloys in at least two ways. First, chlorine concentrations in the ppm range have been shown to increase the propensity of oxide scales (e.g., Al_2O_3 and Cr_2O_3) on alloys to crack

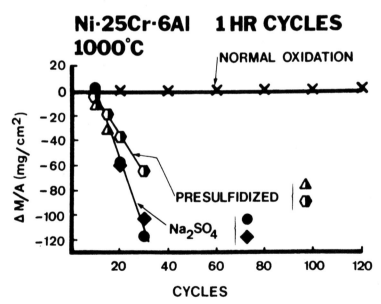

Fig. 14. Comparison of the cyclic oxidation data obtained for Ni–25Cr–6Al alloys that were coated with Na_2SO_4 to those presulfidized in an H_2S-H_2 gas mixture. Approximately 5 mg/cm² Na_2SO_4 was added to specimens after 5 h exposure up to 20 h and then after every 10-h interval beyond 20 h. Presulfidation was performed at the same intervals that the Na_2SO_4 was applied, and the sulfur picked up was equivalent to that in a 5mg/cm² Na_2SO_4 deposit.

and spall.[45] Such a condition must result in alloys progressing to the propagation stage of degradation after shorter exposure times.

It has also been observed that larger concentrations of chloride in deposits causes aluminum and chromium to be rapidly removed from alloys.[10,35] Typical results are presented in Fig. 7, where the attack can be seen to be more severe as the chloride content in increased. Chloride in the deposits results in the development of pores or channels in alloys whereby the chromium and aluminum is preferentially removed, and totally nonprotective oxide scales of these elements are formed on the surfaces of alloys. This process has many features similar to the selective leaching of zinc from copper–zinc alloys during dezincification.[46]

There can be no question that the presence of chloride in deposits can result in more severe hot corrosion. When the phenomenon of low-temperature hot corrosion was first observed in gas turbines, many investigators proposed that it was caused by chloride in deposits. While chloride effects can be severe, the preponderance of data show that low-temperature hot corrosion can proceed without any chloride in the deposit. It is caused by a relatively high SO_3 pressure in the gas phase and nonprotective oxides are formed as a result of the acidic nature of the liquid deposits. A point can still be made, however, that much of the data obtained for Type II hot corrosion has involved burner rig tests and laboratory tube furnace tests under conditions different from those present in gas turbines.[47] Hence, the effects produced by chloride cannot be disregarded. When the conditions are present to permit chloride-induced effects, this attack is significant[48] at temperatures as low as 700 °C (1290 °F) and up to approximately 1000 °C (1830 °F)

EFFECTS PRODUCED BY VARIOUS ELEMENTS ON HOT-CORROSION PROCESS

In attempting to discuss the effects of different elements on the hot-corrosion process, it is useful to first summarize the various propagation modes and to indicate the experimental conditions for which these modes are important. In Fig. 15 the various propagation modes are summarized, and in Fig. 16 the temperature range and gas compositions over which these modes are important are indicated schematically. It is important to emphasize that hot corrosion at the higher temperatures is preceded by longer initiation periods than at the lower temperatures, and therefore a comparison of corrosion rates must be done with care. In Fig. 16 the rates used to compare the various mechanisms are the rates for degradation in the propagation mode. It can be seen that the hot-corrosion attack is more severe at the lower temperatures when SO_3 is in the gas phase. This occurs because deposits are often not liquid if SO_3 is not present, and therefore the increase in attack due to the presence of SO_3 is much greater at the lower temperatures. The sulfidation mechanism extends over the entire temperature range, but again it does not produce as much attack at the lower temperatures when a liquid phase is not formed.

The temperature ranges over which the other mechanisms are important have been selected based on previous discussions in this chapter. The effects of sodium

**MECHANISMS OF PROTECTIVE SCALE BREAKDOWN FOR ALLOYS
DEVELOPING HIGH TEMPERATURE OXIDATION AND HOT CORROSION
RESISTANCE VIA SELECTIVE OXIDE BARRIER FORMATION**

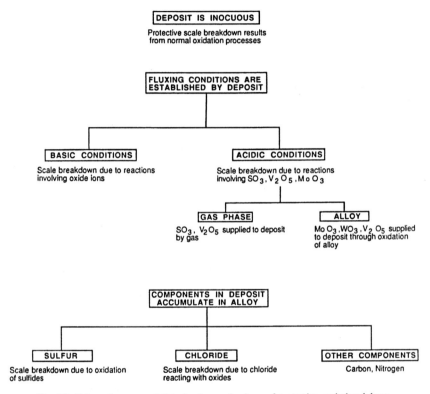

Fig. 15. Schematic representation for the mechanisms of protective scale breakdown.

chloride have not been included in this schematic. Chloride-induced attack is equivalent to the other propagation modes at the higher temperatures. At lower temperatures it definitely will cause more attack than Na_2SO_4 if SO_3 is not present in the gas. If SO_3 is present, the NaCl will be converted to Na_2SO_4; however, the attack produced by NaCl in air may be as great as Na_2SO_4 in SO_3 at low temperatures.

Type II hot corrosion has a specific microstructure (Figs. 1d and 13). Hot-corrosion attack at low temperature can proceed via gas phase acidic fluxing, sulfidation, or chloride-induced effects. Currently the Type II microstructure is taken as that which develops at temperatures between 650 and 850 °C (1200 °F–1560 °F) when SO_3 is present in the gas. These conditions cause attack via gas phase acidic fluxing or sulfidation depending on the alloy composition. High-temperature hot corrosion, or Type I, occurs between 800 and 1000 °C (1470 °F–

1830 °F) and has a microstructure such as that shown in Figs. 1b, 1c, or 4. It can occur via basic fluxing, sulfidation, or alloy-induced acidic fluxing.

In the following, the effects of a number of elements on the hot corrosion of alloys is discussed. These effects are presented under headings of the relevant propagation modes summarized in Fig. 15. As should be apparent from the preceding discussion, the particular propagation mode that is followed depends on test conditions.

Basic Fluxing

Both nickel and cobalt are susceptible to attack via basic fluxing. There is no significant difference between the two.[9,49] No studies have been done on the hot corrosion of pure iron. At temperatures above about 650 °C (1200 °F) iron oxidizes very rapidly even in the absence of hot-corrosion conditions. The effects produced by other elements will be discussed in regard to the effects that they cause in nickel-cobalt-, or iron-base alloys.

Chromium inhibits the onset of basic fluxing in nickel-, cobalt-, and iron-base alloys. It causes the oxide ion concentration to be decreased to levels at which reaction with NiO, CoO, and probably iron oxides is not possible. If the chromium concentration is high enough to form a protective Cr_2O_3 scale, the resistance to hot corrosion is markedly increased.

Aluminum inhibits basic fluxing when its concentration is great enough to result in the formation of continuous Al_2O_3 scales. It does not prevent attack of NiO and CoO scales by oxide ions as in the case of chromium.

Titanium does not produce any significant effects on the degradation of nickel-, cobalt-, or iron-base alloys via basic fluxing. However, studies to examine the effect of this element have not been extensive.

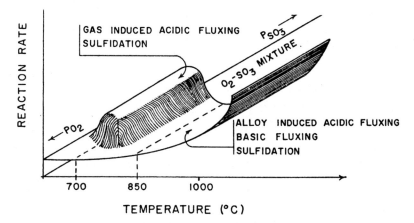

Fig. 16. Schematic showing the temperature and gas compositions over which the hot-corrosion propagation modes are important.

The refractory elements molybdenum, tungsten, and vanadium can inhibit basic fluxing.[50] The oxides of these elements react with oxide ions and hence inhibit the reaction of oxide ions with Al_2O_3, NiO, CoO, and so on. The effect is similar to that of chromium, but the danger of using molybdenum, tungsten, and vanadium to inhibit basic fluxing is that these elements can cause acidic fluxing.

Tantalum may also inhibit basic fluxing, but it is not as effective as chromium. However, it does not cause acidic fluxing, in contrast to the other solid-solution strengtheners such as molybdenum, tungsten, and vanadium.

Other elements that are frequently encountered in superalloys or their coatings do not evidently significantly affect basic fluxing hot corrosion. In the case of coatings on which silica scales are formed during exposure, such scales are extremely susceptible to basic fluxing.

Acidic Fluxing: Gas Phase Induced

Virtually all alloy systems are susceptible to gas-phase-induced acidic fluxing. Cobalt-base systems may be slightly more susceptible than either nickel or iron due to the greater stability of $CoSO_4$ compared to that of $NiSO_4$ and $FeSO_4$. However, this difference does not have any practical value.

Preformed Al_2O_3 scales inhibit the onset of gas-phase-induced acidic fluxing, but once it is initiated, the rate of hot corrosion is not significantly influenced by the aluminum concentration. However, microstructural examination of exposed alloys shows that the attack has preference for the aluminum-rich phases.[36]

Increasing the chromium concentration of alloys does not increase the initiation time for gas phase acidic corrosion, but the rate of this propagation mode is decreased.[51] Cobalt–chromium alloys are reported to have excellent resistance to low-temperature hot corrosion.[34]

Some oxygen-active elements (e.g., Y) that are present in alloys to improve the adherence of oxide scales to the metallic substrates do serve as sites at which molten desposits can penetrate the oxide scales.[52] The oxides of such elements frequently form as continuous stringers extending through the Al_2O_3 scales to react readily with the acidic molten deposits. There are, however, other sites at which the salt can penetrate oxide scales,[52] and the presence of oxygen-active elements in alloys does not shorten the initiation times substantially.

Most of the other elements present in superalloys and their coatings on these alloys do not significantly affect gas phase acidic fluxing. Silica scales formed on coatings are extremely resistant to acidic fluxing. This oxide has a low solubility in acidic melts (Fig. 12), and protective scales develop on many alloys with sufficently high silicon concentrations. A problem with using such scales as part of a protective system, however, is their high susceptibilty to degradation via basic fluxing.

Acidic Fluxing: Alloy Induced

Alloy-induced acidic fluxing of superalloys is caused by the presence of certain refractory elements, specifically molybdenum, tungsten, and vanadium. These elements must be kept at appropriately low concentrations to prevent this form of hot-corrosion attack. The exact concentration that may be used depends on use conditions. There is no significant difference evident in the attack of nickel-, cobalt-, and iron-base alloys containing such elements. With the exception of chromium, other elements do not exert any notable effect on alloy-induced acidic fluxing. However, in order to produce this form of hot corrosion, substantial quantities of the refractory elements must be oxidized. Hence, all elements that promote the selective oxidation of aluminum or chromium in superalloys, in a sense, inhibit alloy-induced acidic fluxing.

Sulfur-Induced Hot Corrosion (Sulfidation)

Sulfur-induced hot corrosion often follows certain forms of basic fluxing since oxide ions are formed in the melt due to sulfide formation in the alloys. Certain nickel-base alloys are much more susceptible to this form of hot corrosion than cobalt-base alloys, as shown in Fig. 3. Such results have given rise to the conclusion that

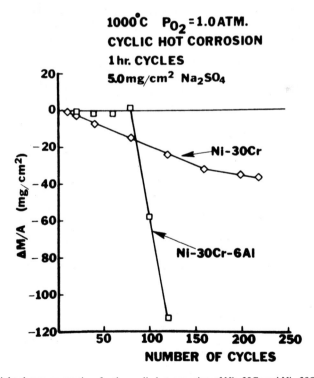

Fig. 17. Weight change versus time for the cyclic hot corrosion of Ni–30Cr and Ni–30Cr–6Al alloys. The Al initially causes the Ni–30Cr–6Al to be more resistant than Ni–30Cr, but after longer times it causes more severe attack.

Table 1. Alloy Systems Susceptible to Hot-Corrosion Attack with Principal Propagation Modes and Procedures to Inhibit Such Degradation

Type of Attack	Susceptible Systems	More Resistant Systems	Procedure for Inhibiting Attack	Contributory Factors
BASIC FLUXING	• Ni, Co, Ni–Al, and Co–Al alloys; Ni–Cr–Al alloys with Cr and Al contents below that required to form external scales of Cr_2O_3 or Al_2O_3 Example: Ni–8Cr–6Al	• Ni–Cr, Co–Cr, and Fe–Cr alloys; Co–Cr–Al, Fe–Cr–Al, and Ni–Cr–Al alloys with Al contents of 10–12% Example: Co–20Cr–12Al–0.5Y	• High Cr, high Cr, and Al • No Al in Ni-base system or Al well above 10% • If low concentrations of Al are required, replace Ni with Co	• Chloride-induced alloy depletion • Carbon-induced oxygen depletion
SULFUR-INDUCED DEGRADATION	• Ni-base alloys containing about 1–6% Al Example: IN-792, U-700	• Ni–Cr, Co–Cr, and Fe–Cr alloys; Co–Cr–Al, Fe–Cr–Al, and Ni–Cr–Al alloys with Al contents of 10–12% Example: TD NiC, X-40, HA-188, HAST-X, SS-304, CoCrAlY coatings	• High Cr • Use Co in place of Ni in alloys with Al contents between 1 and 6% • In Ni-base alloys containing Al, keep the Al content above 10%	• Chloride-induced alloy depletion (especially effective for high-Cr and high-Al alloys) • Carbon-induced oxygen depletion
ALLOY-INDUCED ACIDIC DEGRADATION	• Alloys containing Mo, W, or V Example: B-1900, MAR-M200, IN-100, WI-52, NX-188	• IN-738, HA-188, HAST-X	• High Cr • Lower refractory metal content	• Chloride-induced alloy depletion (HA-188 and HAST-X extremely susceptible) • C-induced oxygen depletion
GAS-PHASE-INDUCED ACIDIC FLUXING	• Virtually all alloys depending on SO_3 pressure or amount of V_2O_5 deposited	• High-Cr alloys, silica formers (Hot pressed Si_3N_4)		• Chloride-induced alloy depletion • Carbon-induced oxygen depletion

cobalt-base alloys are more resistant to hot corrosion than nickel-base alloys. This statement is not generally true and is valid for only a certain set of hot-corrosion conditions. It is worth noting that this difference between nickel and cobalt-base alloys is evident at high test temperatures and with alloys containing chromium and aluminum (Fig. 3). Increasing the chromium or aluminum concentrations in nickel and cobalt alloys increases the initiation time for sulfidation. However, in the case of the nickel-base alloys, as the aluminum concentration drops below 6 wt. %, the alloys become very susceptible to attack. It appears that these compositions are conducive to the rapid removal of sulfur from the deposits accompanied by very severe attack of the resulting sulfide phases (Fig. 17).

Elements other than chromium and aluminum in superalloys do not have substantial effects on the sulfur-induced hot corrosion. There is data available to suggest that the titanium–aluminum ratio may be important,[53] but the role played by titanium is not clear, and the observed results may be due to the effects produced by aluminum.

Other Propagation Modes

In Table 1 the effects of the important elements on the hot corrosion of alloys are summarized as a function of the hot-corrosion propagation mode. In the case of the propagation mode that occurs because of chloride deposits, the only elements that have been observed to produce beneficial effects are titanium and platinum.[54]

There probably are other modes for hot-corrosion propagation. For example, some investigators believe that carbon particles striking the surfaces of gas turbine hardware can play a significant role in the hot-corrosion degradation process.[55]

COMPARISON OF HOT-CORROSION ATTACK OF SOME SUPERALLOYS

It is desirable to attempt a comparison of the hot-corrosion resistances of superalloys. Hot-corrosion attack of superalloys is controlled by alloy composition and other factors defined by the test or use conditions. A comparison of the hot-corrosion attack of superalloys can be attempted by comparing superalloy performance for fixed-use conditions. Difficulties are still encountered, however, because different initiation times are required to produce the hot-corrosion attack. For example, IN-738 is considered to be more hot corrosion resistant than B-1900. Inspection of the data shows that the increased resistance of IN-738 does not arise because this alloy has a lower rate of hot corrosion in the propagation mode than B-1900 but is rather due to the longer times required to initiate attack of the IN-738. It is reasonable to propose that as hot-corrosion attack of superalloys reaches the propagation mode, the rates of attack for any of the propagation modes are so large that all superalloys must be removed from service. Hence, a parameter that has meaning in comparing the resistance of superalloys to hot corrosion is the time required to initiate attack in these alloys via a propagation mode. Unfortunately, much of the data in the literature on the hot corrosion of superalloys does not consider the times to initiate hot-corrosion attack. On the other hand, gas turbine manufacturers are aware of this factor; they utilize their own data and select alloys and coatings appropriately.

Finally, before attempting to compare the hot corrosion of superalloys, it is important to emphasize that as hot corrosion attack is initiated in these alloys, very often a sequence of propagation modes is followed. Such conditions have been documented for B-1900[29] and IN-738.[41] In the case of B-1900 the propagation modes consist of basic fluxing followed by alloy-induced acidic fluxing. Results obtained with an alloy containing the same chromium, aluminum, and molybdenum as B-1900 are consistent with the propagation modes observed for B-1900. A sequence of propagation modes probably exist during the hot corrosion of most superalloys. In comparing the hot corrosion of superalloys by comparing the times to initiate attack via a propagation mode, the propagation mode sequence is not a factor requiring definition.

The most important element in developing resistance to hot-corrosion attack for superalloys is chromium (Table 1). This element inhibits the onset of attack regardless of the hot-corrosion mechanism, but its value in the case of chloride-induced attack is not very great. The next most important composition variable is the refractory metal concentration. It is necessary to keep the concentration low. However, higher concentrations of these elements can be used if the chromium concentration is also increased (e.g., Hastelloy X, Haynes 188). Tantalum does not adversely affect the hot-corrosion resistance as does molybdenum, tungsten, and vanadium. The third important consideration in designing for hot-corrosion resistance is the aluminum concentration. In coatings the aluminum concentration can be taken to 10% and greater, and for such concentrations it has produced beneficial effects. In the case of structural superalloys, however, the aluminum concentrations cannot exceed about 6%. At concentrations between about 2 and 4% aluminum nickel-base superalloys are extremely susceptible to the sulfidation form of hot-corrosion degradation.

There is not a marked difference in the resistance of superalloys to low-temperature hot corrosion. Those alloys with high chromium concentrations (e.g., ~20%) may have somewhat better resistance, but the difference is not great. In the case of resistance to high-temperature hot corrosion, the superalloys with high chromium concentrations and low refractory metal concentrations are better. Moreover, in the case of nickel-base alloys the aluminum concentration should not be high when the chromium concentration is low. For example, IN-738, IN-657, IN-939, IN-597, X-40, and Waspalloy have better resistance to high-temperature hot corrosion than IN-100, IN-713, B-1900, MAR-M 200, and WI-52.

FUTURE PROSPECTS FOR OBTAINING SUPERALLOYS WITH IMPROVED RESISTANCE TO HOT CORROSION

In order to obtain the required mechanical properties in superalloys, compositions must be used that are not highly resistant to hot corrosion. This situation cannot be expected to change in the future even with increased emphasis being placed on single-crystal superalloys. The most likely approaches for developing greater resistance in superalloys to hot-corrosion attack involve the use of coatings and knowing the type of hot-corrosion degradation process that must be withstood. Significant improvements in the hot-corrosion resistance of superalloy-coating systems can be obtained by selecting the superalloy that is most resistant to the form of hot corrosion that is important in the application being considered. A coating is then selected, or developed, to increase the system's resistance to this specific form of hot-corrosion degradation.

REFERENCES

1. W. T. Reid, *External Corrosion and Deposits in Boilers and Gas Turbines*, Elsevier, New York, 1971.
2. *"Hot Corrosion Problems Associated with Gas Turbines,"* ASTM, STP 421 American Society for Testing and Materials, Philadelphia, PA, 1967.
3. P. Hancock, *Corrosion of Alloys at High Tempertures in Atmospheres Consisting of Fuel Combustion Products and Associated Impurities*, Her Majesty's Stationery Office, London, 1968.
4. J. Stringer, *Ann. Rev. Mat. Sci.*, **7**, 477 (1976).
5. R. L. Jones, "Hot Corrosion in Gas Turbines," in *Corrosion in Fossil Fuel Systems*, The Electrochemical Society, Princeton, NJ, 1983, p. 341.
6. D. A. Shores and K. L. Luthra, "Hot Corrosion of Metals and Alloys," in *High Temperature Oxidation*, W. Worrell (ed.), Academic, New York, in press.
7. H. V. Doering and P. A. Bergman, *Mat. Res. Stand.*, **9**, 35 (1969).
8. R. R. Dils and P. S. Follansbee, *Corrosion*, **33**, 385 (1977).
9. J. A. Goebel and F. S. Pettit, *Met. Trans.*, **1**, 1943 (1970).
10. V. Nagarajan, J. Stringer, and D. P. Whittle, *Corr. Sci.*, **22**, 407 (1982).
11. P. A. Bergman, C. L. Sims, and A. N. Beltran, "Hot Corrosion Problems Associated With Gas Turbines," ASTM Special Publication No. 421, Philadelphia, PA, 1967, p. 380.
12. S. Y. Lee, S. M. DeCorso, and W. E. Young, ASME Preprint 70-WA/CD-2, New York, 1970.
13. D. K. Gupta and R. A. Rapp, *J. Electrochem. Soc.*, **127**, 2194 (1980).
14. W. W. Liang, and J. F. Elliot, in *Properties of High Temperature Alloys*, Z. A. Foroulis and F. S. Pettit (eds.), The Electrochemical Society, Princeton, NJ, 1976, p. 557.
15. A. Rahmel, M. Schmidt, and M. Schorr, *Oxid. Met.*, **18**, 195 (1982).
16. A. Rahmel, *Werkst. Korros*, **28**, 299 (1977).
17. J. Stringer, in *High Temperature Corrosion*, R. A. Rapp (ed.), NACE, Houston, TX, 1983, p. 389.
18. T. Huang, E. A. Gulbransen, and G. H. Meier, *J. Met.*, **31**, 28 (1979).
19. J. M. Quets and W. H. Dresher, *J. Mat.*, **4**, 583 (1969).
20. P. D. Jose, D. K. Gupta, and R. A. Rapp, *J. Electrochem. Soc.*, **132**, 73 (1985).
21. Z. S. Zhang and R. A. Rapp, *J. Electrochem. Soc.*, **132**, 734,2498 (1985).
22. R. A. Rapp, *Corrosion*, **42**, 568 (1986).
23. M. L. Deanhardt and K. H. Stern, *J. Electrochem. Soc.*, **129**, 2228 (1982).
24. N. S. Bornstein and M. A. DeCrescente, *Trans. Met. Soc. AIME*, **245**, 1947 (1969).
25. N. S. Bornstein and M. A. DeCrescente, *Corrosion*, **26**, 209 (1970).
26. N. S. Bornstein and M. A. DeCrescente, *Met. Trans.*, **2**, 2875 (1971).
27. R. A. Rapp and K. S. Goto, "The Hot Corrosion of Metals by Molten Salts," in *Symposium on Fused Salts*, J. Braunstein and J. R. Selman (eds.), The Electrochemical Society, Pennington, NJ, 1979, p. 159.
28. D. A. Shores, "New Perspectives on Hot Corrosion Mechanisms," in *High Temperature Corrosion*, R. A. Rapp (ed.), NACE, Houston, TX, 1983, p. 493.
29. G. C. Fryburg, F. J. Kohl, C. A. Stearns, and W. L. Fiedler, *J. Electrochem. Soc.*, **129**, 571 (1982).

30. J. A. Goebel, F. S. Pettit, and G. W. Goward, *Met. Trans.*, **4**, 261 (1973).
31. K. L. Luthra and D. A. Shores, *J. Electrochem. Soc.*, **127**, 2202 (1980).
32. K. L. Luthra, *Met. Trans.*, **13A**, 1647,1843 (1982).
33. K. L. Luthra, *J. Electrochem. Soc.*, **132**, 1293 (1985).
34. K. L. Luthra and J. H. Wood, *Th. Sol. Films*, **119**, 271 (1984).
35. R. H. Barkalow and F. S. Pettit, "On the Mechanisms for Hot Corrosion of CoCrAlY Coatings in Marine Gas Turbines" Proceedings of the 14th Conference on Gas Turbine Materials in a Marine Envirnoment, Naval Sea Systems Command, Annapolis, MD, 1979, p. 493.
36. K. J. Chiang, F. S. Pettit, and G. H. Meier, "Low Temperature Hot Corrosion," in *High Temperature Corrosion*, R. A. Rapp (ed.), NACE, Houston, TX, p. 519.
37. K. P. Lillerud, B. Haflan, and P. Kofstad, *Oxid. Met.*, **21**, 119 (1984).
38. K. P. Lillerud and P. Kofstad, *Oxid. Met.*, **21**, 233 (1984).
39. A. K. Misra, *Oxid. Met.*, **129**, 25 (1986).
40. A. K. Misra, *J. Electrochem. Soc.*, **133**, 1038 (1986).
41. G. C. Fryburg, F. J. Kohl, and C. A. Stearns, *J. Electrochem. Soc.*, **131**, 2985 (1984).
42. E. L. Simons, G. V. Browning, and H. A. Liebhafsky, *Corrosion*, **11**, 505 (1955).
43. A. U. Seybolt, *Trans. TMS-AIME*, **242**, 1955 (1968).
44. J. A. Goebel and F. S. Pettit, *Met. Trans.*, **1**, 3421 (1970).
45. J. B. Johnson, J. R. Nicholls, R. C. Hurst, and P. Hancock, *Cor. Sci.*, **18**, 543 (1978).
46. M. G. Fontana, *Corrosion Engineering*, 3rd ed., McGraw-Hill, New York, 1986, p. 86.
47. P. Hancock, *Vanadic and Chloride Attack of Superalloys*, to be published by the Institute of Metals.
48. A. K. Misra and D. P. Whittle, *Oxid. Met.*, **22**, 1 (1984).
49. J. A. Goebel and F. S. Pettit, "Hot Corrosion of Cobalt Base Alloys," performed by Pratt & Whitney Aircraft for Wright Patterson Air Force Base, ARL TR 75-0235, NTIS, Clearinghouse, Springfield, VA, 1975.
50. H. Morrow, D. S. Sponseller, and E. Kalns, *Met. Trans.*, **5**, 673 (1974).
51. F. S. Pettit and G. W. Goward, "High Temperature Corrosion and Use of Coatings for Protection," in *Metallurgical Treatises*, J. K. Tien and J. F. Elliot (eds.), The Metallurgical Society, Warrendale, PA, 1981, p. 603.
52. S. Hwang, G. H. Meier, G. R. Johnston, V. Provenzano, and J. A. Sprague, "The Initial Stages of Hot Corrosion Attack of CoCrAlY Alloys at 700 °C," in *High Temperature Protective Coatings*, S. C. Singhal (ed.), The Metallurgical Society, Warrendale, PA, 1983, p. 121.
53. P. C. Felix, "Use of Simulation Rig Tests to Determine Corrosion Resistance of Alloys," in *Deposition and Corrosion in Gas Turbines*, A. B. Hart and A. J. B. Cutler (eds.), Applied Science Publishers, London, 1973, p. 331.
54. R. H. Barkalow and F. S. Pettit, "Marine Gas Turbine Hot Corrosion Dependence on Ingested Salt Levels," Naval Research Laboratory Final Report N00173-77-C-0206, April 1979.
55. J. E. Restall, "Influence of Salt and Carbon Particles on the Behavior of Hot Superalloy Components in Gas Turbine Engines," in *Proceedings of the 1974 Gas Turbine Materials in the Marine Environment Conference*, J. W. Fairbanks and I. Machlin (ed.), Battelle Memorial Institute, Columbus, OH, MCIC-75-27, 1974, p. 195.

Chapter 13

Protective Coatings

JOHN H. WOOD and EDWARD H. GOLDMAN

Gas Turbine Division, General Electric Company, Schenectady, New York, and Aircraft Engine Business Group, General Electric Company, Evendale, Ohio

A coating for use at high temperatures on a superalloy substrate can be defined as a surface layer of material, either ceramic or metallic or combinations thereof, that is capable of precluding or inhibiting direct interaction between the substrate and a potentially damaging environment. This damage can either be metal recession due to oxidation/corrosion or a reduction in substrate mechanical properties due to the diffusion of harmful species into the alloy at high temperature. Coatings used on superalloys do not function as inert barriers. Rather, they provide protection by interacting with oxygen in the environment to form dense, tightly adherent oxide scales that inhibit the diffusion of damaging species such as oxygen, nitrogen, and sulfur into the substrate. Coatings must therefore be rich in those elements (such as Al, Cr, or Si) that readily participate in the formation of these protective scales. Essentially, they are reservoirs of those elements; the supply is continually being used to reform new scale to replace that which spalls as a result of thermal cycling or mechanical damage. Thus, by the nature of its protective mechanism, the usable life of a coating is governed by its ability to form the desired protective scale and to retain or replace that scale as needed.

By far the largest use of coatings on superalloys is on components in the hot-gas section of turbine engines, that is, combustors, blades, and vanes. The need for such coatings surfaced in the aircraft engine business in the 1950s when it became apparent that substrate compositional requirements for improved high-temperature strength and optimum high-temperature environmental protection were not compatible. Increasing operating temperatures caused excessive oxidation of the high-strength nickel- and cobalt-base superalloys being used for turbine blades and vanes. This led to the development of the simple aluminide diffusion coatings that solved the oxidation problem. Several of these aluminide coatings are still in use today.

Hot corrosion first became a serious problem in those larger industrial and power generation gas turbines that burned low-quality fuels contaminated with sulfur, sodium, and other impurities, or that were located in areas where deleterious species could be ingested through the air intakes, for example, marine or desert environments. The aluminide coatings developed for aircraft engines to solve the oxidation problem were not effective in inhibiting severe hot-corrosion attack. This sparked development of other classes of coatings aimed specifically at combatting hot corrosion. More recently, another distinct mechanism of corrosion, known as low-temperature hot corrosion, was identified. Its successful inhibition has required coating compositions different from those developed for resistance to classical hot corrosion. Thermal barrier coatings (TBCs), which utilize a ceramic layer to reduce the temperature seen by the superalloy component, have been developed to permit substrate materials to be used at engine operating temperatures that might otherwise exceed their capability. Thus, different classes of coating compositions, and processes for applying them, have evolved to meet the differing needs of various applications.

The factors affecting coating selection are numerous. Obviously, environmental protection is the primary reason for using a coating, and this is governed by the design and application of the part. Possible effects of the coating or coating process on the mechanical or thermal properties of the superalloy to be coated must be considered, including the effects of interdiffusion between the coating and the substrate during high-temperature service exposure. Part geometry may govern what coating process is to be used since some application techniques are line-of-sight. Finally, the cost of the coating is always an important consideration and is often the controlling factor in selection.

It is the intent of this chapter to describe the types of coatings and coating processes that have been developed to provide protection for superalloy substrates at high temperatures. The focus will be on the developments of the past 10–15 years, a particularly active period in the history of superalloy coatings development.

COATING PROCESSES

Just about every known method for altering the surface of a metal has been used or considered for the protection of superalloys. In some cases problems with the protection of superalloy components have actually driven the development of new coating processes or the advancement of existing technologies. Superalloy coating processes are often divided into two main categories: those that involve alteration of the substrate outer layer by its contact and interaction with selected chemical species (*diffusion coating processes*) and those that involve deposition of protective metallic species onto the substrate surface, with adhesion provided by a much smaller amount of elemental interdiffusion (*overlay coating processes*).

Diffusion Coatings

Processes in which aluminum is diffused into a surface are the most widely used in the aircraft gas turbine industry. The first methods were slurry-fusion and pack cementation; the latter remains the most widely used process today. Recently, chemical vapor deposition processes have been shown to have advantages for certain applications.

Pack Cementation. In pack cementation, itself a type of vapor deposition process, both the component to be coated and the reactants that combine to form the vapor are contained in the same retort. The reactants, collectively known as the "pack," consist of an aluminum-containing powder (other elements may also be present), a halide that serves as a chemical activator, and an inert filler such as alumina. On heating in an inert atmosphere, the metal powder and activator react to form a vapor that in turn reacts with the surface of the component, enriching it with aluminum. For nickel-base alloys the phases of interest are Ni_3Al, $NiAl$, and Ni_2Al_3; $CoAl$ and $FeAl_2$ form on cobalt- and iron-base alloys, respectively. The reaction is controlled by the concentrations of the pack constituents and the temperature; these, along with time at temperature and post-coating heat treatment, determine the morphology of the resultant coating.

Diffusion aluminide coatings are classified as either "inward" or "outward" types. An inward coating is produced when the aluminum activity is high with respect to nickel [e.g., high Al and/or activator content in the pack, 1400–1800 °F (760–982 °C) reaction temperature]; the aluminum then diffuses inward faster than the nickel outward through the nickel–aluminum intermetallics that initially form at the surface. When aluminum activity is low with respect to nickel [low pack Al/activator contents, 1800–2000 °F (982–1093 °C) reaction temperature], an outward coating forms; outward diffusion of nickel is favored for subsequent reaction with aluminum in this case. Figure 1 shows typical morphologies of inward and outward

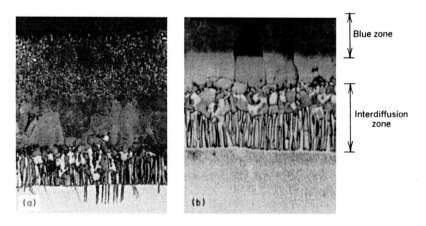

Blue zone

Interdiffusion zone

Fig. 1. Microstructures of (a) high-activity (inward) and (b) low-activity (outward) diffusion aluminide coatings on a Ni-base superalloy.[1] 1000×.

diffusion aluminide coatings.[1] It is interesting to note that both inward and outward coatings may be present on a single component, owing to temperature or composition gradients in the pack or to variable substrate geometry or surface condition.

Both coating types contain the high-melting-point β-NiAl phase. Since NiAl is stable for aluminum contents from about 45 to 60 a/o,[2] pack parameters are often adjusted to produce a more oxidation-resistant hyperstoichiometric NiAl in the outer layer. This "stuffing" of extra aluminum atoms into the NiAl structure causes the usually brown-colored β phase to take on a bluish tint when viewed microscopically. This "blue zone" shows up as a dark layer in Fig. 1b. The solubility in NiAl of most other superalloy substrate elements is small; they are therefore largely rejected from the NiAl outer layer and usually precipitate out as carbides ($M_{23}C_6$, M_6C, MC), metals (e.g., α-Cr), or topologically close-packed phases (e.g., σ, η). Figure 1 shows concentrations of these precipitates in a discrete "interdiffusion zone" between the outer layer and the substrate; in the case of inward coatings precipitates are also present in the NiAl outer layer (Fig. 1a). More detailed discussions on the formation of diffusion aluminide coatings can be found in the first edition of this book[3] and elsewhere.[1]

Because of the compositional complexities of superalloys, a given set of processing conditions will produce different coatings on different substrate alloys. For example, a given coating will usually be thinner on a cobalt-base alloy than on a nickel-base alloy because of the lower diffusivity of aluminum in cobalt. Even when applied to nickel-base superalloys of different compositions, the "same" coating can have different characteristics, particularly with respect to the phase structure of the interdiffusion zone. Single-crystal alloys, for example, generally contain no grain boundary modifiers (C, B, and Zr) because of the absence of grain boundaries. The nature of the interdiffusion zone is altered accordingly; base metal elements present in concentrations that exceed their solubility limits in NiAl must necessarily find means other than carbide formation by which to adapt to the coating phase structure. Parallel alloy/coating development is helpful in generating a desired coating microstructure.

Chemical Vapor Deposition.   In the chemical vapor deposition (CVD) process a vapor of predetermined composition, produced independent of the coating step, is introduced into the coating chamber where it reacts with the surface of the part. The major advantage of CVD over pack cementation is its ability to coat serpentine internal cooling passages of film-cooled airfoils. The vapor can be pumped through the internal passages, providing a fairly uniform coating throughout a very complicated geometry. (During pack cementation a small amount of coating vapor does leak into internal passages through cooling holes, but the "throwing power" is extremely limited.) Another advantage of CVD is its compositional flexibility since the thermodynamics of vapor formation are separated from the thermodynamics of the metal-vapor reaction.

Both pack cementation and CVD processes have been used to deposit other elements, such as chromium and silicon, as well as aluminum. Duplex processes, in which noble metals such as platinum and palladium are plated onto the substrate surface prior to aluminiding, have also been used successfully.

Overlay Coatings

Overlay coatings differ from diffusion coatings in that interdiffusion of the applied surface layer with the substrate is not required to generate the appropriate coating structure or composition. Rather, a prealloyed material having the composition required to form an adherent, protective oxide scale is applied to the surface by any of several methods that require interdiffusion only to ensure that the coating remains attached to the substrate. The overlay coating processes of primary importance today are physical vapor deposition and plasma spraying.

Electron Beam—Physical Vapor Deposition (EBPVD). Physical vapor deposition emerged in the 1960s as the primary overlay coating production technique. The term physical vapor deposition (PVD) refers to deposition of metals by transport of vapor in a vacuum without the need for a chemical reaction.[4] Today, the electron beam (EB) evaporation process is the most commonly used for coating turbine airfoil components. An ingot of the appropriate composition is vaporized in a vacuum using a focused electron beam. The parts to be coated are manipulated within the vapor cloud with the metal condensing out on the preheated substrate surface. The composition of the deposited coating will often be different from that of the starting ingot, due to differences in vapor pressure of the elements in a typical coating alloy; the composition of the ingot must be adjusted accordingly. The technology has progressed to where elements with a broad range of vapor pressures can be simultaneously evaporated from a single source. More detailed descriptions of the process and of the effects of various parameters such as evaporation/deposition rate, substrate preheat temperature, and so on, are presented elsewhere.[5]

Figure 2a shows a typical as-deposited microstructure of a CoCrAlY (Co–19Cr–12Al–0.3Y) coating on a nickel-base superalloy. Post-coating heat treatments

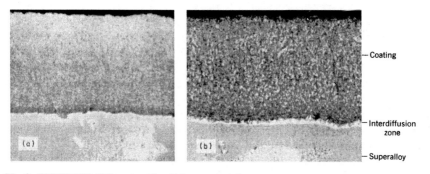

Fig. 2. EBPVD MCrAlY coating (Co–19Cr–12Al–0.3Y) on a Ni-base superalloy substrate (Courtesy of Temescal Division of the BOC Group, Inc.). (a) As coated. (b) After diffusion heat treat. 500×.

are used to ensure good bonding of the coating to the substrate; the microstructure following heat treatment is shown in Fig. 2b. The high aluminum content of this coating causes the precipitation of β-CoAl in the cobalt solid-solution matrix. Limited interdiffusion does occur during coating and post-coating heat treatment, but the structure and composition of these coatings are essentially constant through the thickness and are changed significantly only within the small interdiffusion zone.

Due to the nature of the deposition process, the structure of the as-deposited coating is typically oriented perpendicular to the substrate surface. Separations between adjacent colonies of deposited coating, known as "leader defects," are often present, particularly on convex curved surfaces. Techniques such as glass-bead peening and laser glazing have been utilized to close these defects in order to prevent premature environmental attack and thermal fatigue cracking.

Plasma Spraying.   Plasma spraying has been used as a coating process for many years. The process involves the injection of the coating material, usually in the form of prealloyed powder, into a high-temperature plasma gas stream that has been created inside a plasma gun. Here, the powder particles are melted and accelerated toward the substrate. The molten metal "splats" against the substrate and spreads out in a direction parallel to the substrate surface. The parameters of the process, such as the amount of preheat, plasma gun characteristics, gun to workpiece distance, and so on, all influence the structure and properties of the deposit. For details concerning the plasma spraying process and plasma jet technology, the reader is referred to refs. 6, 7, and 8.

While plasma spray coating is not a new technology, the use of the process in low-pressure vacuum chambers is relatively new. With many of today's coatings being rich in reactive elements such as aluminum and chromium (i.e., MCrAlY), plasma spraying of such coatings in low-pressure environments minimizes the formation of oxide defects within the as-deposited coating structure. The advantages of the low-pressure process also include higher powder particle velocities and broader spray patterns.[9] Coatings can also be applied under the cover of an inert gas shroud. In either case the objective is to deposit a clean, defect-free coating of a desired thickness in a reproducible manner. As with the EBPVD process, adherence of the coating to the substrate is accomplished by means of a post-coat heat treatment.

The microstructure of a typical CoCrAlY plasma sprayed coating is shown in Fig. 3. In the as-coated condition (Fig. 3a) the "splat" interfaces parallel to the substrate surface can be clearly seen. Following diffusion heat treatment (Fig. 3b), individual "splat" layers are no longer visible, and the structure has assumed the two-phase nature that was apparent in the EBPVD coating (Fig. 2b). Typically, the surface finish of plasma-sprayed coatings is rougher than that of EBPVD coatings, and finishing operations (e.g., abrasive slurry and controlled vapor blasting) may be performed to meet the required aerodynamic specifications.

Relative to the EBPVD process, plasma spray offers a significant advantage in terms of compositional flexibility because the vapor pressures of the coating elements are not a concern. Any material that can be produced as a powder of appropriate

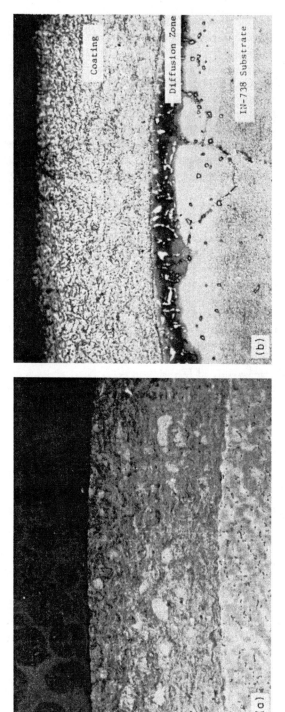

Fig. 3. Low-pressure plasma-sprayed MCrAlY coating (Co–29Cr–6Al–0.3Y) on a Ni-base superalloy substrate. (a) As sprayed; 250×. (b) After diffusion heat treat; 400×.

size fraction can be sprayed through the plasma gun, although variations in the melting temperature of the starting powder may require modifications to gun design for optimum particle melting.

A major disadvantage of plasma spray and EBPVD relative to diffusion coatings is that they are both line-of-sight processes. With the complex shapes common in turbine blade and vane designs, this limitation inevitably causes problems with coating thickness control due to "shadowing"—the blockage or partial blockage of one part of the component being coated by another part that protrudes into the straight-line path of the coating material being deposited. This problem has largely been overcome by sophisticated manipulation of the part being coated (and of the plasma gun in the case of plasma spraying), although this adds complexity and cost to the process.

The internal passages of film-cooled airfoils cannot be coated using either overlay process, however. A hybrid coating consisting of an overlay coating on the external surfaces and a vapor aluminide on the internal surfaces can be used to provide complete protection in this case. Hybrid coatings constitute an increasingly important part of the superalloy coating business. They are composed of two or more compositions that can be applied by one or more processes. In addition to overcoming the processing limitations of overlay coatings, hybrid coatings can also be used to inhibit undesirable interdiffusion. For example, if the surface of a CoCrAlY coating is enriched with silicon, improved hot-corrosion resistance may be achieved without the deleterious effects of the interdiffusion of silicon with the nickel-base substrate. Also, many turbine blade designs have sufficiently broad temperature profiles so as to permit multiple mechanisms of corrosion to be active during operation. Use of a hybrid coating with layers of differing composition can offer some protection against all mechanisms. A special case of a hybrid coating, the thermal barrier coating, is discussed below.

Miscellaneous Coating Processes

Coatings similar to those formed by pack cementation and CVD have been produced by the spraying or dipping and subsequent high-temperature fusion of metal-containing slurries. Electrophoretic deposition processes have also been used to electrodeposit fine metal particles of the desired composition around complex shapes prior to fusion.[10] Success has been reported in producing MCrAlY overlay coatings by a similar technique called occluded plating.[11]

Controlled composition reaction sintering is a related technique that was developed as a low-cost alternative for applying MCrAlY-type overlay coatings.[12] The MCrY components of the coating are applied by slurry spray, which is followed by a controlled reaction with aluminum and sintering in an aluminide pack of appropriate activity. Success with the reaction sintering approach has been reported with NiCoCrAlY[13] and NiCrSi[14] type coatings.

Sputtering and ion plating fall within the broad definition of physical vapor deposition, but the techniques for producing the vapor cloud differ significantly from EBPVD. In sputtering, a target of the coating material is bombarded by high-energy ions. Target atoms are ejected by momentum transfer and subsequently condense on a substrate positioned in the chamber. Ion plating is a hybrid of sputtering and vacuum evaporation. Prior to evaporation ion bombardment cleans the substrate surface. During evaporation the higher energy of deposition leads to better adhesion of the coating. Sputter ion plating of MCrAlY coatings on turbine blades has been reported and is claimed to be at an advanced stage of development.[15]

In the cladding process the desired alloy composition is fabricated into a thin sheet of the required thickness that is then diffusion bonded onto the substrate surface under high temperature and pressure. Although it has been demonstrated that this process does work,[16,17] the difficulties attendant on fabricating some of the more corrosion-resistant, low-ductility alloys into a thin sheet, make it unlikely that such a process would be used on a large scale. Examples of clad turbine blades with many hours of field service are available.[18]

Surface chemistry modifications can also be achieved by ion implantation and ion beam mixing techniques, which involve the high-energy injection of desired elemental species into a thin surface layer of the substrate.

While all of the techniques mentioned in this section may have achieved important commercial status within some segment of the coating industry, they have not yet become widely established as processes for the coating of superalloy turbine components. Therefore, the following sections of this chapter will concentrate only on those coatings produced by the diffusion aluminide, EBPVD, or plasma spraying process.

COATING EVALUATION

Coatings must be evaluated to ensure that they produce the desired benefits in environmental resistance without unacceptable compromises in the mechanical or physical properties of the coated superalloy. Such tests must be as realistic as possible; in the wrong application or with lack of consideration for a critical parameter, coated part life can be less than uncoated life, even though surface stability may be enhanced. On the other hand, tests that are too severe may overpenalize an otherwise acceptable coating. There is, as usual, a trade-off between the need to test coatings realistically and the cost and duration of those tests.

Since the laboratory tests performed on coatings to determine environmental resistance and effects on substrate mechanical properties are similar to those performed on uncoated superalloys, they will not be discussed in detail here. However, it should be emphasized that the coating and substrate to be used in an intended application must always be considered as a materials system and tested accordingly, since interdiffusion of substrate and coating elements with extended time at temperature can dramatically alter performance.

In addition to the performance testing of coatings, coated components must be evaluated for proper thickness, composition, microstructure, and adherence of the applied coating. This is largely done by destructive optical metallography on coated hardware; considerable effort is currently being expended to identify nondestructive techniques such as radiography, ultrasonics, and thermoelectric probe for coating quality assurance.

Although laboratory tests provide extremely useful information about the relative behavior of different coating/substrate systems and, in fact, provide the data required for the design of components using that materials system, it is factory or field engine operation that provides the ultimate test. Such tests include all the variables of stress, strain, temperature, and environment that cannot possibly be duplicated in controlled tests. Long-time exposure of development coatings in the field, in so-called rainbow rotor programs,[18,19] has become particularly useful in sorting out the behavior of various coatings.

COATING PERFORMANCE

Oxidation/High-Temperature Performance

The success of a coating in high-temperature applications as in aircraft engines, is measured by its ability to remain in place, to resist oxidation, and to avoid cracking. In general, the aluminides are most often limited by oxidation behavior, while overlay coatings are more susceptible to thermal fatigue cracking in highly cyclic applications. The major factors that influence the performance of these coatings are described in more detail below.

Aluminide Coatings.   Aluminide-coated superalloys oxidize similarly to the more oxidation-resistant superalloys themselves, except that the reservoir of aluminum at the surface is larger. As described in Chapter 11, oxygen combines with aluminum at the surface, and eventually a continuous, protective Al_2O_3 scale is formed. When the scale cracks and spalls due to thermal cycling, aluminum from the coating

Table 1. Results of X-Ray Diffraction Studies on Oxide Scales Scraped from Coated Oxidation Specimens of a Nickel-Base Superalloy

Coating	Time at 2075 °F (1190 °C), h[a]	Estimated Scale Composition (vol. %)			
		Al_2O_3	TiO_2	$NiAl_2O_4$	Other
Aluminide	240	95	5	—	—
	340	60	20	20	—
	→ 790	60	20	20	—
	940	40	10	50	—
	1290	5	5	80	10 NiO
Platinum-aluminide	350	95	5	—	—
	700	80	10	10	—
	→1100	70	15	15	—

[a] Arrow denotes time at which coating penetration was observed based on metallographic evaluation.

AS-COATED EARLY EXPOSURE

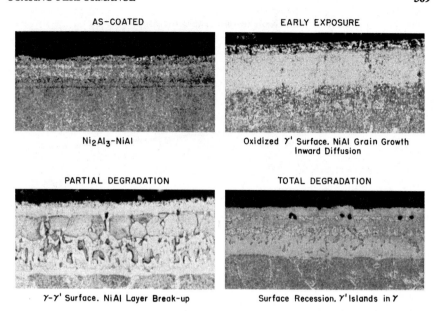

Ni$_2$Al$_3$-NiAl

Oxidized γ' Surface. NiAl Grain Growth
Inward Diffusion

PARTIAL DEGRADATION TOTAL DEGRADATION

γ-γ' Surface. NiAl Layer Break-up

Surface Recession. γ' Islands in γ

Fig. 4. Decomposition of a high-activity aluminide coating on a Ni-base superalloy.[20] 250×.

diffuses to the surface to reform the protective scale. Throughout the exposure, however, aluminum also diffuses from the coating into the base metal. As aluminum is depleted from the coating by diffusion, β-NiAl converts to γ'-Ni$_3$Al and eventually to γ-Ni solid solution. When the aluminum level drops to below about 4–5 wt. %, the continuous Al$_2$O$_3$ scale can no longer be formed (Table 1), and more rapid oxidation occurs. Figure 4 shows a typical degradation sequence from a microstructural standpoint.

Substrate composition plays a major role in determining coating performance. Table 2 shows the oxidation lives of a typical diffusion aluminide coating on several superalloys. The large differences in life are attributed to substrate aluminum level, which affects the rate at which aluminum will diffuse out of the coating, and to the concentration of certain substrate elements that may enhance (Cr and, to some degree, Hf and Ta) or degrade (Ti, V, W, and Mo) oxidation resistance upon diffusion to the coating surface during post-coat heat treatment and service exposure.

The concentration of refractory elements in the diffusion zone may produce other unusual effects, particularly when thermal fatigue cracking occurs. Normally, the diffusion zone is not directly involved in the oxidation process. If the coating cracks, however, this concentration of refractory elements can be directly exposed to the oxidizing environment and may oxidize rapidly. This condition can be reproduced by introducing cracks into oxidation specimens prior to testing. For an experimental vanadium-containing substrate alloy, coating life dropped precipitously when such cracks were introduced; a concentration of vanadium in the diffusion zone of the

Table 2. Oxidation Lives of Diffusion Aluminide Coating (CODEP) on Several Superalloys

Alloy	Coating Life at 2075 °F (1190 °C), h[a]
X-40	23
CMSX-3	85
René 80	100
René 125	300

[a] Mach 1.0 gas velocity, air, cycled once per hour; life determined by visual and metallographic assessment of coating penetration.

coated alloy, while inconsequential in the normal oxidation process, resulted in catastrophic oxidation when the coating was cracked (Fig. 5a). Figure 5b shows this ballooning or burrowing oxidation as it occurred on an aluminide-coated vane from a field-tested aircraft engine. This effect must be considered when selecting coatings for highly cyclic applications.

Another important aspect of the high-temperature performance of diffusion aluminide coatings that is related to the nature of the interdiffusion zone is the temperature at which incipient melting occurs. Even though the melting point of NiAl is around 2900 °F (1593 °C) and that of most superalloys greater than 2300 °F (1260 °C), incipient melting has been observed in the diffusion zone of aluminide-coated superalloys at temperatures as low as 2050 °F (1121 °C).[21] On a macroscopic scale, this can result in wrinkling and, in severe cases, peeling of the coating before oxidation penetration occurs. Again, owing to differences in the nature of the phases in the diffusion zone, the temperature at which melting occurs varies for the same coating on different alloys.

Overlay Coatings. The oxidation scenario for overlay coatings is basically the same as that for diffusion aluminide coatings. The presence of chromium and active elements such as yttrium improve oxidation resistance by increasing the activity of

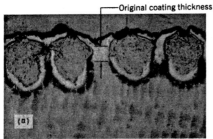
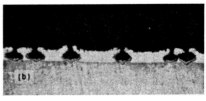

Original coating thickness

Fig. 5. Ballooning oxidation in the interdiffusion zone of a diffusion aluminide–coated Ni-base superalloy. (a) Precracked oxidation test specimens after exposure; 100×. (b) Field-tested turbine vane; 50×.

Table 3. Oxidation Resistance of Coatings on a Nickel-Base Superalloy

Coating	Coating Life at 2075 °F (1190 °C), h[a]
Aluminide	100
Platinum-aluminide	250
NiCoCrAlY	>1000

[a] Mach 1.0 gas velocity, air, cycled once per hour; life determined by visual and metallographic assessment of coating penetration.

aluminum and by improving the spallation resistance of the Al_2O_3 scale, respectively (see Chapter 11). Thus, MCrAlY coatings have been designed that significantly outperform diffusion aluminides in oxidation tests (Table 3). NiCoCrAlY compositions are the most widely used for oxidation protection; the addition of cobalt to the basic NiCrAlY composition, while affording some improved environmental resistance, was also found to improve coating ductility.[22]

As a NiCoCrAlY coating oxidizes, grains of the aluminum-rich β phase gradually convert to islands of γ'; eventually only the less-resistant γ matrix phase remains (Fig. 6). Originally, it was thought that overlay coating composition and performance would be independent of substrate composition. This is not totally accurate, particularly in high-temperature applications where diffusion rates are high. The large grain boundary area present in fine-grained overlay coatings affords a large network of diffusion paths for base metal elements, which can be present in small amounts in the as-deposited and diffusion heat-treated coating (Fig. 7a). With exposure, the amounts of these elements at the coating surface increase (Fig. 7b). As shown in Table 4 and elsewhere,[23] the oxidation resistance of an ovelay coating can indeed be affected by substrate composition, just as are the diffusion aluminides.

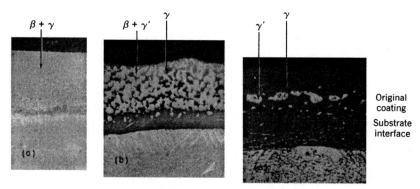

Fig. 6. Typical microstructural changes in a NiCoCrAlX coating on a Ni-base superalloy exposed in a cyclic oxidation test at 2075 °F (1135 °C), air, Mach 0.05 gas velocity, 1 cycle/h. (a) As coated. (b) Exposed 150 h. (c) Exposed 450 h. 250×.

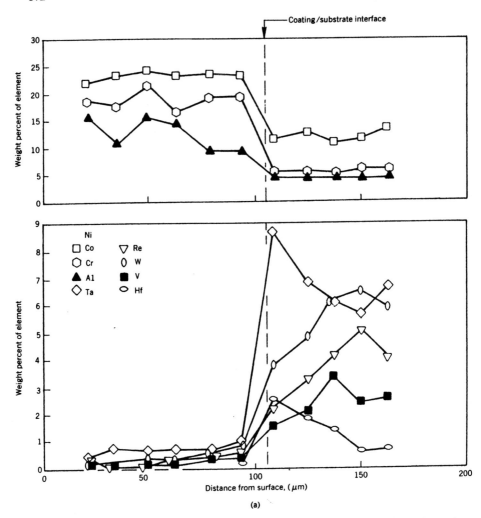

Fig. 7. Electron microprobe traces across a NiCoCrAlY coating on a Ni-base superalloy before and after high-temperature exposure: (a) as coated and diffusion heat treated.

In addition to superior oxidation resistance, a major advantage of MCrAlY coatings over diffusion aluminides for high-temperature applications is a higher melting point that is essentially independent of substrate influence. Melting does not occur in the interdiffusion zone at temperatures significantly lower than the melting point of the bulk overlay composition. Compared to incipient melting points of 2050–2200 °F (1121–1204 °C) for most diffusion aluminides, overlay coatings have survived exposures of 2350 °F (1288 °C) with no evidence of melting. However, the high melting points of the overlay coatings must be balanced against their very

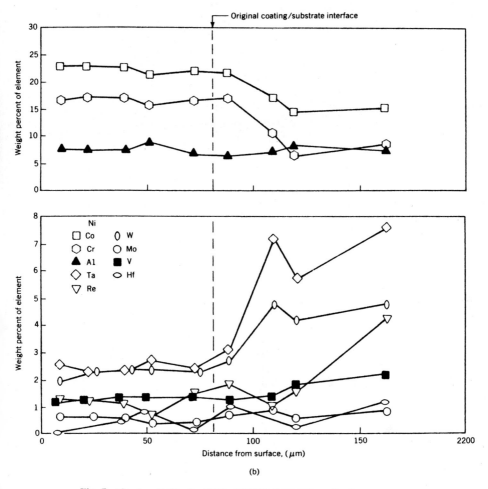

Fig. 7. *(Continued)* (b) after 750 h, 2075 °F (1135 °C), cyclic air exposure.

low strengths at high temperatures, which can result in thermal fatigue cracking in highly cyclic applications.

MCrAlY overlay coating development has largely focused on elemental additions that improve oxidation resistance. Modified NiCoCrAlY containing silicon, tantalum, and/or hafnium, for example, has exhibited improved resistance to oxidation, though ductility is usually reduced.[23] Coating compositions are now being developed that limit interdiffusion between the coating and substrate and increase the high-temperature strength of the coating in order to improve thermal fatigue resistance. The improved understanding, currently being developed, of the strain–temperature cycles experienced by superalloys and coatings in thin-walled, internally cooled turbine airfoil config-

Table 4. Oxidation Resistance of NiCoCrAlY Coating on Various Superalloy Substrates

Substrate	Coating Life at 2075 °F (1190 °C), h[a]
René 80	125
MM-200	200
Advanced Ni-base alloy	500

[a] Mach 1.0 gas velocity, air, cycled once per hour; life determined by visual and metallographic assessment of coating penetration.

urations will also provide direction for future overlay coating development. The almost unlimited compositional flexibility afforded by the low-pressure plasma spray process will continue to provide opportunities for tailoring an overlay coating composition to provide optimum performance on a particular substrate alloy in a service application of interest.

Hot-Corrosion Performance

The need for hot-corrosion-resistant coatings exists in the marine and industrial gas turbines. Here, the thermal cycles are generally not as severe as those in aircraft applications, and thus the limitations on overlay coating use may not be as strict. Often, the potential for hot corrosion can be reduced by proper fuel cleanup and adequate air filtration, but coatings are still required to prevent catastrophic corrosion failure in the event such systems are not available or not performing properly.

Diffusion Coatings. Use of the simple aluminide diffusion coatings in applications where hot corrosion was the primary problem generally led to unsatisfactory results unless the corrosion conditions were relatively mild.[24] Commercial availability of the platinum-aluminide coatings in the early 1970s[25] provided a marked improvement in the hot-corrosion resistance of the aluminide coatings. Other precious metal aluminides, which utilize less expensive elements such as rhodium or palladium in place of platinum, have been developed, but in general, they do not have the capability of the platinum-bearing versions. However, cost considerations can make these coatings, as well as various silicide and duplex chromium–aluminum diffusion coatings, more attractive for use in less severe environments.

The mechanism by which platinum enhances the corrosion resistance of aluminide coatings is not clearly understood. Platinum does not participate directly in scale formation. Rather, it appears to increase the activity of aluminum at the surface and/or influence scale adherence or the rate of scale formation. Degradation of the platinum–aluminide coatings occurs with the depletion of aluminum available to form the protective scale. As was discussed in Chapter 12, the presence of molten alkali metal salts can hasten the destruction of the alumina scale and thus accelerate the consumption of aluminum. A $PtAl_2$ phase (if it is present initially) goes into solution as aluminum is utilized in scale formation. Eventually, as sufficient aluminum is consumed, the β matrix phase is transformed to γ', and effective corrosion protection ceases. In the advanced stages of attack the formation of chromium-rich internal sulfides in the substrate or the interdiffusion zone between the substrate and coating indicates that the usable life of the coating has been exceeded. Micrographs illustrating the different stages of this degradation process are shown in Fig. 8.

The platinum–aluminide coatings are somewhat less ductile than the simple aluminides and therefore may be more restricted in their use in some aircraft engines with severe cyclic operating patterns. However, service experience has generally been very favorable for both land-based and aircraft applications. Satisfactory performance to 40,000 h and beyond has been reported for land-based turbines operating in corrosive environments that would cause the destruction of uncoated superalloys in relatively short times.[18]

Overlay Coatings. Several MCrAlX overlay coatings have also been shown to possess excellent hot-corrosion resistance, both in burner rig testing and in field service.[26,18] The best of the MCrAlX coatings for severe hot-corrosion environments are cobalt based with relatively high chromium–aluminum ratios. The nickel- (or NiCo) and iron-base coatings are effective in oxidation and in relatively mild hot-corrosion environments. A good hot-corrosion MCrAlY coating is exemplified by the Co–29Cr–6Al–0.3Y composition, which is an alumina former despite its relatively high chromium level. Generally, as the chromium–aluminum ratio increases, the hot-corrosion resistance increases at the expense of some oxidation resistance. Whether or not the active element addition (i.e., Y and/or Hf) is as effective in hot corrosion as it is in oxidation is not clearly established, but since many applications on turbine airfoils inevitably involve oxidation as well as hot corrosion, overlay coatings developed for hot-corrosion resistance usually contain an active element.

Other approaches to developing hot-corrosion-resistant overlay coatings have included the use of silicon as either an outer layer in a duplex or graded composition[27] or as the main scale-forming ingredient in a NiCrSi type of coating.[28] The use of noble metals such as platinum in CoCrAlXs has been reported to produce excellent hot-corrosion resistance in marine environments.[29] In general, it would appear that the most effective hot-corrosion-resistant overlay coatings are those alumina formers in which chromium has been increased to the maximum level consistent with the desired mechanical properties, and to which elements such as yttrium, silicon, platinum, and hafnium have been added for optimized performance in a given environment or with a given substrate.

The degradation of MCrAlY coatings in hot corrosion is characterized by the presence of internal sulfides as well as oxides within the coating (Fig. 9). Typically, these chromium-rich sulfides precede the internal oxides, much as in the attack of a bare superalloy. Ultimately, however, it is the depletion of the aluminum, as well as the chromium, needed to form the protective scale that leads to coating failure.

Field experience with some early-generation MCrAlY coatings has shown long-time performance at least equivalent to that of the platinum–aluminide coatings in terms of rate of attack.[18] Given the increased coating thicknesses possible with the overlay coating processes, improved corrosion life of the coated part can be achieved.

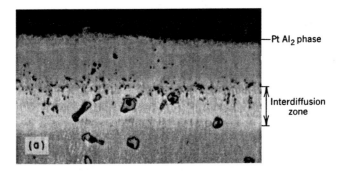

Pt Al₂ phase

Interdiffusion zone

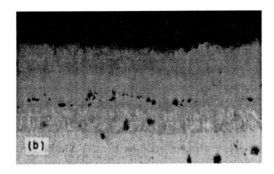

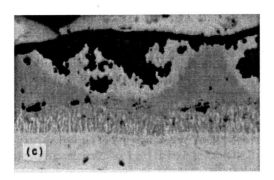

Fig. 8. Pt–Al coating degradation in a hot-corrosion environment. (a) As coated. Note surface layer containing light-colored PtAl₂ phase. (b) Approximately 10,000 h field service in low-corrosion-potential environment. PtAl₂ phase has been solutioned, leaving gray β-NiAl phase with minor surface scaling. (c) Approximately 10,000 h field service in a relatively severe hot-corrosion environment. Gray β-NiAl phase has converted to light-colored γ-Ni-Al or perhaps γ-Ni solid solution. Fine gray sulfides are present in light outer layer, but attack has not penetrated through diffusion zone into substrate. 500×.

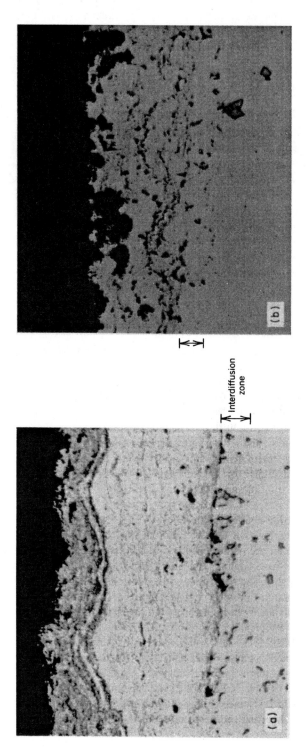

Fig. 9. MCrAlY overlay coating degradation in hot-corrosion environment. (a) Approximately 28,000 h field service in low-corrosion-potential environment. Coating shows internal oxidation with small β-depletion layer but very few, if any, sulfides. (b) approximately 1,000 h in severe hot-corrosion laboratory test environment at 1600 °F (871 °C). Coating shows heavy external scaling, complete β-phase depletion, internal oxides (large dark-gray islands), and internal sulfides (lighter gray particles) in both coating and diffusion zone. 400×.

377

The use of MCrAlX overlay coatings may therefore prove to be more cost-effective, particularly in those hot-corrosion applications where thermal cycling is not overly severe.

Low-Temperature Corrosion Performance

Experience has shown that chromia and possibly silica scales, rather than alumina scales, are required for optimum protection against low-temperature hot corrosion. Since most of the coatings developed for hot-corrosion and oxidation protection depend on alumina scales, these have generally proven ineffective in combatting low-temperature corrosion.[26] Therefore, this section will include discussion of those chromia- or silica-forming coatings that have most recently been utilized for low-temperature corrosion protection.

Diffusion Coatings.   Several commercially available chromide diffusion coatings have proven to be effective against low-temperature corrosion. The ability to rapidly form a continuous, adherent Cr_2O_3 scale appears to be the main requirement for protectiveness. Most pack diffusion chromide coatings are relatively thin [1.5–2.0 mils (0.038–0.051 mm) at most] due to the limitations of the process. Fortunately, the amount of interdiffusion that occurs in a lower temperature environment is small; coating thickness is less of a concern than at higher temperatures where protective elements can be rapidly lost to the substrate. A thinner coating is also more desirable from a mechanical property standpoint, since high chromium compositions tend to be less ductile and therefore more prone to cracking.

Simple silicide diffusion coatings, while probably very effective in combatting low-temperature hot corrosion due to the formation of a SiO_2 scale, are not used in turbine engines because of the deleterious effects of silicon interdiffusion with nickel-base superalloys. The use of more complex silicides (e.g., a Ti–Si composition[30] or Si-rich outer layers in duplex or graded coatings) are the preferred ways of utilizing silicon in the coating of superalloys.

Overlay Coatings.   Overlay coatings with relatively high chromium contents (> 30 wt. %), including MCrXs (see ref. 31) and MCrAlYs (ref. 32), have been reported in the literature; all are predominantly chromia formers. High-chromium cobalt-, nickel-, and iron-base coatings are all effective against low-temperature hot corrosion. However, since the temperature range encountered in many turbine airfoil applications will require protection against both mechanisms of corrosion, the use of cobalt-base coatings with high chromium levels is preferred.[26]

The microstructure of a typical high-chromium CoCrX plasma-sprayed coating is similar to that of the CoCrAlY shown earlier (Fig. 3b) except that the second phase is a cobalt–chromium σ phase rather than the β CoAl phase. Its burner rig test performance relative to that of a more "standard" CoCrAlY coating (i.e., Cr level less than 30 wt. %) is shown in Fig. 10. Clearly, the higher chromium CoCrX coating has vastly superior resistance to low-temperature pitting corrosion in this test. Some investigators[31] have indicated that aluminum may detract from the ability

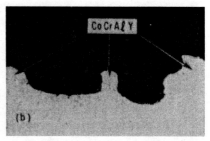

Fig. 10. Typical coating degradation observed on two overlay coatings, CoCrX (> 40 wt. % Cr) versus CoCrAlY (<30 wt. % Cr) in low-temperature hot-corrosion burner rig testing at 1350 °F (732 °C). Fuel: diesel oil + 1% S + 125 ppm Na as sea salt + 784 cm^3/min SO_2; air/fuel = 60, 1 atm, gas velocity = 70 ft/s. (a) CoCrX (>40% Cr), 1021 h, 250×. (b) CoCrAlY (<30% Cr), 64 h, 100×.

of such coatings to resist low-temperature corrosion, while others have found high-chromium CoCrAlY coatings to be acceptable for certain applications.[32] This apparent contradication may reflect differences in the expected environment and thus the test parameters. Suffice it to say that low-temperature corrosion resistance is promoted by the presence of chromia or silica scales, with the alumina-forming coatings that have been developed for hot corrosion and oxidation being less appropriate for this purpose.

Many newly developed low-temperature corrosion coatings are now undergoing field tests, with the results to be reported in the next few years. The identification of low-temperature corrosion as a separate mode of attack, and the coating development work spawned by that realization, is recent enough that as of this writing the record of field performance has still to be written.

THERMAL BARRIER COATINGS

Thermal barrier coatings represent perhaps the most promising and exciting development in superalloy coatings research in recent years. Given the driving force to increase the efficiency and/or output of gas turbines (which inevitably means an increase in turbine inlet temperature), any mechanism by which temperature limits can be raised by overcoming hot-section material restraints is of significant interest. Thermal barrier coatings offer this potential.

Description

A thermal barrier coating, or TBC, is a multilayer coating system that consists of an insulating ceramic outer layer (top coat) and a metallic inner layer (bond coat) between the ceramic and the substrate. In most cases the top coat and bond coat are applied by plasma spraying; sputtering and EBPVD have also been used. Typically, the ceramic top coat is 5–15 mils (0.127–0.381 mm) thick while the metallic bond

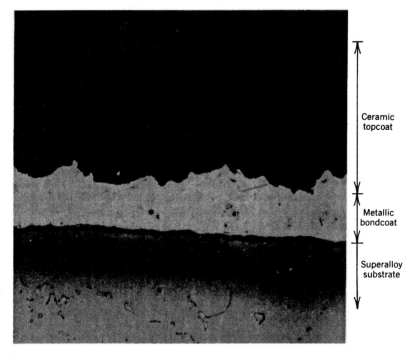

Fig. 11. Microstructure of typical thermal barrier coating on Ni-base superalloy substrate. 200×.

Ceramic topcoat

Metallic bondcoat

Superalloy substrate

coat is 3–5 mils (0.076–0.127 mm) thick. The microstructure of a typical system is shown in Fig. 11.

The function of the ceramic layer is to insulate the metallic substrate from higher surface temperatures than it might otherwise be able to tolerate. Depending on the thermal conductivity of the ceramic, coating thickness, and the heat flux created by the design and cooling configuration of the particular component, temperature gradients of several hundred degrees can be created through the coating. Zirconium oxide (ZrO_2) has been the material of choice because of its very low thermal conductivity and its relatively high (for ceramics) coefficient of thermal expansion. When heated to above about 2140 °F (1170 °C), however, the ZrO_2 structure changes from monoclinic to tetragonal; the accompanying volume change of 4–6% can result in severe spalling of the ceramic layer. Stabilization of the tetragonal phase to room temperature or below can be accomplished by the addition of MgO, CaO, Y_2O_3, or other rare-earth oxides to the ZrO_2. Typical state-of-the-art TBCs utilize ZrO_2 partially stabilized with 6–8 wt. % Y_2O_3.

While the zirconia top coat provides an excellent barrier to heat, it is a sieve with respect to oxygen transport. A major function of the metallic bond coat is thus to impart environmental resistance to the substrate, since the gross formation of oxides at the metal–ceramic interface can cause spallation of the ceramic. The roughness of a bond coat applied by plasma spraying aids in adhesion of the plasma-sprayed ceramic top coat by providing some mechanical interlocking. Air plasma-sprayed MCrAlY's were originally used for most TBCs; low-pressure plasma spray is also used today. An excellent summary of recent literature related to TBCs can be found elsewhere.[33]

Performance/Reliability

By their nature, metal oxides are relatively strain-intolerant. Unfortunately, sources of strain abound in TBCs, resulting from residual stresses from the coating process, thermal expansion mismatch between the ceramic and metal layers, oxidation/corrosion of the bond coat, phase transformations in the ceramic layer caused by thermal cycling, and thermal gradients typical of hot-section components in service. Some components also see mechanically induced strains. Consequently, the ceramic layer is prone to spalling which most often occurs just adjacent to its interface with the bond coat.[34]

In response, much recent work has focused on processing techniques that produce a more strain-tolerant ceramic structure and on the development of bond coats with improved environmental resistance, mechanical properties, and metallurgical stability. Approaches for the development of more strain-tolerant structures have included closer control of the as-deposited phase structure and the intentional incorporation of defects during processing. The as-deposited phase structure, which is critical to the performance of the top coat, has been shown to be very sensitive to the composition and structure of the starting powder[35] as well as to variations of plasma spray parameters (substrate temperature, gun-to-workpiece distance, etc.). The introduction of defects into the ceramic layer has been accomplished by careful control of these parameters in order to produce a controlled amount of porosity and/or microcracking in the deposit.[36] Post-deposition processing, including annealing and quenching, has also met with some success in producing the desired defect structures.[37]

Sputtering[38] and EBPVD[5] have been used to produce segmented structures consisting of a number of fine cracks perpendicular to the substrate surface. This division of the ceramic layer into a network of small individual segments is thought to improve its strain tolerance; improved cyclic lives have been reported for such structures.

Use of improved oxidation- and corrosion-resistant bond coats also leads to improved performance, as do bond coats applied in protective environments. As an aid to protection of the bond coat, techniques such as laser glazing of the ceramic or adjustment of plasma spray parameters near the end of the coating process have been used to produce a dense surface layer that precludes the absorption of corrosive salts.[39]

Current Applications

Commerical application of TBCs in turbine engines has thus far been restricted primarily to the stationary components of the combustion and exhaust systems (e.g., burner cans and transition pieces). Significant reductions in metal substrate temperatures have been achieved, leading to the elimination of creep deflection problems in those components. TBCs are also in limited use on stationary components in the turbine section. A very high potential payoff exists if TBCs can be used successfully on stationary and rotating turbine airfoils.

The future of TBC use on turbine airfoils depends on the success of current efforts to improve characterization capabilities and to enhance reliability. If maximum performance advantage is to be gained, airfoil surface temperatures (i.e., the surface of the ceramic outer layer) will be above the maximum allowable metal temperature of the underlying substrate, making the continued presence of the ceramic top coat critical. Thus, the need for realistic cyclic testing must be carefully addressed and engine testing will be at a premium. To date, there has been little information available on field testing of airfoil TBC coatings. This will be a subject of intense interest in the next few years.

SUMMARY

The development of coatings for superalloys has progressed from the simple diffusion aluminides of the early 1950s through the MCrAlYs to the thermal barrier coatings of the present. Metallic coatings such as the aluminides and high-aluminum MCrAlYs provide oxidation resistance at high temperatures, and TBCs will allow higher turbine-firing temperatures than would otherwise be achievable. Concurrently, higher chromium MCrAlX and MCrX compositions, along with precious-metal aluminides, provide greater hot-corrosion or low-temperature corrosion resistance where required. Several coating techniques are now capable of reproducibly depositing high-quality coatings containing highly reactive elements.

Superalloy coating development will continue to be very active. Efforts to develop reliable TBC coatings for turbine airfoil applications are likely to intensify. As operating temperatures continue to increase, coatings more resistant to oxidation and thermal fatigue will be required, and the advent of large land-based turbines burning coal-derived fuels may well require development of a new class of coatings. New process techniques such as laser melting/cladding and sputter ion plating will continue to be developed, but EBPVD, low-pressure plasma spraying, and pack aluminiding are likely to remain as the major commercial coating processes.

Emphasis will continue to be placed on optimizing coating/substrate combinations such that surface protection is maximized with minimal effect on substrate mechanical properties. This will be driven by the new classes of substrate materials coming into use, such as fiber-reinforced superalloys, oxide-dispersion-strengthened alloys, and so on, which will require that interaction with a coating not affect the stability of the strengthening phases. And finally, continuing, if not increasing, attention must be paid to the testing of newly developed coatings. Particularly in the case of relatively fragile coatings such as TBCs, the thermomechanical test cycles used to evaluate cyclic life must be realistic but not overly severe so as to nullify the possible gains to be realized. As always, engine testing under actual operating conditions will remain the ultimate test of coating system integrity.

REFERENCES

1. G. W. Goward and D. H. Boone, *Oxid. Met.*, **3**, p. 475 (1971).
2. M. Hansen, *Constitution of Binary Alloys*, McGraw-Hill, New York, 1958.
3. S. J. Grisaffe, in *The Superalloys*, C. T. Sims and W. Hagel (eds.), Wiley, New York, 1972.
4. D. G. Teer, Coatings for High Temperature Applications, Ed. by E. Lang, Applied Science Publishers, p. 79 (1983).
5. D. H. Boone and J. W. Fairbanks, *Specialized Cleaning, Finishing and Coating Processes, Conference Proceedings, Los Angeles, CA*, ASM, Metals Park, Ohio, February 1980, p. 357.
6. P. R. Dennis et al. (eds.), *Plasma Jet Technology*, NASA Technology Survey SP-5033, October 1965.
7. E. Muehlberger, "A High Energy Plasma Coating Process," 7th International Thermal Spray Conference, London, England, 1973.
8. D. Apelian et al., *Int. Met. Rev.* **28**(5) (1983).
9. W. F. Schilling, "Low Pressure Plasma Sprayed Coatings for Industrial Gas Turbines," NATO Advanced Workshop, Coatings for Heat Engines, Italy, April 1984.
10. M. J. Barber et al., Report AFWAL-TR-83-4056, Wright Air Development Center, Ohio, 1983.
11. R. Pichoir and J. E. Restall, "Protective Coatings for Superalloys in Gas Turbines," in ref. 9.
12. W. G. Stevens and A. R. Stetson, U.S. Air Force Materials Laboratory Techical Report AFML-TR-76-91, Wright Air Development Center, Ohio, August 1976.
13. L. Hsu and A. R. Stetson, *Metallurgical Coatings 1980, Conference Proceedings, San Diego, CA*, Vol. II, J. N. Zemel (ed.), Elsevier Sequoia, Lausanne, April 1980, p. 419.
14. E. Fitzer et al., *Materials and Coatings to Resist High Temperature Corrosion, Conference Proceedings*, D. R. Holmes and A. Rahmel (eds.), Applied Science, 1977, p. 313.
15. J. E. Restall, *Superalloys 1984*, M. Gell et al. (eds.), The Metallurgical Society of AIME, Warrendale, Pa., 1984, p. 721.
16. A. M. Beltran and W. F. Schilling, *Superalloys 1980*, J. K. Tien et al. (eds.), ASM, Metals Park, Ohio, 1980, p. 413.
17. J. H. Wood et al., "Protective Claddings and Coatings for Utility Gas Turbines," Final Report EPRI Project 1460-1, Palo Alto, CA, November 1983.
18. J. H. Wood et al., ASME Paper No. 82-GT-99, April, 1982.
19. N. R. Lindblad et al., ASME Paper No. 79-GT-47, March 1979.
20. S. R. Levine, NASA TMX-2370, 1971.
21. R. J. VanCleaf, unpublished work.
22. R. J. Hecht, G. W. Goward, and R. C. Elam, U.S. Patent 3,928,026, 1975.
23. R. J. Pennisi and D. K. Gupta, "Sprayed MCrAlY Applications," NASA Cr-165234, 1981.
24. F. S. Petit and G. W. Goward, in ref. 4, p. 341.
25. Canadian Patent No. 877897, August 10, 1971.
26. J. H. Wood et al., ASME Paper No. 85-GT-9, March, 1985.
27. J. R. Vargas et al., in ref. 13, p. 407.
28. M. Villat and P. Felix, *Sulzer Tech. Rev.*, p. 97 (March 1976).
29. J. J. Grisik et al., in ref. 13, p. 397.
30. G. H. Marijnissen, "A Titanium Silicon Coating for Gas Turbine Blades," AGARD CP 317, Maintenance in Service of High Temperature Parts, Noor-wijkerhout, September 1981.
31. K. L. Luthra and J. H. Wood, *Metallurgical Coatings 1984, Conference Proceedings, San Diego, CA*, Vol. II, R. C. Krutenat and J. N. Zemel (eds.), Elsevier Sequoia, Lausanne, Vol. II, April 1984, p. 271.
32. L. F. Aprigliano, Department of the Navy Report No. DTNSRCD/SME-81/10, June 1981.
33. W. J. Lackey et al., ORNL/TM-8959, 1984.
34. R. A. Miller and G. E. Lowell, *Metallurgical Coatings 1982, Conference Proceedings, San Diego, CA*, Vol. I, J. N. Zemel (ed.), Elsevier Sequoia, Lausanne, April 1982, p. 265.
35. D. Chuanxian et al., *Metallurgical Coatings 1984, Conference Proceedings, San Diego, CA*, Vol.

I, R. C. Kristenat and J. N. Zemel (eds.), Elsevier Sequoia, Lausanne, April 1984, p. 467.

36. D. L. Ruckle, in ref. 13, p. 455.

37. G. Johner and K. K. Schweitzer, in ref. 35, Vol. II, p. 301.

38. J. W. Patten et al., in ref. 13, p. 463.

39. S. R. Levine, "Thermal Barrier Coatings Research at NASA-Lewis," paper presented at the 1983 Metallurgical Coatings Conference, San Diego, CA, April 1983.

PART FIVE

PROCESS METALLURGY

Chapter 14

Melting and Refining

LOUIS W. LHERBIER

Cytemp Specialty Steel Division, Titusville, Pennsylvania

BACKGROUND

During the past 40 years superalloys for critical gas turbine parts have become more complex in terms of composition and have been subjected to increasingly sophisticated processing procedures. In fact, the 1980s are often referred to as the "Age of Processing." However, the process on which everything else depends is melting. The performance of a gas turbine engine, through the parts that make up that engine, is ultimately dependent on the starting superalloy ingot. The melting processes fundamentally control the ultimate quality of a disk, shaft, blade, burner can, and so on. No alloying change, no control in forging, or any modification in heat treatment can produce a reliable part from a poor quality ingot. The melting process(es) have been and will remain the dominant feature of superalloy processing technology.[1]

Early superalloys, essentially modifications of the austenitic stainless steels, were satisfactorily melted in electric arc furnaces (EFM). The discovery that reactive element additions could improve strength at elevated temperatures necessitated the development of vacuum induction melting (VIM) with major contributions by General Electric Company, Special Metals, and Universal Cyclops in the 1950s.

Increased demand for superalloys and better vacuum equipment resulted in larger VIM furnaces and larger ingots. The stronger and thus more highly alloyed materials made possible by VIM showed their resistance to man's wishes by exhibiting extensive macrosegregation and microsegregation in static cast products. This prevented direct utilization of these more highly alloyed materials and resulted in what is known today as "duplexing"—the remelting of VIM cast electrodes by vacuum arc remelting (VAR) or electroslag remelting (ESR). This combination of melting processes minimized solidification problems.

Most superalloys are produced by either the VIM/VAR or VIM/ESR melting processes developed in the 1950s and 1960s. Improvements have been made in the remelting processes; process control has allowed good control of macrosegregation and minimized microsegregation. As engine designers have continued to desire quality improvements, material cleanliness (which has been demonstrated to have a profound effect on the reliability of rotating parts) has further improved. To allow still greater process control of major melting processes, development efforts are now aimed at determining the potential of vacuum arc double-electrode remelting (VADER), electron beam cold hearth refining (EBCHR), and plasma melting (PMR). These new developments involve a combination of the various melting processes to achieve maximum quality.

Figure 1 illustrates the interrelationship of the major superalloy production melting techniques and some of the more promising development melting processes. The primary melting processes have been and remain VIM and EFM in combination with argon oxygen decarburization (AOD). The VIM process is used to melt most superalloys, particularly those precipitation-hardened nickel-base alloys containing significant amounts of reactive elements. Other superalloys such as the solid-solution nickel-, nickel–iron-, and cobalt-base materials can be produced by the EFM/AOD process. The principal remelting processes are VAR and ESR. Interest is being shown in EBCHR for potential cleanliness improvement and in VADER for a finer grain cast structure. The melting processes and combinations illustrated

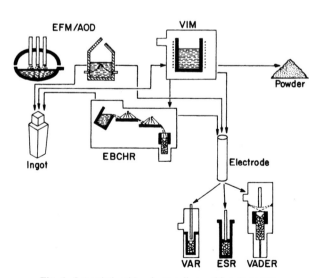

Fig. 1. Interrelationship of superalloy melting processes.

Table 1. Major Melting Processes Applied to Superalloys

Melting Process	Alloys	Use
VIM	B-1900, 713C, MAR-M 246, R-41, IN-718, IN-100, R-95	Mostly remelt for castings or for powder atomization
VIM/VAR	IN-718, Waspaloy, IN-901, R-41, A-286, U-700, U-500, D-979	Generally used for forgings for stock and rolled shapes
VIM/ESR	HX, H-188, N-155, L-605, IN-625, A-286	Primarily cast as slab ingots for flat rolled product
EFM/VAR	HX, L-605, N-155	Usually applied to bar and
EFM/ESR	In-625, A-286, S-816	flat rolled product for economic reasons

in Fig. 1 will be the subject of more detailed discussion in separate sections of this chapter. Melting processes used for some of the well-known superalloys are listed in Table 1.[2]

VIM, the major superalloy melting process, can yield a variety of cast products. The primary output is electrodes for remelting by VAR or ESR with limited amounts for remelting by VADER. Ingots obviously can be cast but are minimal in volume and generally confined to slab ingots destined for flat rolled products or remelt stock to be used in the investment casting industry. In addition, VIM material can be atomized into powder or subjected to EBCHR to achieve improved quality in respect to segregation and/or cleanliness. The other primary superalloy melting process, EFM/AOD, yields cast ingots and electrodes. Direct-cast superalloy ingot product is limited in volume. Most material is cast as electrodes for remelting. All remelting processes produce round ingots. Only ESR is capable of producing slab as well as round ingots. Remelted ingots are forged, rolled, or hot worked in some manner to an intermediate billet, slab, or bar product. Superalloy sheet, strip, and plate are often produced from ESR slab ingots.

There obviously are a number of superalloy melting processes, often used in combination, that can produce economical and reliable product. They are both versatile and complex. It is the purpose of the remaining sections in this chapter to discuss these processes with respect to equipment and procedures, metallurgical reactions, advantages and limitations, and trends based on current and future materials requirements.

PRIMARY MELTING PROCESSES

Vacuum Induction Melting

Equipment and Procedures.   Vacuum induction melting (VIM) is simply the use of an induction furnace placed in a vacuum chamber and so operated that the melting, the tapping of the furnace, and the casting of the metal are accomplished without breaking the vacuum. Apart from the need for a special lining to withstand the vacuum conditions and the remote-control facilities necessary, the melting operation is very little different from that involved in a high-frequency air induction furnace.

Vacuum induction melting furnaces range in size from less than 1 ton to 60 tons, and yet their basic components are essentially the same. The arrangement for a production furnace is shown in Fig. 2. The furnace is contained in a water-cooled metal tank whose configuration is dependent on the production requirements. Most large-scale furnaces today employ a three-chamber arrangement: one to enclose the melting furnace, one to house the molds and their handling mechanism, and one to handle charging of raw materials. In smaller furnaces the pumping system consists of mechanical pumps to rough down the chamber from atmosphere to 1000 μm; then ejector-type diffusion pumps reduce the pressure to the operating range of approximately 10 μm. As furnaces increase in size, steam ejectors are used to handle the larger gas loads, and installation of twin six-stage systems is not uncommon. Since the raw material packing factor prohibits preloading the full charge into the furnace before melting, a means to make major additions is required while melting under vacuum. This is most easily accomplished by means of a separate vacuum chamber for bulk charging. For small furnaces this is not a major problem, but for large furnaces with a wide range of charger configurations, the bulk charging rate becomes a critical item; it must keep up with the melt rate of the induction melting system.

Since most superalloys require the complex control of 8–20 elements to tight ranges or to maxima, great care is required in furnace charge calculations, weigh-out, and expertise in retention of various elements. A number of important minor elements, such as carbon, zirconium, and boron, also have to be controlled as the specified ranges are limited for various reasons. It is important to select the proper

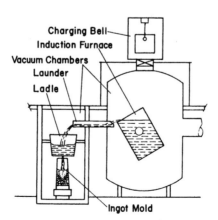

Fig. 2. Schematic diagram of vacuum induction melting furnace.

balance of virgin, scrap, and remelt (recycled) materials for charging that will facilitate melting to the required specification.

Treatment and recycling of scrap has become a business of its own. Scrap dealers routinely collect, clean, and specially pack superalloy scrap for remelting by VIM. Scrap is carefully segregated and recycled internally by the superalloy melter. Some producers even prerefine their scrap, converting it to a more suitable form for charging. Scrap charges are carefully controlled when blended with virgin material, and for some critical gas turbine applications scrap is restricted to specific amounts. Careful quality control procedures require statistical sampling and vacuum melting of representative scrap samples. As a consequence of these requirements, the scrap industry has developed a high degree of sophistication with regard to melting and chemical analysis equipment.

Virgin materials include electrolytic nickel, iron, columbium (niobium), chromium, cobalt, and manganese, "Armco Iron," nickel pellets, tungsten, and molybdenum roundelles. Certain ferroalloys are used for superalloys, allowing a sufficiently high iron residual.[2] The principal ferroalloys used in vacuum melting are ferrochromium, ferromolybdenum, and ferrocolumbium. In certain cases pressed and sintered briquettes of metals such as molybdenum can be used. Loose powder charges are almost never used due to the danger of additive loss and off-analysis (chemistry) as well as potential damage to mechanical vacuum pumps.

The selection of raw materials and the melting practice to be followed are based on the technical and economic objectives for the material being produced. A typical melt practice procedure involves first pumping down to the desired vacuum level and achieving a satisfactory leak rate. Power is turned on and melting commences. The initial virgin charge will contain the nonreactive elements including sufficient carbon for deoxidation during meltdown. In a scrap charge the components must be balanced to achieve the end-point chemistry, and the quantity of scrap containing reactive elements must be controlled. If a mixed heat (virgin/scrap) is to be made, the presence of reactive elements in the charge must be taken into account. The virgin base charge should be refined first before clean scrap is added, followed by alloying additions. Continual leak rate checks serve to assess the progress and completion of the deoxidation and refining reactions.

After completion of base charge refining, volatile elements are added. A partial pressure of inert gas may be utilized to improve recovery. Chemical checks are taken to ensure that the elements are within specification. It is common practice to make minor additions to achieve very close chemistry control. The melt is adjusted to the pouring temperature and the heat is tapped. Depending on the furnace size, the ensuing product can vary from large electrodes to be remelted by VAR and/or ESR to small-size ingots in a wide variety of investment casting dimensions.

Metallurgical Reactions.   In VIM the major means of deoxidation is provided by the available carbon reacting with oxygen to form carbon monoxide gas that is removed by the vacuum.[3] This "carbon boil" proceeds vigorously through the boiling stage into the desorption stage where the CO pressure is insufficient to nucleate bubbles. Hence, CO will form only at the melt surface where it desorbs into the vacuum. The final oxygen level is dependent on hold time at the desorption stage and the stability of the crucible material. Oxygen levels under 0.002 wt. % (20 ppm) are achievable.

In the absence of strong nitride formers and with careful selection of raw materials in the primary charge, rapid removal of nitrogen occurs during the vigorous carbon boil and then slowly declines and levels. Although the solubility of nitrogen in molten alloys is low, the superalloys often contain chromium, titanium, aluminum, columbium, and vanadium, which form stable nitrides and render nitrogen removal by vacuum treatment ineffective. In superalloys nitrogen levels under 0.009 wt. % (90 ppm) are attainable. Increasingly lower levels require longer hold times, which become impractical.

Theoretical computations indicate that sulfur should not evaporate under vacuum conditions. Removal through volatile compounds is possible, but the reaction rates are extremely slow. Very low oxygen contents improve the desulfurization rate. Lime (CaO) has been used and is effective; however, poor refractory life, sulfur reversion, and the presence of the slag layer, which acts as an effective refining barrier, make this an unattractive method in VIM.[4] Addition of manganese and rare-earth elements has proven to be effective in removing sulfur. Residual levels of these elements, however, must be carefully controlled to prevent undesirable effects on hot workability or mechanical properties.

VIM is ideal for the removal of detrimental trace elements such as lead, selenium, cooper, bismuth, and tellurium, which have high vapor pressures. Inductive stirring brings reactants to the melt–vacuum interface where refining reactions can proceed. Fortunately, most of the harmful trace elements have relatively high vapor pressures and are distilled under vacuum. Trace elements such as arsenic, tin, and antimony are not removed under vacuum and again must be controlled by raw material selection.

The introduction of oxygen (or oxides) into the melt from melt reactions with the refractory lining of the furnace is an intrinsic problem. A significant pickup of oxygen in VIM heats can occur as a result of disassociating (decomposing) refractories. This oxygen is then recombined with the reactive elements in the melt to form primary oxide inclusions. One approach to reduce the decomposition of refractories and the accompanying pickup of both oxygen and metal atoms by the liquid alloy is to use the most stable refractories available. However, even the most stable oxides have some drawbacks. Hence, MgO, ZrO_2, Al_2O_3, and mixes continue to be used in practically all installations. The only "solution" to this problem at present is strict adherence to good practice. These include minimizing contact time with the molten metal at high temperatures, avoiding corrosive slag layers and films, using tight brick tolerances, and selecting refractories carefully for quality and high density.

Electric Furnance Melting

Equipment and Procedures. Electric furnace melting (EFM) is the primary process for production of stainless and specialty alloy steels, and a limited number of superalloys. It utilizes selected scrap and raw materials to achieve high-quality product. The furnace consists of a refractory-lined shell holding a charge that is heated by an electric arc generated between graphite electrodes and the charge. For to lower the carbon content. Fluxes are added to remove impurities and to recover alloying elements from the liquid slag covering the bath. Impurities are removed as gases or as liquid slag. The molten metal is cast into ingots for processing to mill products or into electrodes for remelting for further refinement and/or improved ingot structure. EFM alone does not produce the quality required for critical superalloys. In combination with AOD it can in some cases provide a suitable product for subsequent remelting.

AOD was developed as a supplement to EFM. Meltdown of the bulk of the charge is done by EFM, and the molten charge is transferred by ladle to the AOD vessel for refining and bringing the melt into precise chemistry under optimum conditions. Upon transfer to the AOD vessel and the addition of lime, the carbon blow (starting with a mixture of one part argon to three parts oxygen) is begun. This is done by introduction of argon and oxygen through nozzles arranged in the sidewall near the bottom of the vessel. Nitrogen may be used in place of argon for some grades. Carbon blow length and the ratios of argon to oxygen during different positions of the blow are based on initial composition, carbon removal efficiencies, and temperature requirements. These gases are in intimate contact with the entire heat and produce a "rolling-over" motion and scrubbing of slag and metal during the blow.

The specific advantage for stainless steels is that AOD encourages oxidation of carbon in preference to chromium due to reduced partial pressure of CO as it forms in the presence of argon. Nearly complete recovery of chromium and other oxidized elements from the slag is realized through silicon and lime additions and stirring with argon. A final chemistry is taken. The temperature is checked and the heat is tapped.

The vacuum oxygen decarburization (VOD) process, like AOD, shifts the equilibrium in the competition of chromium and carbon for oxygen toward carbon oxidation by lowering the CO partial pressure via the vacuum. Generally, oxygen is blown onto the surface of the bath under vacuum to effect decarburization. An auxiliary means of heating the bath is sometimes provided. Bubbling argon through the bath through a porous plug during and after oxygen blow while under vacuum is still another variation. However, duplexing of EFM is clearly toward AOD for production of most chromium-bearing grades. The VOD process may offer advantages when low hydrogen levels or unusually low carbon plus nitrogen is needed.

Advantages and Limitations

Vacuum Induction Melting. The most important metallurgical justification for VIM is that, of all known methods, it provides the greatest degree of control over composition from the standpoint of desired alloy additions and undesired impurities. The melt is prevented from coming into contact with hydrogen, oxygen, and nitrogen from the atmosphere. Because of low pressure, reactions can take place or proceed to completion faster than they would at atmospheric pressure. Besides homogenizing the melt, induction stirring continually brings reactants to the melt–vacuum interface, which allows refining reactions to proceed as required. Volatilization of gaseous impurities and trace elements improves the mechanical properties of most superalloys.

VIM has some limitations that necessitate, for most end products, a remelting operation. This is primarily to minimize segregation and control solidification structure. Refractory errosion also is a problem since reaction with the melt can detrimentally affect cleanliness due to increased oxide inclusion content.

Electric Furnace/Argon Oxygen Decarburization. The major advantage of the EFM/AOD process is its ability to utilize higher carbon ferrochromium (charge chrome). The process upgrades EFM product quality, lowers ingot costs, and boosts production capacity. It is more than merely a decarburization process because of its ability to produce heats with low sulfur, nitrogen, hydrogen, and oxygen contents. The intimate mixing of slag and metal during stirring is the technical basis for removing these undesirable elements. Consequently, cleaner steel is produced and the high level of internal quality, as measured by microcleanliness standards, provides improved formability, surface quality, and corrosion resistance. Compositional uniformity throughout the heat is insured by the highly effective stirring action of the injected gases. These advantages of control and uniformity permit very precise design of a heat chemistry for a specific application.

The EFM/AOD process has a limitation with respect to superalloys: precise control of reactive elements because of exposure to the atmosphere. Thus, EFM/AOD is excluded for the production of superalloys that have significant amounts of reactive elements and/or require low gas content. Refractory life can also be a problem, with some practices necessary for the production of superalloys.

REMELTING AND REFINING PROCESSES

Vacuum Arc Remelting (VAR)

Equipment and Procedures. The purpose of VAR, the oldest commerical remelting process, is to produce a high-quality ingot by controlling solidification and improving cleanliness. The market created by the gas turbine engine in the 1950s provided the impetus for the rapid development of VAR as the advantages of vacuum purification and cold-crucible solidification became apparent. Continued growth of the superalloy market coupled with continuous refinement of the process has made it the premier remelting process in the world.

Starting material for the VAR process consists of cast or forged electrodes produced initially by EFM/AOD or VIM. A schematic of the process is illustrated in Fig. 3. The ingot builds progressively in a water-cooled copper mold as the electrode slowly melts. Heat required by the process is supplied by electrical energy. Direct-current power is supplied by rectifiers with observed voltages of 20–30 V when operating with currents in the range of 5000–30,000 A. Straight polarity is the norm, with the electrode being the negative pole and the crucible ground positive. Melting is

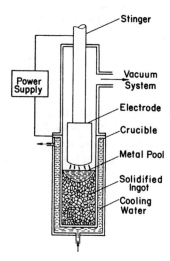

Fig. 3. Schematic diagram of vacuum arc remelting furnace.

usually done in a vacuum environment, although on occasion a partial pressure of nitrogen or argon may be used to retain a gas or high-vapor-pressure element in the material.

VAR usually is initiated by striking an arc into a small quantity of metal chips placed on the crucible base. Power is increased as melting progresses until it reaches a predetermined level based on the desired melt rate. Melt rate conditions and solidification of the molten metal vary with alloy and ingot size. The solidification rate is much slower than conventional ingot static casting because of the desire to control segregation. Near the end of the melt power is gradually reduced in order to hot top the ingot and minimize the size of the shrinkage cavity. Upon completion of melting, the entire assembly is allowed to cool and is removed from the melt station and the mold stripped from the ingot. Depending on alloy and size the ingot may be air cooled, slow cooled, or annealed.

The major components of the VAR process include the power supply, the crucible assembly, the vacuum pumps, and the control systems.[5] Either saturable reactors or silicon-controlled rectifiers can be used to supply dc power supplies for VAR. In either case the objective is to operate the furnace at maximum desired power while maintaining stable arc characteristics. Melting proceeds quite often in a near-short or short condition and the system must have the ability to clear a momentary short, reestablish the arc, and return to the desired preset operating conditions.

For most VAR installations mechanical pumps assisted by a blower reduce the pressure to approximately 200–500 μm after which diffusion pumps further reduce the pressure to a range of 1–5 μm for melting. This sequence is normally controlled automatically and is maintained to remove gases liberated during melting. Abnormally gassy electrodes can occasionally overload the pumping capacity and destabilize the arc, causing a "glow" condition that can be potentially damaging to quality.

An important aspect of control in VAR is maintaining a constant known arc gap between the electrode and the ingot.[6] The arc gap significantly influences heat loss, pool shape, and ingot surface. Therefore, the electrode drive mechanism should be as sensitive as possible so as to maintain the arc gap and prevent sudden changes in electrode position. Generally, arc gaps are maintained at 0.75 in. (19 mm) or less during melting. This type of attempted control frequently results in metal droplets bridging between the electrode and the ingot. The voltage drop that occurs when this bridging takes place is known as "drip-short" and when suitably integrated with time and frequency leads to the most used measure of arc length at short arc gaps. Thus, VAR can be an unstable process, and while normally run as a controlled short circuit, it can be very sensitive to electrode quality, pressure variations, and control system stability.

Two additional factors that must be controlled in VAR are magnetic fields effects and the briefly mentioned electrode melt rate. Since the process uses dc power, it is not uncommon to observe strong stable magnetic fields. These fields, which can be concentrated by the supporting steel framework, interact with the current in the molten pool to create liquid metal movement and influence arc stability. Both can produce ingot solidification defects. Every attempt possible is made to design the furnace to distribute the current coaxially and to minimize the magnetic field.

While the importance of melt rate as a measure of solidification rate has long been recognized, attempts to control it by measuring electrode travel have proven unreliable and erratic. Early melting control strategy consisted of controlling electrode position on the "drip-short" count and then controlling the electrode burn-off rate by regulating the amperage based on the load cell reading. The problem with this approach centered on the nonuniform density caused by the pipe cavity, a condition observed in most cast electrodes. Electrodes cannot be hot topped effectively in VIM, and unusually small diameters and long lengths produce significant shrinkage, pipe, and porosity. To counteract this problem, most new VAR furnaces have been equipped with load cells that directly measure electrode weight.

Metallurgical Reactions.   Superalloys have complex chemical compositions and often contain as many as 20 controlled elements. The performance of these materials ultimately depends on achieving the proper levels of these elements. Consequently, VAR raises questions as to how it will affect the chemistry of the starting VIM electrode. Many years of experience has shown that VAR has little or no effect on the primary elements contained in superalloys. Extensive chemical analyses of VAR ingots by position and location show consistency and uniformity throughout the ingot for major elements such as nickel, chromium, molybdenum, tungsten, and columbium. Similarly, the VAR process has shown insignificant changes in the reactive elements aluminum and titanium and elements contained in minimal quantities such as silicon, sulfur, phosphorus.

Since VAR is conducted under vacuum, one must be concerned about the potential loss of high-vapor-pressure elements. However, many of these are considered to be tramp elements that can have detrimental effects on properties if present in sufficient quantities; removal of lead, bismuth, tin, arsenic, and zinc is beneficial. On the other hand, the removal of the high-vapor-pressure elements manganese and copper in alloys containing specified quantities of these elements requires some modification of the melt practice. This involves melting under a partial pressure of nitrogen or argon or adjusting primary electrode chemistry. It is important to understand that VAR was not designed to remove high-vapor-pressure elements. While beneficial in certain respects, one must keep in mind that these elements reduce arc stability and when combined with the heavy condensate on the mold wall will significantly reduce the surface quality of the VAR ingot.

Gases such as oxygen and hydrogen (and in some instances nitrogen) are considered to be detrimental when present in remelted superalloy and steel products. Fortunately, VAR does an excellent job of lowering gases, particularly oxygen and hydrogen. While the role of CO evolution in VAR is complex and not well understood, there is no doubt that some carbon deoxidation will lead to a reduction in oxygen content. Hydrogen, because of its chemical nature and melting conditions present, is removed easily. Nitrogen also is lowered but not to the same degree as the other two. Nitrogen can form stable nitrides that prevent much, if any, from being removed as a gas. The reduction of nitrogen in superalloy VAR ingots is associated with flotation of the nitrides on the surface of the molten pool and gradual rejection to the ingot surface.

Inclusions have been present in VAR material since inception of the process, are present in current product, and undoubtedly will be present in future product. Inclusions are unwanted since it has been shown that they are detrimental to the properties of the material. This has been emphasized recently as more work has shown the negative effect of inclusions on low-cycle-fatigue behavior. Experience has shown that VAR does improve material cleanliness by removing inclusions during the remelt cycle. However, some large inclusions traceable to the electrode manufacturing process occasionally are encountered. Since the metal is exposed to low pressure and high temperature for only a short period of time, most oxide-type inclusions appear to be physically removed at the pool surface rather than by oxide decompositon. It is believed that oxide inclusions that float are ultimately rejected to the edge of the pool in the region referred to as the "crown." The crown is a metal buildup at the mold wall and has been shown to contain agglomerated inclusions of both oxides and nitrides. While recognized as a process that reduces inclusions, VAR is not a cure-all, and the degree of improvement in cleanliness is a function of the cleanliness of the starting electrodes.

Sulfide inclusions are a special case, and since VAR does not remove sulfur, the problem must be attacked by selecting optimum raw materials during primary VIM melting. The goal is to keep electrode sulfur levels low so as to prevent detrimental property effects, particularly high-temperature ductility. While processes have been developed in VIM to lower sulfur, the best procedure appears to be selecting low-sulfur raw materials and/or fixing the sulfur with late additions such as magnesium or various rare-earth elements.

Electroslag Remelting (ESR)

Equipment and Procedures. The aim of ESR is to produce high quality through a combination of chemical refining and controlled solidification. ESR has roots dating back to the 1930s. However, it was not until the late 1950s that limited production (one producer) became available and offered as an alternative to VAR material for the gas turbine market. Its success was minimal, and further development awaited Russian work in the mid-1960s. It was not until the late 1960s and early 1970s that EST became a viable process for the production of superalloys. Its importance and volume has been increasing since that time. Most of the capacity, approximately two-thirds, still resides in the Soviet Union.

A schematic of the ESR process is illustrated in Fig. 4. It consists of producing an ingot from a cast or forged electrode immersed in a molten slag with heat supplied by electrical energy. Melting is done usually in an air environment. The ingot builds progressively in a water-cooled copper mold as the electrode slowly melts in the molten slag. Heat is generated by the electrical resistance of the slag. The power mode for ESR is versatile and can be direct current with straight or reversed polarity or alternating current single or multiphase. Most modern furnaces are ac single-phase units for quality reasons. For the ac mode power at the furnace is typically 40–50 V with currents ranging from 5000 to 30,000 A.

ESR is initiated by pouring hot liquid slag into the copper crucible or by causing arcing between the electrode and metal chips on the mold base, thereby melting an initial slag volume placed in the mold. These procedures are referred to as hot and cold slag starts, respectively, and both are used commercially on a production basis. The melting is conducted at a controlled voltage with the current linked to the melt rate and electrode feed rate linked to voltage controls. As melting of the electrode nears completion, the ingot is hot topped to prevent pipe and stripped, after allowing enough time for the slag to solidify. Depending on alloy and size, the ingot may be air cooled, slow cooled, or annealed.

Principal components for the ESR process include the power supply, the crucible assembly, and the control systems.[5] The atmosphere and slag are not components in the sense of equipment, but they must be taken into consideration and controlled. A normal 60-Hz continuously variable voltage power supply for an ESR furnace

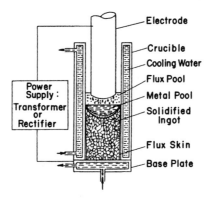

Fig. 4. Schematic diagram of electroslag remelting furnace.

consists of a thyristor-powered saturable core reactor operating the primary of a single-phase water-cooled stepdown transformer. Such a transformer typically supplies a single-phase voltage varying between 15 and 80 V at the bus terminals. Current can range from 5000 to 80,000 A. Few, if any, production furnaces now use dc power except for some converted VAR furnaces.

For the more common single-electrode ESR process, the maximum heat flow into the mold occurs at the slag–metal interface. It has been estimated that most of the total heat generated by the process is removed by the mold-cooling circuit. ESR mold design, therefore, is extremely important. The mold–water interface is very critical in the ESR process, and small changes in heat flux, mold temperature, water temperature, and so on, reportedly influence the surface heat transfer condition by changing the mode between boiling and nonboiling. In the Soviet Union and Europe many ESR systems spray cool molds to minimize water flow and simplify design. In the United States most ESR furnaces use an annular water jacket or channel cooling, which prevent boiling by providing a high surface water velocity.

Molds of two basic geometric shapes are used in the ESR process, cylindrical and rectangular. Because of the severe thermal cycling encountered, most static crucible furnaces use an open-ended copper mold resting on a fixed separate base plate. The purpose of this design is to minimize the stress field caused by thermal cycling. Alloyed copper molds can be used for increased resistance to deformation and fatigue failure. The cylindrical static mold may be manufactured from an extruded or forged tube, an electroplated form, or a rolled plate with a longitudinal weld seam. Large molds must be weld fabricated. Thermal distortion is minimized by using a thin wall approximately 1 in. (2.54 cm) thick, employing a high water velocity and using clean recycled water.[5] Slab molds for ESR are of two basic types, a continuous copper liner formed and welded or a four-piece (book-style) mold. Slab molds use nonboiling water systems with high velocity flow in a water jacket. Slab molds tend to have heavier wall thickness because of the safety factor it provides and for greater ease of attaching restraining bolts from the water jacket.

Most commercial ESR furnaces operate in an open air environment with fume collection equipment to remove vaporized calcium fluoride (CaF_2). Emissions are usually quite small and do not present an environmental problem. A typical slag consists primarily of CaF_2 with additions of Al_2O_3 and CaO. The specific composition of the slag depends on the alloy being melted, its electrical characteristics, and its viscosity.

ESR melt rate is controlled by the current, while voltage characteristics are used to control electrode feed or electrode position. Since the furnace has open access, it is relatively simple to monitor electrode melting by load cells on the electrode holder. For many furnaces melting is automatically controlled by load cells working in conjunction with a preprogrammed melting rate. Electrode position in the slag is very important. Electrode immersion is shallow and "rides on top" of the slag bath. This imposes a constant-voltage constraint as the depth of the slag varies little throughout the melt. The immersion depth must be accurately controlled because of heat balance and current flow requirements and is usually sensed by the width of the recorded voltage trace. Voltage width variations are generated as the electrode begins to come out of the slag bath. This minor voltage variation ("swing control") is analogous to the previously discussed drip-short for electrode positioning in the VAR. Hot topping is accomplished with programmed power reduction. Because of the thermal heat capacity of the slag, the ingot top is usually very sound and generally better than a VAR ingot.

Metallurgical Reactions. ESR is done under a slag blanket in an air or gas environment. Interactions among the molten metal, slag, and atmosphere result in a more complex process than VAR and can consequently affect the final product to a much greater extent. Recognizing these potential interactions coupled with slag composition and such physical factors as viscosity, resistivity, and conductivity, it becomes clear that the complexity of ESR is greater than VAR and that care must be exercised in order to produce a satisfactory product. This is particularly true for precipitation-hardened nickel-base alloys. However, the ESR process, although complex, is also more flexible or "forgiving." For solid-solution superalloys as well as other steels, a wide range of slag compositions and melting parameters may be chosen.

ESR has a greater effect on more elements than VAR during the remelting cycle. One of the major differences involves those elements that have a high affinity for oxygen, particularly aluminum, titanium, and silicon. Only through careful control can ESR ingots be produced with consistent levels of these elements throughout the ingot.[7] Less reactive elements such as nickel, molybdemum, and tungsten are not affected. ESR has the potential for stripping minor elements from superalloys during the remelt cycle. The most representative element is sulfur. The beneficial effect of removing sulfur in ESR can be accomplished with slag containing high lime contents.

The effect of ESR on removal of tramp elements is a controversial issue. Since melting is not done under vacuum, vaporization of such elements as manganese, bismuth, and lead does not occur. Any lowering of tramp elements must be accomplished by molten metal reacting with the slag. Reports are mixed; some melt shops have found reduction of these elements, while others indicate no change during remelting. It would appear that more definitive work is needed to clarify this issue.

ESR does not have the same capacity for reducing gas in the ingot as VAR. Nevertheless, there are conflicting reports concerning the effect of ESR on oxygen, nitrogen, and hydrogen levels. While both increases and decreases have been reported, the consensus is that oxygen and nitrogen levels are little affected if proper melting parameters are used. Hydrogen in austenitic superalloys presents no problem, with low levels being the norm.

Superalloy ingots made by ESR exhibit good cleanliness if properly melted and rival VAR cleanliness ratings. Controversy exists as to which is cleaner. The mechanism for inclusion removal is different from that in VAR, with the slag playing a major role. It is believed that a thin film on the electrode–slag interface provides a site in which oxide inclusions are removed by dissolution. This mechanism may account for the fact that large oxide inclusions traceable to the electrode manufacturing process are seldom found in ESR ingots. Oxides in ESR material are thought to be precipitated during the freezing process and are characteristic of the slag oxides rather than the electrode oxide content. This mechanism is supported by data on production ingots that shows ESR material to contain more and smaller inclusions than VAR material, which has fewer but larger inclusions.

Other Processes

The commercial use of remelting and refining processes other than VAR or ESR is very limited. Active development is underway, however, on a number of other processes. The objective of these developments is to obtain some advantage over either VAR or ESR material in terms of quality or life-cycle costs. Prominent processes include electron beam cold-hearth refining (EBCHR), vacuum arc double-electrode remelting (VADER), and plasma melting (PMR). These processes, in terms of equipment and procedures, will be discussed briefly.

Electron Beam Cold-Hearth Refining (EBCHR).   The purpose of the EBCHR process as applied to superalloys is to provide material with improved purity and cleanliness. The original use of electron beam heating under vacuum involved drip melting and casting of refractory metals. Initial efforts to produce superalloys by this technique gave unsatisfactory results because of unmelted raw material constituents in the cast ingot. The EBCHR process was developed to solve this problem. The first large unit was constructed in the early 1960s but was only used intermittently, and even then mainly for titanium.[8] Two new large units have been installed more recently, and while still used primarily for titanium production, they can also refine superalloys. The superalloy work, however, is still developmental in nature.

The EBCHR process consists of a hearth melting operation of solid or particulate feedstock charged into one end of the furnace, melting with one or more electron beams, traversing the molten metal along the hearth by gravity, and pouring into a mold where solidification is controlled by another electron beam. Refer to Fig. 1. The key element of this process is the presence of an electron beam gun along with a high-voltage supply. The guns generate and control a beam of high-energy electrons and focus them on the material to be melted. Intense heat is generated as the stream of high-energy electrons impinges the material, transforming kinetic energy into thermal energy. The generation and control of electron beams is based on established principles of physics and electron optics. In addition to the guns, the EBCHR process requires a vacuum chamber with appropriate pumping capacity, liquid nitrogen cooling capability (primarily for the copper hearth), and a casting station with appropriate vacuum locks.

A major advantage of electron beam guns is the ability to control the dimensions and shape of the beam pattern that impinges the target. This flexibility allows the use of a small intense beam for melting and a large diffuse beam for refining and hot topping.[9] It is important to maintain high vacuum in this equipment in order to adequately control electron beam generation and control. Melt rate is primarily dependent on beam power, material being melted (type and form), and the degree of refining desired. Actual melt rates can vary from 120 to 1980 lb/h (54–898 kg/h) and sometimes higher under proper conditions.

The development of more highly alloyed superalloys with improved performance has accentuated the need for increased purity and cleanliness. The purpose is to minimize oxide-type inclusions and severely segregated phases. EBCHR (which is carried out under high vacuum) allows the separation of melting and refining from the solidification of the resultant ingot. The process provides sufficient resident time for volatilization reactions to occur independent of melt rate while preventing insoluble constitutents in the molten metal from flowing into the ingot. Vaporizing of unwanted residual and tramp elements to almost undetectable levels is achieved, but significant chromium losses occur because of the high vacuum level and must be compensated for in the feed material.[10] Condensation of vaporized chromium can be a problem. Reactive elements are unchanged for the most part. Oxygen and nitrogen levels are reported to be significantly lower. Nonmetallic inclusions such as oxides can be mechanically removed by water-cooled skimmers or dissociated under the intense heat of the electron beam. Hence, cleanliness of EBCHR material is reported to be better than other melting processes.

Plasma Melting (PMR). Plasma melting is being explored to a limited extent for both the primary melting and remelting of superalloys. Temperatures well above those obtained by most processes, control over the melting atmosphere, and high productivity in combination with high thermal efficiences are major considerations.[9] Plasmas are generated by passing an electric current through a gas. Ionization of the gas must take place to make the gas electrically conducting. A conducting path between a pair of electrodes is established, and the passage of an electric current through the ionized gas causes gaseous discharges. Average gas temperatures range from 5400 to 10,800 °F (3000–6000 °C) and arc temperatures from 10,800 to 36,000 °F (6000–20,000 °C).

The principal component of PMR is the plasma arc torch. Other components include an ac or dc power supply, a gas system to start the torch, a water system to provide cooling to the electrodes and shroud, and a control panel that allows the plasma arc column to be initiated and sustained. There are two major types of plasma arc torches.[11] The one used for most metallurgical melting applications is referred to as the transferred arc torch. The furnace charge becomes the anode. A high-frequency starter is used to start the discharge between the cathode and the nozzle, and then the discharge is tranferred to the charge.

PMR provides intense heat for melting and can be performed under a wide variety of gas compositions. Since melting is not under vacuum, loss of either major or minor elements occurs. Specific data are minimal or lacking on tramp element and gas removal from the feedstock. It is unlikely that the PMR will produce a cleaner material than EBCHR. However, PMR is a cheaper process and it avoids the loss of certain alloying elements that occurs through boil-off in EBCHR. Hence, it is being used to remelt (refine) VAR electrodes of such alloys. No known published data are available.

Vacuum Arc Double-Electrode Remelting (VADER). Vacuum arc double-electrode remelting is relatively new compared to other superalloy melting processes, having been introduced in the late 1970s. Development was based on a potential cost-effective alternative to VAR or powder metallurgy processing of difficult-to-process superalloys such as IN-100, René 95, AF2-IDA, and so on. While details of the process are restricted, VADER essentially consists of dc vacuum arc melting of two consumable electrodes to yield semimolten drops that are solidified in a rotating mold or a withdrawal mold.[12] Casting rates of approximately 20 lb/min (8 kg/min) are possible with lower energy comsumption than standard VAR. Melt rates are about three times those of VAR.

The inherent characteristics of the process make it unlikely that any major changes will occur in the chemistry of either major or minor elements. The same is true of tramp elements contained in the electrode. Since we are dealing with semiliquid droplets in VADER, there is insufficient time for oxide inclusions to dissociate. Consequently, the cleanliness of the remelted ingot will mirror that of the two electrodes. This problem is being circumvented by producing very clean starting electrodes through filtered VIM material, which is subsequently put through the EBCHR process.

Advantages and Limitations

Vacuum Arc Remelting (VAR). The primary advantages of the VAR process are its inert nature with respect to reactive elements and the controlled solidification structure that is obtained in the ingot. Additionally, it was the first major remelting method for the production of superalloys. Its early development and acceptance compared to other processes has provided the time necessary to develop standard practices for many superalloys. The dominance of this process is likely to continue considering the cost necessary to qualify material for gas turbine engines, especially critical rotating parts.

Superalloy ingot size for the VAR process on a production basis can range from 12-in. (30.5-cm) rounds to 30-in. (76.2-cm) rounds. Most are in the 20–25-in. (50.8–63.5-cm) round range, with an average weight of 10,000–12,000 lb (4576–5443 kg). Ingots produced are invariably round; this can be considered a disadvantage, particularly for production of sheet or strip. Shapes other than rounds create arc control problems that have not been adequately solved.

One of the more important disadvantages of VAR ingots that translates into higher costs is surface quality. Concentration of impurities from the pool and condensed volatile elements on the ingot surface frequently necessitate grinding of the ingot prior to hot working. Failure to do so can impair hot workability and significantly increase yield losses. Either way, VAR ingots suffer in comparison to ESR ingots. While hot workability practices have been well established for most superalloys, VAR hot workability is believed by many to be inferior to ESR. While some of this apparent difference may be surface related, subsurface metallurgical structure also seems to play a part.

VAR was originally developed to improve the structure and minimize segregation in superalloy ingots so that maximum properties could be obtained from the chemistry being melted. Consequently, temperature gradients and solidification rate must be controlled as much as possible during the course of melting. The heat supplied by the arc is balanced against the heat transferred through the ingot to the crucible walls and base and radiated from the pool surface, giving the pool its familiar hemispherical shape. It is the shape and depth of this pool that determines dendrite growth and spacing and macrostructure and microstructure.[13] The VAR process has the advantage of being able to control this pool to predetermined limits, particularly with the use of helium cooling along the ingot–mold interface. However, as ingot diameter increases, the parameters required to maintain the appropriate pool size and depth become more difficult to maintain and ultimately limit the size to which a satisfactory VAR superalloy ingot can be produced.

The microstructural features of a VAR ingot only constitute part of the requirements for premium quality. Unlike equilibrium structures that are dependent on solidification rate, melt instabilities and interrupted heat flow can produce defects that may be more harmful than microstructural inhomogeneities. One major defect, freckles, can be present in varying degrees. This defect is originated as a highly segregated channel produced by fluid flow in the liquid–solid region of the ingot. Freckles become more prominent in alloys with large liquidus–solidus zones, typified by Alloy 718. The defect can be initiated by sudden liquid metal movements in the molten pool or by unidirectional pool rotation. Consequently, melt rate, melt rate stability, the maximization of thermal gradients and the minimization of external magnetic fields must be closely controlled in the VAR process.

Additional defects that can manifest themselves in VAR ingots include tree rings and "white spots." Tree rings are associated with microsegregation and result from dendrite growth direction change. They are frequently attributed to mechanically unstable pool shape during the melt sequence.[5] Usually observed as differentially etched rings in the ingot or forged billet, tree rings can be minimized or eliminated, as are many microstructural variations, by homogenization treatments. Recently, a major concern and disadvantage associated with VAR ingots has been the observation of white spots. A number of theories for the formation of these defects have been proposed. Several of the proposed mechanisms are inherent to the VAR process. While infrequently observed in VAR product, their presence raises concern because of demonstrated detrimental effect on quality.[7]

Electroslag Remelting (ESR).   ESR is at a disadvantage to VAR as a superalloy production process because of its late arrival on the scene (VAR enjoys a strong capital investment position) and the complexity of the process attributed to numerous potential reactions between the atmosphere, slag, and molten metal. However, this same complexity offers opportunities to control chemistry and cleanliness to a much greater extent than the VAR process. ESR is being actively pursued as a potential replacement for VAR material in some gas turbine applications. Recent work has shown improved material performance attributed to improved cleanliness.

Ingot size for the ESR process for superalloys varies from 12- to 30-in. (30- to 76-cm) rounds. One major advantage of ESR in this respect is the ability to melt ingots other than circular cross sections, such as slabs as large as 12 × 48 in. (30 × 122 cm), which facilitate the manufacture of superalloy sheet and strip product. Many flat rolled superalloy products use ESR as a standard method of production. Most of these alloys are the solid solution-hardened alloys.

Another advantage exhibited by ESR ingots is surface quality. With the proper melt rate and slag, a smooth surface can be produced that in most cases requires no conditioning prior to hot working. The smooth surface results in improved hot workability and higher yields. It should be pointed out that a number of producers believe that the improved hot workability of ESR ingots is due to advantageous subsurface grain alignment or minor chemical modifications.

Temperature gradients and solidification rates must be carefully controlled during ESR. For many years ESR was thought to possess higher cooling rates than VAR. Melting rates were much higher than VAR for most ESR ingots. However, alloy type varied significantly from VAR product, and recent results indicate that melting rates for the two processes are similar when remelting the same alloy and ingot size.[13] This is especially true for the precipitation-hardened nickel-base alloys. ESR ingots are limited in size by heat flow and the resultant pool size and depth, just as VAR.

Melting rate and the resulting pool profile determine the microstructural and macrostructural features of an ESR ingot and offer no significant advantages over VAR. VAR defects such as freckles and tree rings can occur also in ESR ingots, and the same precautions with electrode stability and magnetic fields must be observed. Perhaps the most important advantage of ESR beside that of inclusion morphology is the freedom from white spots. Several of the proposed mechanisms for the formation of this defect in VAR do not exist in ESR. An example is the absence in ESR of condensation crown on the mold sidewall.

Other Processes.   Major advantages of EBCHR include the potential for use of a wide range of raw materials, removal of nonmetallic inclusions, high gas removal in vacuum refining, alloy addition capabilities, and high throughput rates.[8] These advantages are due to the high vacuum environment, independent casting of the electrode or ingot away from the melting and refining hearth area, and no direct relationship between power and melt rate. Of the advantages noted, the one most sought after in superalloys is cleanliness. The presence of a quiet pool to which the feedstock is added allows for separation of both high- and low-density inclusions as well as thorough deoxidation of the bath. Disadvantages include the loss of high-vapor-pressure elements such as chromium, uncontrolled solidification structures in the electrode or ingot, and higher costs associated with initial investment and maintenance of the equipment. Charge materials must be adjusted for elemental losses, and most product is usually combined with VAR or ESR processes to produce a satisfactory ingot. The size of cast product is small when compared with other production melting or remelting processes and basically has the same size limitations imposed by the inherent segregation characteristics of superalloys.

The major advantages of PMR include the ability to melt under a variety of atmospheres at high pressures, use of a variety of raw materials, capability to achieve a high measure of deoxidation, and the potential for use of slags.[9] These advantages are due to the high-temperature capability of the process, no direct relationship between power and melt rate, high throughput rates, and high thermal efficiency. Because of good atmosphere control, minumum contamination and loss of volatile elements is observed with PMR. Beyond lack of experience in melting superalloys, the chief disadvantages of PMR are limited reduction or elimination of gases that can cause cleanliness problems and solidification problems associated with desired size superalloy ingots and electrodes. The product of PMR would almost invariably be subject to subsequent VAR or ESR for structure control reasons.

Unlike EBCHR or PMR, the VADER process concentrates on controlling the cast structure as opposed to the melting and refining of the alloy composition. It competes with VAR and ESR. The advantage of VADER is its ability to cast fine-grained ingots of sophisticated superalloys that are difficult to hot work into useful shapes. The falling droplets formed by this process, thought by some to be at a temperature between the solidus and liquidus of the alloy, nucleate equiaxed fine grains throughout the entire ingot or electrode. This can be accomplished at a throughput of approximately three times VAR and with much less energy. The product is essentially segregation-free when compared to VAR or ESR product. Disadvantages include a dependence of power and melt rate, no refining capability, and little or no deoxidation possibilities. The most serious problem lies in the fact that the quality of the starting electrodes, which can contain inclusions, white spots, and segrated primary phases, are carried over to the VADER product. Consequently, the ultimate quality other than structure of the final product can be no better than the starting electrode.

CURRENT AND FUTURE TRENDS

Materials Requirements

The first superalloys possessed that which was needed most for gas turbine engines, high-temperature strength. The strength was first measured by tensile properties. This requirement quickly changed as operating temperatures rose. Stress rupture life and then creep properties became the measure. As time passed, fracture toughness and oxidation resistance were added to this list. Recently, the limiting factor for superalloy use in gas turbine applications has been low-cycle fatigue (LCF) or thermomechanical fatigue life, requiring material with low crack initiation potential and with low crack growth rates.

Significant work over the last 10 years has established the importance of cleanliness in superalloys and the detrimental effect of inclusions on LCF life. Inclusion size, shape, and type have variable effects on LCF life, but all are negative. The criticality of inclusions increases as maximum inherent strength levels are approached, and the effect can be catastrophic. The problem is accentuated if the inclusions in the material are associated with melt defects such as white spots and solidification defects such as freckles. Increased sensitivity levels of ultrasonic testing are being used and specific cleanliness evaluation tests such as the electron beam button test are being developed to assess these ultraclean materials.[14]

Melting Development

Individual melting processes applied to superalloys were developed for specific reasons both in the early days of superalloy development and in today's "process"-oriented environment. The VIM process was developed to control chemistry and permit increased hardener content. VAR and ESR were developed to control structure. VADER is an extension of these two processes aimed at producing a fine-grained ingot with acceptable hot workability. The EBCHR and PMR processes are alternatives aimed at improving the cleanliness of these materials. While each has its particular advantage, none of the melting processes alone can yield the type of material desired for today's advanced technology engines.

It is clear to the producers of superalloys that combinations of melting processes are needed to provide the type of quality needed for this market. Circumstances in the United States made VIM/VAR the selected combination. VIM/ESR eventually arrived for the solid solution-hardened superalloys. The appearance of white spots in VIM/VAR material coupled with growing knowledge concerning the effect of inclusions on product quality resulted in "triple-melt" material for critical high-pressure turbine disk applications.[7] This melting process combined VIM, ESR, and VAR in that order for the purpose of minimizing inclusion level and hopefully white spots. The results have been reasonably successful in providing a material with better cleanliness and improved LCF properties. Costs associated with triple-melt material have limited its application.

Various reports on VIM/ESR material show improved cleanliness in this product when compared to VIM/VAR particularly with respect to inclusion size and type. VIM/ESR has not yet been observed to show the presence of white spots. These two important assets have resulted in considerable effort directed toward the manufacture and evaluation of VIM/ESR superalloys for a variety of gas turbine engine rotating parts. Because of the greater complexity of the ESR process, considerable upgrading of equipment is required, and process control variables must be tightly monitored and continuously measured.

Because of the potential for greater cleanliness than VIM, VAR, and ESR or combinations thereof, the EBCHR and PMR processes are being looked at in a variety of development programs. The lack of mechanisms for white spot formation and the potential for superior process control in combination with the cleanliness makes EBCHR of interest for the future. However, the potential of this process has been present for some time and has yet to be realized. It will almost assuredly have to be subsequently VAR, ESR, or VADER processed in order to achieve minimal segregation and sufficent hot workability in a commercial size product.

SUMMARY

Current trends of superalloy production indicate that the quality and reliability depend to a large extent on the initial melting of these materials. While many other factors influence the performance of these materials, they can only be as good as the initial ingot. Mounting evidence of the relationship between structure/cleanliness and performance confirms this premise.

No single-melt process has the capability of supplying the material desired for today's critical applications. As we have seen throughout the history of superalloys, combinations of melting processes will continue to be used and modified as necessary to manufacture these materials. VIM with evolutionary improvements will continue to be the primary melt process well into the next century. The reasons are its ability to control chemistry, remove gases, and permit the use of large quantities of highly reactive elements. Other primary melting processes pale by comparison.

The widespread acceptance of VAR will dictate its continued use to produce ingots with controlled structure and minimal segregation. Its well-known capabilities will be improved with better quality starting electrode material that may come from the ESR or EBCHR process. Independently the ESR process is exhibiting the potential for rivaling the VAR process for the production of superalloys. It is currently used to produce most of the solid solution-hardened superalloys. The ESR process potential for improved cleanliness and minimization of other defects will make it the subject of considerable development programs in the immediate future.

Other melting processes such as EBCHR, PMR, and VADER will be continuously examined to determine what, if any, place they may take in the melting of superalloys. Each of these processes offers some potential for improving quality. However, "new" processes for commercial production of these materials will only become technological realities if they show an economic advantage over the "old" processes or provide a technological advantage that will ultimately improve the performance/ cost trade-off of the gas turbine or other equipment in which they are being used.

REFERENCES

1. L. W. Lherbier, MiCon '78, ASTM STP 672, 514, Houston, TX, 1978.
2. R. S. Cremisio, *The Superalloys*, Wiley, New York, 1972, p. 373.
3. R. Schlatter, *Electric Furnace Steelmaking*, AIME, Warrendale, PA, 1985, p. 175.
4. D. W. Coate, *Proceedings of the Third International Conference on ESR and Other Special Processes, Pittsburgh, PA*, Mellon Institute, Pittsburgh, PA, 1971, p. 31.
5. A. Mitchell, Electric Furnace Steelmaking, TMS-AIME, Warrendale, PA, 1985, p. 191.
6. L. W. Lherbier, *Refractory Alloying Elements in Superalloys*, ASM, Metals Park, OH, 1984, p. 55.
7. J. T. Cordy, S. L. Kelly, and L. W. Lherbier, *Proceedings of the Vacuum Metallurgy Conference, AVS*, TMS-AIME, Warrendale, PA, 1984, p. 69.
8. C. d'A. Hunt, J. H. C. Lowe, and S. K. Harrington, *Electron Beam Melting and Refining Conference*, Reno, NV, 1985, Bakish Materials Corp., Englewood, NV, p. 58.
9. D. Apelian and C. H. Entrekin, Jr., *Electron Beam Melting and Refining Conference*, Bakish Materials Corporation, Bakish Materials Corp., Englewood, NV, 1984, p. 18.

10. A. Mitchell and K. Takagi, *Proceedings of the Vacuum Metallurgy Conference, AVS,* TMS-AIME, Warrendale, PA, 1984, p. 55.

11. S. L. Comacho, *Proceedings of the Vacuum Metallurgy Conference, AVS,* TMS-AIME, Warrendale, PA, 1984, p. 27.

12. Microchip Melting: A Case for Distributed Control, 33 Metal Producing, February 1983, p. 50.

13. K. O. Yu and H. D. Flanders, *Proceedings of the Vacuum Metallurgy Conference, AVS,* TMS-AIME, Warrendale, PA, 1984, p. 107.

14. C. E. Shamblen, S. L. Culp, and R. W. Lober, *Proceedings of the Vacuum Metallurgy Conference, AVS,* TMS-AIME, Warrendale, PA, 1984, p. 145.

Chapter 15

Investment Casting

WILLIAM R. FREEMAN, JR.

Howmet Turbine Components Corp., Greenwich, Connecticut

There are a number of casting processes available to provide "near net shape" superalloy cast parts by economical means. Essentially all such products are produced by the "lost wax," or investment casting, process. The characteristic physical and mechanical properties of investment casting with its related complex hollow shape-making capabilities developed over the past 30 years have made it ideal for amplifying the unusual high temperature properties of superalloys.

With wrought superalloys alloying must be restricted to preserve hot workability. Cast superalloy compositions are not so confined, and alloys with much greater strengths, consistent with other design restraints, are possible. Creep and rupture properties of a given superalloy composition are maximized by the casting and heat treatment processes. Ductility and fatigue properties generally are lower in castings than for their wrought counterpart of similar composition. Although the gap is being reduced by new technological developments to eliminate casting defects and to refine grain size, to some degree these differences are expected to remain.

In the early days of the superalloy casting industry, the unavailability of preformed ceramic cores severely restricted the capability to produce hollow airfoils for turbine blade and vane applications, which is essential for air-cooling needs. Shapes were limited to those that could be developed when the mold slurry flowed around the wax pattern defining the part shape. Airfoil design requirements pushed foundry technology, and significant advances in both core and shell technology were forthcoming (Fig. 1).

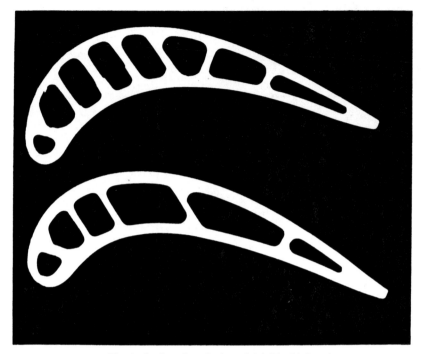

Fig. 1. Sections through air-cooled turbine blade.

Gas turbine airfoils became the first important application for superalloy investment casting. Initially produced in cobalt-base alloys and cast by indirect arc furnaces without a protective atmosphere, a rapid transition to vacuum induction processes occurred in the late 1950s. At the same time interest in microstructural control, particularly grain size and shape, emerged, and by the mid-1960s directionally solidified castings were being produced experimentally.

Integrally cast nozzles and wheels were developed to serve the emerging aviation small gas turbine industry. Initially, cost dictated this move, but ultimately air-sealing requirements and weight savings became increasingly important in design decisions. For stationary components limitations were size and complexity; however, for rotating components mechanical properties such as low-cycle fatigue became a major limitation. Quite recently structural castings have been employed for the highly loaded shell of the turbine engine, both large and small. Castings in excess of 40 in. (102 cm) diameter with pour weights of over 1000 lb. (455 kg) now are common.

BASIC PROCESS

The basic investment casting process has remained unchanged for centuries; however, major refinements have taken place since the late 1920s when production of dental components sparked innovations that ultimately led to the first superalloy applications.

The first step is to produce an exact replica or pattern of the part in wax, plastic, or combinations thereof. Pattern dimensions must compensate for wax, mold, and metal shrinkage during processing. If the product contains internal passages, a preformed ceramic core is inserted in the die cavity around which the pattern material is injected. Except for large or complex castings, a number of patterns may be assembled in a cluster and held in position by the "plumbing" necessary to channel the molten metal into the various mold cavities. Design and positioning of these runners and gating is critical for the achievement of sound, metallurgically acceptable products. Today the molds are produced by first immersing the pattern assembly in an aqueous ceramic slurry. A dry, granular ceramic stucco is applied immediately after dipping to strengthen the shell. These steps are repeated several times to develop a rigid shell. After slow, thorough drying the wax is melted out of the shell and the mold fired to increase substantially its strength for handling and storage. An insulating blanket is tailored to the mold configuration to minimize heat loss during the casting operation and to control solidification.

To make equiaxed grain castings, the mold is preheated to enhance mold filling, control solidification, and develop the proper microstructure. For vacuum casting the alloy charge is melted in an isolated chamber before the preheated mold is inserted and the pressure maintained at about 1 μm for pouring. After casting, exothermic material is applied as a "hot top" for feeding purposes, and the mold is allowed to cool.

Directionally solidified (i.e., polycrystal and single-crystal) castings are produced using vacuum casting equipment yielding substantially different metallographic

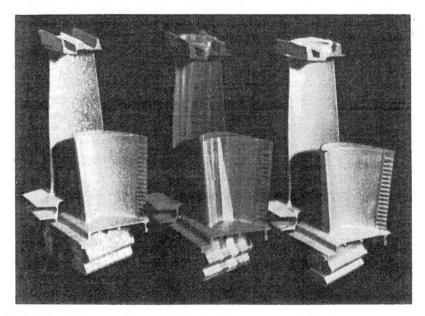

Fig. 2. Turbine blades produced by equiaxed grain (left), directionally solidified polycrystal (center), and directionally solidified single-crystal (right) casting processes.

features (Fig. 2). Although the melting operation is similar to that for equiaxed grain products, the crucible may be of alumina rather than zirconia because of its more refractory nature at the higher temperatures involved. Situated below the melting crucible is an induction-heated graphite susceptor or resistance heater that surrounds the mold at the start of the solidification process. The mold, open at the bottom, rests on a water-cooled copper chill; it is preheated to a specified temperature before being inserted in the heated susceptor. After pouring, the chill and mold are withdrawn from the aforementioned heat source to maintain a thermal gradient during the alloy solidification. Premature solidification above the liquid–solid interface is avoided by maintaining the bulk liquid and surrounding shell at temperatures above the liquidus by means of equipment design, metal-pouring temperature, mold preheat, and susceptor temperature. Details of the process are given in Chapter 7.

Once the casting has cooled, the shell and core are removed from the metal components by mechanical and chemical means, and the individual castings are separated from the gating. From this point a number of finishing operations interspersed with appropriate inspection operations are conducted to produce a deliverable product. Depending on the alloy and casting requirements, the product may be heat treated or densified by hot isostatic pressing (HIP).

PATTERNS

Complicated shapes with close tolerances and intricate core configurations dictate the use of highly specialized pattern materials that will reproduce fine detail, remove easily from the ceramic shell, be dimensionally stable, and produce a smooth surface. Blends of natural and synthetic wax combined with various hydrocarbon resins are the most commonly used pattern materials. If the final part has noncritical surface and dimensional requirements, a low-cost, heavily filled wax blend is selected. Plastics (usually polystyrene) are used selectively and offer high strength and impact resistance, long shelf life, dimensional stability, and relatively low cost. Although urea-base compounds, low-melting-point metals, and some salts are used for pattern making, wax blends and plastics remain the principal choices. A good pattern material has the following:

1. *Strength.* Withstand removal from the pattern die and stresses applied during cluster assembly and shell application.
2. *Fluidity.* Reproduce detail, avoid damaging fragile ceramic cores during pattern injection, remove easily.
3. *Low Ash Content.* Avoid casting inclusions.
4. *Compatibility.* With ceramic shells, cores, and alloy (trace elements).
5. *Dimensional Stability.* Minimal pattern distortion during in-process storage.
6. *Low Expansion/Contraction.* Compatible with shell.
7. *Environmental Safety.* Employee health and waste disposal.
8. *Economical.* Initial cost; reclamation potential.

Wax compounds have the best combination of these requirements and remain the preferred pattern material; however, specially formulated plastic pattern materials are stronger than waxes and exhibit greater dimensional stability with much less stringent ambient temperature and humidity control requirements. Thus, plastic is used whenever strength and stability are paramount, such as for the very thin airfoils of integrally cast turbine wheels.

The majority of wax patterns are produced by low-temperature liquid injection and semisolid extrusion presses. Reciprocating screw injection molding machines generally are employed to produce plastic patterns. Postinjection pattern fixturing and gauging as well as reforming of the patterns have become common because of the closer dimensional tolerance requirements on thin-walled, intricately cored parts. Increasingly, tighter tolerances on finished castings has reduced the ability to make corrections through mechanical straightening, emphasizing the necessity of rigorous control of the pattern. Cluster assembly of wax patterns normally is done by hand using appropriate joining methods.

CERAMIC CORES

Ceramic cores are used in investment casting to reduce component weight, to form the intricate internal passages utilized for transferring fluids, or for cooling air in the case of turbine airfoil components (Fig. 3). Classically, the ceramic used has been silica based, but in recent years other oxides such as alumina have been employed on a developmental basis. The core is embedded initially within the wax pattern during injection and must remain fixed in a preengineered position within the shell after wax removal. Since the coefficient of expansion of the shell and core differ, it must be taken into consideration, usually by single-point attachment. After the metal is cast into the dewaxed cavity around the core and the shell removed, the core is removed from the casting, leaving a cavity having the configuration of the original ceramic core.

The core must be strong enough to withstand the forces of wax injection, chemically compatible with the alloys that will be poured around it (at very high temperature), and highly refractory to retain its shape during pattern removal, preheat, and the metal-pouring operation when subject to the flow of molten metal. Cores also must exhibit excellent thermal shock resistance to withstand the stresses generated during preheat and casting operations without cracking; finally, the core must be easily removable from the cast part. Both chemical and mechanical methods are used for core removal. Thus, cores currently are limited to materials containing a high percentage of silica, which is readily leachable by common bases (e.g., KOH, NaOH), and/or those acids that do not attack superalloys. Intergranular attack of superalloys by caustic can cause loss of stress rupture life and of thermal fatigue cracking resistance. This is controlled by proper selection of container materials, control of caustic chemistry, and proper selection of operating parameters. Large ceramic cores are removed mechanically by blasting with sand or small glass beads.

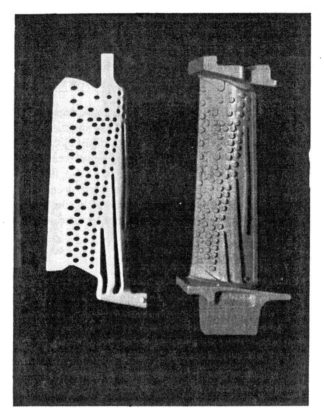

Fig. 3. Typical ceramic core configuration and cutaway of air-cooled turbine blade.

The majority of preformed ceramic cores utilized today are fabricated by either injection or transfer molding. In general, injection molding is utilized to prepare the large, less complex, cores. The process involves heating a ceramic–thermoplastic binder mixture to the softened or fluid state and forcing the mixture by means of a press system into a die maintained at ambient or lower temperature. The material is held in the die until the binder hardens, after which the parts are removed from the die, the binder removed by low-temperature baking, and the core sintered.

Transfer molding starts with a mixture of the ceramic components and a thermosetting binder. The mixture is heated to soften the binder and is then pressed in a heated, closed die. The material remains inside the die under predetermined conditions of time, temperature, and pressure until the binder cures. The parts then are removed from the die cavity and sintered to the desired properties by a controlled thermal cycle that decomposes the binder to silica. Because of the excellent green strength of transfer molded cores, significantly more complex shapes can be made by this process than by the injection technique.

Current practice permits casting parts having 0.015-in. (0.04-cm) thick walls and/or 0.015-in. (0.04-cm) thick slots. Holes of 0.020 in. (0.05 cm) diameter are cast using either quartz tubes or preformed ceramic cores. Cores in excess of 15 in. (38.1 cm) length currently are being produced. Special precautions must be observed during wax injection and the casting operation to avoid core shift, bow, or twist, which will affect casting wall thickness. As superalloy melting temperatures increase, silica-base cores become less suitable because of their reduced mechanical and chemical stability, which leads to rapidly decreasing casting yields due to core deflection and breakage. There is significant development activity in alternate oxide systems (e.g., Al_2O_3) to satisfy the requirements of higher pouring temperature, particularly for single-crystal superalloys that must remain in the molten state for relatively long periods. Use of these new materials will be paced by the development of compatible leachants for removing cores from castings.

SHELL MOLD SYSTEMS

The key to the modern investment casting process is the ceramic shell mold system, which must reproduce the intricacies of the pattern, have sufficient strength to prevent cracking during the pattern removal process, and withstand the thermal and mechanical stresses during the casting operation. On the other hand, excessive shell mold strength makes shell removal difficult and can lead to fracture of the casting during the solidification process, producing defects known as "hot tears." This results from the stress produced by the thermal mismatch between the metal and ceramic shell acting on the casting during solidification and cooling when the alloy has both low strength and ductility. Thus, ideally, the shell mold must be strong enough to shape and hold the dense metal but weak enough to deform or fragment as the metal freezes and shrinks. Advanced casting processes such as directional solidification in addition subject the shell mold to high temperatures for particularly extended periods. These high temperatures distort the mold and aggravate mold–metal interaction with the more reactive alloy constituents (e.g., Al and Hf).

The shell mold[1] is produced by alternate applications of a ceramic slurry and dry refractory grain. The ceramic slurry commonly consists of a fine-grained refractory flour, usually 200 mesh (74 μm) or finer, in a binder of ultrafine silica. The slurry is applied to the wax assembly by dipping followed by draining to remove excess slurry. Precise manipulation of the assembly during application evenly distributes the slurry coating (Fig. 4). While the slurry is still wet, refractory grain, or "stucco," is applied either by allowing the grains to rain onto the assembly or by immersing the assembly in a fluidized bed of refractory particles. After application of the stucco the binder is cured by chemical reaction or controlled drying so that additional coats may be applied without disrupting previous layers. A typical shell mold consists of 5–10 layers. Increased mold size, larger parts, and need for consistency during the shell molding process has promoted the automation of the process.

Mold construction materials vary and the initial dip, or "facecoat," is the most critical since it establishes the surface finish of the casting, reproduces the intricacies

Fig. 4. Automated application of facecoat slurry to wax assembly.

of the pattern, and is directly exposed to the molten metal. For equiaxed grain superalloy castings the facecoat usually includes 3–8% of a nucleating agent such as cobalt aluminate, silicate, or oxide.[2] The stucco applied to the first coat is finer than the other stucco materials, ranging from 70 to 120 mesh (210–125 μm), minimizing the possibility that the stucco particles will penetrate through the slurry coating. Stucco for the second and the third coat is typically coarser, with a particle ranging from 50 to 90 mesh (297–160 μm). From the third to the final coat the objective is to develop shell thickness for strength. These succeeding layers or "backup" dips use a coarser refractory powder in the slurry and a larger particle size stucco, 10–40 mesh (2000–420 μm). A cross-sectional view of a completed shell mold (Fig. 5) shows also the integral ceramic pouring cup and the "cover dip" to facilitate handling and avoid the stucco particles entering the mold cavity.

Selection of a shell material is dependent on a number of parameters such as alloy, temperature, casting process, and cost. Some of the typical refractory materials and maximum use temperatures are as follows:

Alumina (Al_2O_3)	3340 °F (1840 °C)
Silica (SiO_2)	3050 °F (1680 °C)
Zircon ($ZrSiO_4$)	3000 °F (1650 °C)
Aluminosilicate (variable)	2850 °F (1560 °C)

Although these refractory materials are important to the high-temperature properties of the shell mold, the binder is critical to the stability and rheology of the ceramic slurries, green strength of the shell, and high-temperature shell strength. Although phosphate and aluminate binders have been used for shell molds, most superalloy investment shell molds today use either colloidal silica or ethyl silicate as the primary binder. Both provide an extremely small particle size that is evenly distributed around the ceramic aggregate. The small particle size provides a very large surface area, which, because of its high reactivity, produces excellent bonding in both the green and fired states. Small amounts of organic materials are added to the binder system in slurries to improve pattern wetting, reduce foaming, and increase shell green strength.

Rate and uniformity are important parameters in the drying of the mold. Temperature and humidity are critical since variations during drying can cause the pattern to distort. Drying cabinets are used to maintain stable conditions despite the cooling associated with the evaporation of water.

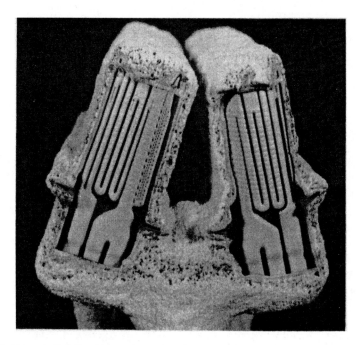

Fig. 5. Cutaway of shell mold revealing multilayer dip coating and core orientation after dewax.

Completion of the shell-molding process requires the removal of the pattern material. Removal of wax pattern materials by solvents is slow. Preferably the wax is removed by either steam autoclave or the flash dewax process. Both of these start with a thoroughly dried mold. In the steam dewaxing process the mold is placed in a vessel and high pressure [90–150 psi (620–1035 kPa)] and high-temperature steam [300–360 °F (150–185 °C)] are introduced; the wax removed from the mold by this process can be reclaimed. Residual wax is removed during subsequent mold firing.

For flash dewaxing the molds are placed in a furnace and preheated to 1800–2000 °F (980–1090 °C). The one-step pattern removal and shell firing offers an economic advantage, but reuse of the wax is sacrificed. Also, thermal shock of the shell may occur with this process depending on the thermal expansion characteristics and complexity and size of the shell. This process is required for the removal of plastic patterns.

Both dewaxing processes are dependent on the rapid transfer of heat to the mold to melt quickly the surface of the pattern to allow space for expansion of the remainder of the wax as its temperature increases; failure to do so many produce sufficient internal pressure to crack the mold.

Then, prior to casting, for equiaxed grain castings the molds are either placed in a metal can surrounded by ceramic grog or wrapped with a ceramic insulating blanket. Both techniques minimize heat loss when transferring the molds between the preheat and casting furnace aid control of the solidification and the cooling processes.

CASTING ENGINEERING

The wax pattern(s) must be incorporated into a cluster containing gates, runners, sprues, and a pouring cup arranged precisely to control the flow and solidification behavior of the liquid metal. The ultimate objective for a superalloy casting is to achieve the designed geometric shape to acceptable limits with optimized, reproducible properties. The latter is achieved by control over soundness, grain configuration, microstructural features, and freedom from inclusions. Soundness depends on controlling solidification such that there is an ample supply of molten metal to feed previously solidified interdendritic areas. This is accomplished by ensuring that the last area to solidify is in the gating outside the finished casting. Lower solidification rates achieved by increasing mold and metal temperature generally produce a more sound casting. This can be compromised by local hot spots in the mold caused by metal impingement on the shell wall during the pouring operations, creating an undesirable "reverse temperature gradient" leading to surface-connected porosity.

Hot spots are minimized by redirecting the flow of incoming liquid metal to reduce impingment on, and heat transfer to, the shell. Primary gating on turbine blades is through the root attachment area and is patterned on past experience with similar shapes. Airfoil tip shrouds usually have separate gating to assure soundness particularly at the junction of the shroud and airfoil where operating stresses are high. For low-pressure blades in large commercial engines airfoil gating may be required for local feeding; however, this is undesirable for a number of reasons, including the cost of gate removal, potential alteration of the desired airfoil dimensions during finishing, and undesirable metallurgical features such as columnar grains that generally are present in the airfoil under the gate. Insulation originally was wrapped around the shell mold to help maintain mold preheat temperature; however, added layers are used today to control solidification direction and enhance feeding. Silica-based fiber roll board and blanket materials have replaced asbestos as the preferred insulating materials.

Microshrinkage in castings is a natural phenomenon relating to the contraction of alloys on solidification. Ideally, and for castings that will be highly stressed in service, microshrinkage is forced to the centerline of the casting section so that closure by HIP is feasible. Centerline shrinkage will occur when the mold temperature stays below the solidus of the alloy, causing inward movement of the solidification front. The heat flow patterns are affected by the location of gates and cores as well as by changes in section thickness. Heat transfer considerations are of fundamental importance in any mathematical modeling of the casting process, which is expected to predict locations with significant microshrinkage. To the degree acceptable, microshrinkage should be uniformly dispersed and represent less than 0.5% (measured metallographically) to avoid significant degradation of mechanical properties.

With increasing microshrinkage, rupture properties are reduced more rapidly at intermediate temperatures [around 1400 °F (760 °C)] than at high temperatures (Fig. 6). Low ductility and the notch effect of the porosity are important factors at the

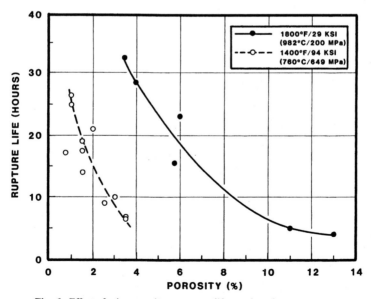

Fig. 6. Effect of microporosity on rupture life at selected temperatures.

lower temperatures. As microshrinkage increases, the effective section size becomes the predominant factor.

Control of grain size is an important means for developing and maintaining both physical and mechanical properties. Generally, a number of randomly oriented equiaxed grains in a given cross section is preferred to provide consistent properties, but often this is difficult to achieve in thin sections. To meet this objective, the mold facecoat nucleant, mold and metal temperature, and other parameters are chosen to accelerate grain nucleation and solidification.

Finer grain size generally improves tensile, fatigue, and creep properties at low-to-intermediate temperatures (Fig. 7). The finer grain size produced by relatively rapid solidification is accompanied by a finer distribution of γ′ particles and a tendency to form blocky carbide particles. The latter morphology is preferred to the script-type carbides produced by slow solidification rates, particularly for a fatigue-sensitive environment. Under these conditions the carbide particles do not contribute to superalloy properties. As the service temperature increases, they impart important grain boundary strengthening provided continuous films or "necklaces" are avoided.

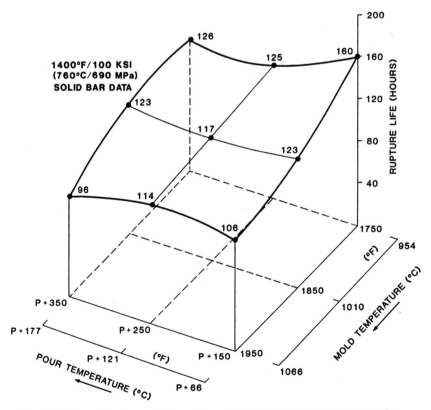

Fig. 7. Influence of casting variables on intermediate-temperature creep rupture properties.

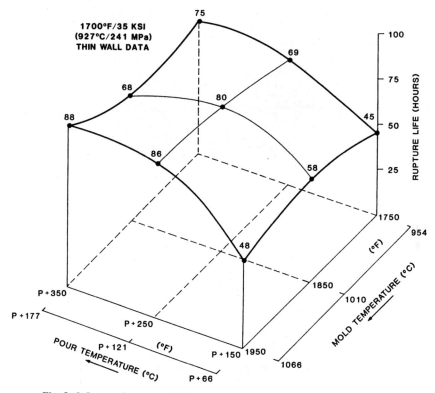

Fig. 8. Influence of casting variables on high-temperature creep rupture properties.

For high-temperature rupture performance, slower solidification and cooling rates are preferred to coarsen both the grain size during solidification and the γ' precipitated during cooling (Fig. 8). While benefitting high-temperature strength through a reduction in grain boundary content, more property scatter can be expected due to (random) crystallographic orientation effects. For turbine blades the desired micro-structure is difficult to achieve since the thin airfoils operating at the highest temperatures should have coarse grains and the heavier section root attachment area, being less rupture dependent, should have a fine-grain microstructure. Where conventional practice fails, a gate, or gutter, along the airfoil edges may be employed through which metal is caused to flow, thereby creating deliberate hot spots to retard the local solidification rate (Fig. 9). This approach also may be used to eliminate columnar grains growing from airfoil edges. When casting hollow airfoils, the core defining the internal cavity loses heat less rapidly than the surrounding shell, causing the grains to nucleate from the external surface and grow toward the core, often producing single grains through the wall.

Thin-walled castings, especially turbine blades with equiaxed grains, generally possess reduced high-temperature properties resulting from a combination of physical

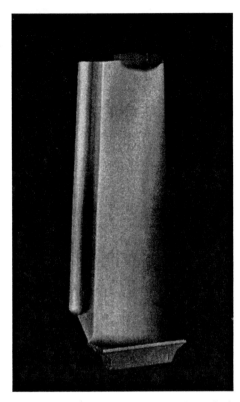

Fig. 9. Trailing edge gating employed to achieve uniform fine grain throughout casting.

size, grain orientation, grain boundaries, environment, and related coatings[3] (Fig. 10). The combined effect is to initiate cracks sooner and propagate them faster while concurrently being influenced by oxidizing and related alloy depletion effects. As the elastic modulus may vary from 18×10^6 to 48×10^6 psi (124×10^6 to 331×10^6 GPa) depending on orientation, the few (e.g., one to two) grains in the thin-wall section may share the applied stress field unequally. Also, a single grain boundary oriented normal to the stress field provides an opportunity for rapid crack propagation. Coatings to retard oxidation effects, while beneficial, are not load bearing, and may represent a major fraction of the thin-wall cross section, thus increasing stress levels. Concurrently, coatings may be brittle at intermediate temperatures, thus providing the opportunity for stress risers and associated property reductions. Section size plays a lesser role in directionally solidified castings, which have neither transverse grain boundaries nor unfavorable grain orientation effects.

As the grain refinement produced by facecoat nucleants is two-dimensional, a columnar structure may be created for some distance below the surface. This can affect properties significantly for components such as integral turbine wheels where heavy sections are involved (Fig. 11a). To avoid this condition, the mold can be

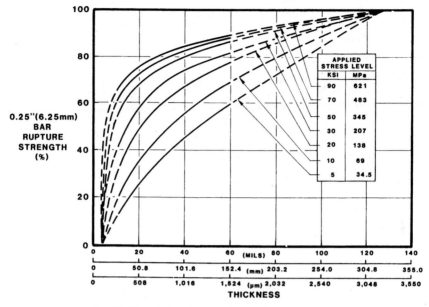

Fig. 10. Degradation of rupture life in thin cast sections.

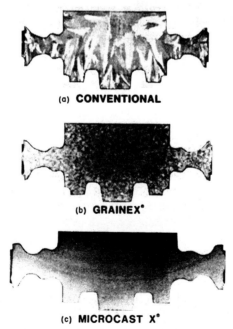

(a) CONVENTIONAL

(b) GRAINEX°

(c) MICROCAST X°

Fig. 11. Refinement in grain size achieved by mold agitation (Grainex) and pouring temperature control (Microcast-X) to improve mechanical properties and isotropic characteristics.

425

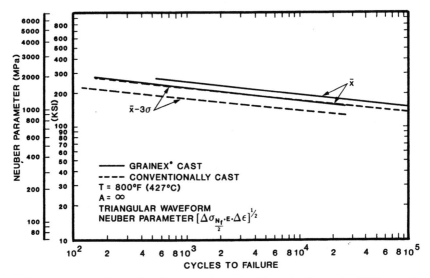

Fig. 12. Improvement of low-cycle fatigue properties achieved by grain refinement and HIP processing.

agitated during solidification of the more massive sections. This causes the advancing dendrites at the liquid–solid interface to fragment, thus acting as nucleation sites within the remaining unsolidified metal (Fig. 11b). Unfortunately, feeding during solidification is disrupted and a more porous casting results, but HIP processing provides the solution. Although the average properties are not altered substantially by this particular process, scatter is reduced significantly and design minimum properties are increased accordingly (Fig 12). Grain refinement and random orientation also reduce crack growth rates (Fig. 13).

THE CASTING PROCESS

Most superalloys are cast in vacuum to avoid oxidation of reactive elements in their compositions. Some cobalt-base superalloys are cast in air using induction or indirect arc rollover furnaces. Vacuum casting of equiaxed grain products is usually in a furnace divided into two major chambers, each held under vacuum and separated by a large door or valve. The upper chamber contains an induction-heated reusable ceramic crucible in which the alloy is melted. Zirconia crucibles are commonly employed; single-use silica liners may be specified when alloy cleanliness is especially critical.

The preweighed charge is introduced through a lock device and is melted rapidly to a predetermined temperature, usually 150–300 °F (85–165 °C) above the liquidus temperature. Precise optical measurement of this temperature is crucial. Metal temperature during casting is much more critical than mold temperature in controlling grain size and orientation; it also strongly affects the presence and location of microshrinkage. When the superheat condition is satisfied, the preheated mold, having been heated to 1600–2300 °F (870–1260 °C), is rapidly transferred from the preheat furnace to the lower chamber that is then evacuated. The mold is raised to the casting position and the molten superalloy poured rapidly into the cavity; speed and reproducibility are essential to achieve good fill without cold shuts and other related imperfections. Precise mold positioning and pour rates also are imperative. For maximum consistency, melting and casting are automated with programmed closed-loop furnace control. The filled mold is lowered and removed from the furnace.

Shrinkage during solidification is minimized in part by maintaining a head of molten metal to "feed" the casting; this is achieved by adding an exothermic material immediately after mold removal from the furnace.

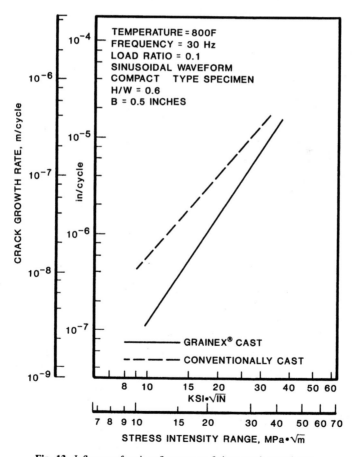

Fig. 13. Influence of grain refinement on fatigue crack growth rate.

Directionally solidified single or polycrystal castings are produced in equipment especially designed for the purpose. A unique feature of the furnace is a heater that surrounds the mold, maintaining it at a temperature above the liquidus of the alloy being cast. The gradient within the open-bottomed mold is achieved by placement on a water-cooled copper chill; the mold is withdrawn from the heater at a rate of 4–15 in./h (10–38 cm/h). To maintain the continuity of the grain structure, it is common to program variations in the withdrawal rate.

Due to thermal expansion differences, the shell mold usually fractures on cooling, facilitating its removal by mechanical or hydraulic means. Prior to grit- and sand-blasting operations, the individual castings are separated from the cluster by abrasive cutoff. After shell removal the cluster is checked by one of several commercially available emission or X-ray fluorescence instruments to verify the alloy identification.

A major portion of casting cost is in the finishing operations, which remain labor intensive. Superficial surface defects are blended out abrasively within specified limits, and the castings may require mechanical straightening operations before and after heat treatment to satisfy dimensional requirements.

Thermal Processing

Microshrinkage is eliminated by the simultaneous application of gas pressure and high temperatures, called HIP. In practice, the highest possible temperature is chosen consistent with avoiding incipient melting and, where possible, with heating above

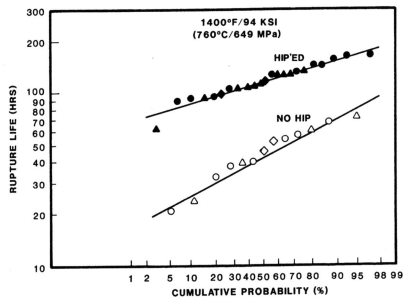

Fig. 14. Elimination of microporosity substantially raises rupture life.

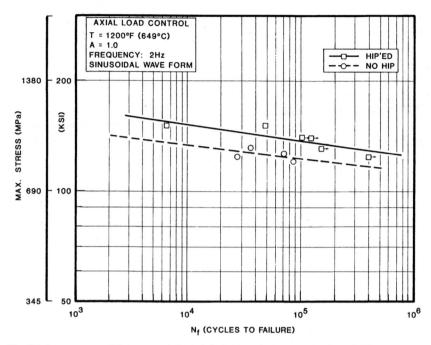

Fig. 15. Improvement of fatigue properties by elimination of microporosity through HIP processing.

the γ' solvus. For most superalloys the range is 2200–2225 °F (1200–1220 °C) with a pressure of 15 ksi (103 MPa). Four hours at temperature is sufficient under these conditions. When hafnium is an alloying constituent, the melting point is depressed and a lower temperature [i.e., 2165 °F (1185 °C)] is selected with a correspondingly higher pressure [e.g., 25 ksi (174 MPa)]. Higher strength alloys such as MAR-M 247 and IN-792Hf do not respond to the (economically preferable) lower pressures in reasonable time cycles. With HIP, scatter in rupture properties is reduced substantially, especially at intermediate temperatures where the stress-rising effects of porosity acting as metallurgical notches is significant (Fig. 14). Fatigue property improvements are noted consistently (Fig. 15). However, if shrinkage is surface connected, the internal shrink cavities are pressurized, and void elimination is impossible.

Heat treatment of cast superalloys in the traditional sense was not employed until the mid-1960s. Before the use of shell molds the heavy-walled investment dictated a slow cooling rate with its associated aging effect on the casting. As faster cooling rates developed with shell molds, aging response varied with section size and the many possible casting variables. This, coupled with significant alloying additions of titanium as a strengthener, provided an opportunity to minimize property scatter by heat treatment and later to effect an overall improvement in rupture life to the extent that the γ' produced after solidification and cooling could be resolutioned and aged. The cooling rate from the solution temperature can have a significant effect on rupture properties, especially when the procedure follows HIP processing. At temperatures above about 2100 °F (1149 °C), the less stable MC carbides decompose, forming $M_{23}C_6$ and/or M_6C carbides, as in the case of IN-100 and MAR-M 246, respectively. When this occurs, the intermediate temperature rupture ductility following post-HIP heat treatment is reduced significantly.

The presence in the microstructure of serrated grain boundaries benefits rupture properties above the equicohesive temperature by retarding grain boundary sliding. This feature may exist in the as-cast structure and be lost upon subsequent HIP or heat treatment operations if the cooling rate is not managed properly. This structure is achieved by emphasizing a cooling rate from above the γ' solvus so that the formation and movement of grain boundary γ' will cause the displacement of local grain boundary segments.[4] A prerequisite for this occurrence is that the γ' solvus must exceed the carbide solvus temperatures.

Casting Defects

Commonly occurring defects in superalloy castings include inclusions, hot tears, cold shuts, and microstructural features that are peculiar to certain alloy systems. Brittle phases (e.g., laves and η) and particular grain patterns (e.g., chill and columnar) frequently are unacceptable, particularly in airfoils. The concern for casting defects varies inversely with their detectability by nondestructive methods.

Nonmetallic inclusions (Fig. 16) are more readily detectable and represent primarily economic problems of scrap loss and rework, except when present in very heavy sections where detection sensitivity is reduced. Defects may be present from several sources, including the alloy manufacture, remelting, or mold facecoat debris. Chemical analysis of the inclusion (by *in situ* energy-dispersive analysis in a scanning electron microscope) can point to its source.

Depending on the alloy used, the vacuum levels during remelting and casting, and casting furnace operator practice, dross may become another type of inclusion, especially with alloys containing hafnium. Dross can be made to adhere to the remelt crucible wall immediately before pouring, but this depends on operator skill. In recent years there has been significant effort to reduce inclusions from various causes. During alloy manufacture most superalloys are filtered through reticulated ceramic foam containing 10–20 cells per inch. In certain instances filtering is employed additionally during casting. Generally, alloys such as INCO 718 containing many reactive elements require filtering routinely on critical castings.

Hot tears are created when the casting is plastically strained by the mold at high temperatures, causing the recently solidified metal to separate. While this is influenced by part geometry (e.g., where thin sections join thick sections), constraints by the significantly different expansion characteristics of the shell and core material often are the cause. Although tears usually are surface connected, they can be difficult to detect nondestructively when filled with oxide formed as the casting cools. Frequently, parts are thermally cycled to about 2000 °F (1095 °C) in vacuum or hydrogen to disrupt the oxide and aid in detectability by fluorescent penetrant

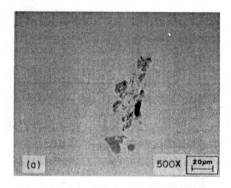

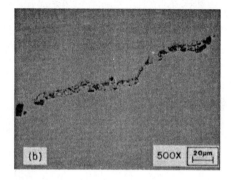

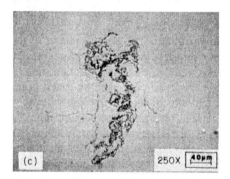

Fig. 16. Typical inclusions found in superalloy castings.[4]

inspection. Cold shuts occur when two advancing fronts of liquid metal meet and do not bond metallurgically because of a thin oxide film at the interface and solidification of the metal locally. Detection of this type of defect is extremely difficult, and when a problem is recognized, the casting process is reengineered.

There are several undesirable phases that may occur in superalloy castings. These include TCP (topologically close-packed) phases, which usually are observed in regions last to solidify. Prior to the mid-1960s a number of superalloys had been

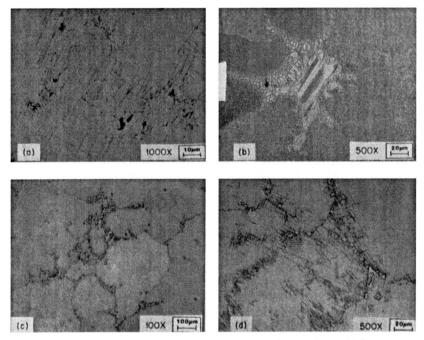

Fig. 17. Phases present in certain superalloys: (a) σ; (b) η; (c) Laves; (d) δ.

developed that tended to form σ phase (Fig. 17a) either during solidification or in service, thereby degrading rupture life. IN-100 is one of the widely-used alloys that required chemistry refinement to prevent this occurrence. More recently developed alloys have avoided this condition using Phase Computation (PHACOMP) analysis techniques. The η phase [Ni$_3$Ti,Cb,Ta)] may be found in heavy section castings (Fig. 17b), such as in the hubs of integral turbine wheels, and also in relatively thin sections of some newer superalloys containing high levels of titanium and tantalum (e.g., IN-792). As the compound is very brittle with little strain tolerance, cracking through normal engine operation or during machining is a major concern. The η phase cannot be dissolved by solution treatment and must be avoided or minimized by alloy chemistry control and foundry practice.

IN-718 was developed as a wrought disk material with good weldability and excellent strength capabilities up to about 1200 °F (650 °C). With upgraded commercial practice, including premium raw materials, vacuum melting, and filtering, IN-718 is now the preferred alloy for major engine cases and other large structural components that are cast, HIPed, and heat treated (Fig. 18). Laves phase (Fig. 17c) is present in the cast structure and must be minimized to develop required properties. This can be accomplished by homogenizing at 2050 °F (1120 °C) or above, with time determined by the degree of segregation present. Either the HIP temperature exposure or a separate homogenization step has the undesirable effect of dissolving the normal δ (Ni$_3$Cb) component of the microstructure (Fig. 17d), producing a notch-rupture-sensitive product. Recent research[5] indicates that reformation of δ at 1600 °F followed by the desired solution treatment resolves the notch problem with no detrimental side effects.

Many parts are "grain size controlled"; on airfoils a minimum as well as a maximum size may be specified. On thin edges there is a tendency to form chill grain—an extremely fine (ASTM 3–5) size without a typical dendritic structure (Fig. 19). As would be expected, the local creep rupture performance is significantly reduced (Fig. 20). Likewise, columnar grains emanating from airfoil leading and trailing edges affect thermal fatigue life adversely. Both grain problems are resolved by adjustment of casting parameters and the gating arrangement.

Nondestructive Testing

Due to the critical nature of most superalloy castings, nondestructive means for detecting defects are important and, combined with scrap and rework, can represent 20% of casting cost. Aside from visual and dimensional inspection, film-type radiography and fluorescent penetrant are the most common procedures in use today. Recent improvements in real-time radiography have demonstrated sensitivity essentially equivalent to film techniques. This has been made economical by improvements in imaging technology (i.e., cameras, screens) and computer enhancement.

Fig. 18. Typical large structural casting.

Fig. 19. Enlarge view of airfoil trailing edge depicting chill grain formation.

Fluorescent penetrant inspection methods vary significantly depending on sensitivity requirements established by the user. Detection requires the defect to be surface connected and of sufficient dimension to accommodate entry of the penetrant.

In contrast to the practice for wrought materials, ultrasonic inspection is not used generally for defect detection due to the relatively coarser grain size of castings. However, with technological developments of finer grain castings, increased usage is expected. A summary of these and other inspection methods and the type defect being sought is shown in Table 1.

FUTURE DEVELOPMENTS

Casting technology developments have been instrumental in the improved performance of superalloy products and in shape-making capability and have provided opportunities for new alloys. Current casting process developments include single-crystal and fine-grained equiaxed grain technology, *in situ* eutectic composites, and bicastings. The latter process is considered when two or more cast components, usually of different alloy composition, are to be assembled in intimate contact by a mechanical, or metallurgical joint. Of particular interest are turbine wheels and nozzles. The process involves assembling castings (e.g., vanes and blades) in a ceramic mold that provides an appropriate cavity (e.g., shrouds and disks) for the introduction of the alternate molten alloy.

Due to the savings potential, further application of superalloys in large, complex structural castings will generate, providing low-cycle fatigue and tensile properties can be improved further. To accomplish these goals in section thicknesses ranging from 0.030 in. (0.76 cm) to 0.5 in. (1.27 cm), much development effort is being undertaken in the direction of refining grain size by lowering superheat during the casting operation (Fig. 11c). This conflicts with the desire to raise pouring temperature to improve fluidity for mold-filling purposes, particularly in thin sections. Where these conditions can be satisfied with a cast ASTM-3 or finer grain size, wrought properties will be approached closely (Fig. 21).

Cyclic fatigue performance today tends to dominate superalloy component life more than creep rupture behavior. This is true for both wrought and cast products. Oxide inclusions have become a major initiation site for these fatigue failures. In the late 1970s a major effort was undertaken[6] to produce clean superalloy powder for subsequent consolidation into P/M wrought, ultrahigh-performance disks. It was concluded that while conventional vacuum induction melted (VIM) superalloy could be made substantially cleaner by electron beam cold-hearth refining (EBCHR), oxide skins formed on the powder particles during the powder-making process, thereby losing some benefit from prior alloy-cleaning operations. Concurrently, superalloy producers of casting remelt stock introduced the use of reticulated ceramic filters with 10–20 cells/in. (4–8 cells/cm) to reduce the inclusion count in VIM materials. While an improvement, EBCHR material is being investigated as remelt stock for critical applications or products with a low casting yield due to retained inclusions. While cleaner alloy is desirable, the full benefit will not be realized until improved remelt crucibles and shells are available that do not contribute to

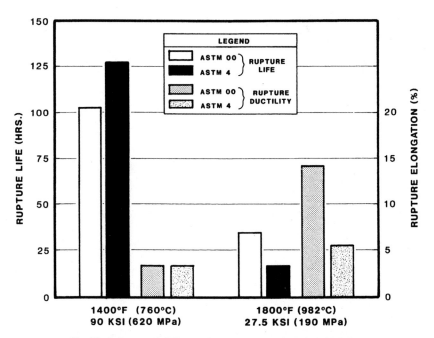

Fig. 20. Influence of chill grain on rupture properties (MAR-M 247).

Table 1. Methods of Inspection for Various Types of Casting Defects

Inspection Method	Defect Type[a]											
	Micro-shrinkage	Gas Porosity	Cold Shuts	Hot Tears	Inclusions	Residual Core	Wall Thickness	Grain Size	Cracks	Freckles	Misorientation	Recrystallized Grain
X radiography	2	1	5	4	1	3	—	—	5	5	—	5
Neutron radiography	—	—	—	—	—	1	—	—	—	—	—	—
CT radiography	—	—	—	—	—	—	2	—	—	—	—	—
Fluorescent penetrant	3	—	4	2	2	—	1	—	1	—	—	—
Eddy current	—	—	2	2	—	—	—	—	1	—	—	—
Ultrasonic	—	—	—	—	—	—	3	—	3	—	—	—
Chemical etch	—	—	—	—	—	—	—	1	—	1	3	1
Acoustic emission	—	—	—	—	—	—	—	—	1	—	—	—
X-ray diffraction	—	—	—	—	—	—	—	—	—	—	1	—

[a] Reliability: 1, high (best available); 5, low; —, currently not used.

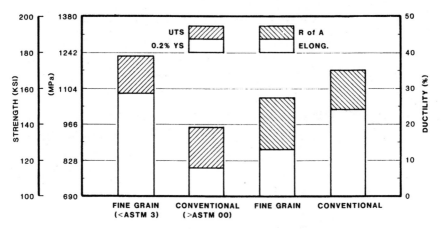

Fig. 21. Comparison of IN-718 tensile properties as influenced by grain size.

the "dirt" problem. Skull furnaces such as used for titanium casting for many years represent one attractive approach that is being used both for structural and, in modified form (i.e., electron beam), for directionally solidified castings.[7]

Directionally solidified polycrystal and single-crystal airfoils are in quantity production for both high- and low-pressure airfoil applications and will be extended to large land-based industrial gas turbine buckets. Innovations can be expected both in processing and alloys. Current single-crystal castings require $\langle 001 \rangle$ alignment with $\pm 15°$ of the airfoil stacking axis but no secondary orientation limitations. This is accomplished readily with the use of the "pigtail" crystal selector.[8] Where secondary orientation (e.g., chordal) is of importance for stiffness or fatigue purposes, a seed crystal approach is employed.[9] By this means any primary orientation may be developed with associated variations in rupture properties (Fig. 22) obtainable in certain superalloys.[9]

Early single-crystal alloys are characterized by the relative absence of grain boundary strengtheners, (i.e., carbon, boron, zirconium, and hafnium) and high levels of chromium, tantalum, and aluminum.[10] Major shortcomings were the high intrinsic value of the alloying elements, particularly tantalum, and the small solution treatment "window" between the γ' solvus and the incipient melting point. More recent alloy developments have addressed these issues, reducing tantalum content by as much as 6 wt. % and by increasing the solution treating range from 15 °F (8 °C) to 85 °F (47 °C). Further improvements in cast single-crystal alloys can be expected from the judicious addition of rhenium[11] for strength (Fig. 23), and small additions of hafnium and yttrium for improved oxidation resistance. As the latter two elements are very reactive and reduce traditional mold facecoat materials (e.g., Al and Zr), more stable refractories are required to maintain these alloying elements at the low, preferred levels.

Both government and industry support is increasing to automate more in the investment foundry. Improvements in quality and economics already are realized

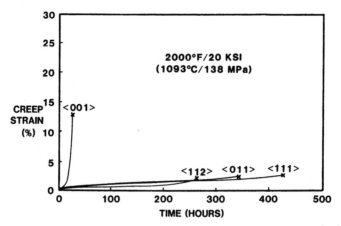

Fig. 22. Influence of primary crystallographic orientation on creep behavior of single crystals.[9] Reprinted with permission from Superalloys II, edited by R. P. Dalal, M. Gell et al., The Metallurgical Society, 420 Commonwealth Dr., Warrendale, PA.

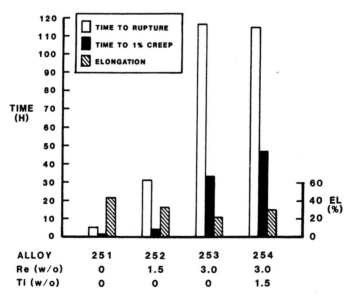

Fig. 23. Improvement in high-temperature rupture properties with small additions of rhenium. Reprinted with permission from Nicholas J. Grant Symposium "Processing and Properties of Advanced High-Temperature Alloys," 16–18 June 1985, Massachusetts Institute of Technology, Cambridge, MA. Proceedings published by ASM *Advanced High-Temperature Alloys: Proceeding and Properties* (ISBN 0-87170-222-3).

from such efforts in the mold making and directionally solidified casting areas. Fully automated equiaxed grain vacuum casting furnaces are nearing reality, with certain functions already programmable in existing equipment. The key impediment at this time is the issue of precise melt temperature measurement. As suggested earlier, rapid advances in the automation of nondestructive inspection techniques will be implemented early in the factory automation cycle. There is significant interest developing in computer simulation of the solidification process. Due to the significant number of variables affecting product characteristics and the current tolerances required, it is too soon to predict the degree to which this endeavor will be successful, particularly when considering the shape complexity of most superalloy castings and the present unavailability of alloy physical property data. However, it would be quite a useful engineering tool if initial casting trials could be engineered and analyzed by the computer, leaving fine tuning to existing methods.

ACKNOWLEDGMENTS

The author expresses his appreciation to a number of colleagues who made significant contributions to this chapter. Messrs. Kenneth Canfield, John Miller, and Thomas Ernst and Dr. Thomas Wright provided the narrative covering the wax and ceramics sections, and Mr. George Strabel provided the graphics and photographic material. Dr. Frederick Norris and Mr. Gregory Bouse participated in the editorial and support activities, and Jay L. VanderSluis acted as a consultant on all technical matters.

REFERENCES

1. T. Operhall et al., U.S. Patent 2,961,751, 1960.
2. R. C. Feagin, U.S. Patent 3,259,948, 1966.
3. P. Linko, private communication, 1986.
4. A. K. Koul et al., *Met. Trans.* **16A**, 17 (1985).
5. E. E. Brown, private communication.
6. E. E. Brown et al., "Manufacturing Technology for Improved Quality Remelt Stock for Powder Atomization," AFWAL-TR-84-4187, Air Force Wright Aeronautical Labs, Wright Patterson AFB, Ohio, 1985.
7. H. A. Hauser, private communication, 1986.
8. J. S. Erickson et al., U.S. Patent 3,724,531, 1973.
9. R. P. Dalal et al., *Superalloys 1984*, M. Gell et al. (eds.), TMS-AIME, Warrendale, PA, 1984, p. 185.
10. M. L. Gell et al., U.S. Patent 3,567,526, 1971.
11. L. E. Dardi et al., "Metallurgical Advancements in Investment Casting Technology," ASM Metals/ Materials Technology Series, Metals Park, Ohio, 8519-002.

Chapter 16

Wrought Alloys

WILFORD H. COUTS, JR. and TIMOTHY E. HOWSON

Wyman-Gordon Co., North Grafton, Massachusetts

Other chapters in this book describe advances in performance that have been achieved in wrought superalloys. These advances have not come easily; superalloys are difficult to deform and easy to crack during deformation. This chapter will describe wrought alloy processing, the equipment used, and the most important parameters that need to be understood for successful deformation processing.

Specification of a thermomechanical process requires the definition of a goal. For sheet metal the goal may be some balance of strength, formability, and weldability. For forged disks the goal is some compromise among the strength needed for overspeed, creep resistance to prevent tip rub, cleanliness and fine grain for resistance to crack nucleation in low-cycle fatigue, and slow crack growth for longer service between overhauls.

Because testing of these mechanical properties is costly and slow, micro- and macrostructure are frequently used to demonstrate reproducibility of a thermomechanical process. Metallography may also be used to search for stringering of brittle phases that create the ductility problems demonstrated in Fig. 1 and Table 1. Much of this chapter will be phrased in terms of microstructure.

The extent of testing, both mechanical and metallographic, is immense. Its purpose is to demonstrate process control. Processing must be controlled in almost every step of manufacture of these history-sensitive materials. The processes that must be reproducibly controlled include primary and secondary melting, homogenization, ingot conversion, hot and cold (if any) deformation, and heat treatment.

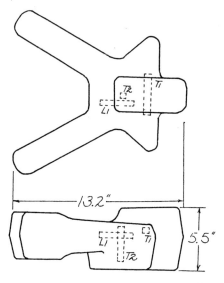

Fig. 1. Contour shape hammer forged in René 41. The starting bar axis was parallel to L1. Tensile properties are reported in Table 1.

A process that is controlled operates within specified limits, but usually the optimum limits that are adequate for both quality and cost-effectiveness are not known. Because of this, the industry is restricted to the current extensive testing of the product.

Disk quality is vital to flight safety as well as to economical operation. More understanding of the factors that affect disk behavior is required, based on understanding

Table 1. Effect of Orientation on Tensile Properties of René 41 Contoured Forging[a]

Orientation	0.2% YS (ksi)	UTS (ksi)	Elongation (%)	RA (%)
	Room Temperature			
L1	125.0	174.6	30.0	32.8
T1	126.2	163.6	16.0	18.2
T2	129.3	162.5	16.0	16.8
	1400 °F (700 °C)			
L1	108.1	142.0	17.0	17.5
T1	104.0	139.7	14.5	16.8
T2	104.0	135.2	14.0	18.8

[a] Heat Treatment was 1975 °F (1080 °C) for 4 h, oil quench (OQ), age at 1400 °F (760 °C) for 16 h, air cool (AC). Abbreviations: YS, yield strength; UTS, ultimate tensile strength; RA, reduction in area.

of alloy physical metallurgy or on empirical results if that is all that is available. The efforts of many researchers in government and at engine manufacturers, forge vendors, billet vendors, and universities are directed at disk behavior. Only a small sample of this effort is included in the references.

Despite the lag in basic understanding, there have been strides in the state of the art. Some will be described in this chapter. Progress is hampered by (1) proprietary limitations, (2) the huge cost of experiments that thoroughly explore the interrelations, and (3) inadequate tools to measure the variations of strain, strain rate, and temperature within the metal being deformed. Within these limits, the authors have sought to describe thermomechanical processing as it is currently practiced.

Finally, some predictions are offered about the directions of technical progress that will occur in the next 5–10 years, along with some of the research that must be accomplished to achieve this progress.

CURRENT INDUSTRIAL METALWORKING PROCESSES

Billet Cogging

Superalloy ingots melted by VAR or ESR (Chapter 14) are converted to billet prior to forging. The goal is to refine the cast structure. This hot working, called cogging, is usually accomplished on relatively fast acting manually controlled hydraulic presses of about 2000 ton [1815 t (metric tons)] capacity using open dies. The conversion process may include upsetting on open flat dies, with alternating upsetting and cogging steps and wash heats (reheating of the billet) as needed.

In 1972, at the time of the first edition of this book,[1] the goals of billet cogging were to refine the structure for macroetch and sonic inspection, seal microporosity, and remove ingot surface defects. Two major problems were (1) nonuniform billet structure due to double-ending (the practice of cogging one end only, reheating, and then cogging the other end with no reduction on the first end) and (2) misshapen cross sections that decreased the ability to sonic inspect, to cut pieces (or mults) at the desired weight, and to uniformly fill the die cavity. The main goal of cogging is still the same, that is, to refine the cast structure. However, what has changed is that both a finer and more uniform grain structure is now required for many applications.

One driving force for improved microstructure is better LCF properties. Because of factors such as lower forging temperatures, nonhomogeneous strain, die chill, die lock, and friction, the forged microstructure in parts of a forging can contain remnants of the as-cogged microstructure that carry through the forging deformation. Thus, as LCF and grain size control requirements in forgings have become more demanding, the requirements for billet structural uniformity and quality have become more stringent. In response to these demands, new thermomechanical cogging practices have been developed, and the industry has installed automated equipment such as the GFM[2] that reproducibly convert the full length to accurate shape.

A second driving force for improved billet structure and improved surface finish is better sonic inspectability. A need for billet inspectability beyond the current capability has been created by fracture mechanics design where allowable stresses depend on the largest nondetectable defect. Billet cogging is receiving unprecedented attention, but despite its importance, the literature on cogging of superalloys is limited.[3,4]

Extrusion

Extrusion is used extensively for the production of seamless tubing[5] for applications such as reactor heat exchangers, for conversion of ingot to billet, and for powder consolidation. The widespread use of glass lubrication has facilitated the application. Available equipment ranges up to in excess of 20,000 tons (18,144 t). Some of the extrusion presses control ram speed throughout the stroke.

Tamura[6] has published the results of an extensive investigation of hot deformation and extrusion of nickel-base alloys as a function of temperature, strain rate, and composition. The steady-state load required during extrusion was reduced by slower extrusion speed, but peak breakthrough load was not changed. By use of a carbon steel can, the peak load at breakthrough was eliminated, and only the tonnage needed for steady-state extrusion was required.

Die shape is receiving new attention. Extrusion of aerospace materials has utilized shear dies or conical dies for the most part. New die geometries are being studied such as the streamlined die that has a smooth entry and exit and does not create sharp velocity discontinuities. A streamlined die can potentially increase product yield and enhance uniformity. Gegel et al.[7] have studied optimization of streamlined die geometry using the results of analytical modeling of metal flow.

As in other forming processes, the force needed to accomplish the deformation is sensitive to the starting microstructure. Using HIPed MAR-M 200 powder billet, Kandeil et al.[8] demonstrated that thermal cycles to coarsen the grain size and the γ' size reduced the pressures required for cold hydrostatic extrusion.

Extrusion conversion from ingot to billet is extensively used for the more richly alloyed, crack-prone chemistries such as Astroloy. It is much less common for the INCO 718–Waspaloy types. However, extrusion has become almost universal for powder processing,[9] and the availability of more well-instrumented extrusion equipment may lead to more extrusion of cast-wrought alloys.

Rolling

Powerful equipment is required for rolling of superalloys.[10] Fast handling is mandatory to eliminate edge cracking. This section focuses on sheet rolling, but bar and ring rolling are subsets of the rolling process. Sheet is frequently used as rolled and annealed. This gives greater importance to dimensional control, surface finish, formability, and weldability in production than for any other manufacturing process. Hurlbatt[11] describes heating cycles, mill specifications, number of passes, pickling, camber control, and yield for the Nimonics.

Analytical tools are needed for equipment purchases and the design of rolling sequences. Kelley[12] studied the resistance to deformation characteristics during hot rolling of Waspaloy, Hastelloy X, and several other alloys. Reductions per pass of 2–30% were investigated, which probably covers the range of commercial capability. Using regression analysis of data from an instrumented laboratory mill, relationships for a resistance to deformation parameter were obtained for each alloy. For example, for Waspaloy,

$$K = 1842.0 + 291.5\dot{\epsilon} - 1.378T - 0.228T\dot{\epsilon}$$

where

K = resistance to deformation parameter

$\dot{\epsilon}$ = strain rate

T = temperature

Although no microstructural data were reported, the relationships are useful for input to Sim's rolling equation for predicting rolling loads (as described in ref. 12).

Microstructure had been the missing link but is the subject of recent work by Dinis-Ribeiro and Sellars.[13] By control of initial structures and measurement of the adiabatic heating, they have rationalized the interrelations of starting temperature, rolling reduction, quench delay, and starting and finishing microstructures in Nimonic 80A, 90, and Waspaloy. At temperatures above 1832 °F (1000 °C) they propose that static recrystallization begins quickly after rolling but is then halted by carbide precipitation at boundaries. As a result, variations in rolling reduction and temperature over the range investigated [1740–2160 °F (950–1180 °C)] had relatively small effect on the degree of recrystallization.

These authors[13] also report an inconsistency between plane strain compression tests in which dynamic recrystallization was observed and rolling tests of 10–40% reduction in which static but not dynamic recrystallization was observed. The difference was attributed to local strains being more heterogeneous in plane strain compression than in rolling. Dynamic recrystallization began in the plane strain test when local, but not the average, strains exceeded a critical value. During rolling, however, because dynamic recrystallization may not commence until much higher average strains (reductions) are reached than in the plane strain test, only static recrystallization was observed for the rolling reductions investigated. This work illustrates the careful documentation and detailed analysis required for proper interpretation of thermomechanical experiments.

Forging

To forge superalloys, a wide variety of equipment is utilized from small hammers to large presses. An important variable is rate of die closure, which may be as fast as 25 ft/s for hammers and as slow as 0.001 ft/s for isothermal hydraulic presses.

Most hammers are controlled for force and timing of blows by the operator. Some presses are manually controlled, but newer ones have microprocessor controls on head speed as well as data logging.

Die chill may be another variable. Most forgings are made in steel tooling heated to 400–800 °F (204–427 °C). There now exists considerable isothermal forge capacity that uses molybdenum alloy tooling heated to the same temperature as the forging. Hot superalloy dies heated to the 1200–1800 °F (649–982 °C) range are also used. These terms are not well defined. Figure 2 demonstrates the temperature differentials.

Isothermal and hot die tooling are expensive. However, they can be used to produce near-net shapes that reduce both raw material input and machining time. Initially these methods were applied only to costly materials. However, the drive for greater quality assurance is accelerating the use of these near-net methods because they reduce the amount of metal affected by die chill and thereby increase the uniformity of the microstructure in the finished component.

Rotating components of jet engines usually require closed dies since the periphery of a pancake made on open flat dies does not receive the required amount of working. Lack of definition of strain gradients has been a limiting problem. While the percentage of reduction in height has been commonly used as a criterion up until now, it is inadequate and even misleading. New tools of computer simulation of metal flow with finite-element method-based codes can be used to predict the strains in a forged shape, as well as strain rates and stresses during forging. The capability of simulating forging on a computer will provide an important tool for the forging designer and metallurgist.

Examples of some problem areas are shown in Figs. 3–5. Folds or laps that can occur are often attributed to poor die design and to conditioning gouges (Fig. 3). The improved geometric control on the new cogging presses (described previously[2]) will alleviate this problem. It will also allow increased upset ratios (original height

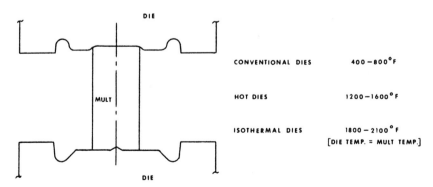

Fig. 2. Die temperatures corresponding to the definitions of conventional, hot die, and isothermal forging of superalloys.

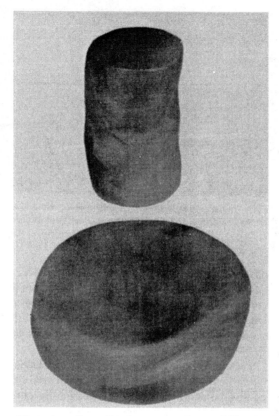

Fig. 3. Illustration of the tendency for conditioning gouges to become folds or laps in the stiffer nickel-base alloys such as René 95. 38×.

to original diameter) because of reduced susceptibility toward folding. In addition, taller, smaller diameter mults have a reduced friction cone (a volume of metal affected by friction between workpiece and die) in which recrystallization may not occur. Figure 4 illustrates the effect of friction and die chill in a Waspaloy pancake upset on conventional tool steel flat dies. This friction cone is also sensitive to die contour, as demonstrated in Fig. 5 for a contoured disk that was forged by direct upset from ingot. Inhibition of metal flow caused primarily by die shape is called die lock.

In the past, manufacture of a component usually included two to four sets of sequential forge operations. Now the improved billet surface, microstructure quality, and modeling allow many single-forge operations. This also fits everpresent goals of cost control and process control. Of course, many other factors such as the complexity of the forge geometry, workability, die wear, and lubrication must be considered.

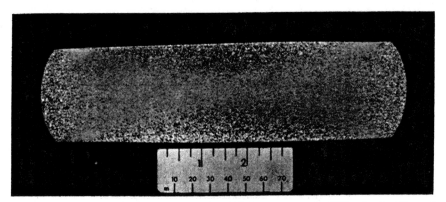

Fig. 4. Cross section of a Waspaloy pancake upset 43% from coarse-grained billet showing effects of friction and die chill on top and bottom surfaces and effect of lower strain at the periphery.

Use of air-atmosphere furnaces with radiation heating is still dominant, and a control of ±25 °F (14 °C) is still the norm in the forge shop. Controlling temperatures and times within narrow limits is desirable when one is trying to optimize many competing requirements, for example, by achieving a microstructure for good tensile and creep properties and resistance to crack nucleation, but much heating equipment is not adequate for the need.

Details of forge practice vary tremendously depending on phase equilibria, desired compromise of properties, available equipment, and quality and size of billet. These details are often considered proprietary, so published information is virtually nonexistent. However, very good reviews have been published of the variety of alloys, microstructures, and heat treatments and the multiplicity of mechanical property goals in forged components.[14-16]

Fig. 5. Turbine wheel of René 95 forged by direct upset of ingot showing some retention of cast structure due to die lock. 50×.

PROCESS VARIABLES REQUIRING DEFINITION

The Goal

In 1972 (i.e., in *The Superalloys*) the superalloy forger usually pursued one of two microstructural goals.[1] The older microstructural goal was really no goal at all, with the forger being guided by "geometry in forging, properties in heat treat." The second goal was to control the as-forged grain size by keeping the grain size as fine as possible for better LCF resistance. Now there are still two microstructural goals, but they have both changed. One is to attain a grain size of ASTM 10–14, primarily for formability, tensile strength, ductility, and resistance to crack nucleation in LCF. The other goal is a grain size of ASTM 4–8, primarily for creep strength and resistance to crack propagation.

In all cases uniformity of microstructure is important because nonuniformity seriously lowers allowable design levels. Although unrecrystallized grains were earlier not included within grain size specifications, the trend now is to include *all* grains in the rating. This reinforces the search for uniformity.

Since most current heat treatments include subsolvus "solution" treats, the grain size must be achieved and controlled in the hot-work operations. There are some applications that require supersolvus solution heat treatments or necklaced structure. The necklace structure, shown in Fig. 6, is a partially recrystallized structure consisting of a bimodal distribution of grain sizes—large grains surrounded by smaller recrystallized grains. But in the majority of applications the grain size requirements fall in the two uniform categories described.

Clearly, all the mechanical properties cannot be optimized at once. In each case one set of properties is slightly subordinated to the other. Dual property disks, with a fine grain structure in the hub for tensile and LCF properties and a coarser grain structure at the rim for creep resistance, have been considered. One approach to dual property disks has been to bond material for the rim to other material (not necessarily the same alloy) for the hub. A second approach has been to achieve a variation in grain structure by clever design of forge shapes and thermomechanical cycles. At this time, however, dual property disks are not yet in use for man-rated engines.

Homogenization of Alloy Chemistry

The goal of microstructural uniformity can depend strongly on the degree of chemical uniformity. The kinetics of precipitate nucleation, particle coarsening, and boundary mobility are all sensitive to the variations in chemistry that exist in cast material (Fig. 7). To reduce chemical segregation in the as-cast VAR or ESR ingot, conversion usually includes a static homogenization cycle. The temperature selection is limited on the high end by the onset of incipient melting and by the temperature fluctuation in the furnace in which the homogenization will occur. Sometimes an additional purpose of homogenization is to eliminate brittle nonequilibrium phases, such as Laves in INCO 718.

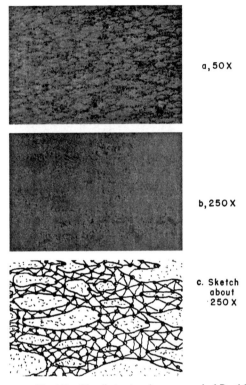

a, 50 X

b, 250 X

c. Sketch
about
250 X

Fig. 6. Partially recrystallized "necklace" structure in warm-worked René 95.

Typically, bulk diffusion in the homogenization cycle cannot completely eliminate segregation in the cast material. Grain boundary migration can help to homogenize the chemistry, but the cast grain structure is already very coarse. Homogenization after some deformation may result in grain boundary mobility, which helps to achieve a more uniform chemistry.[17]

Cooling from the homogenization temperature can result in heavy and undesirable precipitation of carbides and other phases at grain boundaries and other interfaces. Sometimes a slow cool to precipitate and agglomerate the secondary phases will be necessary.

Resistance to Metal Flow

Knowledge of the energy required to deform a superalloy is essential for selection of a forging cycle that is compatible with available equipment, die materials, and the final specification requirements. The factors that affect the flow stress are temperature, strain rate, total strain, starting microstructure, and chemistry. The dependence of flow stress on temperature, strain rate, and starting microstructure has been documented in the literature for many superalloys such as Waspaloy, INCO 718, IN-100 and René 95.[18-21] Flow data for superalloys have been generated by many different types of test including compression, tensile, torsion, and Gleeble. Today, the compression test is most widely used to generate flow data and also data for workability information and constitutive models.[22,23]

Since the flow stress decreases with increasing temperature, the forging temperature is chosen as high as allowed by other considerations, such as the microstructural goal and workability. For fine-grained billet flow stress is quite sensitive to strain rate—it decreases as the deformation rate decreases. Slow strain rate may be desired for minimum load on the tooling or to allow achievement of superplastic deformation. At slow strain rates the temperature of the tooling becomes an overriding factor. To prevent temperature loss from the workpiece, it is essential that the die temperature be maintained at or very near the workpiece temperature. Conventional tool steel and superalloy tooling are ruled out, and a molybdenum alloy, TZM, which requires a vacuum or inert-gas environment, is used today. The choice of strain rate (or forging equipment) can be influenced by other factors. For example, productivity

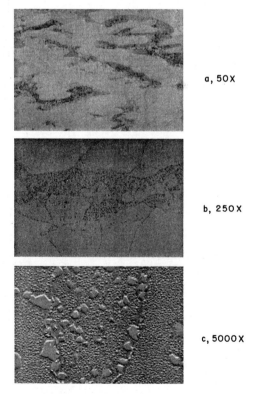

a, 50X

b, 250X

c, 5000X

Fig. 7. Inhibition of grain boundary migration in Astroloy by γ′ particles as a result of chemical heterogeneity.

can be increased by forging at higher strain rates. However, this will cause higher loads on the tooling, may cause excessive adiabatic heating, and strain localization may be encountered. Starting microstructure can be particularly important when superplastic deformation is the goal. At a given temperature the maximum strain rate at which superplasticity is achieved will increase as grain size decreases.[21]

Tamura[6] reported an analysis of the chemical factors affecting resistance to extrusion in nickel-base alloys. Of the elements evaluated, columbium was most potent, tungsten next, and then molybdenum. Chromium was slightly strengthening, and most other elements were weakening. A correlation was observed with the diffusion constant in nickel at 2100 °F (1150 °C), and it was theorized that tantalum would be most potent of all. The result was a regression equation that could predict the force needed to extrude alloys with new chemistries.

In production of forgings new chemistries happen infrequently, but new shapes are encountered every day. Prediction of the force required has been an experience-based art, but that is changing. Techniques such as slab analysis are used to predict the profile of loading on the tooling and now finite-element-method-based software for metal flow simulation yields very accurate information on forces on the dies. An important factor that is out of the realm of this discussion is friction (or lubrication).

Hot Workability

The two characteristics that influence hot workability are strength and ductility. At the high temperatures at which superalloys are worked and forged, workability is better than at lower temperatures because of lower strength and higher ductility. The good ductility is due to the dynamic recovery and dynamic recrystallization that occur during the deformation.[24,25] With sufficient strain, a steady state may be reached in which the flow stress is constant. However, in industrial hot-working processes steady-state conditions are most likely only achieved in extrusion or superplastic forging.

The importance of understanding the regions of good workability and the operative mechanisms is illustrated by Guimaraes and Jonas.[19] They studied dynamic recovery and dynamic recrystallization in Waspaloy and INCO 718 over ranges of temperature and strain rate. In the regions where dynamic recovery occurs during deformation (but not dynamic recrystallization), pores form at triple points that would be detrimental to continued deformation.

Techniques for developing maps for use in choosing hot-working parameters have been developed. A processing map that delineates regions in temperature and strain rate space in which various damage mechanisms operate and the regions that are safe for warm and hot working has been published.[26] However, a processing map for a superalloy has not been published.

A different kind of map derived by Gegel et al.[22] shows, in temperature–strain rate space, the efficiency of power dissipation during plastic deformation from metallurgical processes such as dynamic recrystallization or internal fracture. The map defines the combinations of temperature and strain rate that yield high efficiency and the combinations that yield low efficiency. Each map is derived for a specific material and starting microstructure. Flow stresses are measured at a given strain for a number of tests over a range of temperatures and strain rates. The strain rate sensitivity of the flow stress is then obtained, and the efficiencies are defined (see ref. 22). A map such as this can offer a guide for choosing hot-working parameters as long as the metallurgical process (or processes) corresponding to the efficiencies in different parts of the map is characterized. It is quite possible that the microstructural goal in an alloy would be obtained not at the temperature–strain rate combination corresponding to the highest efficiency but rather at a combination corresponding to a much lower efficiency.

Figure 8 shows a map of constant efficiency contours for −150-mesh extruded René 95.[27] The data were generated by compression testing at temperatures of 1900–2075 °F (1038–1135 °C) and strain rates of 10^{-3}–10 sec^{-1}. From the flow curves, the strain rate sensitivity and the efficiency are defined at each test temperature and strain rate at a given strain. The map in Fig. 8 was constructed using flow stresses measured in the compression tests at 0.5 strain.

In addition to being able to specify the temperature–strain rate combinations for good workability, it is also essential for the forger to understand the phase equilibria of the alloy, the microstructures of the input material as a function of temperature, the effect of the amount of strain, the deformation mechanisms that operate, the flow curves the material exhibits, and the microstructures achieved at the different strain rate–temperature combinations. For example, the initial microstructure depends on the prior thermomechanical history and on whether the working temperature is established by heating from a lower temperature or by cooling from a higher temperature.

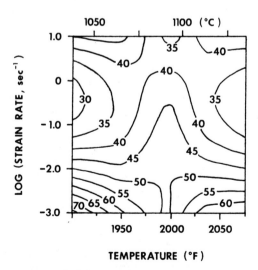

Fig. 8. Map of constant contours of efficiency of power dissipation through metallurgical processes for −150-mesh extruded René 95, measured in compression specimens at 0.5 strain.

The successful forger must ask and answer satisfactorily at least the following questions:

1. Are the secondary phases agglomerating, dissolving, or precipitating?
2. How long will it take for these processes to occur and to possibly affect other aspects of the microstructure such as grain structure?
3. To what extent is the structural response heterogeneous due to chemical segregation?
4. Will the strain in all areas of the forging be sufficient to achieve the microstructural goals for the temperature and strain rate chosen?
5. If not, is a change in temperature and strain rate needed or is a modification of the forging preform dictated?

The conditions for best workability are not necessarily the same as the conditions that will yield the desired microstructural goals.

There is currently an extensive effort to generate the data needed for important production alloys. The data are being used for basic flow stress information, for generation of "processing maps," and also for use in modeling of hot-forming processes using numerical methods.

Furnace heating time now receives careful attention for two reasons. The first reason is energy costs. Numerical analysis of heat transfer allows minimum heating times to be specified very precisely. Cooling rates in quenching after forging or in heat treatment can also be obtained. The second reason is control of the final microstructure. One must apply knowledge of the phase equilibria to the selection of temperature and time to control the microstructure as heated prior to deformation if one is to produce the desired microstructure at the end of deformation. Much of this understanding is empirical, but new tools are becoming available. For example, the coarsening rate of γ' can be predicted in a well-homogenized alloy,[28] and a technique for the study of carbide dissolution using computer simulation has been demonstrated.[29]

An important point is that the metal temperature in the furnace is not exactly the same as metal temperature during deformation. Heat loss by radiation, conduction to the tooling, and adiabatic heating are variables sensitive to the environment, the shape of the part, insulation, lubrication, the rate of deformation, and so on. It is only now that the tools that can analyze metal deformation and predict the temperature within the metal during deformation are becoming available.

But what about the development metallurgist who must determine a region of good workability and define a working cycle for a new alloy but who has limited material and/or computer capability? With a microstructural goal in mind, he might read the description of hot working and warm working by Kear et al.[30] and then identify the most critical temperature characteristics for his alloy (i.e., γ' solvus, $M_{23}C_6$ solvus, etc.) and forge a few small samples in the temperature regime of interest. Use of isothermal tools or hot dies will minimize cracking if the optimum conditions have not yet been determined.

After each forge operation in a sequence, the microstrucuture should be examined. Waspaloy has been studied extensively for the microstructural changes that occur as a result of changes in strain, strain rate, and temperature. Detailed descriptions by Livesey and Sellars[18] and McQueen et al.[24] may be used as roadmaps to adjust subsequent forge operations according to the phase equilibria of the experimental alloy as one progresses toward the final goal. These references present the best available clues for the recognition of strain hardening, dynamic recrystallization, or static recrystallization in experimental superalloy chemistries.

Heat Treatment

The solution heat treatment temperature must usually be specified within a narrow range of temperatures. If the temperature is too high, grain growth may occur in some areas of the forging where, due to the nonuniform chemistry that carries through from the cast structure, solutioning of grain growth-inhibiting phases such as γ' or δ takes place. If the temperature is too low, the grain size may be finer than desired, and the γ' may over-age, both of which may cause the tensile and creep strengths in service to be unacceptably low. An understanding of the furnace capability is always needed. Use of workpiece thermocouples and adjustment of furnace controls should allow a ± 15 °F (9 °C) range as a maximum in heat treat.

Quenching after solution treatment is almost universal. Quenching immediately after forging is sometimes practiced. Water is an economical and safe quenchant for dilute alloys or sluggish γ' formers, but water quenching is likely to cause cracking in richer alloys such as Astroloy. Oil or hot salt are common alternatives. Polymer quenching is attractive due to environmental controls. At this time selection of shape to be quenched, surface preparation, control of transfer time, quench medium agitation, and so on, are empirically chosen, but computer analysis is beginning to be helpful.

Due to great improvements in process control, direct aging cycles have become possible. This cycle uses the heating for finish forge as a "solution heat treat." In this case time and temperature in the forge shop must be controlled as precisely as in the heat treat department. Improved tensile and LCF properties may be obtained with little or no loss in creep resistance at disk-operating temperatures. Little information has been published about minimum deformation in finish forge, role of adiabatic heating, or optimized age cycles for direct age.

THE NEXT DECADE

Thermomechanical processing is still an experience-based art, although a very sophisticated one. New influences outside the forge shop are changing this situation. Computer software for metal flow simulation[31] is one such influence. Application of the finite-element method to solve problems in plasticity coupled with the availability of fast, large memory computers have magnified the capability of the die designer and forge shop metallurgist. This tool can eliminate costly time-consuming die tryouts, improve material utilization, and define the strain-strain rate–temperature history internal to the forging and thereby allow better control of the microstructure. Figure 9 shows the range of variation of strain revealed by analysis of a simple preform forged to a simple shape.

Statistical process control (SPC) works backward. It takes the results and searches for sources of problem areas. The benefit for superalloys is that it can focus research and development on the manufacturing areas that need it most. It can minimize the random application of effort that has characterized the past.

Advances in elevated-temperature fracture mechanics analysis are influencing billet cleanliness, which is a function of melt or powder conversion practice. As discussed, fracture mechanics needs have also led to a more uniform, finer grained billet for improved sonic quality. These advances have resulted in improved forgeability and will give more credence to the predictions obtained by computer simulation. Increased understanding of crack behavior should impact the goals of both the as-forged microstructure and the as-heat-treated microstructure.

Simulation, SPC, billet control, deformation mapping, and equipment instrumentation are powerful tools. Now they must be applied to the development of manufacturing cycles for specific alloys, specific microstructures, and specific desired properties. Motivated by customer demands for performance upgrades, improved reliability, and cost control, these tools should enable the industry to convert the finish forge to a science-based operation within five years. Heat treatment should follow a similar schedule. Preliminary operations such as upsetting, cogging, and homogenization may not convert to a scientific foundation for another five or more years. Solidification control of either powder or ingot is also needed as an integral partner to thermomechanical processing.

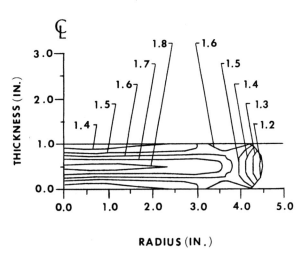

Fig. 9. Effective strain contours in a Waspaloy pancake predicted by computer simulation.[31] The starting mult (preform) was upset 80% under isothermal conditions at 1950 °F, with a friction factor of 0.2.

There is no perceivable requirement for radical change in the size range of the forgings to be needed. Thus, no major changes in ingot size, billet size, or extrusion size are anticipated. If a market for larger disks, that is, 60–100 in. (1525–2550 mm) diameter, should materialize, it would be serviced by conventional experienced-based processing.

The battle of the last decade between cast-wrought and powder billet should continue. The competition has stirred melt vendors into process improvements, and the product should be able to capture some of the applications that are now powder. New and/or better powder billet should capture the newer, most demanding applications.

Since powder billet is costly and can usually be superplastically formed due to the fine grain size, most of it is forged on isothermal dies. Isothermal forging produces the maximum in shape refinement. For well-defined shapes in production quantity, isoforge should continue to be the preferred process.

In contrast, cast-wrought billet is usually not superplastic and averages lower cost per pound. For cast-wrought alloys, conventional forge, hot die forge, and isoforge have competed on tooling cost, delivery, machining envelope weight, quantity, uniformity of product, follow-on order possibilities, and so on. If process advances continue to occur in all three forge practices, one may expect continued competition. However, any process that ceases to advance will wither.

REFERENCES

1. C. T. Sims and W. Hagel (eds.), *The Superalloys*, Wiley, New York, 1972.
2. Anonymous, "Forging the Future of Specialty Metals," CarTech, Reading, PA, 1984.
3. J. G. Wistreich and A. Shutt, *JISI*, **193**, 163 (1959).
4. T. W. Allen and L. J. Cartmell, *JIM*, **96**, 321 (1968).
5. H. Aigner, H. P. Degischer, and E. Hertner, *Nuc. Tech.*, **66**, 54 (July 1984).
6. M. Tamura, *Proceedings of Japan–U.S. Seminar on Superalloys*, Japan Institute of Metals, December 1984, p. 151.
7. H. L. Gegel, J. C. Malas, J. S. Gunasekera, J. T. Morgan, and S. M. Doraivelu, "Computer-Aided Design of Extrusion Dies by Metal Flow Simulations," Advisory Group for Aerospace Research and Development, Neuilly-sur-Seine, France, AGARD-LS-137, 8-1, 1984.
8. A. Y. Kandeil, W. Wallace, J-P. A. Immarigeon, and M. C. de Malherbe, *Can. Met. Quart.*, **19**, 245 (1980).
9. D. R. Chang, D. D. Krueger, and R. A. Sprague, in *Superalloys 1984*, M. Gell et al. (eds.), TMS-AIME, Warrendale, PA, 1984, p. 245.
10. Anonymous, *Met. Bull. Mon.*, **142**, 43 (1982).
11. J. D. Hurlbatt, *Met. Tech.*, **2**, 326 (1975).
12. E. W. Kelley, in *Superalloys 1980*, J. K. Tien et al. (eds.), ASM, 1980, p. 141.
13. N. Dinis-Ribeiro and C. M. Sellars, "Strength and Structure During Hot Deformation of Nickel-Base Superalloys," presented at Superalloys Conference, Araxa, Brazil, April 9–12, 1984.
14. N. A. Wilkinson, *Met. Tech.*, 234 (July 1977).
15. A. E. Marsh and G. Oakes, "Forging High-Value Aerospace Materials for Cost Effective Control of Product Shape and Properties," Hot Working and Forming Processes Conference, Sheffield, England, July 1979.
16. F. Turner, in *The Development of Gas Turbine Materials*, Applied Science Publishers, Barking, England, 1981.
17. K. Smidoda, C. Gottschalk, and H. Gleiter, *Met. Sci.*, 146 (March–April 1979).

18. D. W. Livesey and C. M. Sellars, *Mat. Sci. Tech.*, **1**, 136 (1985).

19. A. A. Guimaraes and J. J. Jonas, *Met. Trans. A*, **12A**, 1655 (1981).

20. J-P. A. Immarigeon, "The Role of Microstructure in the Modelling of Plastic Flow in P/M Superalloys at Forging Temperatures and Strain Rate," Advisory Group for Aerospace Research and Development, Neuilly-sur-Seine, France AGARD-LS-137, 4-1, 1984.

21. T. E. Howson, W. H. Couts, Jr. and J. E. Coyne, in *Superalloys 1984*, Gell et al. (eds.), TMS-AIME, Warrendale, PA, 1984, p. 275.

22. H. L. Gegel, Y. V. R. K. Prasad, J. C. Malas, J. T. Morgan, K. A. Lark, S. M. Doraivelu, and D. R. Barker, "Computer Simulations for Controlling Microstructure During Hot Working of Ti-6-2-4-2," 1984 Pressure Vessels and Piping Conference and Exhibition, PVP-Vol. 87, ASME, New York, p. 101.

23. J-P. A. Immarigeon, A. Y. Kandeil, W. Wallace, and M. C. de Malherbe, *J. Test. Eval.*, **8**, 273 (1980)

24. H. J. McQueen, G. Gurewitz, and S. Fulop, *High Temp. Tech.*, **1**, 131 (1983).

25. H. J. McQueen and J. J. Jonas, *J. Appl. Metalwork.*, **3**, 233 (1984).

26. R. Raj, *Met. Trans. A*, **12A**, 1089 (1981).

27. S. Jain, General Electric Co., private communication, 1986.

28. D. McLean, *Met. Sci.*, **18**, 249 (1984).

29. W. E. Voice and R. G. Faulkner, *Met. Sci.*, **18**, 411 (1984).

30. B. H. Kear, J. M. Oblak, and W. A. Owczarski, *J. Met.*, **24**, 25 (June 1972).

31. S. I. Oh, *Int. J. Mech. Sci.*, **24**, 479 (1982).

Chapter 17

Powder Metallurgy

STEVEN REICHMAN and DAVID S. CHANG

Wyman-Gordon Company, North Grafton, Massachusetts,
and Aircraft Engine Business Group, General Electric Company, Cincinnati, Ohio

The potential advantages of powder metallurgy processing for superalloys became apparent in the late 1960s.[1] Superalloy users and producers began to view it as a potential route to low-cost, high-performance aerospace components.[2,3] In fact, by the mid-1970s one could almost envision an all-powder turbine engine with near-net-shape processed blades, disks, and structural components. However, as with any "new" technology, this was overapplied, and a share of disappointments and disasters came with the successes.

Present applications (estimated to consume about 1,000,000 lb/yr of consolidated power product[4]) establish the boundary conditions for the use of powder metallurgy processes to superalloy components for turbine engines. Powder-based superalloys are used when "conventional" cast or wrought components will not meet the performance requirements of the application, the achievement of which will have a high engine payoff. The failure of conventional material is usually a result of segregation, which results in low or inconsistent properties and reduced thermomechanical response. So, far from being a panacea, powder-based processes are applied when other processes (usually preferred) cannot do the job.

The attributes that make powder-based processing appropriate for turbine products are intrinsic to the powder process; that is,

1. reduced segregation due to rapid solidification rates, resulting in smaller intermetallic particles and reduced interdendritic spacing not achievable in practical-sized components through casting processes;

2. forgeable microstructures to achieve very high levels of mechanical properties (the uniformity of the consolidated powder is on the "micro"-scale, so that property uniformity and inspectability can be outstanding); and

3. the ability to achieve unique structures for special environments [e.g., oxide-dispersion-strengthened (ODS) superalloys are not achievable by any process other than powder metallurgical.]

The factors that limit the use of these materials over a broader application base are a combination of performance limitations and cost considerations. They are:

1. for highly stressed components such as fracture-critical or fatigue-limited parts, the extreme sensitivity of these alloys to low defect levels arising from contamination, and attendant performance degradation (this susceptibility is alloy related, not processes related);

2. the need for thermomechanical processing for critical components to assure the neutralization of contaminant defects;

3. the need for very fine powders (reduced yield) to control the maximum defect size;

4. in the case of product, very careful and stringent process requirements both in powder production and thermomechanical processing;

5. high inherent cost due to expensive processing operations necessary to assure product integrity; and

6. a further limit to the near-net shape potential of P/M processing by limitations in nondestructive inspection (NDI).

The current application of P/M processing to superalloy turbine components is in 1000–1400 °F (540–760 °C) compressor and turbine disks (and associated rotating hardware). While most of the current cast-wrought and even cast compositions have been evaluated as powder-processed materials, only the alloys that fulfill a high-performance payoff to the engine are considered cost-effective; the cost-effectiveness of powder processing is considered beyond the acquisition cost of the component itself. These components are produced from prealloyed atomized powder. The powder (for the majority of applications) is consolidated by hot isostatic pressing or hot extrusion. These billets or preforms are for the most part subsequently forged to near sonic shape.

The other application of P/M processing of superalloys is for static vanes in the turbine section, although there is good potential for combustor cases from sheet product. These are produced from ODS alloys. In this process elemental or master alloy powders, including the very fine oxide powder, are mixed together in the solid state and form homogenous powder particles by mechanical alloying. This powder is then consolidated to full density by hot extrusion and is subsequently directionally recrystallized to a specific crystallographic texture, which results in stable but anisotropic structures. These materials are used up to about 2000 °F (1100 °C). The oxide dispersion provides strength through dislocation pinning

(Orowan strengthening) at temperatures above those where γ' strengthening is effective. these alloys also show very good environmental resistance through generation of adherent oxide coatings.

The balance of this chapter will discuss the processes used for powder manufacture, consolidation processes, thermomechanical processing, physical and mechanical properties and design criteria, and future applications and trends.

POWDER-MAKING PROCESSES

During the development and application of superalloy powder products, virtually every conventional and unconventional technique in particulate production has been tried. However, due to the reactive nature of the alloying elements in this family of materials, only the processes that are conducted in inert atmosphere (either gaseous environment or vacuum) have prevailed. It was recognized early that low oxygen and nitrogen contents were needed, and powder particle surfaces free of oxides, nitrides, and carbides were required in order to effect sound interparticle bonds in the consolidated product.[5] The processes that fulfilled these requirements and also lent themselves to the production environment were inert and soluble gas atomization (IGA and SGA), the rotating electrode process (REP and PREP), and centrifugal atomization [the so-called rapid solidification process (RSP)]. Oxide dispersion powders are quite different in their requirements and are produced by the mechanical alloying process.

The atomized superalloy powders are generally spherical in nature and are used in rather fine-mesh cuts (-100 mesh, 150 μm to -325 mesh, 43 μm) in order to minimize contamination effects on critical flaw size considerations.

Inert-Gas Atomization and Soluble-Gas Atomization

Inert-gas atomization (IGA) and soluble-gas atomization (SGA) are the most widely used processes for powder production. In the basic IGA process vacuum-refined alloy is remelted in an inert-gas environment and poured into a tundish through which the molten metal exits as a stream and is metered through a gas nozzle. The nozzle delivers a continuous circumferential stream of inert gas at high pressure to the molten metal stream, which is disintegrated into spherical particles. These particles cool at rates approximately 10^2 °C/s.[6] The powder is collected at the outlet of the atomization chamber. Figure 1 shows a typical argon atomizer.

The soluble gas process involves atomization upward from an induction melting crucible (see Fig. 2). The melt environment (above atmospheric pressure) contains a partial pressure of gas that is soluble in the alloy (usually H_2). The atomization takes place by immersing a tube (usually ceramic) into the melt; the open end of the tube connects to the upper atomization chamber, which is under vacuum. The molten metal is drawn up the tube, and the combination of the pressure to vacuum effect and the dissolution of the soluble gas causes the metal to be atomized. Cooling rates of 10^3 °C/s[1] are typical as measured from dendrite arm spacing.

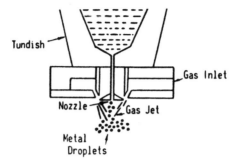

a

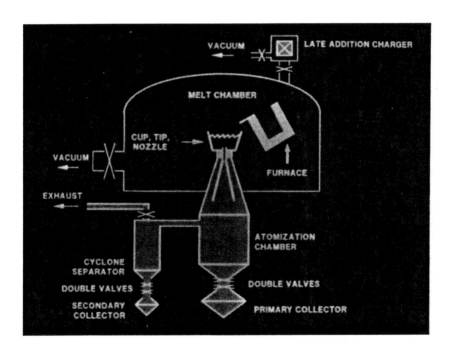

b

Fig. 1. Gas atomization system for superalloy powder production: (a) atomization nozzle; (b) typical system.

While there are variations on the above processes among the producers, the basic mechanisms are common. All current atomization processes utilize refractory systems for melting and molten metal transfer, which is a major source of contamination and a limitation to the full realization of maximum property performance.

Rotating Electrode (REP) and Plasma Rotating Electrode (PREP) Processes

The REP process was used in early production of IN-100 powder but is presently not used due to technical and economic considerations. In both processes (REP and PREP) an electrode of the alloy is rotated at high speed in an inert chamber. In REP a nonconsumable tungsten electrode is used to strike an arc to the surface of the alloy electrode. The surface melting of the rotating electrode centrifugally removes the molten metal surface, which forms spherical powder particles in the chamber. PREP is similar to REP except that instead of a tungsten electrode a plasma arc is used to melt the superalloy electrode surface, as shown in Fig. 3. Cooling rates as high as 10^5 °C/s for IN-100 powder have been seen for PREP

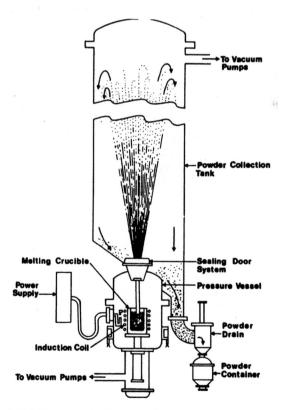

Fig. 2. Soluble gas atomization system for producing superalloy powder.

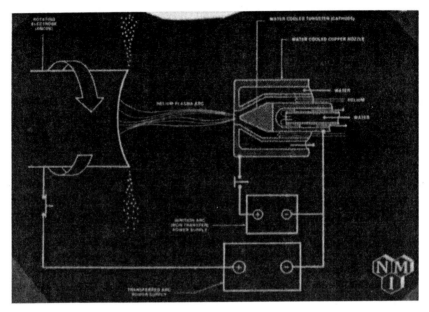

Fig. 3. Principle of plasma rotating electrode process (PREP). The process is carried out in a vacuum tight chamber.

titanium powders and similar rates are expected for nickel-base powders,[7] though lower cooling rates (10^2 °C/s) may be appropriate for larger particle diameters.

These processes, while not in active production, offer the potential for noncontaminating powder production (especially PREP). However, electrode segregation, which persists through powder production, and certain volatiles caused by the high-temperature arc or plasma are technical problems that have yet to be adequately resolved.

Centrifugal Atomization

Another powder production process, pioneered by Pratt and Whitney,[8] that is in limited superalloy production is centrifugal atomization. This process offers very rapid cooling rates (1-8 × 10^6 °F/s)[6] and an extremely narrow particle size distribution in the product. In the centrifugal atomization process a stream of molten metal, usually from an induction furnace, is very carefully metered onto the surface of a rapidly rotating disk, as shown in Fig. 4. The molten metal forms a liquid sheet on the disk, and as it accelerates off the circumference, it forms spherical particles that are further atomized by vertical jets of an inert gas (helium). This secondary atomization results in the very high cooling rates. The retention of the benefits of the very high cooling rates on the microcrystallinity of the powder is unclear in light of the time–temperature excursion during conventional consolidation.[9] (The application of these powders in unique alloy compositions for turbine airfoils will be discussed later in this chapter.)

The types of alloys to which atomization processes have been applied are high-volume γ' fraction and refractory element compositions that show unacceptable segregation in conventional cast/wrought processing. For the most part these alloys as powders are similar in overall composition to the conventional alloys, except for reduced carbon levels, since carbides are not necessary for grain size control and can in fact cause powder surface decoration, which subsequently causes bonding problems [prior powder boundary (PPB) effects] in the consolidated product. The alloys commonly used are René 95, low-carbon IN-100, MERL-76, and low-carbon Astroloy. Other alloys specifically tailored for powder are are under development and should see production in the future. Typical chemistries of superalloy powders are shown in Table 1.

Processes for ODS Alloys

Compared to the above processes, ODS alloys made by mechanical alloying take a completely different approach to producing a homogenous powder particle. The mechanical alloying process is a solid-state process (i.e., no melting takes place) in which elemental or master alloy particles and oxide particles in predetermined proportion are mixed together in a high-energy ball mill. The particle size of the master alloy mix is in the 2–200-μm range. The oxide particles are initially <10 μm in size.[10] During the milling the energy from the balls is dissipated either as

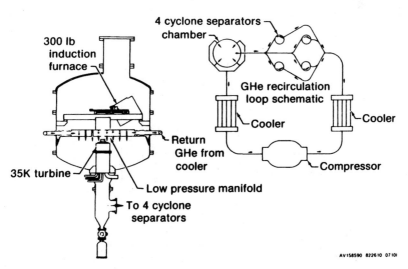

Fig. 4. Centrifugal atomization system for rapid solidification processing of superalloy powder. Low-pressure manifold provides the secondary quench (He).

Table 1. Typical Chemical Compositions for P/M Superalloys (w/o)

	Ni	Cr	Co	Mo	W	Cb	Al	Ti	Hf	Ta	C	B	Zr	V
René 95	Bal	13.0	8.0	3.5	3.5	3.5	3.5	2.5	—	—	0.05	0.01	0.05	—
IN-100	Bal	12.4	18.5	3.2	—	—	5.0	4.3	—	—	0.07	0.02	0.06	0.8
MERL-76	Bal	12.5	18.5	3.2	—	1.4	5.0	4.4	0.4	—	0.04	0.2	0.6	—
APK1	Bal	15.5	17.0	5.0	—	—	4.0	3.5	—	—	0.03	0.025	—	—
NiALMo (Range)	Bal	0.10	—	10–14	6–10	—	7.8	—	—	0–6	—	—	—	—

Table 2. Typical Chemical Compositions for ODS Alloys

	Fe	Ni	Cr	Mo	W	Al	Ti	Ta	B	Zr	C	Y_2O_3
MA-754	—	Bal	20.0	—	—	0.3	0.5	—	—	—	—	0.6
MA-956	Bal	—	20.0	—	—	4.5	0.4	—	—	—	—	0.5
MA-6000	—	Bal	15.0	2.0	4.0	4.5	2.5	2.0	0.01	0.15	0.05	1.1

heat or in collisions with the powders. These collisions cause interparticle welding, plastic deformation, and fracturing. The process is carried out in an inert environment so that the welding and fracturing take place between atomically clean surfaces. The weld fracture mechanism occurs over an extended period of time (up to 24 h) so that a sufficient number of events occur until the particles are homogenous on a fine scale. X-ray analysis of properly milled powder reveals only one crystal structure intermediate to the elemental constituents.[11] The incorporation of very fine (200–400 Å size) inert oxide particles (usually Y_2O_3–Al_2O_3) provide the basis for the dispersion strengthening of the subsequently consolidated powder. The oxide particles represent about 1 v/o of the alloy.

This powder, as contrasted to the atomized powder, is not spherical in nature but is rather acicular or platelike.

ODS processing has been applied to either solid-solution austenitic (MA-754), ferritic (MA-956), or γ'-strengthened austenitic (MA-6000) alloys. MA-6000 offers good intermediate-temperature strength, a shortcoming of MA-754 and MA-956. Typical compositions are shown in Table 2. It is reported that at least 4 w/o aluminum is needed in these alloys to aid in oxide scale formation.[12]

Characterization of Powder

The atomized powders are generally spherical in shape and have oxygen contents, depending on particle size, of about 100 ppm (proportional to specific surface area), a Gaussian-type distribution of sizes, and structures either microcrystalline or dendritic depending on particle size and cooling rate.[13] Figures 5–7 show typical powder morphologies, and Fig. 8 shows particle size distribution versus production process.

Atomized powders show varying degrees of satelliting (clustering), splats, and shells depending on the peculiarities of the producers' system (Fig. 5). These features, while not shown to be detrimental to the end product, can have an influence on yield of product.

It is interesting to note in Fig. 8 that the two rotating processes (PREP and centrifugal atomization) both show a significantly larger average particle size than the atomization processes (80 vs. 45 μm). The physics of the processes limit the fineness of the powders produced by these distinctly different processes. (The liquid metal sheet shears at the solid–liquid interface at higher speeds.) The break in the centrifugal atomization powder size distribution is probably a result of the two atomization mechanisms superimposed (secondary atomization).

Several authors[14,15] have shown or predicted the segregation of various alloying elements on powder surfaces due to the solidification kinetics for powder particles. The degree of segregation is dependent on the cooling rate and particle size. Aubin et al.[15] shows that the combination of high bulk carbon and oxygen content can significantly reduce the ductility and toughness of the consolidated product through PPB (prior particle boundary) decoration.

Oxidized powder particles likewise are shown to be deleterious to properties such as ductility and LCF life.[16] These particles are generated during atomization as a result of transient air leaks in the system. Their incidence in a lot of powder

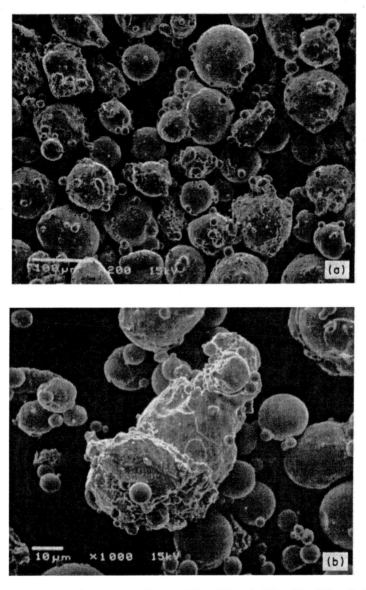

Fig. 5. Typical Ar-atomized powder (R95): (a) −150 + 325 mesh; 200×; (b) −325 mesh; 1000×.

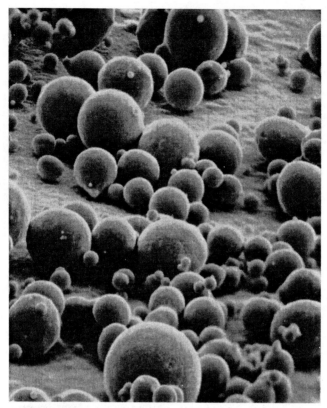

Fig. 6. Typical powder morphology from soluble-gas atomized powder (IN-100); HMI powder; 100×.

Fig. 7. Centrifugally atomized IN-100 powder; 500×.

469

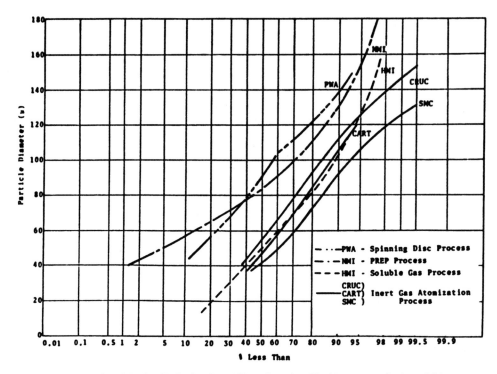

Fig. 8. Typical particle size distribution for −100-mesh produced by inert-gas atomization soluble-gas atomization and two centrifugal atomization processes. PWA, Pratt & Whitney Aircraft; NMI, Nuclear Metals, Inc.; HMI, Homogeneous Metals, Inc.; CRUC, Crucible Compaction Metals; CART, Carpenter Technology; SMC, Special Metals Corp.

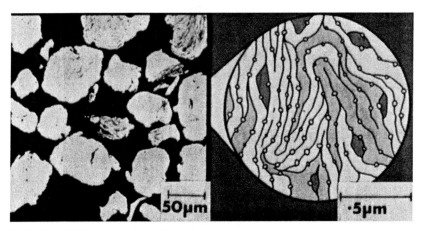

Fig. 9. Typical fully processed mechanically alloyed powder: (a) microstructural cross sections; (b) schematic showing oxide particle delineating boundaries between the solid-state alloyed elements.

is low and therefore difficult to detect and remove. Their presence influences minimum property design criteria.

Hollow particles are also found in atomized powders[17] arising from either entrapped argon or solidification shrinkage. These pores generally close during consolidation, and only the entrapped argon variety can open during subsequent high-temperature exposure. One production test called the thermally induced porosity (TIP) test is routinely applied to consolidated powder (especially HIPed powder). This test involves a high-temperature [2200–2250 °F (1200–1230 °C)] thermal exposure after which the density change and microstructure are observed. There is a limit to the allowed density change (e.g., <0.2%) arising from pore generation. This test can also show the presence of a consolidation process that has had a container leak. The presence of triple-point porosity as opposed to particle center porosity is indicative of a breached container.

Inorganic or oxide contamination as well as organic contamination of the powder is usually characterized or quantified by an elutriation procedure. A sample powder is elutriated with deionized water, and the low-density particles are collected, characterized, and counted. Acceptability levels are established for type and amount of contamination.

Contamination from other alloys (cross-contamination) produced in the same equipment is a source of defect generation that can limit the minimum properties of the component. Cross-contamination by a lower hardness alloy or an alloy with a different γ' solvus can result in a few scattered coarse grains or carbide–powder boundary problems in the consolidated powder.

Mechanically alloyed powder shows significantly different powder characteristics than atomized powder. Figure 9 shows typical MA-754 powder. When looked at on a microscopic scale, the powders show a banded structure with spacing of about 0.5 μm with Y_2O_3 oxide particles of 200–400 Å spaced at about 0.1 μm. The banding can be eliminated in the powder by annealing at 2191 °F (1200 °C), which causes recrystallization to irregular equiaxed grains.[11] This type of thermal excursion occurs during densification. The typical particle size of the mechanically alloyed powder is platelike disks about 200 × 100 μm by 50–100 μm thick.

POWDER CONSOLIDATION

As with powder production, current powder consolidation evolved from attempts to densify superalloy powders by virtually every P/M process, pressing and sintering, vacuum hot pressing, explosive compaction, hot isostatic pressing (HIP), direct forge consolidation, and extrusion.

Because of the necessity to achieve purity in the product at a reasonable price, the methods currently employed in powder consolidation are hot isostatic pressing and hot compaction of the powder to near full density followed by hot extrusion.

Prior to densification and due to the need for cleanliness (minimization of contamination), most of these powders are stored under inert or at least controlled environment conditions.

Various studies have shown that the desorption of surface gases occurs over a range of temperatures, and the deleterious species, that is, H_2O, CO, CO_2, and CH_4, evolve from surfaces from 100 °C for H_2O to about 600 °C for CO (peak desorption).[14] These investigations show the detrimental effect these species have on PPB retention and attendant properties. The most direct approach to solve this problem is to avoid contamination by adsorbing species through all inert processing.

There have been a number of approaches to particulate defect removal[20] using various magnetic and electrostatic techniques that reportedly use the differences in dielectric and magnetic characteristics of the different contaminants. It has also been suggested[18] that a preconsolidation deformation operation such as powder rolling could be effective in reducing ceramic defect size. There are no reports of these processes being used in production, and a general approach to avoid generating defects in the first instance is preferred at present.

All densification involves the containerization of the powders under vacuum. (The container is disposed of subsequent to densification of the powder.) The powder is usually loaded into its consolidation container (metal) under a dynamic vacuum. This can be accomplished either warm or cold. The evacuated container filled with powder is then sealed and ready for consolidation.

Contamination of the powder up to this point (postatomization) by so-called reactive defects is by far the most insidious type of contaminant. These defects in the consolidated product can be many times larger than the initial contaminant due to decomposition during thermal processing prior to densification and the generation of reactive species that form PPB networks of oxides and carbides. These defects can arise from many sources that contribute material to the powder mass such as cloth, rubber, grease, human hair, and low-stability oxides such as mill scale.

Hot Isostatic Pressing

In hot isostatic pressing (HIP) the containerized powder is heated to an elevated temperature (either above or below the γ' solvus depending on the desired grain size of the product) under a high external gas pressure (approximately 15,000 psi). The combination of temperature and pressure on the container is transferred to the powder mass and affects densification.

The general structural goal of HIP consolidation is to have a fully dense product without retention of prior PPBs. The conditions for successful HIP consolidation require sufficient time, temperature, and deformation during the cycle, as shown by Tien et al.[19] Varying HIP cycle parameters can influence whether plastic deformation or creep is the predominant densification mode, which can in turn influence the PPB retention in the microstructure.

Some components are used in the HIPed and heat-treated condition. These applications are evidently successful, but one would envision extreme control of the processing of assure product integrity. Examples of HIPed structure are shown in Fig. 10.[21]

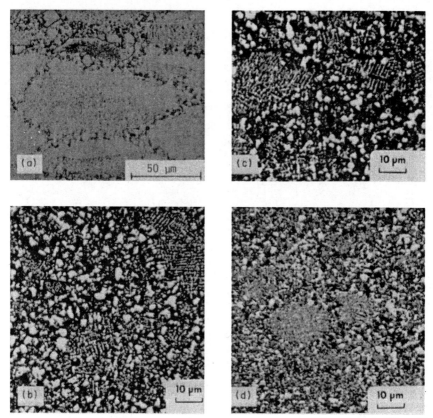

Fig. 10. René 95: (a) cast and wrought; (b) as HIP (−100 mesh); (c) HIP + forge (−100 mesh); (d) as extruded (−150 mesh).

Hot Compaction and Extrusion

The other commercial consolidation process (in fact, the predominant method) is subsolvus hot compaction followed by extrusion. In this process containerized powder is hot compacted (usually in a closed die, although HIP can be used). The consolidated powder (> 95% dense) is subsequently extruded at a ratio of about 6:1 to fully dense billet.

The structure of subsolvus extruded billet is characterized by a fine recrystallized grain structure with almost complete elimination of prior powder boundaries and any dendritic structure carryover from the starting powder (Fig. 10).[21] These structures are dependent on the starting powder size and powder production process.[20] The aforementioned structures are all from −150-mesh powder.

Consolidation of ODS Powders

Hot extrusion is the typical consolidation process. These materials require directional thermomechanical processing to achieve proper structure and properties. The powders are not necessarily handled inertly and can be cold compacted to about 80% density. This can be done by uniaxial pressing or cold isostatic pressing (CIP). The nonreactive

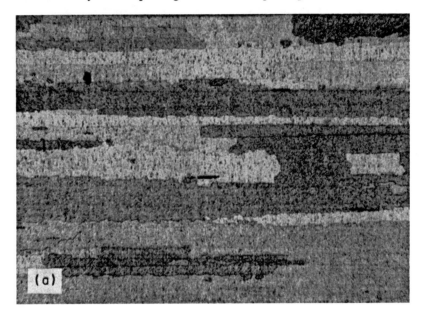

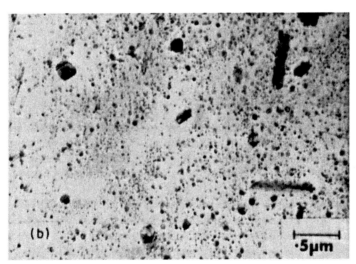

Fig. 11. MA-754 alloy extruded and recrystallized: (a) longitudinal section 100×; (b) TEM showing dispersed oxides/carbides and twins.

nature of these powders lends them to this processing. The cold compacted mass or loose powder is then containerized, heated to a high temperature, and extruded to full density. The resultant product (due to the oxide dispension) is extremely fine grained (1 μm) (Fig. 11). The extruded bar may be used in the post-consolidation heat-treated condition or further worked (forged). Regardless, the final step in processing is a recrystallization step (often directional) that results in elongated (high-aspect-ratio) grains that are very stable due to the inert oxide pinning. This step in processing is typically performed at about 2300 °F (1260 °C). Often each grain can be several inches long. This grain size and directionality contributes to the exceptional high-temperature stability and strength of these materials by minimization of transverse grain boundaries.[22] During the consolidation operation, and probably due to excess free oxygen, $Y_2O_3-Al_2O_3$ (yttria aluminate) is reported to form.

THERMOMECHANICAL PROCESSING

The goals of thermomechanical processing with specific application to P/M superalloys are several-fold. In order of priority they are:

1. Hot deformation to neutralize the effect of contamination-related defects.[23,24] These can be either intrinsic to the powders (e.g., ceramics, argon pores, and oxidized powder particles) or extrinsic, such as reactive and metallic contamination occurring during powder handling (e.g., rubber, cloth, mill scale, and cross-contaminated alloys). (For the sake of the foregoing we define *intrinsic* to mean defects that arise from the powder-making process, which includes melting and atomization. *Extrinsic* refers to defects arising from powder handling, can fabrication, etc.)
2. Cost reduction by conservation of material usage to the part. These materials show good formability due to their superplastic behavior, so near-sonic-shape forgings are possible.
3. Property generation through development of structural features such as grain flow, necklace structures, or critical deformation in the case of ODS alloys.

In general, the properties of as-consolidated superalloy powders are quite high and uniform. There is little if any improvement in static properties through forging itself, though it tends to break up defect structures due to metal flow and dynamic recrystallization. Thermomechanical processing does, however, increase certain minimum dynamic properties determined by defect content.

Forging of P/M superalloys takes several approaches, the majority of processing being isothermal forging (tooling and metal at the same temperature). For instance, for René 95 the alloy powder is consolidated to a fine grain size (subsolvus) in the 2–5-μm range and exhibits superplastic behavior[25]; that is, very low flow stresses are required at low deformation rates. The response to deformation is to reach a peak stress followed by flow softening with increasing strain at constant strain rate.

Typically, forging takes place at 1 in./in. min (range 0.1–2 in./in. min) rates with flow stresses of about 7–15 ksi at 2000 °F (1200 °C) depending on strain rate. These alloys exhibit strain rate sensitivities ($M = \partial\dot{\epsilon}/\partial\sigma$) of about 0.5, so that strain rate control during the forging operation is critical. Forging can also be carried out at higher strain rates (up to several in./in./min) on hot dies; however, yield (shape-making potential) and tooling life can suffer. Subsequent to forging these alloys are usually given a partial (subsolvus) solution heat treatment and single or multiple aging cycles.

There has been considerable work on a P/M low-carbon version of Astroloy (APKI) whose composition is shown in Table 1. This material is HIP consolidated at supersolvus conditions to yield a relatively coarse-grain structure (ASTM 5–6). The thermomechanical processing of this alloy is performed on conventional forging equipment at high strain rates (up to 1000 in./in./min.), and the resultant micro-structure consists of the so-called necklace characterized by a deformed coarse grain surrounded by a necklace of recrystallized fine grain. It is claimed that this structure shows a good combination of crack initiation and crack growth rates. This approach was attempted early with René 95 powder but was later abandoned in favor of a fully recrystallized fine-grain structure.[26,27]

The rapidly solidified alloys made by the RSP centrifugal atomization process have taken a somewhat different direction (than the R-95, IN-100, MERL-76, and LC Astroloy materials) towards their end use by Pratt & Whitney Aircraft. These materials (NiAlMo series; see Table 1) are being evaluated for turbine airfoils.[23] The powders are consolidated as above but are subsequently directionally recrystal-lized much like ODS alloys to produce high-aspect-ratio textured and aligned mi-crostructures. These materials reportedly show equivalent or superior properties to cast airfoils now used in advanced turbine engines.

The thermomechanical processing of ODS materials follows standard superalloy procedures since they are fine grained when consolidated and show superplasticity in the γ'-strengthened MA-6000 composition. Processing takes place in the 1500–2000 °F (815–1100 °C) range. The major requirement in processing is to provide critical warm deformation for subsequent recrystallization. The deformation process is usually performed on hot tooling to reduce cracking, minimize the material needs in the component, and provide necessary process control of the deformation. Sub-sequently, these materials are recrystallized at high temperatures (>2300 °F) to achieve the proper grain growth and often in a thermal gradient to enhance the desired directionality ($\langle 100 \rangle$ cube on edge texture). The MA-6000 alloy type is further heat treated like a γ'-strengthened superalloy for γ' size distribution and carbide size distribution.

MECHANICAL PROPERTIES

The monotonic mechanical properties of P/M superalloys, such as tensile, creep, and stress rupture, are directly related to alloy composition and structure. The structure, as discussed previously, is strongly influenced by powder particle size, consolidation processing, and heat treatment. Data from literature on current production alloys will be referenced for information and is shown in Table 3.[24]

Because of the continued demand for improved gas turbine engine performance and reliability, a great deal of emphasis has been focused on understanding and controlling the cyclic mechanical properties of superalloys, especially low-cycle fatigue (LCF). This critical design property in current P/M superalloys is largely controlled by the presence of defects in the material. For the purposes of this discussion, a defect is defined as any material heterogeneity that results in initiation of fatigue cracks by any mechanism other than crystallographic initiation.

Because of unique processing and application requirements, the ODS material properties will be discussed separately.

Structural Effects: Grain Size

One of the positive attributes of powder superalloy structures is the fine uniform grain size. Typical grain size of current production P/M superalloys is in the range of ASTM 7–12. Figure 10 illustrates the microstructure of typically cast and wrought René 95, HIPed René 95, and extruded + isothermally forged P/M René 95. The HIPed P/M material (ASTM 8) has finer and more uniform grain size than the c & w structure (ASTM 3–6), and thermomechanical working after extrusion further refines the grain structure (ASTM 11–12). A mechanical property comparison of these three materials is shown in Table 4.[25] Tensile strength and ductility increase as grain size becomes finer, in full agreement with Petch.[26]

Heat Treatment

The effect of grain size is even more pronounced when coupled with heat treatment. Table 5 shows the effect of heat treatment and grain size on P/M René 41 made by the Rotating Electrode Process (REP) and extruded at a reduction ratio of 12:1.[27] The rupture life increased fivefold as the solution temperature was raised from 2050 °F (1120 °C) to 2200 °F (1200 °C). The 0.2% yield strength, on the other hand, dropped approximately 18%. Another example (Table 6) is for René 95 heat treated above the γ' solvus temperature (2110 °F), showing that both the grain size and stress rupture life increase while tensile yield decreases. The increase in grain size after supersolvus heat treatment is expected since the dissolved grain boundary γ' permits grain growth.

The properties of γ'-strenthened superalloys such as René 95 also are sensitive to cooling rates from the solution temperature. The cooling rates are controlled by quench media and material section thickness. Chang et al.[28] showed that cooling rate affects γ' size and tensile and creep rupture properties, as in René 95 disks heat treated at three conditions:

1. 2050 °F/h salt quench + 1600 °F/h air cool + 1200 °F/24 h air cool.
2. 2000 °F/h oil quench + 1400 °F/16 h air cool.
3. 2050 °F/h air cool + 1400 °F/16 h air cool.

Table 3. Mechanical Properties of PM Superalloy[a24]

| Alloy | Tensile | | | | | Stress Rupture | | |
	Temperature (°F)	UTS (ksi)	0.2% YS (ksi)	Elongation (%)	R/A (%)	Temperature (°F)	Stress (ksi)	Average Life (h)
U-700[b]	1400	149	148	20	28	1400	85	25
IN-100[c]	1300	184	154.5	20	21	1400	95	35
René 95[d]	1200	218	165	13.5	14.9	1200	150	54
Astroloy (L.C.)[e]	1200	192	142	25.6	25.9	1400	92	89

[a] Abbreviations: UTS, ultimate tensile strength; YS, yield strength.
[b] HIP + heat treated.
[c] Minus 100 mesh, extruded and gatorized.
[d] Minus 150 mesh, 2050 °F HIPed and heat treated.
[e] At 1150 °F HIP, 4 h/1975 °F oil quench + 24 h/1200 °F air cool + 8 h/1400 °F air cool.

Table 4. Heat Treatments, Grain Size, and Tensile Properties of René 95 Forms[a]

Alloy Form	E&F[b]	HIP[c]	C&W[d]
Heat treatment	1120 °C/1 h AC +760 °C/8 h AC	1120 °C/1 h AC +760 °C/8 h AC	1220 °C/1 h AC +1120 °C/1 h AC +760 °C/8 h AC
Grain size (μm)	5 (ASTM 11)	8 (ASTM 8)	150 (ASTM 3–6)
100 °F tensile properties			
0.2% yield strength (ksi)	165.4	162.4	136.4
Ultimate tensile strength (ksi)	226.3	226.3	175.5
Elongation (%)	8.6	16.6	8.6
Reduction of area (%)	19.6	19.1	14.3
1200 °F tensile properties			
0.2% yield strength (ksi)	165.4	159.5	134.7
Ultimate tensile strength (ksi)	217.6	217.6	181.3
Elongation (%)	12.4	13.8	9.0
Reduction of area (%)	16.2	13.4	13.0

[a]From ref. 25. AC = air cool.
[b]Processing: −150 mesh powder, extruded at 1900 °F (1070 °C) to a reduction of 7 to 1 in area isothermally forged at 2012 °F (1100 °C) to 80% height reduction.
[c]Processing: −150 mesh powder, HIP at 2050 °F (1120 °C) at 15 ksi (100 MPA) for three hours.
[d]Processing: cross-rolled plate, heat treated 2225 °F (1218 °C) for 1 h.

Table 5. Elevated-Temperature Mechanical Properties of REP René 41 [a]

Extrusion Temperature	AMS5713B	Solution Treatment 2050 °F, 0.5 h AC				Solution Treatment 2125 °F, 2 h AC			Solution Treatment 2200 °F, 4 h AC
		1850 °F	1900 °F	1950 °F	2000 °F	1850 °F	1900 °F	1950 °F	2000 °F
1400 °F Tensile									
UTS (ksi)	135	144	142	144	143	146	143	143	142
0.2% YS (ksi)	105	136	134	136	134	123	123	126	111
Elongation (%)	5	21.9	23.6	15.1	20.6	28.5	28.2	20.6	25.5
RA (%)	8	35.3	37.1	19.6	27.3	34.7	35.4	27.0	33.9
Stress rupture (1650 °F, 25 ksi)									
Life (h)		11.7	4.9	4.7	12.0	51.8	34.0	29.6	68.6
Elongation (%)		29.4	25.0	31.6	20.4	15.9	18.3	18.8	11.5
RA (%)		36.8	33.4	39.4	29.9	30.9	29.9	27.9	30.7
ASTM Grain Size		12	11	11	9.5	7	8	9	6.5

[a] All samples aged 1650 °F, 4 h AC. Abbreviations: AMS, Aerospace Material Specification; UTS, ultimate tensile strength; YS, yield strength; RA, reduction of area; AC, air cool.

Table 6. PM René 95 Properites

| Alloy | Grain Size | 1200 °F Tensile | | | | Stress Rupture (h) 1200 °F, 140 ksi |
		UTS (ksi)	0.2% Yield	Elongation (%)	RA	
HIPed René 95[a]	ASTM 8	218	165	13.8	15	87,150
HIPed René 95[b] with supersolvus heat treatment	ASTM 6	205	145	16	19	244,298

[a]Processing: −150-mesh powder, HIPed at 2050 °F, 15 ksi/3 h 2050 °F/1 h + 1600 °F/1 h + 1200 °F/24 h.
[b]Processing: −150-mesh powder, HIPed at 2050 °F, 15 ksi/3 h 2200 °F/1 h + 2050 °F/1 h + 1600 °F/1 h + 1200 °F/24 h.

The "solution" temperature in each of these heat treatments is below the γ' solvus (approximately 2110 °F). This prevented the extensive grain coarsening associated with complete solutioning of the γ' phase. Figure 12 shows TEM replica micrographs that illustrate primary (eutectic) and intermediate γ' features. The primary γ' sizes were similar and ranged from 1 to 3 μm in size. The volume fraction of primary γ', however, increased with decreasing solution temperature. Intermediate γ' forms on cooling from the solution temperature. The more rapid the cooling, the finer the size and distribution. Heat treatment C (Table 7) air cooled from the annealing temperature, resulted in the largest intermediate γ', measuring approximately 0.2 μm in size. The quenched materials had smaller intermediate γ', measuring approximately 0.05 μm in size. Tensile test data (Fig. 13) show that slow cooling results in lower strength but higher ductility.

Trends in rupture and creep strength (Table 7) are consistent with that observed from the tensile study; slow cooling rate results in lower creep and rupture properties.

Defects and Fatigue

While P/M processing provides property improvement and cost-effectiveness over cast and wrought processing, the low-cycle fatigue (LCF) capability of P/M materials has been limited by the presence of defects introduced during various processing steps. Their nature, size, and distribution are dependent on the powder size and consolidation technique.

Defects in P/M alloys have been extensively studied for HIP-consolidated René 95. This experience has principally resulted from a large-scale study of LCF specimen fatigue origins.[25,27] Table 8 summarizes the characteristics of four major types of defects found at the failure initiation sites of HIP René 95 LCF specimens produced from −150-mesh powder and processed similarly to Table 4 condition.

Types 1 and 2 ceramic inclusions are illustrated in Fig. 14. Type 1 defects exist as discrete and often chunky particles, while type 2 are agglomerates of fine particles. Both types originate from the melting crucible, pouring tundish, or atomizing nozzle. Aluminum and hafnium (in Hf-bearing alloys) are usually the major metallic elements

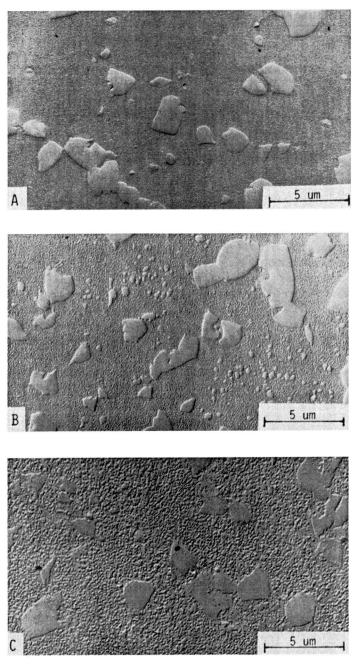

Fig. 12. TEM replications of René 95 showing the effect of heat treatment on γ′ size: (a) 2050 °F/ 1 h/ salt +1600 °F/1 h/AC +1200 °F/1 h/AC; (b) 2000 °F/1 h/oil +1400 °F/16 h/AC; (c) 2050 °F/ 1 h/AC +1400 °F/16 h/AC

482

Table 7. EXT + Iso-René 95 Disk Heat Treat Study Creep and Rupture Test Data[a]

Forging S/N	Location/ Orientation	Temperature (°F)	Stress (ksi)	Total Strain at 50 h (%)	Rupture Life (h)	Heat Treatment[b]
1	Rim-tang.	1200	150	—	85.2	A
1	Rim-tang.	1200	123	0.02		A
2	Rim-tang.	1200	150	—	107.7	B
2	Rim-tang.	1200	123	0.02		B
3	Rim-tang.	1200	150	—	18.6	C
3	Rim-tang.	1200	123	0.16		C

[a] From ref. 28.

[b] Heat treatments: A = 2050 °F/1 h salt quench, 1600 °F/1 h air cool + 1200 °F/24 h air cool. B = 2000 °F/1 h oil quench, 1400 °F/16 h air cool. C = 2050 °F/1 h air cool, 1400 °F/16 h air cool.

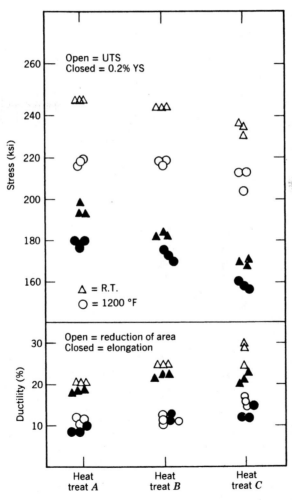

Fig. 13. EX+F René 95 disk heat-treat study tensile results.

Table 8. Comparison of Defect Type versus Size at LCF Initiation Site[a] HIP versus H+F versus EX+F[b] René 95

Temperature (°F)	Total Strain Range (%)	Defect Type[c]	Defect Size (mils²)					
			HIP		H+F		EX+F	
			Average	Maximum	Average	Maximum	Average	Maximum
1000	<0.75	1	6.7	36	5.3	16.0	4.8	9.8
		2	9.5	50	6.1	11.2	8.9	12.1
		3	18.7	242	None	None	None	None
		4	6.3	10.3	None	None	None	None
1000	>0.75	1	11.2	50	3.9	12.0	4.8	8.1
		2	13.1	111	4.8	13.0	7.1	16.0
		3	19.3	83.6	4.2	4.5	None	None
		4	5.3	6.8	0.8	1.4	1.0	1.5
750	0.75	1	7.5	9.1	6.4	13.7	4.1	8.7
		2	8.6	13.1	4.2	9.3	5.2	10.2
		3	6.4	6.4	None	None	None	None
		4	None	None	1.1	1.5	0.5	0.7

[a] From ref. 28.
[b] HIP, as HIP; H+F, HIP + isoforge; EX+F, extrusion + isoforge.
[c] 1, ceramics; 2, ceramic clusters; 3, prior particle boundaries; 4, voids.

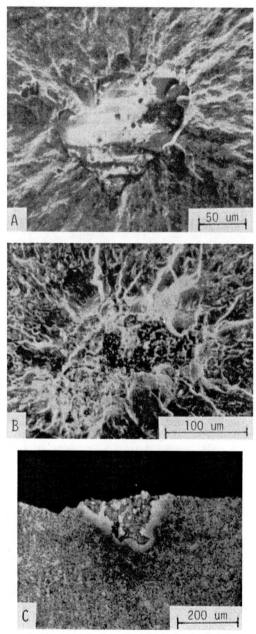

Fig. 14. Micrographs showing ceramic: (a) type 1 SEM fractograph showing fatigue initiation site; (b) type 2 SEM fractograph showing fatigue initiation site; (c) type 3 optical micrograph from metallographic section.

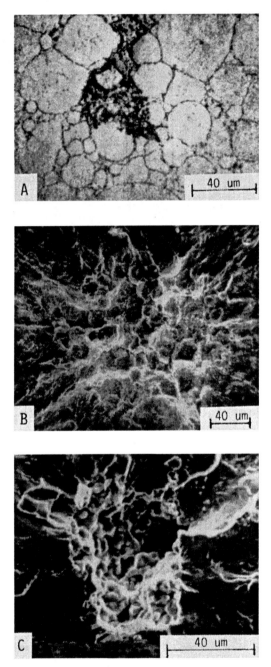

Fig. 15. Micrographs showing (a) type 3 (PPB) optical micrograph from a metallographic section; (b) type 3 (PPB) SEM fractograph showing fatigue initiation site; (c) type 3 (void) SEM fractograph showing fatigue initiation site.

with zirconium, magnesium, and calcium also present in various amounts, depending on the type of melt refractory used. The defect sizes are limited by the screen opening through which the powders have been sieved and are not altered significantly by reaction during the HIP cycle. As indicated in Table 1, ceramic defects average about 10 mils^2, with a very few reaching 50–100 mils^2.

Type 3, or prior particle boundary (PPB), defects are "diffused" and generally occupy a much larger volume of material than type 1 or 2 defects. As shown in Fig. 15, type 3 defects are characterized by semicontinuous network of fine oxide or carbide particles around the host powder surfaces. A residual core of the initial contamination source may be present at the center of the affected area. Figure 15a illustrates an example of a PPB defect with a core of refractory remaining. PPB defects generate ball and socket features as shown in the fatigue origin fractograph of Fig. 15b. Formation of type 3 defects appears to require a source of oxygen or carbon that reacts with powder surfaces during the HIP cycle. Depending on the nature and amount of the contaminant, PPB defects can range from several mils squared to orders of magnitude larger than the maximum size of 250 mils^2 shown in Table 8. Consequently, PPB defects are the most detrimental to dynamic properties of P/M superalloys.

Since the powders are produced by argon atomization, argon is occasionally entrapped in the powder particles, giving rise to voids or type 4 defects. An example of a type 4 defect is shown in Fig. 15c. These defects seldom exceed 10 mils^2 from −150-mesh material. The voids could be significantly larger when powder size increases and could cause much more scatter in LCF life.

The effect of defects on LCF life depends on defect size and location in the LCF specimen. This is illustrated in Fig. 16 for HIPed René 95 tested at 1000 °F (540 °C). Note that for a given size a defect is less detrimental when situed internally than when it is located at the surface. The less severe effect of internal defects is largely due to the absence of adverse environmental interactions.

Effect of Thermomechanical Working on LCF

It has been observed that forging of consolidated P/M René 95 is beneficial to LCF life.[25,27] Miner and Gayda[25] show that at high strain ranges, the LCF behavior of HIPed, extruded plus forged (EX + F), and cast plus wrought René 95 are similar. While at strain ranges less than 1.0% both P/M René 95 show longer LCF life than cast and wrought René 95, which is attributed to grain refinement. However, the extruded and forged material showed the longest life, as shown in Fig. 17.[27] The beneficial effects of forging are twofold: the defects are dispersed during processing and their size may be reduced, and the grain size is further refined. By proper control of the TMP process, the effects of the most detrimental PPB defect can be substantially reduced or eliminated. This is evident by the results of failure analysis at LCF origins shown in Table 8. The data show the reduced average defect size and lack of PPB-type defects in the thermomechanically processed material. The LCF life of a thermomechanically processed P/M material now appears to be limited by the presence of small ceramic defects.

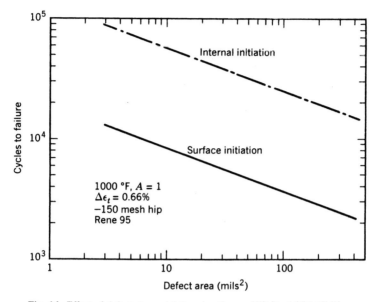

Fig. 16. Effect of defect size and failure location on HIP René 95 LCF life.

Crack Propagation

Since the fatigue life of current P/M superalloys is limited by the presence of defects, the investigation of crack propagation of aerospace engineering materials has increased. Most of the studies assume a certain starting defect size and investigate the crack propagation rate and residual fatigue life. Work by Miner and Gayda[25] and Van Stone et al.[29] have shown that grain size plays a predominant role in determining the Stage II crack propagation rate (see Fig. 18). Coarse grain material typically has lower Stage II crack propagation.

Properties of ODS Material

ODS material is strengthened at high temperature by an ultrafine dispersed oxide such as Y_2O_3, where the conventional γ'-strengthening phase becomes unstable. The mechanical properties of engineering ODS materials such as MA-754 and MA-6000 are obtained by controlled thermal mechanical processing. These TMP processes are designed to create a stable, recrystallized grain structure that is coarse and highly elongated in the direction of hot working. Grain aspect ratios may be as high as 10:1. The highly directional structure results in varying degrees of anisotropy in mechanical and physical properties. For applications that require good thermal fatigue resistance, such as gas turbine vanes, MA-754 is given strong texture with a ⟨100⟩ crystallographic direction parallel to the working direction. The texture results in a low modulus of elasticity in the longitudinal direction. The low modulus improves thermal fatigue resistance by lowering stresses for a given thermal strain.

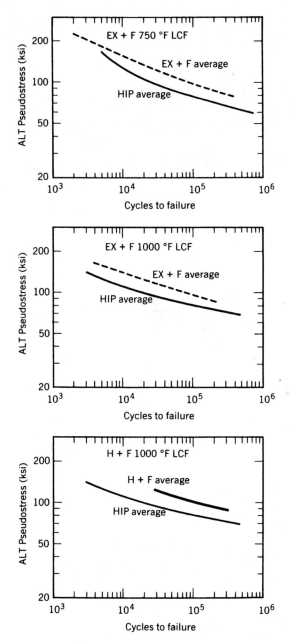

Fig. 17. Comparison of average LCF lives of HIP versus HIP and forge and EX+F René 95.

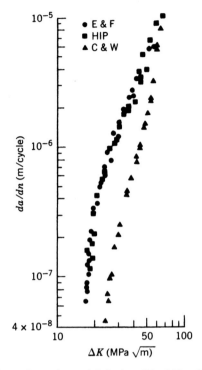

Fig. 18. Cyclic crack growth behavior of René 95 at 540 °C.

Table 9. ODS Material Properties[a]

| Alloy(s) | 2000 °F Tensile | | | | Rupture Capability of Temperature (Stress to Produce 1000 h Life) (ksi) |
	UTS (ksi)	0.2% Yield (ksi)	Elongation (%)	RA	
MA-754					(2000 °F)
Longitudinal	21.5	19.5	12.5	24	13.6
Long Transver	19	17.5	3.5	1.5	3.5
MA-956					(1800 °F)
Longitudinal	13.2	12.3	3.5	—	9.7
Transver	13.0	12.0	4.0	—	9.1
MA-6000					(2000 °F)
Longitudinal	32.2	27.8	9.0	31.0	16
Transver	25.7	24.7	2.0	1.0	—

[a] From ref. 30.

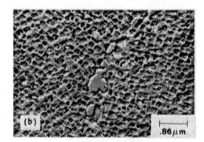
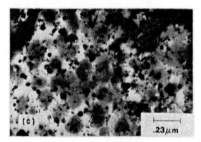

Fig. 19. Combined coarse elongated grain structure, γ' precipitates, and Y–Al Oxide Particles in Inconel alloy MA-6000. (a) Optical macrograph of bar showing elongated grains; (b) replica micrograph showing precipitate; (c) transmission electron micrograph showing oxide dispersion and γ' precipitates.

The anisotropy is apparent in short-time tensile properties but becomes much more pronounced in the time-dependent, creep and rupture properties. The longitudinal and long transverse properties of MA-754 and some other alloys are shown in Table 9. The long transverse data shows about 5% reduction in tensile, $5\times$ reduction in tensile reduction of area and $2-3\times$ reduction in stress rupture strength at 2000 °F.

Other ODS alloys such as MA-956 were developed for high-temperature sheet application. This is achieved by taking advantage of the excellent oxidation resistance of the alloy. MA-6000 was developed to combine the high-temperture strength of an ODS alloy with the intermediate-temperature strength of γ' precipitate. Typical mechanical properties of these alloys are also shown in Table 9, and the structures are shown in Fig. 19.

FUTURE WORK AND SUMMARY

From this discussion it is obvious that in a restricted sense powder metallurgy technology is a cost-effective means to produce highly alloyed compositions with desirable mechanical properties.

For gas turbine applications the limiting property is still defect-degraded fatigue life. Further efforts in processing are needed to eliminate these defects. This can be achieved by using clean melted input stock material followed by a ceramicless powder atomization process. Developments are also necessary for process monitoring, process modeling, and process control in order to assure the quality of the product as produced rather than later inspected. Since the LCF life is also limited by surface-initiated failure, surface enhancement techniques capable of generating a surface compressive stress layer to inhibit surface failure should be applied. This concept has been effectively demonstrated by a number of investigators.[31,32] Alloy development activities to modify composition and improve the intrinsic defect tolerance of the materials are also needed. These compositional modifications must consider and balance all other properties for engine applications. In addition, new innovative consolidation techniques may be necessary to take advantage of the unique structure produced by rapid solidification processes. As powder quality, availability, and manufacturing techniques improve, implementation is expected to increase significantly.

Current applications of powder-based processing are expected to increase as turbine engine requirements become more demanding. The potential for new applications, for example, in dispersion-strengthened systems or new alloys such as the NiMoAl series is good, though the competing technologies and materials such as ceramics and ceramic composites will make the requirement for cost-effectiveness intrinsic.

Powder-processed superalloys have been in production for over 15 years, and many of the current limitations placed on their processing or use were not envisioned then. There was no experience to draw upon 15 years ago. The new applications will be developed from an existing knowledge base that will make realistic assessment of their performance potential possible.

REFERENCES

1. C. T. Sims and W. C. Hagel, *The Superalloys*, Wiley, New York, 1972.
2. K. Moyer, "An Improved Manufacturing Process for the Production of Higher Purity Superalloy Powders," Air Force Materials Laboratory, Wright-Patterson AFB, Ohio, AFML TR-69-21, February 1969.
3. Metal Progress, "Technology Forecast," Powder Metallurgy Section, ASM, Metals Park, Ohio 1969–1975.
4. R. L. Dreshfield and H. R. Gray, in *1984 International Powder Metallurgy Conference*, MPIF, Metal Powder Industries Federation, Princeton, NJ.
5. S. H. Reichman and J. W. Smythe, in *1969 Powder Metallurgy Conference Proceedings*, APMI, p. 829, New York, NY, 1970.
6. N. J. Grant and R. M. Pelloux, in *Rapid Solidification Technology Source Book*, ASM, 1983, pp. 362, 369.
7. P. Roberts, Nuclear Metals, private communication.
8. P. R. Holiday et al., and R. J. Patterson "Apparatus for Producing Metal Powders," U.S. Patent 4,078,873, 1978.
9. R. D. Field, S. J. Hales, W. O. Powers and H. L. Fraser, in *Superalloys 1984*, TMS-AIME, Warrendale, PA, 1984, p. 487.
10. L. R. Curwick, in *Frontiers of High Temperature Materials*, J. S. Benjamin (ed.), Inco Alloy Products, New York, NY, 1981, p. 4.
11. J. Nutting et al., in *Frontiers of High Temperature Materials*, J. S. Benjamin (ed.), Inco Alloy Products, New York, NY, 1981, p. 35.
12. P. G. Bailey, in *Frontiers of High Temperature Materials*, J. S. Benjamin (ed.), Inco Alloy Products, New York, NY, 1981, p. 62.

13. R. D. Field, A. R. Cox, and H. L. Fraser, in *Superalloys 1980*, ASM, Metals Park, OH, 1980, p. 441.
14. C. Aubin, J. H. Davidson, and J. P. Trottier, in *Superalloys 1980*, ASM, Metals Park, OH, p. 353.
15. H. E. Mobius, in *P/M Superalloys: Current and Future*, 1984 International Powder Metallurgy Conference, MPIF.
16. R. D. Eng and D. J. Evans, in *Superalloys 1980*, ASM, Metals Park, Ohio, p. 498.
17. D. R. Chang, D. D. Kruger, and R. A. Sprague, *Superalloys 1984*, TMS-AIME, Warrendale, PA, p. 271.
18. R. T. Thamburaj, W. Wallace, Y. N. Chari, and T. L. Prakash, *Powd. Metall.*, **27**(3), 171 (1984).
19. R. D. Kissinger, S. V. Nair, and J. K. Tien, in *Superalloys 1984*, TMS-AIME, Warrendale, PA, 1984, p. 294.
20. T. E. Howson, W. H. Couts, and J. E. Coyne, in *Superalloys 1984*, TMS-AIME, Warrendale, PA, 1984, p. 280.
21. C. E. Shamblen, R. E. Allen, and F. E. Walker, *Met. Trans. A*, **6A**, 2073 (November 1975).
22. M. W. Cockell and K. A. G. Boyce, *Met. Powd. Rev.*, **40**, No. 3 139 (March 1985).
23. R. J. Patterson II, A. R. Cox, and E. L. Van Reuth, in *Rapid Solidification Technology Source Book*, ASM, 1983, p. 417.
24. *Proceedings of a Metal Powder Conference at Movenpick, Zurich*, November 18–20, 1980, Metal Powder Rept., Shrewsbury, U.K. Papers 11, 16, and 17.
25. R. V. Miner and J. Gayda, *Int. J. Fat.*, **6**(3), 189 (1984).
26. N. J. Petch, *J. Iron Steel Inst.*, **174**, 25 (1953).
27. G. I. Friedman and G. S. Ansell, in *The Superalloys*, C. T. Sims and W. C. Hagel (eds.), Wiley, New York, 1972, p. 427.
28. D. R. Chang, D. D. Kruger, and R. A. Sprague, in *Superalloys 1984*, TMS-AIME, Warrendale, PA 1984 (see ref. 20).
29. R. H. Van Stone, D. D. Kruger, and L. T. Duvelius, in *Fracture Mechanics: Fourteenth Symposium*, Vol. II, STP 792, 1983. American Society for Testing and Materials, Philadelphia, PA.
30. Inco Map Technical Brochure, 1983.
31. W. Renzhi, in *First International Conference on Shot Peening, Paris*, September 1981, Pergamon, Oxford, p. 395, 1981.
32. B. D. Boggs and J. G. Byrne, *Met. Trans.*, **4**, 2153 (1973).

Chapter 18

Joining

WILLIAM YENISCAVICH

Westinghouse Electric Company, Pittsburgh, Pennsylvania

The evolution of modern superalloys has coincided with the evolution of the gas turbine engine for the aircraft industry. The rate of development of new alloys has been rapid and, in many respects, surpassed the development in companion joining techniques. Traditionally, emphasis in alloy development has been on high-temperature strength, stress rupture, and oxidation properties. Joining has been treated as a separate problem by welding and brazing engineers who were usually required to devise procedures and techniques to join each new alloy after it had been developed. This has resulted in exaggerated fabrication costs for some alloys because of unpredictable and uncontrolled metallurgical variations causing such problems as cracking during welding.

This chapter primarily treats the *welding* of superalloys. Space limitations, unfortunately, inhibit an equivalent discussion on *brazing*.

ADVANTAGES OF WELDING

Despite many difficulties, welding has been and will continue to be one of the main fabrication techniques for aircraft and land-based engine components. Welding permits fabrication of economical size subcomponents with essentially no additional weight and at moderate cost. Welded joints cause relatively little reduction in service capabilities, particularly if they are placed in noncritical locations. The tolerances capable of being held on welded components are close if proper fixturing and welding techniques are used.

PROBLEMS ASSOCIATED WITH WELDING

The main problem encountered in the welding of superalloys is cracking and fissuring. A crack is a large planar separation visible to the unaided eye. A fissure, on the other hand, is a small crack that is usually only detectable by metallographic examination. The prevention of defects is one of the most challenging problems in the welding of superalloys. Many superalloys, like the casting grades 713C and B-1900, have such high fissuring sensitivity that it is impossible to make fissure-free fusion welds.

A second problem associated with welding of superalloys is reduction in mechanical properties. Generally, techniques can be used that do not cause a significant reduction in tensile or yield strengths; however, ductility of welded specimens is practically always reduced. This is because the structure of solidified weld metal is segregated and less ductile than an equivalent wrought structure. The segregation that occurs in solidified weld metal could also cause a reduction in oxidation resistance. If high-electron-vacancy elements segregated on solidification, they could cause σ or other embrittling phases to precipitate during welding or after being put in service. Each alloy must be examined individually to assess the degradation in properties that may result from welding. Postweld heat treatments can be helpful in reducing segregation, but effective heat treatments are often difficult to perform on large fabrications.

Regions in the heat-affected zone are exposed to high temperatures and can undergo exaggerated grain growth, solutioning, and reprecipitation of carbides and other precipitates. These changes may cause deterioration of properties, such as corrosion and oxidation resistance, and must also be evaluated individually.

Weld reinforcements, that is, weld overbeads and underbeads, must be avoided in situations where fatigue is the mode of failure. Fatigue strength reduction factors of 2.25–2.50 have been reported for weld underbeads.[1,2] The best method for avoiding a fatigue problem is to put the weld joint in a low-stressed area.

WELDING PROCESSES

There are 45 distinct welding processes listed in the Welding Engineering Data Sheets.[3] In superalloy welding the most common are illustrated in Fig. 1: shielded-metal arc, gas–tungsten arc, gas–metal arc, resistance, and electron beam. The best general references on these processes and on specific details of welding superalloys are the welding handbooks.[4–8]

The purpose of these welding processes is to generate heat in a localized area and thereby cause melting and joining of two pieces of metal. Although processes differ significantly from each other, from an overview they are just different ways

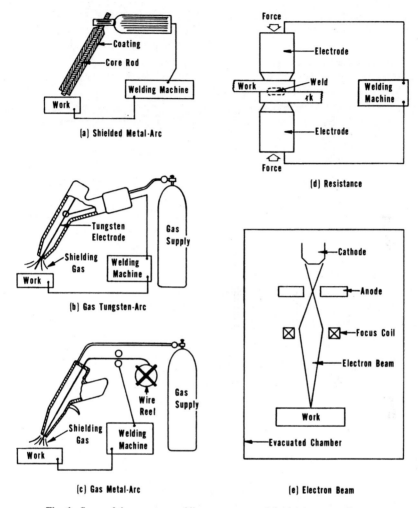

Fig. 1. Some of the common welding processes used for joining superalloys.

of generating localized heating. However, when welding superalloys, a general rule is to use as small a heat input as possible and to keep the interpass temperature (temperature of workpiece before a following weld bead is deposited) as low as possible. This will help prevent cracking.

Shielded-Metal Arc

Shielded-metal arc welding is an arc-welding process where heat is produced with an electric arc between a coated metal rod called an electrode and the workpiece. Shielding is obtained from decomposition of the electrode coating, and filler metal is obtained from the electrode's metal core. Wide use is made of this technique in the welding industry, but it is used relatively less in superalloy fabrication because of the difficulty of flux removal, the difficulty in welding thin sections, and the inability to automate this welding process. In general, the lower limit of sheet thickness that can be welded by shielded-metal arc welding is 0.037 in. for jigged structures and 0.062 in. for welding without jigging or backing. Types of nickel-base electrodes available and the suppliers have been summarized.[9] Cobalt- and iron-base superalloy electrodes can be obtained by contacting suppliers.[10,11]

Gas–Tungsten Arc

Gas–tungsten arc welding is an arc-welding process wherein heat is produced by an arc between a single tungsten electrode and the workpiece. Common gases are argon and helium. Filler metal, if used, is preplaced in the weld joint or fed into the arc from an external source during welding. This is by far the most popular welding technique for superalloys. It is a clean process, and thin sections can be easily welded. A modification of this process called the plasma needle arc[12] allows small stable welding currents and permits welding of 0.010-in.-thick foils. The gas–tungsten arc process is also readily adaptable to automatic welding. Nickel-base filler wires and their suppliers for this process have been summarized.[13] Cobalt- and iron-base filler wire can be obtained by contacting suppliers.[10,11]

Gas–Metal Arc

The gas–metal arc welding process is similar to the gas–tungsten arc process except that a consumable filler metal electrode is used instead of the tungsten electrode. The consumable electrode, usually in coil wire form, is fed through the welding torch and provides filler metal for making the welded joint.

Resistance

Resistance welding processes develop heat at the joint to be welded by short-time flow of low-voltage, high-density electric current through the interface being joined. Force is applied before, during, and after application of current to assure a continuous electrical circuit and to forge the heated parts together. The maximum temperature achieved is usually above the melting point of the base metal. However, for dispersion-strengthened alloys preliminary work has been done where the maximum temperature is kept below melting to prevent agglomeration of the dispersion.

Electron Beam

Electron beam welding usually takes place in an evacuated chamber in which the beam-generating and beam-focusing devices, as well as the workpiece, are in this vacuum environment. Welding in a chamber imposes several limitations but, at the same time, provides a pure inert environment in which metal may be welded without fear of contamination. Electron beam welding is also practiced in medium vacuum and at atmospheric pressure. However, the beam is always generated in high vacuum, and it is only the workpiece that is at these higher pressures.

An outstanding feature of electron beam welding is its ability to make exceedingly narrow, deeply penetrated welds. A weld 0.500 in. deep and 0.060 in. wide is not uncommon in superalloys. The process is adaptable to many applications not feasible with conventional processes, the most important of which are thick-to-thin materials and dissimilar metals. It is also important in complex structures where little distortion can be tolerated, structures where welding is required in deep holes, deep grooves, and other relatively inaccessible places. It is an easily automated process, but it has the highest capital investment for equipment and fixtures of all the welding processes.

AUTOMATION OF WELDING

The shielded-metal arc process using coated electrodes is a manual welding process; no significant attempts have been made to automate this process. Capital investment in equipment is low, and operator skill required is high. The gas–tungsten arc and gas–metal arc welding processes can be operated either manually or automatically. Semiautomatic welding consists of attaching the welding torch to a carrier, which will track it over the weld joint, attaching travel speed controls and electrode arc gap controls. Further additions, such as programmed tracking, programmed voltage changes, and programmed amperage changes, result in a fully automated welding process. Resistance welding and electron beam welding processes, by their nature, are either semiautomated or fully automated welding processes.

The previous automated techniques are based on preprogramming the required information before starting to weld. This technique has been referred to as the numerical control technique. Recently, a new automated approach to welding has been applied, called the adaptive control technique. In this case response parameters such as weld bead width are monitored during the welding process and are used to continually adjust the input parameters, such as welding amperage, to give the preset response. These adaptive control techniques can be of great use to the superalloy fabricator in certain applications.

DESCRIPTION OF A WELD

In their simplest description, welds consist of only two regions: weld metal and heat-affected zone. Savage[14] has expanded the concept of a weld into four regions. These regions of the weld are shown in Fig. 2 and are defined as follows:

1. *Composite Region.* The bulk portion of the weld metal, within which stirring causes the chemical composition to be modified by dilution of the filler metal with material melted from the surrounding base metal.

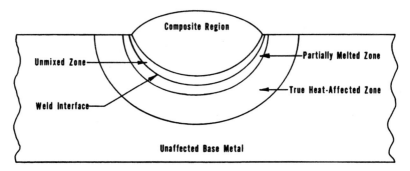

Fig. 2. Proposed terminology and definitions of discrete regions in welds.[15]

2. *Unmixed Zone.* A boundary layer at the outer extremities of the weld metal, consisting of base metal that is melted and solidified during welding without experiencing mechanical mixing with filler metal.

3. *Weld Interface.* The *surface* bounding the region within which complete melting was experienced during welding as evidenced by presence of a definite solidification substructure.

4. *Partially Melted Zone.* That portion of the base metal located just outside the weld interface, within which the portion melted ranges from 0 to 100%.

5. *True Heat-Affected Zone.* That portion of base metal within which all microstructural changes produced by welding occur in the solid state.

RAPID THERMAL CYCLE

Most metallurgical studies on superalloys have been conducted under slow thermal cycling conditions, that is, at or near equilibrium. In the case of welding, however, rapid thermal cycles are the normal course of events, and metallurgical changes occurring during rapid thermal cycles can be considerably different. Knowledge of these changes is necessary for understanding phenomena such as cracking.

Depression of Solidus

First consider the effect of rapid freezing on solidification. The phenomenon of segregation during solidification is well known; that is, the solid will be of a different composition than the bulk liquid from which it is freezing. If the equilibrium partition coefficient is less than unity, the solid will have less solute than the liquid when freezing, and this will, in turn, cause solute enrichment of the liquid. This results in a lowering of freezing temperatures as the remaining liquid is enriched. In alloy systems exhibiting a eutectic, the liquid may be enriched until the eutectic composition is reached and further lowering of freezing temperature is impossible. Hence a liquid that would freeze at a relatively high temperature under equilibrium conditions

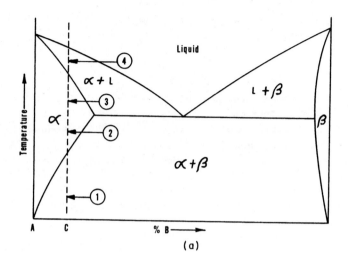

(a)

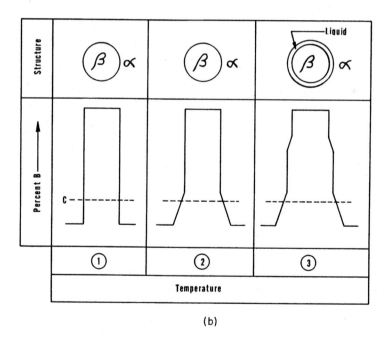

(b)

Fig. 3. (a) Phase diagram for a simple binary alloy. (b) Structures and distribution of B in alloy C at three temperatures shown in (a) when rapidly heated.

could, by solute segregation during rapid freezing, have liquid remaining to much lower temperatures.

Constitutional Liquation

Another phenomenon that can occur during rapid heating is localized melting. Figure 3a shows a phase diagram for a simple binary system A and B. An alloy of nominal composition C is shown on the diagram, and its equilibrium structures are indicated at various temperatures: at temperature 1 two phases $\alpha + \beta$ are present; at temperatures 2 and 3 a single α phase; and at temperature 4 a two-phase $\alpha +$ liquid structure.

Now suppose this alloy is heated rapidly starting at temperature 1. Figure 3b shows that at temperature 1 a two-phase $\alpha + \beta$ structure exists and has a solute distribution of B as shown. At temperature 2 equilibrium would predict only the α phase. However, since the alloy is being heated rapidly, bulk equilibrium is not obtained. Local equilibrium exists at the $\alpha-\beta$ interface, and dissolution of the β particle is in progress. The distribution of solute under these conditions is shown. If temperature 3 is reached before the β particle has gone completely into solution, a liquid phase will form. The phase diagram shows that a liquid phase must exist between α and β at temperature 3. Hence, if the two phases are present because of rapid heating and local equilibrium prevails, a liquid interface must exist between the two phases. This phenomenon of localized melting due to a nonequilibrium distribution of phases during rapid heating is called "constitutional liquation."[16]

MEASURING CRACKING SENSITIVITY

There are numerous methods used to measure cracking sensitivites of different alloys and between heats of the same alloy. These methods can be divided into five basic types: production welds, mockups, weld-restraint crack tests, variable-strain weld tests, and nonwelding tests.

Production Welds

When components being welded have relatively little value, the most practical test is to do actual production welding. If welds crack, the component is scrapped. This approach cannot be used on components where expensive machining or finishing has been done before it is welded.

Mockups

In many cases the critical features of an expensive component can be duplicated in a less expensive mockup. In this case the mockup can be used to develop welding parameters and can also be used to assess cracking sensitivity if it is made from the same heat of alloy as the production component. Mockup tests are most meaningful for predicting production behavior, but they are usually still somewhat expensive.

Weld-Restraint Cracking Tests

A large variety of weld-restraint cracking tests are available and have been summarized.[17] In general, these tests consist of some rigid restraining method that develops stress as a weldment is made. "Crack" or "no crack" is most often the response parameter measured. If the restraint developed in the test selected exceeds that which will be developed in the production application, the test selected can be used as a go/no-go test for incoming materials. For these applications it is very suitable, but on the other hand it gives very little quantitative information on the mechanism of cracking.

Variable-Strain Weld Tests

There are several tests where the strain developed during welding is completely controlled and varied in incremental amounts by the investigator. The most notable of these tests is the Varestraint test developed by Savage and Lundin.[14] Briefly, the test utilizes a plate-type specimen supported as a cantilever beam. A weld is made along the length of the plate, and as the arc passes a predetermined point, the plate is bent to conform to the curvature of the top surface of a removable die block. Blocks of different radii are used to develop different levels of augmented strain. In this test the augmented strain required to initiate cracking can be determined as well as the tendency for crack propagation to occur with increased strain.

Nonwelding Tests

Tests to predict weld-cracking sensitivity or weldability that do not include making a weld are ambitious from the production point of view; that is, a go/no-go welding decision is difficult to make based on mechanical or metallurgical properties. A far easier method for production applications is the mockup or weld-restraint test. On the other hand, to gain an insight into mechanisms of cracking during welding, more sophisticated testing techniques are required. The most usable and best explored effort along this line is the Gleeble test introduced by Nippes and Savage.[18] In this test specimens are heated through simulated welding thermal cycles and tensile tested at various points within the thermal cycle. Strength and ductility results are used to infer weld-cracking behavior.

HOT-DUCTILITY TEST

Numerous experimental tools are available to study the metallurgical changes that occur during welding, but one of the most effective to date is the Gleeble test. The equipment consists of a time–temperature control device constructed to permit duplication of heating and cooling cycles experienced by regions adjacent to an arc weld. A 0.25-in.-diameter specimen is held in the jaws of a high-speed tensile-testing machine. At selected points in the simulated weld thermal cycle, tensile tests are performed and tensile strength, and ductility of the alloy are measured as a function of test temperature. Typical hot-ductility curves obtained on heating and on cooling are shown in Fig. 4. The zero-ductility temperature is defined as that temperature on heating where the ductility drops to zero. Similarly, the zero-strength temperature is defined as that temperature on heating where the strength drops to zero.

In early work with the Gleeble many tests were conducted on heating as well as on cooling. Although useful information was obtained on heating, the maximum decrease in ductility and tensile strength was found on cooling during the drop from a peak temperature close to the melting range. In order to take advantage of these effects, testing is now usually done on cooling. This procedure also appears more consistently correlatable with the cracking mechanism during welding. True heat-affected zone cracks appear to occur during cooling, and composite region cracking certainly must occur during cooling because if cracks cracks formed on heating, they would be healed when the composite region melted.

Criteria for Interpretation

Different criteria have been used by different investigators for interpreting hot-ductility curves. A few of these are discussed briefly. They are minimum arbitrary ductility, recovery rate of ductility, recovery rate of ultimate strength, zero-ductility range (ZDR), and zero-ductility dip in middle-temperature range.

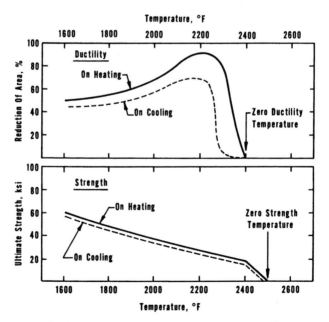

Fig. 4. Typically hot-ductility curves for Inconel 600.[19]

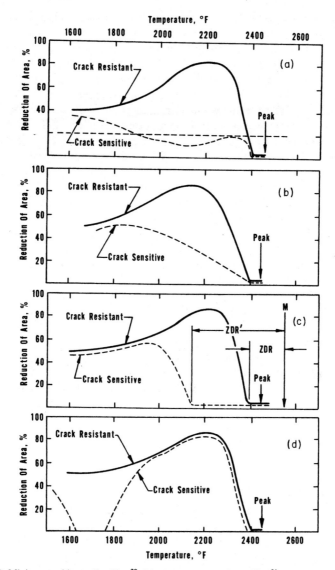

Fig. 5. (a) Minimum arbitrary ductility,[20] (b) recovery rate of ductility,[21] (c) zero-ductility range (ZDR),[22] (d) midtemperature range ductility[22,23] being used as crack sensitivity criteria.

An example of minimum arbitrary ductility is illustrated in Fig. 5a. In this case it is assumed that if the ductility obtained on cooling from a peak temperature is less than 20% reduction of area, the alloy will be crack sensitive when welded. On the other hand, if the ductility exceeds 20% reduction of area, the alloy will be crack resistant. The 20% level is an arbitrary choice and could also have been 10 or 30%.

The recovery rate of ductility, another criterion for separating crack-sensitive from crack-resistant alloys, is illustrated in Fig. 5b. Here it is assumed that if ductility recovers quickly on cooling from a "peak" temperature, the alloy will be crack resistant. However, if ductility recovers at a slow rate, the alloy will be crack sensitive. Consider the case where the recovery rate of ductility for two alloys is identical but the recovery rates for ultimate tensile strength differ. The alloy with the low ultimate strength recovery rate is classed as *crack sensitive*, while the other with the high recovery rate in ultimate strength is classed as *crack resistant*.

Another criterion for discriminating between crack-sensitive and crack-resistant alloys is the ZDR, as shown in Fig. 5c. When cooling from a peak temperature near melting, ductility will remain zero for some temperature range before recovering. Length of this zero-temperature range, measured from the start of melting (M), is called the ZDR. A crack-sensitive alloy will have a large ZDR, while a crack-resistant alloy will have a small ZDR. Another criterion for measuring cracking sensitivity is the middle-temperature-range ductility dip obtained on cooling, as illustrated in Fig. 5d. If ductility dips to zero in the middle-temperature-range the alloy is classed as crack sensitive; however if the ductility remains high, the alloy is crack resistant.

Testing Parameters

Testing parameters can have a significant effect on hot-ductility response and thus on judgments of cracking sensitivity. Variations in peak temperature, cooling rate, and strain rate can be particularly important. The effect of peak temperature is illustrated in Fig. 6, where a three-dimensional representation is made of a weld heat-affected zone in Hastelloy X. For illustration, the size of the heat-affected zone has been greatly magnified compared to the weld pool. Regions in the heat-affected zone are initially heated and then cooled as the weld pool passes. Peak temperature will depend on distance from the weld interface, decreasing as distance from the weld interface increases. A point 0.014 in. from the weld interface will be heated to a peak temperature of 2200 °F (1200 °C) and shows no loss in ductility on heating or cooling. A point 0.010 in. from the weld interface will reach a peak temperature of 2250 °F (1230 °C). At this temperature ductility drops to zero but recovers immediately on cooling. At 0.006 in. from the weld interface peak temperature is 2300 °F (1260 °C). However, ductility drops to zero on heating at 2250 °F (1230 °C) and does not return until a significant amount of cooling below 2300 °F (1260 °C) occurs. Closer to the weld interface, and consequently at higher peak temperatures, ZDR become larger and, when combined, shows the zero-ductility plateau as illustrated. Zero-ductility ranges discussed previously are traces across this zero-ductility plateau. Peak temperature of testing consequently has a significant effect on the value of the ZDR. Longer ZDR are obtained from higher peak temperature.

Structures

An effect of microstructure and segregation on the ductility response of INCO 82 weld wire alloys has been reported.[22] In general, the wrought structure had a smaller

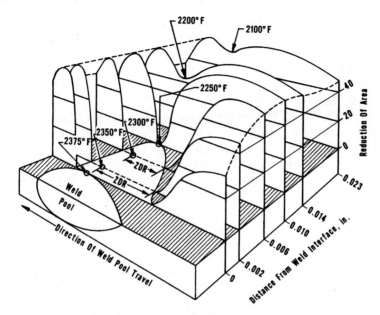

Fig. 6. Three-dimensional representation of the effect of peak temperature on ductility response. (From ref. 24.)

ZDR than welded specimens of the same alloy. Increases of 200–340 °F (110–190 °C) in ZDR were observed between wrought and welded structures of the same heat of alloy. An effect of welding speed was also found in the same INCO 82 alloy welded at 1, 5, and 10 in./min and showed that faster welding speeds produced a finer grain size and also gave smaller ZDR.

Cracking Criterion

A review of some of the criteria for interpreting hot-ductility curves and of the effects of testing parameters and structure revealed that careful and precise testing procedures and techniques are necessary to get meaningful results. Disagreements in results reported by different investigators could easily be attributed to differences in testing conditions.

Let us now take a look at what might be a meaningful criterion for predicting heat-affected zone-cracking sensitivity. To do this, we must first define what causes cracking. Cracking will occur when an inposed strain exceeds the ability of the material to deform; for example, if a material can undergo 10% elongation, a strain in excess of 10% will cause the material to crack. Using coefficient of thermal expansion, Young's modulus, and a welding thermal cycle, it can be calculated that the maximum uniform strain that can be developed during welding is on the order of 0.1%. However, if strain-concentrating phenomena occur, such as in areas of reduced section or because of thermal and yield strength gradients, much larger strains may occur in local areas. These strain-concentrating phenomena, however, will occur only over macroregions, such as through the diameter of a reduced bar, and will not develop to any significant magnitude within microregions. Hence, in the case of a weld heat-affected zone no significant strain concentration would be expected, and the total magnitude of strain should be something less than 1%. Because of such small strains, it is assumed that cracking will only occur when the alloy is in a condition of zero ductility.

Let us now consider a weld pool moving on a 0.5-in.-thick plate and the isotherms around this moving weld pool as shown in Fig. 7. The isotherms show decreasing temperature with increasing distance from the weld pool. If a line parallel to the welding direction is drawn tangent to each isotherm, the locus of these tangent points forms a boundary between heating and cooling. This is illustrated in Fig. 8a. The area to the right, in the direction of travel, of the heating-cooling line is being heated, while the area to the left is being cooled. In the heated area the material is expanding and developing compressive stresses. Once cooling begins, stresses are gradually relieved until they arrive at zero. The concept of a zero-stress boundary is also shown. To the right of the zero-stress line material is under compression, and to the left it is under tension. As material is cooled further (or looking further to the left of the zero-stress line), it eventually reaches its yield strength. This is a difficult line to define precisely because the yield stress also increases with decreasing temperature. Nevertheless, a yield strength line has been drawn in its approximate location.

Adjacent to each moving weld pool is a zero-ductility plateau. In Fig. 8b, the zero-ductility plateau does not intersect the yield strength line. When the yield strength is reached, the material has ductility and no cracking occurs. In contrast,

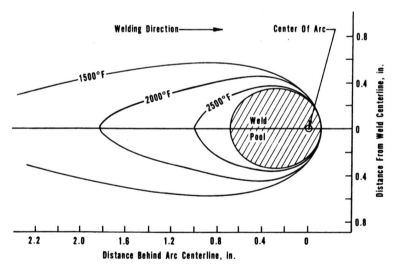

Fig. 7. Isotherms around a moving weld pool in 0.5-in.-thick steel plate welded at 24 V, 208 A, and 3 in./min.[4]

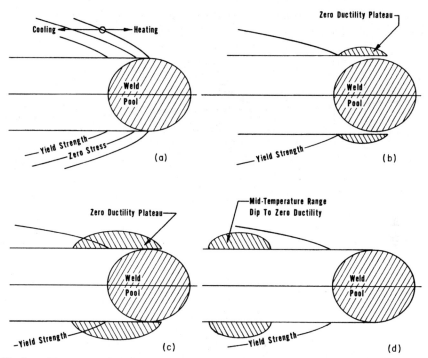

Fig. 8. (a) Zero-stress and yield strength lines in a weld heat-affected zone. (b) Yield strength line and zero-ductility plateau of a crack-resistant alloy. (c) Yield strength line and zero-ductility plateau of a crack-sensitive alloy. (d) Yield strength line and a midtemperature range dip to zero ductility.

Fig. 8c shows a zero-ductility plateau that intersects the yield strength line. Hence, when the yield strength is reached, the material still has zero ductility, and therefore it cracks. Figure 8d illustrates the case where the midtemperature-range ductility drops to zero. Because of the high stresses present in the midtemperature range, one would expect a drop to zero ductility in the midtemperature range to cause gross fissuring.

This schematic method of comparing the yield strength line with the zero-ductility plateau is useful for understanding several basic phenomena:

1. A small ZDR or plateau can exist and not necessarily cause cracking.
2. A variation in welding parameters can shift the yield strength line and also change the size and shape of the zero-ductility plateau. Hence, changes in welding parameters can eliminate or aggravate cracking.
3. A midtemperature-range ductility dip to zero is very detrimental.

Interpretation

A typical hot-ductility curve along with the individual contributions that combine to make up this curve as shown in Fig. 9. Although these contributing curves are

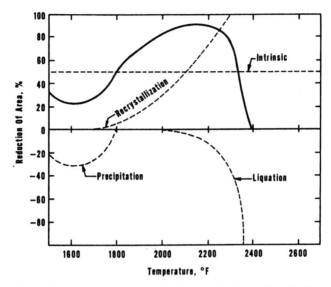

Fig. 9. Individual factors that combine to make the normally observed hot-ductility response.

only treated in a qualitative manner, they provide a basis on which to predict and interpret hot-ductility response. The construction of the hot-ductility curve starts by assuming an arbitrary intrinsic ductility for the alloy, independent of temperature and equivalent to 50% reduction of area. A positive contribution to ductility arises at high temperatures because of recrystallization. As temperature is increased beyond the initial recrystallization temperature, the specimen has a chance to recrystallize and replace deformed metal with new undeformed grains and hence allow the metal to undergo further deformation without rupturing. Continued recrystallization of deforming metal during the test will allow 100% reduction of area to be obtained. Ability of the metal to provide undeformed metal by recrystallization will increase with temperature (Fig. 9). This recrystallization mechanism of increasing ductility during hot-ductility testing has been previously proposed by several authors.[26,27] Two negative contributions to ductility are shown in Fig. 9: precipitation hardening and liquation. Consider precipitation hardening first. Stickler and Vinckier[26] made a study of the causes of the midtemperature-range ductility dip in stainless steel and attributed loss in ductility to hardening caused by precipitation of metal carbides. Annealing at 1575 °F (857 °C) for 1.5 h prior to hot-ductility testing eliminated the midtemperature-range ductility dip by overaging of the precipitate. The Ni–Cr–Fe alloys studied by Yeniscavich[22] showed a similar midtemperature-range ductility dip on heating, which was also attributed to precipitation hardening. The second negative contribution to ductility occurs at higher temperatures and is postulated as grain boundary liquation at temperatures several hundred degrees below bulk melting.[19,22] The model postulates that liquation begins at approximately 2000 °F (1100 °C) as small isolated areas of thin films that occupy only a small fraction of the grain boundary area. As temperature increases, film thickness grows, and so does the fraction of grain boundary area containing the liquid film. When a certain fraction of the grain boundary has liquated, ductility of the specimen will be reduced to zero because of the notching effect caused by distributed liquid films. However, the specimen will still have some tensile strength. When the liquid film extends completely through the grain boundary volumes and reaches a thickness in excess of 2000 Å, tensile strength will also go to zero.

EFFECTS OF MINOR ELEMENTS

Solution-hardened alloys of the cobalt- or nickel-base type should be relatively easy to weld. They have no phase changes or precipitation-hardening reactions, which would normally cause problems during welding. However, because of the long liquidus-to-solidus range normally present in these alloys, hot cracking and fissuring during welding is often a problem.

Minor elements such as sulfur in nickel-base alloys and boron in cobalt-base alloys are notorious for causing dramatic increases in cracking during welding if slightly above their specified values. In these cases the ZDR are greatly increased. However, in many cases crack-sensitive heats of particular alloys appear even with all the minor elements within specification. To explore this problem further, a statistically designed factorial experiment was performed on Hastelloy X.[23] The experiment was conducted to measure effects of eight minor elements on cracking sensitivity of the weld heat-affected zone in Hastelloy X. Cracking sensitivity was increased by high levels of boron, sulfur, phosphorus, and carbon. Silicon and magnesium had a slight detrimental effect, and manganese and zirconium had a slight beneficial effect. However, even for the four elements showing the greatest detrimental effects, the increase due to each individual element was not large. Hence it was concluded that minor elements, when within normal specification limits, do not individually cause significant increases in heat-affected zone-cracking sensitivity. However, when several minor elements are present at nominally high levels, their effects are additive, and significant increases in heat-affected zone-cracking sensitivity can occur. Hence it appears that a total compositional evaluation is necessary to predict effectively cracking sensitivity of superalloys.

γ'-STRENGTHENED ALLOYS

The present status of welding technology for nickel-base superalloys is shown in Fig. 10. Those with low aluminum and titanium contents, shown below the dashed line, are readily weldable. However, as combined aluminum and titanium is increased, welding becomes more difficult. Alloys like René 41 and Waspaloy are borderline; they weld with relatively little difficulty but sometimes crack during postweld heat treating. Casting alloys with high aluminum and titanium, like 713C and IN-100, have low ductility at all temperatures and usually crack during welding.

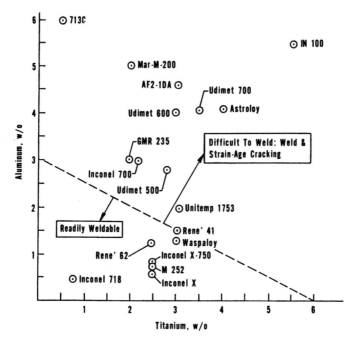

Fig. 10. Separation between γ'-strengthened alloys that are difficult to weld and those readily weldable. (From ref. 28.)

In general, alloys like René 41 will weld without cracking if they are solution annealed, thoroughly cleaned, and special precautions taken to prevent oxygen contamination. However, gross cracking will occur if the weldments are then directly age hardened. To prevent this, the weldments are stress relieved by reheating to the solution-annealing temperature before aging. Then they can usually be age hardened without difficulty. In some cases, however, cracking occurs after welding when heating to the annealing temperature. Large slowly heated fabricated components can show such cracking. Alloys and heats of alloys that exhibit a greater tendency than others to crack when reheating to anneal after welding have been called "strain-age cracking sensitive."

Weld Tests

To duplicate production conditions using smaller quantities of materials, weld restraint tests have been used in the laboratory. Although upward of 25 different types of weld-restraint tests are available, General Electric[29] and Rocketdyne[30] both chose a "circular-patch"-type test. The General Electric approach was to follow Hinde and Thorneycraft[31] and develop the time–temperature relationship for cracking of restrained welds. Isothermal annealing at temperatures between 1100 °F (595 °C) and 1800 °F (980 °C) was used to develop a "C" curve, which showed crack initiation. Typical C curves showing the difference between strain-age cracking sensitive and resistant heats are shown in Fig. 11. The nose of the C curve for the sensitive heat is 4 min. compared with 45 min. for the resistant heat. Since strain-age cracking occurs on heating through the temperature range of 1100–1800 °F (595–980 °C), displacement of the nose of the curve to longer times should be significant in reducing strain-age cracking. Sensitivity ratings on these heats were obtained in actual production fabrication, where the sensitive heat experienced gross strain-age cracking and the resistant heat experienced no strain-age cracking.

Restraint

A high amount of restraint is needed for strain-age cracking. This is borne out in production fabrication of René 41 and is an underlying consideration in its usage; that is, René 41 is avoided in welded, large, highly restrained components. Also, in the restrained weld tests it was found that if restraining welds break, strain-age cracking does not occur. Aside from external restraint imposed by geometry of the component being welded, internal residual stresses are also believed to play a significant role in strain-age cracking. Stress distribution around an unrestrained long butt weld in a thin sheet is given qualitatively in Fig. 12. It can be seen that a tensile stress exists parallel and perpendicular to welding direction in the weld metal and adjacent heat-affected zone and that these stresses are balanced by compressive stresses. In the case of a restrained weld balanced compressive stresses

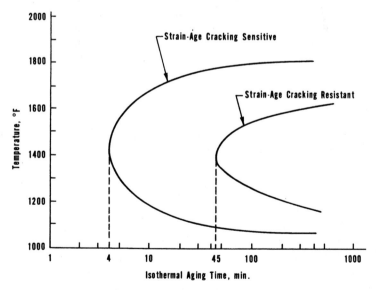

Fig. 11. Isothermal annealing curves showing the initiation of cracking for strain-age crack-sensitive and resistant heats of René 41. (From ref. 29.)

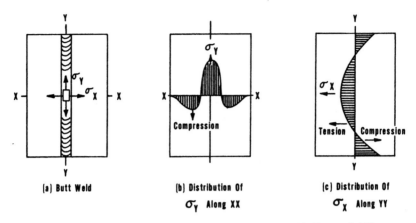

| (a) Butt Weld | (b) Distribution Of σ_Y Along XX | (c) Distribution Of σ_X Along YY |

Fig. 12. Typical distribution of residual stresses in a butt weld. (From ref. 32.)

can lie further away from the weld, that is, in the restraining members. The magnitude of the residual tensile stress in the y direction can be as high as the yield stress.

Strain

In reviewing the strain-age cracking phenomenon, it appears that stress, either residual or applied, is not necessarily the direct cause of the cracking, but rather the strain resulting from these stresses is the important part. In order to calculate maximum strain that can result from a residual weld stress, one can assume that stress is equivalent to the yield stress and that the component is rigidly fastened. Using a yield of 100,00 psi and a modulus of 30×10^6 psi would give a strain of 0.33% to completely relax this stress. The relaxation data of Rocketdyne[30] on simulated heat-affected-zone specimens indicate stress would only relax to 50,000 psi while heating at 25–30 °F/min (14–17 °C/min) to 1600 °F (870 °C); hence, maximum strain would be 0.16%.

Another source of strain during aging René 41 is the volume contraction associated with the aging. Very little quantitative data are available on this contraction; however, it was reported[28] that aging a 40-in.-diameter component resulted in a 0.040–0.050-in. contraction. Hence, a maximum strain of 0.125% could result from volume contraction on aging.

Heat-Affected Zone

The metallurgical changes that occur in the heat-affected zone are also of considerable importance. A study of simulated heat-affected zones in René 41 by Wu and Herfert[33] showed that in regions exposed to temperatures in excess of 2200 °F (1200 °C) the M_6C carbides were taken into solution. During subsequent aging these were precipitated as a continuous grain boundary phase of $M_{23}C_6$ and caused embrittlement. Based on these results, it was concluded that susceptibility to strain-age cracking is limited to those areas in the heat-affected zone that experience thermal cycles above 2200 °F (1200 °C).

Overaging

At General Electric[29] overaging René 41 prior to welding was claimed to be the single most effective method of reducing strain-age cracking. The overaging thermal cycle consisted of: 1975 °F (1080 °C), hold for 0.5 h, cool at 3–8 °F/min (2–5 °C/min) to 1800 °F (980 °C), hold for 4 h, cool at 3–8 °F/min (2–5 °C/min) to 1600 °F (870 °C), hold for 4 h, cool at 3–8 °F/min (2–5 °C/min) to 1400 °F (760 °C), hold for 16 h, air cool to room temperature. The result of this overaging treatment was to shift the nose of the C curve from approximately 4 min in the solution-annealed condition to something in excess of 500 min when overaged, a fantastically large shift in the time to initiate cracking.

Mechanism

Strain-age cracking appears to be caused by a loss in ductility during aging. In the solution-annealed condition René 41 will exhibit 1–4% elongation at approximately 1500 °F (815 °C) before fracture. Hence, solution-annealed material will not normally exhibit strain-age cracking even when fully restrained because volume contraction on aging only produces approximately 0.125% strain.

When a weld is made, several additional factors come into play. First, a residual stress is developed in the heat-affected zone, which may be additive to volume contraction and accelerate embrittling precipitation. Second, weld reinforcement may cause a notch effect and produce a additional localized plastic deformation in the weld heat-affected zone. Third, heat input of the welding cycle causes changes in carbide type and morphology in the heat-affected zone and may also cause solutioning of other precipitates that result in midtemperature-range embrittlement by reprecipitation during aging.

Overaging has and should reduce strain-age cracking for several reasons. First, it removes aluminum, titanium, and carbon from solid solution and reduces the degree of fine precipitation during postweld heat treatment. Second, volume contraction on aging is avoided since the material is already aged.

SUMMARY

With the introduction of the jet engine in 1940 the age of superalloy development was born and has grown at a rapid pace ever since. Joining techniques had a late start and were not initially considered a significant part of this new age until a multitude of cracking problems in welding (and lack of wetting in brazing) focused the necessary effort in this area. A fair understanding of the mechanisms causing these problems has already been developed along with some advanced techniques

516

to study these mechanisms. Continued work will lead to tighter specifications of composition and structure or processing procedure to eliminate these problems and will also lead to the development of new weldable superalloys. The advantages of welding (and brazing) for joining are low cost, low weight, and high strength. These advantages together with other continued improvements in joining technology will make welding (and brazing) the primary techniques for constructing superalloy turbine parts from several components in the foreseeable future.

REFERENCES

1. R. P. Newman and T. R. Gurney, *Brit. Weld. J.*, **11**, 341 (1964).
2. R. R. Gatts and G. J. Sokol, *Effect of Closure Weld Configuration on the Fatigue Strength of Omega Seals*, KAPL-2000-21, Reactor Metallurgy Report No. 24, Metallurgy, 1.1–1.11, 1964.
3. T. B. Jefferson, *Welding Engineer Data Sheets*, 5th ed., Monticello, IL, 1967, p. 1.
4. C. Weisman, *Welding Handbook*, Vol. 1, 7th ed., American Welding Society, 1976.
5. W. A. Kearns, *Welding Handbook*, Vol. 2, 7th ed., American Welding Society, 1978.
6. W. A. Kearns, *Welding Handbook*, Vol. 3, 7th ed., American Welding Society, 1980.
7. W. A. Kearns, *Welding Handbook*, Vol. 4, 7th ed., American Welding Society, 1982.
8. W. A. Kearns, *Welding Handbook*, Vol. 5, 7th ed., American Welding Society, 1984.
9. Specifications for Nickel and Nickel Alloy Covered Electrodes, AWS A5, 11-33, American Welding Society, 1983.
10. Arcos Divison, Hoskins Manufacturing, Mount Carmel, PA.
11. Stoody Company, 16425 Gale Avenue, Box 1901, Industry, CA.
12. Patented welding processes by Linde Division, Union Carbide Corporation.
13. *Specifications for Nickel and Nickel Alloy Bare Welding Rods, and Electrodes*, AWS A5.14-83, American Welding Society, 1963.
14. W. F. Savage and C. D. Lundin, *Weld J.*, **44**, 433S (1965).
15. W. F. Savage, *Weld. Des. Fab.*, **42**, 56 (December 1969).
16. J. J. Pepe and W. F. Savage, *Weld, J.*, **46**, 411S (1967).
17. J. J. Vagi, R. P. Meister, and M. D. Randall, Weldment Evaluation Methods, DMIC Report 244, 1968.
18. E. F. Nippes and W. F. Savage, *Weld J.*, **28**, 5348 (1949).
19. B. Weiss, G. E. Grotke, and R. Stickler, *Weld. J.*, **49**, 471S (1970).
20. J. Heuschkel, *Weld J.*, **39**, 236S (1960).
21. E. F. Nippes et al., *Weld. J.*, **34**, 183S (1955).
22. W. Yeniscavich, *Weld. J.*, **45**, 334S (1966).
23. W. Yeniscavich and C. W. Fox, *Effects of Minor Elements on the Weldability of High Nickel Alloys*, Proceedings of a Symposium Sponsored by the Welding Research Council, 1969, pp. 29–35.
24. D. S. Duvall and W. A. Owczarski, *Weld. J.*, **46**, 423S (1967).
25. W. Yensicavich, *Methods of High-Alloy Weldability Evaluation*, Proceedings of a Symposium Sponsored by the Welding Research Council, 1970, pp. 2–12.
26. R. Stickler and A. Vinckier, Westinghouse Research Laboratory Report, 10-0103-3-R17, 1960.
27. W. N. Platte, Westinghouse Research Laboratory Report 10-0103-3-R5, 1959.
28. M. Prager and C. S. Shira, *Weld. Res. Counc. Bull.*, 128 (1968).
29. T. F. Berry and W. P. Hughes, *Weld. J.*, **48**, 505S (1969).
30. E. G. Thompson, S. Nuney, and M. Prager, *Weld. J.*, **47**, 299S (1968).
31. J. Hinde and R. Throneycroft, *Brit. Weld. J.*, **7**, 605 (1960).
32. K. Masubuchi, *Weld Imperfections*, Addison-Wesley, Reading, MA, 1968, p. 567.
33. R. C. Wu and R. E. Herfert, *Weld. J.*, **46**, 32S (1967).

PART SIX

THE NEXT YEARS

Chapter 19

Alternative Materials

NORMAN S. STOLOFF and CHESTER T. SIMS

Rensselaer Polytechnic Institute, Troy, New York

When considering possible replacements for superalloys in high-temperature applications, two classes of materials suggest themselves:

1. materials that offer higher strength-to-weight ratios and/or better corrosion resistance than superalloys to serve without significantly increasing maximum operating temperatures and
2. materials that, by virtue of higher melting points, concomitant with usable strength and corrosion resistance, offer the capability of operating at higher temperatures, in the 2000–4000 °F (1095–2205 °C) temperature range.

Materials of class 1 include composite alloys based on nickel or cobalt and many intermetallic compounds, especially the aluminides of titanium, nickel, and cobalt; class 1 might also include ceramics. Higher melting intermetallic compounds than the aluminides have been identified, but data that would allow consideration for high-temperature service are either nonexistent or only conjectural for them.

In class 2 one normally thinks of the body-centered-cubic (BCC) refractory metals, especially tungsten, molybdenum, tantalum, and columbium, or of structural ceramics, including ceramic–matrix composites. Another type of material falling into this category is the carbon–carbon composite class.

The principal characteristics of the various types of alternative materials considered in this chapter are summarized in Table 1. We will first consider the lower melting,

Table 1. Characteristics of Alternative Materials

Systems	Ambient Ductility	High-Temperature Strength	Database	Resistance to Oxidation	Corrosion
Cb, Ta	High	Low	Moderate	Fair[a]	Fair[a]
Mo, W	Low	High	Moderate	Poor	Poor
Aluminides	Low	Intermediate	Low	Good	Poor
Ni-base					
DS eutectic	High	High	Moderate	Good[b]	Good[b]
Wire-reinforced					
FeCrAlY	Low	High	Moderate	Good	Good
SiC,Si$_3$N$_4$	Low	High	Moderate	High	Good
C–C composite	Low	High	Low	Poor	Poor
Superalloy	High	High	High	High[b]	High[b]

[a] Must be coated.
[b] Usually coated.

class 1 materials, focusing on intermetallics, DS eutectics, and wire-reinforced superalloys. The second half of this chapter will discuss the refractory metals, monolithic ceramics, and composites (ceramic–matrix and carbon–carbon).

INTERMETALLIC COMPOUNDS

For many years the beneficial effects of long-range order on strength, especially at elevated temperatures, has created interest in possible application of intermetallic compounds in turbines. A list of representative candidates for replacing superalloys at current operating temperatures appears in Table 2. While many higher melting intermetallics exist, little or no data on mechanical or physical properties are available. Aluminides represent a particularly attractive choice because of high oxidation resistance resulting from the protective nature of Al$_2$O$_3$ films. Further, aluminides such as CoAl and NiAl already are in service as coatings, while γ' (Ni$_3$Al) is the principal microstructural constituent (up to 70 vol. %) of advanced nickel-base superalloys. TiAl and Ti$_3$Al have received extensive attention because of their low densities and potential weight savings.

Two severe limitations to use of aluminides also have been identified: lack of appreciable ductility at ambient temperatures and lack of adequate creep resistance at high temperatures. However, with regard to the first problem, there is renewed interest in monolithic Ni$_3$Al and other aluminides as a result of the striking increase in ductility imparted to nickel-rich compositions of Ni$_3$Al by the addition of small amounts of boron (see Fig. 1).[1] The boron segregates to grain boundaries and strengthens them, but it is only effective so long as the aluminum content remains below precisely 25 at. % and so long as the grain size is reasonably small. (In rapidly solidified foils, 25 at. % aluminum material reveals some ductility.[2] However, lack of high-temperature creep strength is of concern, and boron alone does not confer adequate ductility in oxidizing environments at elevated temperatures.[3] Recent work to alleviate these problems is described below.

Strength

Strengths of nickel, cobalt, and iron aluminides generally are not very high at room temperature. However, one of the striking features of the plastic deformation of most ordered alloys, particularly for most of those with the $L1_2$ structure, is the sharp rise in flow stress with increasing temperature. The flow stress peak occurs in single crystals as well as in polycrystals, and its position with respect to temperature in Ni_3Al is a function of crystal orientation and alloy content. Alloying with zirconium and hafnium is especially effective in raising the high-temperature strength of Ni_3Al; yield strengths of these alloys are higher at 1562 °F (850 °C) than those of commercial superalloys (Fig. 2),[4] especially on a density-compensated basis.[5]

Increases in flow stress with temperature also have been noted in long-range-ordered alloys of other crystal structures.[6] The peak strength is sometimes associated with the order–disorder temperature, T_c (e.g., in FeCo), or with the temperature of a transition from one ordered structure to another (e.g., in Fe_3Al); in other cases it bears no apparent relation to a transformation (e.g., Ni_3Al, CuZn, and Ni_3Ge). Therefore, it is unlikely that a single mechanism can explain the effect in all compounds.

Long-range-ordered alloys usually exhibit high strain-hardening rates compared to their disordered or partially ordered counterparts. For $L1_2$ superlattices the strain-hardening rate can double with order at temperatures near 70 °F (22 °C), whereas lesser increments in rate are noted in other crystal structures. High strain-hardening rates induced by long-range order may permit attainment of very high strengths

Table 2. Properties of Intermetallic Compounds

Alloy	Structure	Young's Modulus (10^6 psi)	T_m				Density (g/cm^3)
			°F	°C	°F	°C	
TiAl	$L1_0$	25.5	2260	1460	2660	1460	3.91[a]
Ti_3Al	DO_{19}	21.0	2912	1600	2012	1100	4.2[b]
NiAl	B_2	42.7	2984	1640	2984	1640	5.86[a]
Ni_3Al	$L1_2$	25.9	2534	1390	2534	1390	7.50[a]
FeAl	B2	37.8	2282– 2552	1250– 1400	1250– 1400	2282– 2552	5.56[a]
Fe_3Al	DO_3	20.4	2804	1540	1004	540	6.72[a]
CoAl	B2	42.7	2998	1648	2998	1648	6.14[a]
Zr_3Al	$L1_2$	19.6	2552	1400	1787	975	5.76[a]
Fe_3Si	DO_3	39.4	2318	1270	2318	1270	7.25[a]
MM-246	FCC	30	2400	1316	—	—	8.44

[a]Calculated from lattice parameter data.
[b]Estimated.

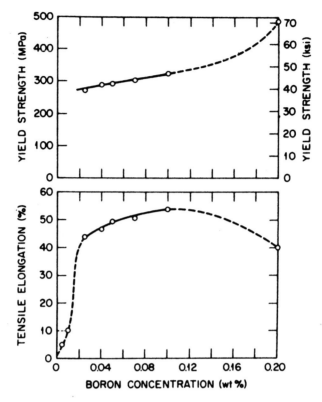

Fig. 1. Yield strength and tensile elongation of boron-doped Ni$_3$Al (24 at. % Al), as a function of boron concentration.[1]

through cold-working operations or thermal-mechanical treatments, as was shown in Fig. 2 for Ni$_3$Al+B.[4] Wear resistance also should be enhanced by rapid strain hardening, permitting possible replacement of cobalt-base alloys for such applications.

Fracture

Polycrystalline intermetallic compounds usually are brittle when tested in tension, although considerable plasticity may be displayed by single crystals or by polycrystals tested in compression.

In the iron–aluminum system sharply reduced ductility is observed as the aluminum content approaches 25 at. %.[7] Alloys with 25–50 at. % aluminum usually have been reported to be completely brittle at room temperature when processed by conventional ingot techniques. However, the ductility of P/M alloys with 23–35 a/o aluminum is 5–7% at room temperature,[8] and fracture is generally transgranular; nevertheless, all of the iron aluminides display extreme notch sensitivity. Ti$_3$Al and TiAl also are brittle at low temperatures.

Polycrystalline NiAl undergoes a ductile-to-brittle transition near 750 °F (400 °C), depending on aluminum content and grain size.[9] Failure is usually intergranular, although some transgranular cleavage cracks also have been noted. Single crystals, on the other hand, exhibit considerable ductility at low temperature. In the case of the aluminides the segregation of impurities to grain boundaries is considered unlikely to be the principal source of brittleness.

Improved Ductility by Microalloying

Considerable effort is being devoted to improving the ductility of several aluminides (FeAl, Fe$_3$Al, NiAl, and Ni$_3$Al) through such diverse techniques as grain refinement through thermal-mechanical treatment, microalloying with boron, and various rapid solidification techniques. Particularly promising is the previously described demonstration of high room temperature ductility of hypostoichiometric Ni$_3$Al doped with small quantities of boron,[1,2,10] as was shown in Fig. 1.[1] Boron segregates preferentially to grain boundaries, even in melt-spun material. The ductility of

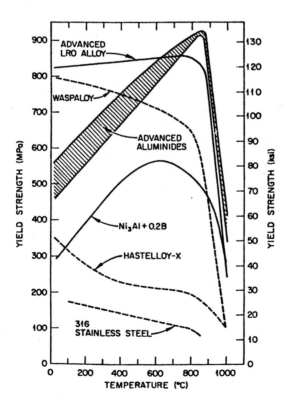

Fig. 2. Yield strengths of advanced Ni$_3$Al and (Fe,Ni)$_3$V-type LRO alloys compared to conventional alloys.[4,5]

polycrystalline Ni_3Al + B decreases substantially with increasing temperature, with a pronounced ductility minimum observed in the range 1112–1472 °F (600–800 °C) when tests are conducted in air.[11] The method of preparation of the alloys has a large effect on ductility at any temperature below 1472 °F (800 °C). The minimum is less severe in vacuum and in alloys containing chromium, leading to a suggestion that dynamic embrittlement by oxygen is responsible.[3] Consequently, hot working of cast Ni_3Al + B is inadvisable; rather, cold working with alternate anneals is preferable. Boron also improves the ductility of melt-spun Ni_3Si.[12] Little success has been obtained through microalloying with respect to the ductility of Fe_3Al and NiAl. In the case of Fe–40 at. % Al, small ductility improvements with boron have been achieved coupled with a change from intergranular to transgranular fracture.[13]

The ability to improve ductility of Ni_3Al by microalloying in conjunction with additions of manganese, hafnium, or iron has led to a breakthrough in alloy development schemes for this system. Advanced alloys based on solid-solution additions of hafnium and iron have demonstrated very high strengths (Fig. 2) relative to commercial alloys, coupled with about 10% lower density.[4] Further improvements achieved by cold work would be useful for low-temperature applications. There exists now an opportunity also to dispersion strengthen Ni_3Al or to utilize Ni_3Al as the matrix of mechanically incorporated composites.

Processing

The small but growing number of ordered alloys that display some ductility [Ni_3Al + B, Co_3Ti, Fe_3Al, and $(Fe,Ni)_3V$] has led to increased attention to fabricability of such alloys. Aluminides in bulk may be consolidated by conventional powder processing, plasma spray, with or without hot isostatic pressing (HIP), and arc melting and casting, followed by either high-temperature extrusion or alternate cold rolling and recrystallization treatments. Drop casting has proven to be a useful means of refining grain size and reducing segregation during solidification of Ni_3Al + B. Other conventional casting techniques, followed by working operations, have proven unsuccessful, as the product is invariably brittle. However, melt spinning can provide ductile ribbons.[2] The addition of several alloying elements to Ni_3Al + B has been shown to have an appreciable effect on ductility. Small amounts of iron (2 at. %) cause severe cracking during cold rolling, while alloys containing large quantities of iron (6–15%) were readily rolled to sheet.[14] Hafnium, zirconium, and manganese as well as 0.1–0.2% carbon also improve the ductility and cold formability of Ni_3Al; the influence of manganese probably is due in part to removal of sulfur from grain boundaries by sulfide formation. However, high ductility at low temperatures is no guarantee of successful fabrication at elevated temperatures.

Rapid solidification rate (RSR) processing via the P/M route also has been employed to produce intermetallic compounds with improved ductility. Koch[15] has summarized representative rapid solidification processing efforts.

Creep

Creep resistance of unalloyed NiAl (ref. 16) and Ni_3Al (ref. 17) tends to be poor relative to that of commercial high-temperature alloys. However, alloying of boron-modified Ni$_3$Al with hafnium and zirconium provides creep resistance comparable to or better than that of Waspaloy.[4] The second-stage (steady-state) creep rate may be expressed by

$$\dot{\epsilon} = A(\sigma/G)^n e^{-Q/RT} \tag{1}$$

where A and n are experimental constants, G is shear modulus, Q is the activation energy for creep, R is the gas constant, and T is absolute temperature. The stress exponent n in equation (1) is 3 for this system. Advanced long-range-ordered alloys [(FeNi)$_3$V type] in which aluminum replaces vanadium also display good creep resistance, although not quite as high as that of Ni$_3$Al + B. These results, which were obtained at a very early stage of development of both classes of alloys, suggest the desirability of further intensive research on these systems.

Preliminary reports also show encouraging improvements in creep resistance of iron aluminides. For example, 6 at. % Mo + Ti added in solid solution to Fe$_3$Al increases rupture life by six orders of magnitude at temperatures near 1292 °F (700 °C).[18] This increase is attributed to a large increase in activation energy for creep resulting from a 360 °F (200 °C) increase in the critical temperature for DO$_3$-type order [from about 1020 °F (550 °C) to 1380 °F (750 °C)]. Increases in creep strength of FeAl-base alloys also have been reported.[19] However, in this system the solutes responsible for hardening, principally columbium and tantalum, form ternary intermetallic second phases. These provide considerable strength at 1520 °F (827 °C).

Fatigue

The supression of cross-slip or reduction in number of available slip systems with long-range order that occurs in most alloys suggests a diminished probability of crack nucleation under cyclic loading. In the few systems for which room temperature fatigue data have been obtained,[6] ordering does, indeed, lead to an increase in high-cycle (stress-controlled) fatigue life that cannot be accounted for by yield stress differences. The stress-controlled fatigue life of alloyed γ' single crystals is independent of temperature to 1472 °F (800 °C).[20] However, fatigue lives of polycrystals drop precipitously above 932 °F (500 °C) even when tests are conducted in 5 × 10^{-5} torr vacuum.[21] The effect is particularly severe in P/M processed material, perhaps due to the higher internal oxygen content relative to wrought stock.

Crack propagation data for ordered alloys are not yet generally available in the literature. However, recent work has shown that the crack growth resistance of Ni$_3$Al + B + Hf and of a (FeNi)$_3$V alloy, LR0-60, is superior to that of conventional alloys, especially at low stress intensity ranges at 78 °F (25 °C) (see Fig. 3).[21] At 1112 °F (600 °C), when tested in argon, LR0-60 displays excellent fatigue crack growth resistance, but Ni$_3$Al + B fails prematurely, apparently due to the same factors that cause a ductility minimum in tension. Fe$_3$Al alloys, on the other hand, reveal inferior crack growth resistance at high ΔK, perhaps to the extreme notch sensitivity of such alloys.

There are few data available on low-cycle fatigue (LCF) of ordered alloys. Generally, LCF resistance increases with increasing tensile ductility, so that most intermetallics will not be expected to display good LCF resistance at temperatures below the ductile-to-brittle transition temperature (DBTT).

Oxidation

The excellent oxidation resistance of aluminides based on nickel, cobalt, and iron has been extensively exploited by using them as coatings on gas turbine hardware. Although brittle, compounds such as NiAl readily form an adherent layer of Al_2O_3, which is protective under oxidizing conditions to temperatures in excess of 1832 °F (1000 °C). In the case of the iron–aluminum system, Kanthal alloys have been utilized for heating elements for many decades.

Interest in utilizing aluminides and other ordered alloys in structural applications has led to several investigations of their oxidation and corrosion behavior. Hafnium-modified Ni₃Al + B displays good resistance to oxidation at 1832 °F (1000 °C) (see Fig. 4).[22] It appears that although slightly more weight gain occurs in the hafnium-containing alloys, oxide scale adhesion is improved. Considerable weight loss and severe flaking of oxide occurred in the iron-containing alloy. The oxidation resistance at 1832 °F (1000 °C) of hafnium-modified Ni₃Al is superior to that of

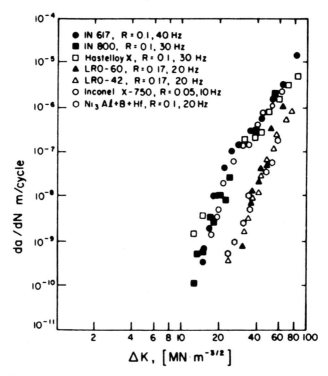

Fig. 3. Crack growth of several intermetallics and commercial alloys at room temperature.[21]

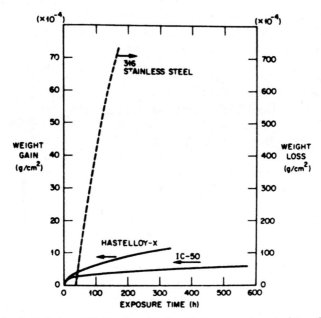

Fig. 4. Oxidation of Ni₃Al + B + Hf alloy IC-50 compared to other alloys.[22]

Hastelloy X and 316 stainless steel, undoubtedly due to the formation of a compact, adherent film of Al_2O_3 on the aluminides. Chromium also improves oxidation resistance while supressing high temperature embrittlement during exposure to air. Hot-corrosion behavior has not yet been reported but would be expected to be inferior to that of superalloys unless adequate chromium is present.

Potential Applications

At their present stage of development alloys based on Ni_3Al, Ti_3Al, and perhaps Fe_3Al may have some potential in the temperature range for some turbine engine structurals and disks, about 1200–1250 °F (627–675 °C). Strength and creep resistance fall rapidly for Ni_3Al at higher temperatures (see Fig. 2), and little is as yet known of the potential for alloying to higher temperatures. However, alloy development is proceeding at a rapid pace, and licensing agreements for commercial applications of nickel aluminides already have been entered into by Oak Ridge National Laboratory.

COMPOSITES

Eutectic Composites

Commencing in the late 1960s considerable effort was devoted to development of eutectic composites for turbine hardware. Prepared by DS techniques, composites offered very high tensile, fatigue, and creep strengths and seemed to offer significant temperature and/or strength advantages over conventionally cast alloys and DS alloys such as MAR-M 200. Property disadvantages of the eutectics were poor transverse and shear strengths, but low solidification rates and cost were the ultimate factors preventing usage. Further, the single-crystal superalloys being developed concurrently generated such good properties that they challenged the eutectics technically and were placed into production as turbine blades. Thus, present nickelbase eutectics offer little prospect of replacing nickel alloy single crystals. However, some strong iron-base alloys based on Fe–Mn–Cr–C have been developed that might find application in high-performance applications where the cost disadvantage is not a critical factor. In view of the current inactive status of metal-based eutectic systems, only a brief survey of their properties will be given here.

Preparation. When eutectic or very near eutectic compositions in a "normal" eutectic system are cooled in a steep temperature gradient, an aligned fibrous or lamellar structure may result. The α and β phases may be alloys, intermetallic compounds, or nonmetals such as carbides. Growth conditions are stringent in that a critical ratio of liquid–solid temperature gradient G to solidification rate R must be maintained. Too low a G/R results in a nonaligned dendritic or partly aligned (cellular) structure. Also important are inert atmosphere and high-purity starting materials in order to maintain a planar solid–liquid interface vital to forming the aligned structure.

Among the advantages of this means of preparing a composite are simplicity and the ability to control strength by altering microstructure, either through altering R or by postsolidification heat treatment to develop a strengthening precipitate in the matrix. In general, the interfiber or interlamellar spacing, λ, is proportional to $R^{-\frac{1}{2}}$.

Strength. A summary of high-strength cobalt-, nickel-, and iron-base eutectics appears in Table 3. Postsolidification heat treatment has been successfully applied to many of the alloys in Table 3, with demonstrable improvements in tensile, fatigue, and creep strengths. In nickel systems strengthening by heat treatment is attainable by adding aluminum to precipitate the γ' in the matrix. The Ni–Al–Mo ($\gamma/\gamma'-\alpha$) alloys are particularly notable for sharp increases in strength after solution treatment (due to refinement of the γ/γ' distribution) and again after aging to precipitate additional strengthening phases (e.g., Mo platelets and Ni_xMo intermetallic particles).[23] Unfortunately, these increases in strength are accompanied by severely reduced ductility.

Mechanical properties also are sharply affected by changes in λ; at low temperatures increased strength is attained at smaller λ, while at temperatures approaching the eutectic melting point degradation of properties may occur at small spacings. This is a consequence of sliding along interlamellar boundaries and/or grain boundaries.

As in all composites, off-axis properties of aligned eutectics are a problem, and in most cases they deteriorate as temperature increases. This is a serious problem for gas turbine applications, although additions of small amounts of boron or carbon

Table 3. High-Temperature Eutectic Alloy Compositions

Alloy	V_f	Ni	Co	Cr	Al	Nb	Ta	C	Other
Fibrous									
Nitac	0.05	69	—	10	5	—	14.9	1.1	—
Nitac 13	—	63	3.3	4.4	5.4	—	8.1	0.54	3.1W, 6.2Re, 5.6V
Cotac 74	—	bal	20	10	4	4.9	—	0.55	10W
Cotac 741	—	bal	10	10	5	4.7	—	0.5	10W
Cotac 3 or 33[a]	0.10	10	56	20	—	—	13	1	—
Cotac 50B3W	0.10	9.5	59	15.7	—	—	12	0.77	3W
γ'/γ–Mo (AG-15)	0.26	65.5	—	—	8.1	—	—	—	26.4Mo
γ'/γ–Mo (AG-34)	0.26	62.5	—	—	6.3	—	—	—	31.2Mo
Cotac 744	—	bal	10	4	6	3.8	—	0.46	10W, 2Mo
Lamellar									
γ/γ'–δ (6%Cr)	0.37	71.5	—	6	2.5	20	—	—	—
γ/γ'–δ (0%Cr)	0.3	76.5	—	—	2.5	21	—	—	—
γ–δ	0.26	66.7	—	—	—	23.3	—	—	—
Ni_3Ta–Ni_3Al	0.35	64.1	—	—	4.9	—	31	—	—
γ/γ'–Ni_3Ta	—	67.6	—	—	3.7	—	28.7	—	—
Ni–Ni_3Ta	—	63	—	—	—	—	37	—	—
γ'–δ	0.44	bal	—	—	4.4	23.4	—	—	—
MM-246	—	bal	10	9	5.5		1.5	0.15	2.5Mo, 10W, 1.5Ti, 0.015B, 0.05Zr

[a] 1300 °C, 2 h; 1000 °C, 24 h, air cooled.

have been utilized to improve transverse strengths of Nitac and γ/γ'–δ-type alloys, respectively.

Creep and Stress Rupture. Aligned eutectics (e.g., γ/γ'–δ and Nitac 14B) are characterized by excellent longitudinal creep and stress rupture properties, as may be noted in Fig. 5.[24] Activation energies for creep of eutectics tend to be higher than for conventional alloys, perhaps due to slowed diffusion in the ordered intermetallic compounds themselves or in the refractory metal carbides which usually are the reinforcing phase.

Apart from increasing the solidification rate, postsolidification heat treatment may be utilized to reduce creep damage. For example, heat treatment to precipitate fine TaC particles between the fibers results in a large improvement in stress rupture resistance of Co, Cr, and Ni–TaC (Cotac) alloys.[25]

Fatigue.   Fatigue resistance is generally outstanding when tests are carried out in tension–tension loading. For example, while nickel-base superalloys generally exhibit fatigue-limit-tensile-strength ratios of 0.25–0.3 at 70 °F (22 °C), ratios of up to 0.62 are noted for Cotac alloys and up to 0.84 for $\gamma/\gamma'-\delta$ (0% Cr). At elevated temperatures the ratio increases for both Cotac and Nitac.

Fatigue data under strain control remain sparse. A comprehensive investigation of low-cycle fatigue in a simple Nitac alloy (Ni, Cr-TaC) has shown that the Coffin–Manson relation between plastic strain range, $\Delta\epsilon_p$, and life, N_f, is observed[26]:

$$N_f^\beta \Delta\epsilon_p = C \qquad (2)$$

The low-cycle fatigue resistance of this alloy was shown to be superior to that of René 80, a cast nickel-base superalloy at 1600 °F (871 °C). Nitac-14B, a much stronger alloy, is superior in LCF resistance to many superalloys, including single-crystal U-700.[27]

Thermal Fatigue.   The carbide-reinforced eutectics exhibit wide ranges of behavior as a result of thermal cycling. Ni–CbC, Ni,Cr–CbC, Co,Cr–CbC, and Co,Cr–TaC are reportedly severely damaged by thermal cycling between 762 °F (400 °C) and 2050 °F (1120 °C) in 2 min by self-resistance heating[28]; Bibring[25] reported, to the contrary, that Nitac and Cotac alloys suffer little damage as a result of cycling between 73 °F (23 °C) and 1832 °F (1000 °C), both with and without simultaneous loading. Heavily alloyed Nitac showed no microstructural changes after 3000 cycles

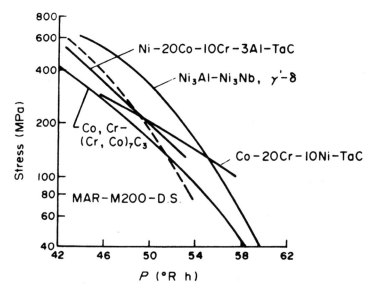

Fig. 5. Larson–Miller parameter $P = T$ (°R) (20R + log t_R) comparing stress rupture resistance of several eutectics with DS MAR-M 200.[24]

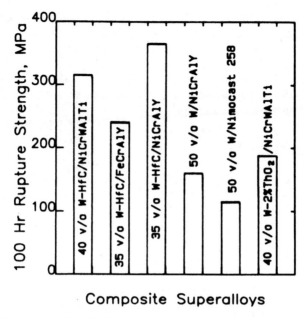

Fig. 6. Rupture strength at 2012 °F (1100 °C) for several fiber-reinforced superalloys.[32,33]

between 762 °F (400 °C) and 2050 °F (1120 °C). Cobalt alloys are inherently less stable to thermal cycling than nickel-base alloys due to the occurrence of the HCP–FCC phase transformation and the extremely planar slip characteristics of the former.[29] Two nickel-base alloys reinforced with CbC fibers, Cotac 74 and Cotac 741, in which the nickel matrices are precipitation hardened by γ′ and solid-solution strengthened by tungsten, demonstrated superior thermal cycling resistance to TaC-reinforced alloys.[30]

Transverse Properties and Blade Design. Off-axis properties, including shear strength, transverse strength, and transverse ductility, may be design limiting for these composites. Shear properties are important in root design, while transverse tensile and stress rupture properties may influence airfoil thermal fatigue lives.

Wire-Reinforced Superalloys

Considerable effort has been devoted to the preparation of superalloys reinforced with ceramic or metallic filaments. The reinforcement phase must be stiff, strong, and stable. Ceramic fibers possess these qualities as well as excellent oxidation and corrosion resistance and low density. Unfortunately, ceramic-fiber-reinforced superalloys suffer from *in situ* fiber surface attack and fiber matrix thermal expansion mismatch, thereby leading to inadequate strengths. Refractory metal fibers based principally on tungsten have proven more successful because of the ability of the fibers to relieve mismatch-induced stresses by plastic deformation. Tungsten-alloy-reinforced superalloys actually have shown significantly better thermal fatigue resistance than superalloys when rapidly cycled to 1100 °C.[31] Representative rupture properties of several wire-reinforced superalloys are shown in Fig. 6.[32,33] Note that W–HfC wires are particularly effective in strengthening nickel-base alloys. Other super-alloy–fiber combinations currently under development include MAR-M 200 with HfN-coated W and FeCrAlY with SiC.

REFRACTORY METALS

Properties of the BCC refractory metals of groups Va (Ta and Cb) and VIa (Mo and W) are summarized in Table 4.[34] Loosely defined as abundant metal elements with melting points in excess of 4000 °F (2200 °C), they melt at substantially higher temperatures than iron, nickel, and cobalt and their alloys and undergo no phase transformations below the melting point. Nevertheless, development of strong high-temperature alloys suitable for atmospheric service has had to face several glaring deficiencies: the open BCC structure (precluding high creep resistance relative to the melting point), lack of low-temperature ductility in the VIa metals, severe lack of oxidation resistance for all, and significantly higher density than superalloys for all save columbium.

A major effort to solve or bypass these problems and to create a new class of "superalloys" occurred in the 1950–1960 era, but with only scant results. The principal unsolved problem then was generation of suitable surface stability and the discovery of surface contamination problems. Columbium-base alloys edged in marginally, since they form solid oxides and showed some surface stability promise, with a Si–Ti–Al-based surface coating. (As described in Chapter 20, there is now one airframe application of Cb alloys.)

Consequently, attention to these alloys in general (except for Cb) has not focused on their attributes for air-breathing services. They are used as structural elements in high-temperature gas-cooled reactors and fusion reactors, for space nuclear-powered devices, electrical applications, thermocouples, and so on; the dominance of tungsten in lamp filaments and of tantalum in capacitors, for instance, is well known in those industries.

However, in the 20 years since the "Campaign of the Sixties" much has been learned about physical metallurgy. Truly startling developments have occurred in the processing of metals, particularly superalloys. At the same time superalloys have now come close to reaching their limits, at least as measured by temperature. Thus, refractory metals, led by columbium, are again being reviewed for turbine or other air-breathing devices in the hope that new metallurgy and new processing can bring them into service.

Alloy Properties and Applications

Molybdenum.   Molybdenum is the most readily available and broadly utilized refractory metal. Applications are based on its high melting temperature, high

Table 4. Comparative Properties of Selected Refractory Metals and Their Alloys

	Group V		Group VI		MM-246[a]
	Nb	Ta	Mo	W	
Pure metals					
Melting temperature, °C	2468	2996	2610	3410	1315
Density (25 °C), g/cm³	8.6	16.6	10.2	19.3	8.44
Thermal expansion (600 °C), 10^{-6}/°C	8.3	6.7	5.8	4.6	8.2
Thermal conductivity (600 °C), cal/cm² s °C cm	0.15	0.16	0.28	0.30	140
Specific heat (20 °C), cal/g °C	0.065	0.036	0.061	0.032	0.095
Young's modulus (25 °C), GPa	97	185	325	400	205
(Approximate) recrystallization temperature, °C	1150	1350	1200	1500	Melting point
Alloys					
Mechanical properties [(0–1000 °C) (32–1832 °F)]					
Tensile/creep strength	Low	Low	Very high	Very high	Very high
Fatigue strength	Good	Good	Good	Good	Very high
DBTT (bending)	<0 °C	<0 °C	≤0 °C	>>0 °C	Melting point
Chemical properties					
Liquid metal compatibility	Good	Good	Very good	Very good	Good
Gas–metal interactions	Fair	Fair	Poor	Poor	Excellent
Fabricability					
Forming	Good	Good	Difficult	Poor	Good
Joining (welding)	Good	Good	Difficult	Very poor	Fair

[a] Properties limited by melting point.

strength and stiffness, high thermal conductivity, and resistance to corrosion in many nonoxidizing environments. A selected group of alloys is listed in Table 5. Mo-TZM is typical; it has come into demand for cores and inserts in die casting of steel, aluminum, zinc, and copper. It is used also for hot-working tools. TZM dies are used for the isothermal forging process known as Gatorizing, developed by Pratt and Whitney Aircraft for large turbine disks. TZM also has been employed recently for hot gas valves and seals for high-temperature gas systems. Molybdenum alloy turbine buckets have been tested in nuclear turbines, successfully for potassium vapor service and unsuccessfully for the (HTR) high-temperature gas-cooled reactor; in the latter case the buckets were embrittled from carburization.

TZM relies on solid-solution strengthening by small quantities of zirconium, titanium, and carbon, dispersion strengthening by precipitation of complex Mo–Ti–Zr carbides, and cold work. The tensile strength of Mo-TZM is significantly greater up to 2550 °F (1400 °C) than that of other commercial molybdenum-base alloys, and it compares well on a strength weight basis with other refractory metals (Fig. 7).[35] It can be processed by vacuum arc melting or powder metallurgy.

A similar alloy, TZC-Mo, also has good creep rupture strength at elevated temperatures. Consolidation of TZC-Mo is by powder-metallurgical processes to avoid cracking in vacuum arc casting. Other molybdenum-base alloys strengthened by carbides of hafnium (Mo–0.2Hf–0.2C) or columbium generate still higher creep strength (see Fig. 8)[36] but are not commercially available.

Table 5. Composition of Representative Refractory Metal Alloys

Alloy Name	Nominal Solutes w%
Columbium Alloys	
Cb-1Zr	1.0Zr–0.005C
Cb1	30W, 1Zr, 0.06C
B88	28W, 2Hf, 0.067C
FS85	28Ta, 10W, 1Zr, 0.004C
C129Y	10W, 10Hf, 0.2Y, 0.015C
C-103	10Hf, 1.0Ti, 0.7Zr, 0.015C
Tantalum Alloys	
T-111	8W, 2Hf
T-222	10W, 2.5Hf, 0.01C
ASTAR 811C	8w, 1.0Hf, 1.0Re, 0.025C
Tungsten Alloys	
Thoriated W	2.0 ThO$_2$
W-Re	25 Re
3D	3 Re
Molybdenum Alloys	
TZC	1.2Ti, 0.25Zr, 0.15C
TZM	0.5Ti, 0.1Zr, 0.02C
Mo-Re	50Re
Mo-13Re	13Re

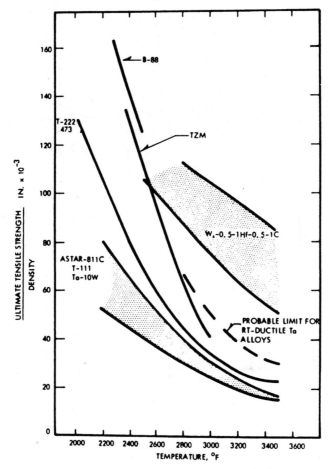

Fig. 7. Comparison of strength/density versus temperature for refractory metals.[35]

Molybdenum alloys can suffer service embrittlement as a result of the characteristic BCC ductile-to-brittle transition. Additions of up to 50% rhenium completely nullify this effect, but the cost and high density of rhenium limit the usefulness of such alloys. A recent development, Mo–13% Re, may combine good creep resistance with improved low-temperature ductility.

Molybdenum alloys suffer from both contamination by oxygen during processing and recrystallization and carbide stringers on grain boundaries parallel to the working direction.[37] As little as 6 ppm oxygen can promote brittle behavior. However, carbon ties up oxygen in the lattice, and as long as the carbon–oxygen ratio is greater than 2:1 the effect of oxygen can be blocked.[37] In TZM the presence of titanium and zirconium ties up carbon as MC carbides, so the situation is more complicated. Thus, among methods utilized to improve the ductility of molybdenum

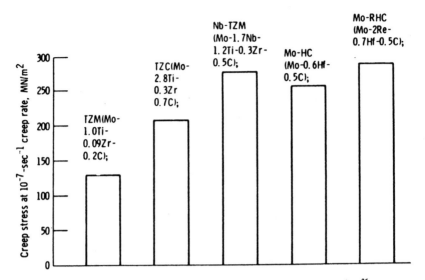

Fig. 8. Creep strengths of several Mo-base alloys at 2400 °C (1315 °C).[36]

alloys is addition of rhenium, removal of surface contamination, control of impurity levels, and removal of carbides from grain boundaries.[38]

In brief review, even though molybdenum-base alloys have seen trial service as gas turbine blades, their application appears limited to special nonoxidizing atmospheres. In other industrial applications at low to moderate temperatures, molybdenum alloys possess thermal-physical properties superior to superalloys for some applications.

Tungsten. Tungsten, the highest melting and most dense metal in the periodic table, also suffers from extreme brittleness at low temperatures unless processed with extreme care. The processing involves very extensive "hot" cold work in the presence of fine particulate matter such as K_2O or ThO_2. A 30% rhenium alloy is ductile, but the expense of rhenium limits application. A W–3% Re alloy called "3D" is used for flash bulbs due to its high resistivity.

Strength in tungsten alloys is generated by extensive cold work plus oxide or carbide dispersions. Strengths of tungsten-base alloys are very high, even when corrected for density differences (see Fig. 7). The possible use of tungsten alloys as high-strength fiber reinforcements for metal–matrix composites has already been cited.

Due to their special high-melting position, tungsten-base alloys have carved for themselves a strong industrial market. However, their extremely high density and lack of oxidation resistance [W alloys develop a gaseous dissociating oxide above about 1000 °C (1800 °F)] preclude success in competition with superalloys.

Columbium. "Columbium superalloys" are possible, as testified by the use of C-103 in turbine engine nozzle flaps (see Chapter 20), but there are still severe problems to overcome if their full potential is to be realized. A summary of the properties offered by columbium alloys when considered for service in oxidizing atmospheres is as follows:

Advantages	Disadvantages
Density is approximately that of nickel	Low Young's modulus
Good lattice solubility for alloying	Open BCC structure
Forms a solid oxide	Embrittlement by interstitial elements

It is important to comment parenthetically here that the solid oxide formed (Cb_2O_5) develops rapidly and is not protective and also that a low Young's modulus can sometimes be beneficial.

Most columbium alloys (Table 5) display good formability, weldability, and moderate strength. To date, alloying of columbium for strength has been straightforward: it is solution strengthened by refractory and high-modulus elements and dispersion strengthened by refractory MC carbides. The substitutional solutes most often added to columbium for creep resistance are tungsten, molybdenum, and tantalum. The reactive metals zirconium and hafnium, together with carbon and nitrogen, produce fine precipitates that further contribute to creep strength. Aluminum and titanium improve base metal oxidation resistance, but they lower the melting point and thus have a negative effect on strength. Alloys are fabricated by electron beam and double-vacuum consummable arc melting and working. Cast columbium-base alloys are not known.

While columbium does form a solid oxide, the unprotective Cb_2O_5 forms at a linear rate. Accordingly, the solution is to coat the alloys and attempt to alloy the base metal to support the coating chemically. To avoid reducing strength, careful alloy balancing is essential. A protective coating is required even for *processing* at temperatures above 800 °F (424 °C) in oxidizing atmospheres in order to minimize solution of oxygen, which can cause embrittlement.

Coatings are generated by surface treatments with aluminum, titanium, and silicon. The protective silicide layer formed is currently limited to about 2700 °F (1500 °C) service.

Columbium alloys are under continuing consideration for nuclear fuel applications (due to low nuclear cross section) and are used for corrosion-resistant chemical service as well as the aircraft engine structural part mentioned. Cb–1 wt. % Zr alloy is used extensively in nuclear systems containing liquid metals operating at 1800–2192 °F (982–1200 °C) because of its low thermal-neutron cross section, moderate strength, and excellent fabricability. It often is identified as the preferred "first wall" material for fusion reactors and is utilized in sodium vapor lamp applications.

Alloy B-88 is among the strongest columbium alloys (Fig. 7), as is the alloy C-1. Alloy C-103 (see Table 5) obviously depends heavily on hafnium for solution strengthening and complex MC carbides for dispersion strengthening. The coating is "on its own," but it works. Thus the coated C-103 system is the first refractory metal system to have generated aircraft engine acceptance. Alloy C-103 also has been utilized for rocket components, which require moderate strength at 2000–2500 °F (1093–1370 °C).

Thus, columbium-base alloys have strength potential to somewhat above 2500 °F (1370 °C) and have demonstrated they can extend the service range of superalloys in air, at least in one structural case. Application of contemporary processing developments (PM, RSR, plasma processing, etc.) combined with demonstrated classic fine-particle and solution strengthening and extensive hot-cold working may well be a route to greater strength and temperature capability.

Tantalum. Tantalum alloys also are readily fabricable and have strength potential, but their high cost, high density, and limited availability have hindered development. Strengthening of tantalum by substitutional solutes generally parallels the trend in columbium. Since tungsten is a more potent strengthener than molybdenum, 7–10% tungsten is present in all tantalum-base alloys. T-111 (Fig. 7) and T-222 are hafnium-containing modifications of Ta–10W (with C) and have comparable fabricability. Protective coatings are required for tantalum and its alloys above 900 °F (482 °C) if they are expected to see service in any oxidizing atmosphere. Tantalum has widespread use in capacitors and, because of its corrosion resistance to acids and chemicals, in certain industrial plant applications.

Summary

Of these four refractory metals, three (Mo, W, and Ta) are not suitable for air-breathing turbine service due to atmospheric reactivity and high density. Depending on other properties, they remain viable market commodities for electrical, chemical, and nonoxidizing mechanical service, some at high temperature.

The fourth element, columbium, while difficult to strengthen and with marginal surface stability capability in air atmospheres, is now in one high-temperature service application; with more study and work it may become an avenue to extend the temperature range of superalloys.

CERAMICS

Defined loosely, "ceramics" encompasses a variety of materials, including monolithic ceramics, ceramic composites, and often carbon–carbon composites. Monolithic ceramics have been under extensive study for about 10 years, with strong emphasis now being directed toward composites. For brevity, we will emphasize monolithic ceramics and carbon–carbon.

Covalently bonded ceramics offer the prospect of useful heat resistance, possibly to temperatures near 3000 °F (1650 °C), coupled with low density and in some cases excellent oxidation and corrosion resistance. A further advantage is the low cost and ready availability of the principal starting materials, silicon, carbon, and nitrogen. Unfortunately, these ceramics also are brittle, prone to thermal shock, and less thermally conductive than heat-resistant metals, leading to severe deficiencies under tensile loading. These are inherent properties determined by the nature of the interatomic bonds. Mechanical properties also are highly variable, depending sensitively on preparation technique, impurities, and surface finish; in ceramics the process basically determines the properties. Nevertheless, the toughness and thermal shock resistance of Si_3N_4 and its ability to form protective SiO_2 layers makes it a candidate for turbine or diesel applications. SiC has similar properties.

Thus, large weight savings and increased efficiencies are potential benefits to be gained by replacing metals with Si_3N_4, SiC, or other ceramics in hot sections. Automotive diesel engines made substantially of Si_3N_4 already have been tested successfully in Japan. Stationary products such as heat exchangers, recuperators, and gaskets for exhaust systems, furnaces, and heaters also are being considered. Two ceramic automotive gas turbines have been developed and demonstrated in the United States.

For other high-temperature applications oxides (MgO, Al_2O_3), silicates (Mg_2SiO_4, Fe_2SiO_4), and glass–ceramic composites also have received attention. (Oxides are principally ionically bonded.) Zirconia (ZrO_2) has been in service as a thermal barrier coating in aircraft combustors on superalloys for many years. However, the oxide-type ceramics tend to be less desirable mechanically than are carbide–nitride ceramics, although they are very stable in oxidizing atmospheres.

Table 6 summarizes physical and mechanical properties of several high-temperature ceramics.[38] These properties are only representative, as wide variations arise from different preparation schemes, even in commercially available materials.

Processing

With ceramics processing is more heavily integrated with properties of the product than for metals. Bulk structural ceramics classically are processed from powders. Property characteristics can be traced to many factors: powder size, shape, purity, and density; existence of second phases; pore size and distribution, grain size, and grain boundary condition; microstructural stability; and the nature and size of critical flaws. As one example, bulk density is a major factor influencing strength of Si_3N_4, with flexure strength increasing linearly with density.[39]

Hot-pressed material is stronger than sintered or reaction-sintered Si_3N_4 for the same bulk density. However, hot processing is not suitable for producing complex shapes, nor is it amenable to high production rates. Much of the scatter in properties among various Si_3N_4 materials arises from the necessity to use oxide additives to aid in sintering of high-purity Si_3N_4 powders. The additive, concentrated at grain boundaries, forms liquid silicates by combining with the silica impurity commonly present in all Si_3N_4 powders. Densification occurs by dissolution and precipitation within the liquid silicate. Properties are then determined by the composition and volume fraction of silicate left at the grain boundaries, and usefulness will be limited by the temperature at which the grain boundary material loses strength; for MgO additives in Si_3N_4 it is about 2400 °F (1300 °C). Rare-earth oxides such as Y_2O_3 also have been investigated as sintering aids; excellent properties are displayed by $Si_3N_4–Y_2O_3$, particularly at temperatures above 2192 °F (1200 °C), but some

Table 6. Physical and Mechanical Properties of High-Temperature Ceramics

Property	Low-Polytype Sialon (Glassy Phase)	Hot-Pressed Silicon Nitride	Reaction-Bonded Silicon Nitride	Sintered Silicon Carbide	Alumina	Partially-Stabilized Zirconia (PSZ)
Room temperature modulus of rupture, MPa(ksi)	945 (137)	896 (130)	241 (35)	483 (70)	380 (55)	610 (88)
Typical Weibull modulus	11	10–15	10–15	10	10	10–20
Room temperature tensile strength, MPa(ksi)	450 (60)	~580 (~80)	145 (21)	299 (42)	210 (30)	466 (67)
Room temperature compressive strength, MPa(ksi)	>3500 (>500)	>3500 (>500)	1000 (~150)	2000 (~300)	2750 (~400)	1850 (~270)
Room temperature Young's modulus, MPa(ksi)	3×10^5 (4.4×10^4)	3.1×10^5 (4.5×10^4)	2.0×10^5 (3×10^4)	4.1×10^5 (6×10^4)	3.6×10^5 (5.3×10^4)	2.0×10^5 (3×10^4)
Room temperature hardness, kg/mm² (VHN, 0.5 kg load)	2000	2200	900–1,000	2500	1600	1500
Fracture toughness (K_{IC}), MPa m$^{1/2}$	7.7	5	1.87	3.0	1.75	9.5
Poisson's ratio	0.23	0.27	0.27	0.24	0.27	0.3
Density, kg/m³(g/cm³)	3230–3260 (3.23–3.26)	3200 (3.2)	2500 (2.5)	3100 (3.1)	3980 (3.98)	5780 (5.78)
Thermal expansion coefficient, 1/K (0 to 1,000 °C)	3.04×10^{-6}	3.2×10^{-6}	3.2×10^{-6}	4.3×10^{-6}	9.0×10^{-6}	10.6×10^{-6}
Specific heat, J/kg/K(cal/g/K)	620 (0.15)	710 (0.17)	710 (0.17)	1040 (0.25)	1040 (0.25)	543 (0.13)
Room temperature thermal conductivity, W/m/K	21.3	25	8–12	83.6	8.4	2
Thermal shock resistance, ΔT_oK	~900	500/700	~500	300/400	200	~500

Source: SAE paper 850521, Lucas Cookson Syalon Ltd.

compositions exhibit phase instabilities at temperatures between 1292 °F (700 °C) and 2012 °F (1100 °C).

Hot isostatic pressing (HIP) is used to make high-density Si_3N_4 5% Y_2O_3. For both SiC (below) and Si_3N_4, cold isostatic pressing (CIP), sintering, and HIP without canning is a promising combination of processes, except that high-temperature sintering prevents microstructural control. HIP of glass-canned ceramics at temperatures above 3632 °F (2000 °C) also has been reported.

Si_3N_4 has been produced near theoretical density without sintering aids if sufficient purity and high processing temperatures are employed. Sintering temperatures between 2900 and 3300 °F (1600 and 1800 °C) and pressures of 150,000–750,000 psi (1–5 GPa) produce a material that is 88% β. The resulting hardness is about 50% greater than for Si_3N_4 doped with 4% Y_2O_3. A Swedish process called nitrided pressureless sintering involves melting silicon with a sintering agent in a matrix of Si_3N_4.[40] A ceramic body is formed and the part is nitrided. The free silicon is converted to Si_3N_4 at temperatures between 2372 and 2552 °F (1300 and 1400 °C). Final sintering is carried out at 3273 °F (1800 °C) to fully densify the material. Linear shrinkage is 8–9% compared to 15–20% by more conventional sintering.

Silicon carbide, which crystallizes as α (hexagonal) or β (cubic) variants, typically is produced by reaction bonding (RB), by sintering, or by hot pressing; the latter process generates the strongest and toughest SiC. SiC ceramics tend to be more oxidation resistant than Si_3N_4. An approach sponsored by the U.S. Army is to produce a fine, uniform-siliconized microstructure from a carbon skeleton made from liquid polymer solutions.[41] This material is much stronger than RB or sintered SiC and is about the same as for hot-pressed SiC. An alloy of SiC with AlN has been produced by carbothermal reduction of silica and alumina in a nitrogen atmosphere. A solid solution is formed when the alloy is hot pressed above 3632 °F (2000 °C), while a mixture of an SiC-rich and an AlN-rich phase is formed by heat treatment at lower temperature.

Composite Ceramics

Chemical vapor deposition (CVD) SiC as a foam is under development for use as heat exchangers and thermal protection insulation. This material can be formed into tubes, shapes, and structures and can be reinforced by ceramic fibers. Typical reinforcements, usually continuous filaments, are graphite or ceramic (SiC, alumina-borosilicate, and Al_2O_3) fibers. Advantages of these materials are cited as low-weight, efficient heat transfer, high-"temperature" resistance, corrosion resistance, and good thermal shock characteristics and fracture toughness. The latter (and vital) property is acquired by partial fiber pull-out, which donates a type of irreversible "deformation," creating the toughness.

Infiltration of carbon fiber woven cloth, mats, or preshaped bodies with molten silicon converts the carbon to SiC, generating a silicon–matrix SiC-reinforced material called "Silcomp." It is capable of producing large structures.

Mechanical Properties

The extreme brittleness of ceramic materials requires most testing to be carried out in three- or four-point bending. Al_2O_3, always a candidate because of its refractory and nonreactive nature, shows strength retention to 2192 °F (1200 °C) but then creeps rapidly due to the weak interatomic bonding of its ionic structure. Covalently bonded SiC is generally stronger than Al_2O_3, especially at temperatures in excess of 1832 °F (1000 °C). Hot-pressed Si_3N_4 (also covalent) displays a flexural strength in excess of 100 ksi (690 MPa) at room temperature and retains its strength well to 1100 °C (2112 °C). At temperatures above 1000 °C both sintered SiC and hot-pressed Si_3N_4 display considerably higher flexural strengths than the tensile strength of cast IN-100. Various additives to Si_3N_4 to achieve full density during hot pressing reduce the high-temperature strength because of the creation of glassy grain boundary phases. Little of this phase is found in RB product. Typical steady-state creep data for SiC and Si_3N_4 are shown in Fig. 9.[39] Note the wide range of properties due to scatter and differing impurity contents.

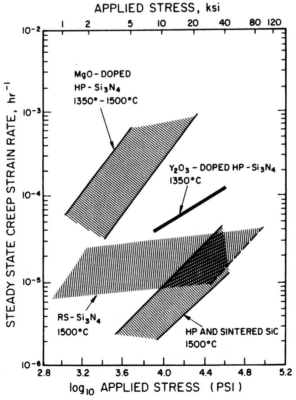

Fig. 9. Steady-state creep in fluxure of several Si_3N_4 and SiC materials.[39] (Reprinted with permission)

Stress exponents for creep in ceramics are much lower than for metals, with values of 1.8–2.3 reported for both tension and compression creep of both forms of Si_3N_4.[42] The low exponents are suggestive of creep deformation predominantly by grain boundary sliding accommodated either by cavity formation or by formation of microcracks as opposed to dislocation motion. The activation energy for creep of Si_3N_4 is variable, dependent on processing and composition.

Si_3N_4 and SiC display clear superiority over other ceramics in thermal shock resistance; lower elastic modulus and thermal expansion coefficient makes Si_3N_4 superior to SiC in all forms; good thermal conductivity in SiC helps, but this cannot overcome the other relevant properties. For example, HP Si_3N_4 vanes have withstood 250 cycle exposures at temperatures of 2500 °F (1370 °C) without cracking. However, wide differences in behavior have been noted, depending on machined surface condition, density, variations, degree of open porosity, and composition. For instance, excessive free silicon is detrimental.

Oxidation

Oxidation behavior of HP SiC and Si_3N_4 depends strongly on porosity.[39] For example, MgO-doped HP Si_3N_4 exhibits more rapid oxidation than Y_2O_3-doped samples in the range 1832–2192 °F (1000–1200 °C). Increased weight gain occurs as additive content increases. MgO-doped Si_3N_4 forms a SiO_2-based scale modified by cation impurities such as magnesium, aluminum, iron, calcium, and manganese in the unexposed substrate; these impurities can form mixed crystalline silicates or dissolve into the SiO_2. The principal scale product is usually $MgSiO_3$. Pits formed by reaction of $MgSiO_3$ with Si_3N_4 become fracture origins in subsequent mechanical loading, resulting in as much as 40% strength loss.

The behavior of RS Si_3N_4 also varies depending on the nature of the porosity, pore size, and pore size distribution.[39] Since these materials generally are much more pure than HP product (typically containing 0.5–1% cation impurities), impurities only have minor effects. The strength of RS Si_3N_4 after oxidation may be either higher or lower than for unexposed Si_3N_4 depending on the relative extent of internal and external oxidation, the nature of the SiO_2 film, and whether or nor thermal cycling has occurred.[40] Generally, oxidation causes a reduction in strength at 2552 °F (1400 °C) for 100 h exposure or more.

SiC oxidizes relative slowly compared to Si_3N_4. Oxidation kinetics are controlled either by desorption of CO gas formed at the $SiC–SiO_2$ interface[43] or by the inward diffusion of oxygen through the surface SiO_2, resulting in parabolic kinetics.[44] This behavior is preserved with additives such as boron, Al_2O_3, and B_4C, but reaction rates are increased.

From the overall surface stability view, SiC materials may be considered more stable than Si_3N_4 in long-term high-temperature applications. Oxidation of SiC results in lower weight gains and thinner scales as well as less alteration of strength distribution relative to Si_3N_4. This may be attributed primarily to the higher purity and greater density of SiC. However, recent forms of HP Si_3N_4 doped with 4% Y_2O_3 as well as SiO_2 show considerable promise of improved oxidation resistance.

Both materials are limited by the inability of SiO_2 to maintain a stable state over about 2700 °F (1500 °C). [Correspondingly, Al_2O_3, while not a "strong" ceramic, maintains stability in oxidizing environments to about 3500 °F (1950 °C).]

Carbon–Carbon Composites

The highest temperature capability of any material seriously considered for high-temperature use is exhibited by carbon–carbon composites, graphite fibers in a carbon–graphite matrix. Carbon–carbon composites are now used for one-time service in rocket-nozzle and missile exit core structures and in turbine aircraft brake shoes; SiC-coated carbon–carbon parts are being used as the nose cap and heating edges of the space shuttle.

A major factor in the strength of carbon-based materials is the exceedingly low self-diffusivity for carbon in graphite [10^{-14} cm^2/s at 1472 °F (800 °C)]. The activation energy for self-diffusion is in excess of 180 kcal/mol. Another outstanding advantage is density, in the range of 1.47–1.7 g/cm^3 for unidirectional composites with 55–65 wt. % fibers. These materials reportedly display useful strength to 4000 °F (2200 °C); strength may actually *increase* somewhat from room temperature.

However, carbon–carbon is not stable in oxidizing environments at temperatures above 800 °F (427 °C), and coatings are absolutely essential. Silicon carbide coatings have been engine tested at 2500 °F (1371 °C) on carbon–carbon cruise missile turbine engine parts, and efforts to develop an oxidation-resistant carbon fiber for high-temperature use is now underway.

Carbon fibers are formed from three different precursors: rayon, acrylic copolymers, and mesophase pitch. Matrix precursors used in fabrication of these composites are based on coal tar and petroleum pitches, several synthetic resins, or CVD carbon. Precursor materials have not been optimized. During carbonization coal tar and petroleum pitches (by-products of the catalytic cracking of crude oil) develop ordered phases that affect mechanical properties. Most synthetic resins transform to brittle vitreous carbon. CVD carbon can assume several morphologies (amorphous, columnar, or laminar), the specific form being controlled primarily through experiment.

SiO_2 film stability and viscosity, of course, are major issues in considering use of SiC or Si_3N_4 coatings for carbon–carbon composites. Addition of second-phase particles is the most effective method of raising the viscosity of SiO_2 to minimize sloughing off of the oxide, but decomposition around 2700 °F (1500 °C) is still a problem.

SUMMARY

This chapter has attempted a concise evaluation of the materials and materials systems that have, in recent years, arisen as potential competitors or substitutes to replace or extend the service of superalloys. The next chapter specifically faces the *future* of superalloys and, in that context, also touches on some of these alternate materials.

To summarize, the four "systems" discussed in sufficient detail here to allow the reader to make some meaningful comparisons are: intermetallic compounds (IMCs); metal–matrix composites, including DS eutectics; refractory metals; and ceramics and ceramic composites. Of these groups and their major subdivisions, the authors submit that only two groups of materials may succeed in the near term. These are (1) titanium aluminides (from IMCs) and (2) columbium alloys (from refractory metals). The primary reasons these two materials subgroups will probably be utilized are as follows:

1. *Titanium aluminides* possess outstanding oxidation resistance and very low density. (Their problems are fabrication difficulties, unknown hot corrosion behavior, and concern for service ductility retention/crack growth limitations.)
2. *Columbium alloys* possess strength potential to beyond superalloys in temperature range and (when coated) acceptable surface stability. (Their problems are the need to further increase strength and to improve protection against surface attack and contamination.)

In the general sense, the overriding failings of the balance of the list can be cited as follows:

Intermetallic compounds, with the exception of the titanium aluminides (TiAl and Ti$_3$Al), at their present stage of development provide only marginal (or no) property advantages against advanced superalloys. Ni$_3$Al is a prime example; its density and melting point are not a *sufficient* advantage. However, iron, cobalt, and nickel aluminides may find advantage in other high-temperature or severe-condition service. In fact, TiAl and/or Ti$_3$Al will be accepted only after some further improvements in fabricability/cost and toughness/reliability parameters.

Eutectic alloys have been shown technically to be engine acceptable. However, unless a major processing breakthrough is achieved to bypass the characteristically slow growth rates required in processing by directional solidification, their cost is too high to allow a rational trade-off against the modest advantages shown over more "conventional" cast single-crystal superalloys.

Refractory metals, columbium alloy systems excepted, clearly are noncompetitive in density and surface stability, as shown by over 40 years of on–off considerations, despite their very great melting point advantage. However, they may very well become integral phases in fiber-strengthened composites.

Ceramics and ceramic composites are the most difficult to evaluate. On strength, density, surface stability, availability, and cost bases, they generate very high interest. Further, many ceramics (and their composites) would appear to have the capability to extend usable strength capacity to much higher temperature levels.

However, their *innate* brittleness must be bypassed by revolutionary mechanical designs and materials design techniques to allow service under conditions heretofore considered impossible. At this time these solutions have not been achieved in monolithic ceramics and their alloys after 10 or more years of intensive work. Ceramists are turning to ceramic composites, which offer the possibility of somewhat improved toughness. Ultimately, service as bearing materials represents, perhaps, the highest potential really achievable.

Carbon–carbon composites, with usable strength to virtually 4000 °F (2200 °C), have such low potential for reasonably reliable surface stability performance that they are limited to very short time or relatively moderate temperature use.

Thus, in conclusion, superalloys are beginning to meet some limited competition from "alternate" materials on a slender and individual basis from highly qualified materials in an atmosphere of strenuous efforts. However, the virtually unique properties of superalloys up to 2000 °F (1200 °C) forestall, in the short run, great inroads into their prime application fields of excellence, such as for gas turbine hot-stage parts. Nevertheless, it must be understood clearly that superalloys are nearing their operating temperature limits. Further improvement in gas turbine efficiency and performance will, therefore, demand continuing efforts to overcome the evident deficiencies cited above for the alternative materials. See also the next chapter.

REFERENCES

1. C. T. Liu, C. L. White, and J. A. Horton, *Acta Met.*, **33**, 213 (1985).
2. A. I. Taub, S. C. Huang, and K. M. Chang, *Met. Trans. A.*, **15A**, 399 (1984).
3. C. T. Liu and C. L. White, Oak Ridge National Laboratory, unpublished.
4. Structural Uses for Ductile Ordered Alloys, NMAB-419, Washington, DC, August 31, 1984.
5. C. T. Liu, in *High Temperature Alloys: Theory and Design*, TMS-AIME, Warrendale, PA, 1984, p. 289.
6. N. S. Stoloff and R. G. Davies, *Prog. Mat. Sci.*, **13**, 1 (1966).
7. M. J. Marcinkowski, M. E. Taylor, and F. X. Kayser, *J. Mat. Sci.*, **10**, 406 (1975).
8. S. K. Ehlers and M. G. Mendiratta, *J. Met.*, **33**, 5 (December 1981).
9. E. M. Schulson, COSAM Program Overview, NASA TN-8300006, NASA, Washington, DC, 1982, p. 175.
10. K. Aoki and O. Izumi, *J. Jap. Inst. Met.*, **43**, 1190 (1979).
11. A. I. Taub, C. Huang, and K. M. Chang in *Proceedings of the Symposium on High Temperature Ordered Intermetallic Alloys*. C. T. Liu, C. C. Koch and N. S. Stoloff (eds.), Materials Research Society, Pittsburgh, PA, 1985, p. 221.
12. A. I. Taub, C. L. Briant, S. C. Huang, K. M. Chang, and M. R. Jackson, *Scripta Met.*, **20**, 129 (1986).
13. K. Vedula, Case Western Reserve University, unpublished.
14. C. T. Liu and C. L. White, in *High Temperature Ordered Intermetallic Alloys*, C. T. Liu, C. C. Koch, and N. S. Stoloff (eds.), Materials Research Society, Pittsburgh, PA, 1985, p. 365.
15. C. Koch, in *Proceedings of the Symposium on High Temperature Ordered Intermetallic Alloys*, C. T. Liu, C. C. Koch and N. S. Stoloff (eds.), Materials Research Society, Pittsburgh, PA, 1985, p. 397.
16. P. R. Strutt and R. A. Dodd, in *Ordered Alloys, Structural Applications and Physical Metallurgy*, B. H. Kear, C. T. Sims, N. S. Stoloff, and J. H. Westbrook (eds.), Claitor, Baton Rouge, LA, 1969, p. 475.
17. P. H. Thornton, R. G. Davies, and T. L. Johnston, *Met. Trans.*, **1**, 207 (1970).
18. R. T. Fortnum and D. E. Mikkola, Abstract Bulletin, TMS-AIME Annual Meeting, February 1986, p. 14.
19. K. Vedula, Abstract Bulletin, TMS-AIME Annual Meeting, February 1986, p. 51.
20. J. E. Doherty, A. F. Giamei, and B. H. Kear, *Met. Trans. A*, **6A**, 2195 (1975).

21. A. K. Kuruvilla and N. S. Stoloff, *Scripta Met.*, 21, to be published (1987).
22. J. C. Greiss, C. T. Liu, and J. H. Devan, Oak Ridge National Laboratory, unpublished.
23. T. Ishii, D. J. Duquette and N. S. Stoloff, *Acta Met.*, **29**, 1467 (1981).
24. N. S. Stoloff, in *Encyclopedia of Materials Science and Engineering*, Pergamon, Elmsford, NY, 1986, p. 1204.
25. H. Bibring, *Proceedings of the Conference on In-Situ Composites*, NMAB-308-II, NRC, Washington, DC, 1973, p. 1.
26. M. F. Henry and N. S. Stoloff, in *Fatigue of Composite Materials*, STP 569, American Society for Testing and Materials, Philadelphia, PA, 1975, p. 189.
27. N. Bylina, and N. S. Stoloff, *Mat. Sci. Eng.*, to be published (1987).
28. E. M. Breinan, E. R. Thompson, and F. D. Lemkey, *Proceedings of the Conference on In-Situ Composites*, NMAB 308-II, NRC, Washington, DC, 1973, p. 201.
29. D. Woodford, *Proceedings of the Second Conference on In-Situ Composites*, Xerox, Lexington, MA, 1976, p. 211.
30. M. Rabinovitch, in *Advances in Composite Materials*, Applied Science, Barking, Essex, England, 1978, p. 289.
31. R. A. Signorelli and J. A. DiCarlo, *J. Met.*, **37**, 41, (June 1985).
32. J. K. Tien and V. C. Nardone, in *Fracture: Interactions of Microstructure, Mechanisms, Mechanics*, TMS-AIME, New York, 1985, p. 321.
33. D. W. Petrasek and R. A. Signorelli, NASA TM-82590, 1981.
34. R. E. Gold and D. L. Harrod, *J. Nucl. Mat.*, **85–86**, 805 (1979).
35. R. W. Buckman, Jr. and R. C. Goodspeed, in *Refractory Metal Alloys*, Plenum, New York, 1968.
36. W. R. Witzke, NASA TM-X-3239, Lewis Research Center, Cleveland OH, 1975.
37. J. Wadsworth, T. G. Nieh, C. M. Packer, and J. J. Stephens, Abstract Bulletin TMS-AIME Annual Meeting, February 1986, p. 60.
38. Lucas Cookson Syalon, Ltd., SAE Paper 850521, reprinted in *Adv. Mat. Proc.*, **2**, 37 (January 1986).
39. D. C. Larsen et al. (eds.), *Ceramic Materials for Advanced Heat Engines*, Noyes, Park Ridge, NJ, 1985.
40. L. M. Shephard, *Adv. Mat. Proc.*, **1**, 39 (September 1985).
41. L. M. Shephard, *Adv. Mat. Proc.*, **1**, 39 (November 1985).
42. M. S. Seltzer, "High Temperature Creep of Ceramics," AFML-TR-76-97, Air Force Materials Laboratory, Wright-Patterson AFB, OH, June 1976.
43. S. C. Singhal, in *Properties of High Temperature Alloys (with Emphasis on Environmental Effects)*, Electrochemical Society, Princeton, NJ, 1977, p. 697.
44. J. W. Hinze, W. C. Tripp, and H. C. Graham, in *Mass Transport Phenomena in Ceramics*, Plenum, New York, 1975.

Chapter 20

Future of Superalloys

G. S. HOPPIN III and W. P. DANESI

Garrett Turbine Engine Company, Phoenix, Arizona

When this chapter was written initially 14 years ago, several technologies were identified as holding promise if research and development were conducted in those areas. Since that time significant strides have been made with development of (1) superalloy overlay and thermal barrier coatings to meet specific alloy/application needs, (2) grain-size-controlled cast and wrought turbine wheel materials, (3) several alloys designed specifically to be produced by the single-crystal casting process, (4) development of a high-volume casting process for the manufacture of superalloy turbocharger wheels, and (5) an AMS specification for the control of tramp elements. As with any forecast, some areas have not been brought to production, although some engine development has been conducted on each: eutectic alloys, refractory alloys, and ceramics. Of the three, ceramics appears to be the closest to a production application. Eutectic alloys have been shown to be engine acceptable but appear to be too costly for production implementation.

Even with these advances, there are tremendous challenges still facing the materials engineer in adapting alloys and processes to meet the demands of current and future engine requirements.

Second-generation jumbo jet engines such as the Pratt and Whitney PW2037, General Electric SNECMA CFM56, and Rolls-Royce RB211-535E4 are now in service in the latest large passenger jet aircraft. These engines are operating at turbine inlet temperatures 500 °F (280 °C) greater than the comparable engines of 1970, with 250 °F (140 °C) of the temperature improvement coming from improvements in superalloy turbine blade temperature capability.[1] Similar material improvements are incorporated in the latest large American military jet engines, GE F404 and PW1120. As was the case with the prior generation of large jet engines, over 50% of the weight of these engines is composed of superalloys. The same types of superalloys used in the large engines are now in use in Garrett's TFE731 business jet engines, Pratt and Whitney of Canada's PW100 commuter turboprops, and Allison's Model 250 helicopter engines. Both large and small turbine engines are now classified as long-life aircraft engines, with times between overhaul typically in the 3000–10,000-h range.

These engines have always been a glamorous application of superalloys. Several other types of turbines are also vitally dependent on superalloys: industrial turbines for electrical power and mechanical drive applications, turbopumps for liquid propellant rocket engines such as the SSME (space shuttle main engine), auxiliary power units (APUs), missile engines, vehicular turbines, and automobile and truck turbosuperchargers.

Industrial "heavy-duty" turbines are required to operate reliably for lives in excess of 100,000 h burning a wide variety of fuels, often in highly corrosive environments such as off-shore oil platforms. Together with steam turbines in combined cycles, these large gas turbines significantly raise the overall thermal efficiency of central station electric power generation systems. Such "STAG" (steam and gas) systems are now becoming common. The "heavy-duty" gas turbine is also integral in plants designed to take energy directly from coal, such as those employing fluidized-bed combustors.

Auxiliary power units are widely used to provide electrical power and compressed air for aircraft starting and air conditioning. They require long life and high reliability and typically operate at lower metal temperatures than long-life aircraft engines. Missile engines have evolved to meet two challenging and dissimilar criteria: low-cost construction and high reliability. Vehicular turbines were originally planned for high-volume production in trucks and automobiles. This market did not materialize due to unanticipated improvements in diesel and gasoline engine efficiency. A new market did, however, develop for vehicular turbines in military-tracked vehicles, such as the U.S. Army's M1 tank, where the Lycoming AGT 1500 turbine is the powerplant. A high-volume production market also evolved for automobile and truck turbochargers, which are the most cost-driven of all superalloy-using turbines.

Because of the wide differences in duty cycles and operational environments, the various types of gas turbines impose significantly different requirements on superalloy components. Requirements for these different classes of turbines include maximum creep resistance for turbine blades and vanes (airfoils), maximum oxidation resistance for airfoils, maximum corrosion resistance for airfoils, long-time metallurgical stability, high-cyclic life (10,000 + cycles) for disks, negligible (less than 100 cycles) disk cyclic life, high tensile strengths for disks, minimum crack growth rates for disks, minimum cost components, and maximum resistance to thermal mechanical fatigue. Since many of these requirements are conflicting, a wide variety of superalloys, coatings, and component manufacturing processes are needed to satisfy the unique requirements of each class of turbine. The future in superalloys lies in providing materials that address all these requirements.

ALLOY DEVELOPMENT

A quick look at where we have been is good preparation for contemplating the future. Considering only stress rupture strength versus time, Fig. 1 depicts progress versus time for the forged and cast superalloys used as turbine blades and vanes. Between 1940 and 1970 temperature capabilities of the alloys shown grew at about 17 °F/yr (10 °C/yr) until an upper asymptote was reached for conventionally cast polycrystalline superalloys. Figure 1 shows that since the early 1960s two families of blade and vane materials have been available. The alloys in the upper band offer maximized stress rupture strength and generally poor hot-corrosion resistance while those in the lower band offer relatively good hot corrosion at lower strength levels. The two families evolved as increasingly higher alloy contents forced a choice between high strength and metallurgical stability at low (8–12%) chromium levels or good corrosion resistance and stability at higher (14–16%) chromium levels.

Fig. 2 presents another way of looking at the data from Fig. 1 and developments of recent years. Three eras become apparent: the weak alloys producible by air melting in the early years, the stronger alloys plateauing about 1970 made by vacuum melting, and the current era of high-strength alloys having directional anisotropic macrostructures and properties.

The directional structure development began with the work of VerSnyder et al.[2] at Pratt and Whitney Aircraft that eventually resulted in the mass production of blades and vanes of directionally solidified (DS) MAR-M 200 + Hf with polycrystalline bundles of grains grown directionally from a water-cooled copper chill in the low-elastic-modulus, high-creep-strength (001) crystallographic orientation. During the 1975–1985 period vigorous research and development competition was waged between several competing material systems with directional structures to become the successor to DS MAR-M 200 + Hf. The competing systems included single-

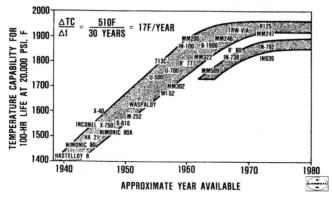

Fig. 1. Temperature capability of superalloys.

THE THREE ERAS OF SUPERALLOY PROCESSING

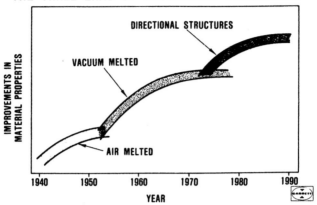

Fig. 2. The three eras of superalloys.

crystal (i.e., single-grain) cast superalloys, directionally solidified eutectic cast superalloys, directionally recrystallized extruded powder superalloys (with and without oxide dispersoids), and fiber-reinforced superalloys. The winner in this competition—both for technical and economic considerations—was the cast single-crystal class of superalloy. These have now been adopted by major aircraft engine producers as materials for turbine blades and vanes in their latest high-performance engines.

The compositional evolution of some representative single-crystal alloys is traced in Table 1. (A detailed account of these alloys appears in Chapter 7.) The high-strength cast alloy MAR-M 247 has been successfully used in production to produce both equiaxed polycrystalline and directionally solidified (DS) castings. Single-crystal castings made of the MAR-M 247 chemistry were found to offer no improvement in stress rupture strength over DS castings, even though a reduction in data scatter was noted. By removing all the grain-boundary-strengthening elements (i.e., C, B, Hf, and Zr) from MAR-M 247, the NASAIR 100 SC alloy was created. It was experimentally determined that NASAIR 100 had about a 50 °F improvement in stress rupture properties over DS MAR-M 247. Tests of NASAIR 100 showed it to be metallurgically unstable, with the μ phase being present as cast and converting to M_6C with subsequent time at temperature. The addition of 5% cobalt to NASAIR 100 solved the stability problem. A further addition of 0.6% hafnium improved coated oxidation resistance and resulted in "Alloy 3," developed on the same NASA research program as NASAIR 100.[3] The ultimate evolution of this work was CMSX-3, a commercial single-crystal alloy similar to Alloy 3, with a lower hafnium content and a modified tantalum–tungsten ratio. Other first-generation single-crystal alloys include PWA1480[4] and SRR-99.[5] All of these first-generation single-crystal alloys have virtually identical creep and stress rupture strengths when properly heat treated, although other critical properties vary.

The next major thrusts anticipated in single-crystal alloy development are the development of alloys containing rhenium for improved stress rupture strength and minor additions of yttrium and/or rare-earth elements such as lanthanum for improved environmental resistance. Rhenium is beneficial for stress rupture strength in that it strengthens the alloy matrix and also retards the coarsening of fine γ' with time at temperature. Yttrium and rare-earth elements in proper proportions stabilize the alumina–chromia scale on the alloy surface and can impart remarkable oxidation resistance that may obviate the need for protective surface coatings on airfoils.[6] The use of rhenium as an alloying element will significantly increase alloy cost. To effectively use rhenium-containing alloys in production, a revert process must be developed to recover the alloy expended in casting gates, risers, and scrap parts. The successful development of "no-coat" alloys containing yttrium and rare earths will require extremely close control of molten metal-ceramic reactions during initial alloy manufacture and in the casting process. This control will be necessary to prevent the active elements from reducing oxides in the ceramic molds and cores during alloy manufacture and casting.

Little will be accomplished in the development of conventional superalloys for disks. No new high-strength disk alloys have been developed since the powder superalloys (modified IN-100 and René 95) were developed in the 1960s. The extremely high tensile strength of these alloys confers the desired maximized low-cycle fatigue strength but at the expense of high-cyclic crack-growth rates. A monumental effort has been expended to minimize the size of intrinsic defects in these alloys and to develop ultrasensitive nondestructive evaluation techniques and equipment to detect small flaws within disks and on the surface at critically stressed areas. The probability of stronger disk alloys being developed is small, since all past experience indicates that the defect sensitivity of stronger alloys would be greater than that of the current alloys. Some intriguing possibilities exist in developing higher strength/density disks by using a γ' Ni_3Al-type alloy as the matrix, with further strengthening being accomplished by some second-phase dispersoid.

Another potential competing alloy system for advanced turbine disks is based on the Ti_3Al intermetallic compound. Alloys derived from this system have significantly lower density than nickel superalloys and retain useful creep strength up to 1250 °F (625 °C). As yet, the tensile strengths of the Ti_3Al alloys are inadequate for turbine disks, and room temperature ductilities are low. Developmental efforts are actively underway to correct these deficiencies.

Table 1. Chemical Compositions of Major Single-Crystal Alloys and Three of Their Precursors

	Chemical Composition (wt. %)										
	Co	Cr	Mo	W	Ta	Al	Ti	Hf	C	B	Zr
MAR-M247	10	8.4	0.6	10.0	3.3	5.5	1.0	1.4	0.15	0.015	0.05
NASAIR 100	—	9.0	1.0	10.5	3.3	5.8	1.0	—	—	—	—
"Alloy 3"	5.1	8.7	8.0	10.0	3.2	5.4	1.1	0.6	—	—	—
CMSX-3	4	7.5	0.5	7.5	6	5.5	0.9	0.1	—	—	—
PWA-1480	5	10	—	4	12	5.0	1.5	—	—	—	—
SRR-99	5	8.5	—	9.5	2.8	5.5	2.2	—	—	—	—

ULTRACLEAN METAL

The major thrust in disk alloys will be in producing extremely clean material and processing it into parts with extremely uniform microstructures maximizing tensile and low-cycle fatigue strength and resistance to crack growth. The use of ultrahigh-strength powder metallurgy disk alloys such as René 95 and Gatorized IN-100 has resulted in major efforts to minimize the size of the largest defect present in finished parts due to relatively rapid crack propagation when these alloys are used at high stress levels.[7] A developing trend that will become stronger in the future is the use of specialized refining processes to produce the cleanest possible starting stock for subsequent powder production. The most promising process identified to date is electron beam cold-hearth refining (EBCHR).[8] The EBCHR process has also been effectively used to produce high-quality superalloy remelt bar for producing vacuum investment cast airfoils and structural parts.[9] This process and other competing high-purity alloy processes such as electroslag remelting (ESR) and plasma arc melting will be increasingly used for improved cleanliness and precise chemistry control of superalloys.

Processes such as EBCHR, ESR, and plasma arc melting will all be used in conjunction with vacuum induction melting (VIM) to effect primary melting. Although the oldest of the vacuum melting processes, VIM is the most versatile in its ability to melt a wide variety of charge materials from 100% virgin elements to 100% alloy scrap. A new approach in VIM to improve alloy purity is the use of ceramic filters to remove oxides during pouring of the molten alloy into the ingot molds.[10] This process is currently in production by one alloy supplier and is being studied by others. It currently appears that VIM + ESR will replace VIM + VAR for a wide variety of alloys due to better workability of ESR ingots. VIM + EBCHR or plasma arc melting will probably be restricted to high-strength disk alloys due to the inherently higher cost of these processes.

To determine "how clean is clean," the electron-beam button-melting process will be employed, with specifications developed to define accept–reject criteria based on the "raft" of oxides remaining on the top of a 1–2-lb "button" after melting.

MINOR AND TRACE ELEMENTS

The advent of modern single-crystal castings resulted in the effective removal of as much boron, carbon, and zirconium as possible from the alloy's chemistry.[4,5] The manufacture of the remelt bar for these alloys has required more stringent chemical controls of these elements than in prior superalloys. The next steps in single-crystal alloy development will involve alloys containing rhenium for improved creep resistance[11,12] and alloys with minor additions of hafnium and yttrium for maximized oxidation resistance.[6] Very close control of both master alloy melting practice and investment casting parameters will be necessary to prevent oxidation of the highly reactive yttrium (or La, which is another logical candidate based on its successful use to maximize oxidation resistance in wrought alloys).[13]

The trace-element problem encountered by the industry around 1970 has been resolved by a specification (AMS 2280) universally accepted in the United States. This specification controls 20 trace elements that can cause degradation of the mechanical properties of nickel-base alloys. This specification will be applied to single-crystal airfoil alloys in the future, probably in a modified version, as the effects of trace elements on the mechanical properties of single-crystal alloys have yet to be determined.

COBALT IN SUPERALLOYS

The world cobalt shortage precipitated by the insurrection in Zaire in 1978–1979 led to an excellent NASA program exploring the role of critical materials in gas turbines.[14] The research performed in this study clearly showed that many cast and wrought nickel-base superalloys contained much more cobalt than was needed to assure alloy producibility and maximized mechanical properties. Specifically, Waspaloy at about 8% cobalt was found to have equivalent properties to the conventional 14% cobalt alloy. The lower cobalt composition has not yet been adopted, since cobalt is no longer in short supply. Cast single-crystal alloy work also has indicated that about 5% cobalt is needed to prevent μ phase formation in a single-crystal alloy made by chemically modifying MAR-M 247.[3] The 5% cobalt level is about half that used in most production cast turbine airfoil alloys. As Table 1 shows, 4 −5% cobalt has been the level selected for the first generation of production single-crystal alloys.

COBALT-BASE ALLOYS

The cobalt shortage of the late 1970s also totally discouraged the development of new cobalt-base alloys. Alloy development had previously been lagging due to the lack of γ'-type strengthening in cobalt alloys. Certain older cobalt-base alloys have survived for use in cast turbine vanes where the alloys' weldability is an asset for component repairability, and alloys such as FSX-414 remain the mainstay choice for virtually all heavy-duty gas turbine nozzle diaphragm castings. However, the trend of 10 years ago to use the cobalt-base sheet alloy HA-188 for combustor liners in all new engines has been halted with greater usage of nickel-base combustor alloys becoming the accepted practice.

ODS ALLOYS

The economic decline of the nickel industry resulted in developmental work on oxide-dispersion-strengthened (ODS) alloys being essentially terminated. The INCO-developed yttria–dispersoid ODS alloy MA-754 has evolved into a high-volume production alloy for turbine vanes that is competitive with single-crystal castings in this application. Usage of MA-754 will grow with time and a renaissance of

ODS alloy development work will occur for turbine blades and disks, primarily because other strengthening mechanisms have been essentially exhausted for these types of superalloys.

COATINGS AND COATING ALLOY DEVELOPMENT

Increasing metal temperatures on blade and vane alloys with lower chromium contents than earlier alloys will make the need for improved coatings imperative to obtain acceptable oxidation and hot-corrosion lives on turbine airfoils. The relatively simple diffusion aluminide coatings will be increasingly displaced by "overlay" coatings of the MCrAlY type. The overlay coatings will be carefully designed to maximize diffusional stability with the substrate alloy and to minimize differentials in the coating and alloy thermal coefficients of expansion. The overlay coatings are applied by plasma spraying metal powders or the electron beam physical vapor deposition processes. A typical overlay coating composition is Ni–23% Co–20%Cr–8.5%Al–4%Ta–0.04%Y, a commercial Alloy Metals, Inc. product.

PROCESS DEVELOPMENT

Improved reliability of superalloy parts will be attained by processes producing unique microstructures, either directional in nature for turbine blades and vanes or extremely isotropic for turbine disks. Powder alloys will gain increasing acceptance for turbine disks and for other parts produced by vacuum plasma deposition of superalloy powders. Hybrid components manufactured by diffusion bonding together two or more pieces to produce a final monolithic part will also be a major processing thrust.

Fig. 3. Fine-grained casting of radial turbine wheel produced by mold agitation process.

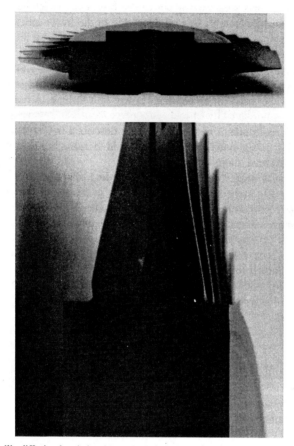

Fig. 4. Bimetallic diffusion-bonded turbine wheel. Upper, cross section of entire wheel. Lower, bond between cast blade ring and fine-grained powder hub.

Directionally solidified and single-crystal alloy turbine blades and vanes will see increasing use in turbines in sizes ranging up to 10–20 in. in length. Automated, computer-controlled processes to produce these castings will improve product quality and reduce the costs of these parts. Single-crystal parts will almost universally be produced by "seeding" techniques to assure that primary and secondary crystal growth directions are rigidly controlled.

Turbine disks for aircraft engines will be produced by total processing technology that will maximize isotropy by uniform grain size control. This will be obtained by true isothermal forging of cast ingots of weaker alloys and consolidated powders of the stronger alloys. Maximized isotropy of properties and uniform grain size are prerequisites to maximizing low-cycle fatigue life while minimizing crack growth rates in highly stressed disks. Fine-grained cast preforms will emerge as viable competitors to prealloyed powders for forging disks of the stronger alloys. Disks and other rotating parts produced by hot isostatic pressing (HIP) powders will find increasing use for moderately stressed disks due to the lower inherent costs when compared to forgings. Disks made by HIP are only limited by the working zone of the HIP vessel, the current maximum diameter being about 4 ft.

Another process that will find increasing application is fine-grained casting for integral turbine wheels in small engines. Fig. 3 shows a cross section of a cast radial turbine wheel used for an auxiliary power unit turbine. A uniform fine grain size structure was produced by a mold agitation process that fragmented the growing dendrites into a multitude of nucleation sites as described in Chapter 15. The major advantage of the fine-grained casting is a fivefold increase in low-cycle fatigue life over the large-grained static casting.

Multialloy parts made by HIP diffusion bonding also appear to have a bright future. The cross-sectioned turbine wheel shown in the top of Fig. 4 is composed of a cast ring of high-creep-strength turbine blades bonded to a fine-grain-size powder metallurgy hub made of a high-tensile-strength disk alloy. The lower photograph of Fig. 4 shows a close-up of the bond between the two alloys. Initial applications of this type of process have been in turbine wheels for small engines, but the process has also been studied for very large industrial turbine blades wherein the airfoil is made of one alloy and the blade root and attachment of a second alloy. Thus, it is anticipated that this type of technology will develop over a wide variety of superalloy component sizes.

COMPETING MATERIAL SYSTEMS

The material systems competing with the superalloys are described in Chapter 19. The principal competing systems are intermetallic compounds, carbon–carbon and metal-matrix composites, refractory metals, and ceramics.

Intermetallic Compounds

In recent years exploration of alloy systems based on the intermetallic compounds Ti_3Al[15], TiAl and Ni_3Al[16] has been actively pursued. The Ti_3Al and TiAl systems are potentially competitive with superalloys in the 1100–1500 °F (600–815 °C) regime, and the Ni_3Al alloys are potentially competitive to 2000 °F (1100 °C). It currently appears that engineering materials for a broad spectrum of gas turbines will be developed from the Ti_3Al system. The other systems are further from engineering application, but the ongoing research merits attention due to potential weight and cost reductions.

Carbon–Carbon

Composite materials of woven carbon fibers infiltrated with pitch or other carbonaceous materials and pyrolyzed have become known as *carbon–carbon*. Parts of carbon–carbon coated with SiC and a glassy infiltrant have been successfully used as the nose cap and leading edges of NASA's space shuttle. In this application parts experience very high heat fluxes and local temperatures for short periods of time during reentry from earth orbit. Since carbon has a density about one-fifth that of superalloys and retains useful strength to temperatures in excess of 4000 °F (2200 °C), the coated carbon–carbon materials are intriguing for use in gas turbines. Similar to the refractory metals, the key problem to solve is oxidation, which becomes a problem at all temperatures above 800 °F (425 °C). Successful use of coated carbon–carbon is thus totally dependent on coating technology as are the refractory metals discussed following. The use of coated carbon–carbon for gas turbines is being actively researched for military applications. Time will determine the success of this endeavor.

Refractory Metals

The four refractory metal elements having alloys of engineering significance are molybdenum, tungsten, tantalum, and columbium. Major efforts in alloy development in these systems occurred in the 1950–1965 period. During this period many engineering alloys of molybdenum, columbium, and tantalum were developed. The Achilles heel of these materials was and is poor oxidation resistance, which in turn stimulated the development of protective coating systems for the alloys. Tungsten and molybdenum and their alloys also exhibit ductile–brittle transition temperature behavior. This drawback had been surmounted by controlled mechanical working to lower the transition to acceptable values. The engineering alloys of columbium and tantalum have found application in both liquid and solid propellant rocket engines. In these applications the poor oxidation resistance has not been limiting since the exposures to very high temperatures are relatively brief and often occur at extremely high altitudes where the partial pressure of oxygen is extremely low.

The first major use of a columbium alloy in a gas turbine engine has been on a production military engine, which has successfully used coated columbium alloy exhaust nozzle flaps for several years. The alloy used for the flaps is C-103 (nominally Cb–10%Hf–1%Ti) and the coating is a fused silicide. Development work is in progress to produce columbium alloys with much greater inherent oxidation resistance than C-103. It is highly probable that coated columbium alloys will find future use in specialized applications where local temperatures would melt superalloys and no cooling air is available.

Experimental silicide-coated turbine wheels of forged TZM molybdenum have been successfully run in a small turbojet demonstrator engine by Williams International for 7 h at a maximum gas temperature of 2450 °F (1343 °C).[17] Molybdenum alloys also have found other high-temperature applications in glass melting and as heating elements in vacuum furnaces. There appears to be some potential uses of molybdenum alloys in short-life missile engines but not in the other categories of gas turbines.

Tantalum and tungsten alloys have not found application in gas turbines. The oxidation problem coupled with the high densities and relatively high costs of these metals will preclude their use in the future as superalloy competitors in gas turbines.

Ceramics

Considerable progress in the manufacture, design, and testing of gas turbine components made of the structural ceramics SiC and Si_3N_4 has been made in the past decade by European, American, and Japanese companies. Much of the work has been heavily government supported. Two demonstrator automotive gas turbines have been developed and demonstrated, the AGT-100 by Detroit Diesel Allison and the AGT-101 by a Garrett-Ford partnership.[18] These demonstrator engine programs have identified a number of deficiencies in current ceramic technology, which has led to a major U.S. Department of Energy program to improve the ceramic technology level.[19] The best existing ceramic materials are capable of operating uncooled at temperatures of about 2500 °F (1371 °C)—well beyond the temperature capabilities of superalloys. The biggest drawback of ceramics—their completely brittle fracture mode—has not been solved, but significant advances have been made in design methodology for these materials, and the consistency of material properties has improved. Research is just beginning in ceramic–ceramic composites, which exhibit pseudo-plasticity in failure and may extend the engineering usefulness of ceramics.

Ceramic material technology has matured to the point where it now appears viable for use in short-life missile engines and vehicular engines of relatively small size. It should be carefully monitored as a competitor to superalloys in the future.

SUMMARY

Superalloys are alive and well and entering the second era of the "age of processing." Some key future trends include

1. use of rhenium as an alloying element to maximize high-temperature creep strength;
2. development of "no-coat" alloys containing controlled amounts of hafnium, lanthanum, and yttrium for oxidation resistance;
3. greater usage of directionally-solifidied and single-crystal turbine blade and vane castings;
4. improved cleanliness to minimize intrinsic defects in disk alloys;
5. greater usage of isothermal forging to achieve uniform grain size in disks;
6. greater usage of disk alloys produced from prealloyed powders;
7. greater usage of fine-grained cast turbine wheels densified by hot isostatic pressing; and
8. widespread development and application of hybrid parts made by diffusion bonding two or more alloys together to form a single part.

The competition from ceramics and intermetallic alloys will become more intense, with these materials finding specialized applications first in short-life engines. The usage of coated refractory metals will be minor, while coated carbon–carbon materials may see specialized military usage.

While the competing materials will make inroads in specialized niches, the γ'-strengthened nickel-base superalloys will remain the mainstay of the gas turbine business due to their proven versatility and amenability to being formed into gas turbine components of all sizes by a variety of casting, forging, and powder metallurgy processes.

Two items discussed in the past[20] are still very true and will be so into the 1990s: (1) the metallurgist is currently faced with a tremendous challenge to advance the applications of superalloys and (2) no other metal systems are blessed with the combination of melting points, corrosion resistance, and inherent precipitates with reversible solubility to accommodate temperature variations as the superalloys.

Although the superalloy systems continue to encounter competing metallic and nonmetallic systems, the superalloys will continue to be the dominant materials in future gas turbine engines.

REFERENCES

1. F. E. Pickering, "A Decade of Progress in Turbomachinery Design and Development," Paper 851989, SAE, October 15, 1985.
2. J. S. Erickson, C. P. Sullivan and F. L. VerSnyder, in *High Temperature Materials in Gas Turbines*, P. Sahm and A. Speidel, (eds.), Elsevier, Amsterdam, p. 315, 1974.
3. T. E. Strangman et al., *Superalloys 1984*, AIME, p. 795.
4. M. Gell, D. N. Duhl, and A. F. Giamei, *Superalloys 1980*, Metals Park, OH, ASM, p. 205.
5. P. A. Ford and R. P. Arthey, *Superalloys 1984*, The Metallurgical Society of AIME, Warrendale, PA, p. 115.
6. G. H. Meier, F. S. Pettit, and A. S. Khan, Rapid Solidification Processing, Principles and Technologies, III, National Bureau of Standards, December, 1982, p. 348.
7. D. R. Chang, D. D. Krueger, and R. A. Sprague, *Superalloys 1984*, The Metallurgical Society of AIME, Warrendale, PA, p. 245.
8. C. E. Shamblen, D. R. Chang, and J. A. Corrado, *Superalloys 1984*, The Metallurgical Society of AIME, Warrendale, PA, p. 509.
9. C. d'A. Hunt, J. C. Lowe, and S. K. Harrington, "The Use of High Quality Superalloy Melt Stock Refined by Electron Beam Cold Hearth Melting in Precision Investment Foundaries," Proceedings of Conference on Electron Beam Melting and Refining, Bakish Materials Corp., November 1985.
10. D. Apelian and W. H. Sutton, *Superalloys 1984*, The Metallurgical Society of AIME, Warrendale, PA, p. 421.
11. L. Dardi, A. Dalal, and C. Yaker, "Metallurgical Advances in Investment Casting Technology, Howmet Turbine Components Corp., Whitehall, MI, 1985.
12. G. Erickson, K. Harris, and R. Schwer, "Development of CMSX-5. A Third Generation High Strength Single Crystal Superalloy," Cannon-Muskegon Corp., Muskegon, MI, February 1985.
13. M. F. Rothman, ASME Paper 85-GT-10, ASME, Philadelphia, March, 1985.
14. J. R. Stephens et al., NASA Technical Memorandum 83006, NASA Lewis Flight Center, Cleveland, OH, October, 1982.
15. H. A. Lipsitt and S. M. L. Sastry, *Met. Trans. A*, **8A**, 1543 (1977).
16. National Materials Advisory Board NMAB-419, *Structural Uses for Ductile Ordered Alloys*, National Academy Press, Washington, DC, 1984.
17. M. Egan et al., *Physical Metallurgy and Technology of Molybdenum and its Alloys* AMAX Materials Research Center, Ann Arbor, MI, 1985, p. 33.
18. J. Kidwell and D. Kreiner, ASME Paper 85-GT-177, ASME, Philadelphia, March 1985.
19. D. R. Johnson et al., American Ceramic Society Bulletin, February 1985, p. 276.
20. W. Danesi and M. Semchyshen, in *The Superalloys*, Wiley, New York, 1972, p. 565.

Appendix A

Phase Diagrams

Compiled by ROBERT L. DRESHFIELD and T. P. GABB

NASA Lewis Research Center, Cleveland, Ohio

This appendix contains ternary and polar phase diagrams of systems important to superalloys. To conserve space, the nickel–iron–cobalt system has been omitted as complete solubility exists in the region of greatest importance to superalloys.

The diagrams are reproduced as they were originally published; therefore, they contain atomic or weight percent as originally used. It should also be noted that the face-centered-cubic phase (Ni, Co, and Fe) is called γ in this compilation.

The polar diagrams shown are examples of diagrams prepared by Sims[1] to show the maximum extent known or expected for the intermetallic phases shown. They are of particular value for assessing the potential stability of alloy compositions. A more complete collection and details on the interpretation of the diagrams may be found in Sims and Hagel.[1]

The diagrams shown for nickel–aluminum–chromium are those of Taylor and Floyd.[2] A recent review of the system was performed[3] and substantially confirmed the diagrams shown here.

The following diagrams are included in this appendix:

TERNARY PHASE DIAGRAMS

Figure Number	System	Temperature, °F (°C)	Reference
A1	Co–Cr–Fe	2200 (1200)	4
A2	Co–Cr–Mo	2200 (1200)	5
A3	Ni–Al–Cb	1380 (750)	6

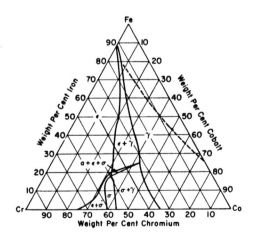

Fig. A1. Co–Cr–Fe system at 2200 °F (1200 °C).

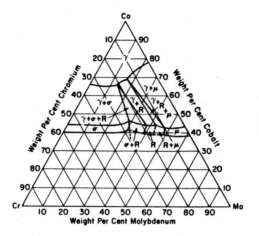

Fig. A2. Co–Cr–Mo system at 2200 °F (1200 °C).

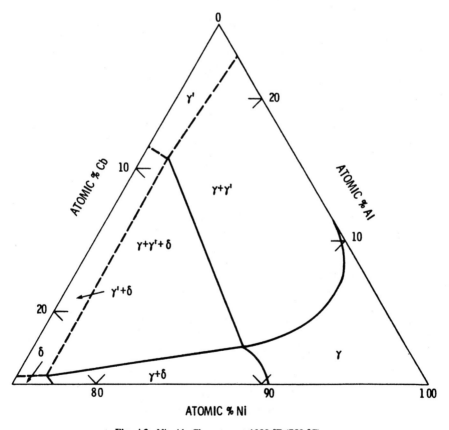

Fig. A3. Ni–Al–Cb system at 1380 °F (750 °C).

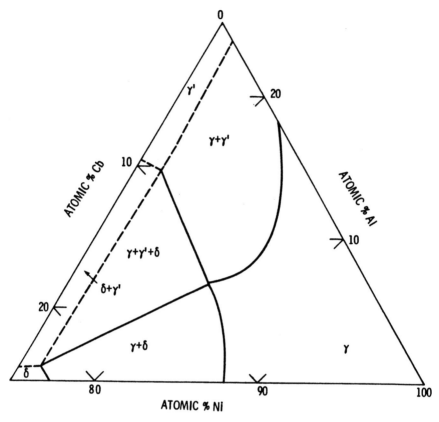

Fig. A4. Ni–Al–Cb system at 2200 °F (1200 °C).

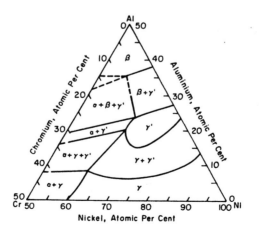

Fig. A5. Ni–Al–Cr system at 1550 °F (850 °C).

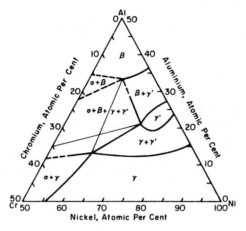

Fig. A6. Ni–Al–Cr system at 1830 °F (1000 °C).

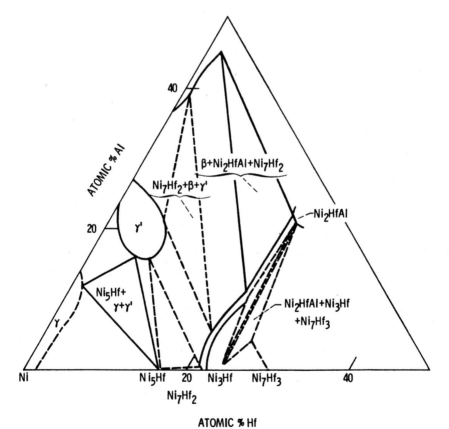

ATOMIC % Hf

Fig. A7. Ni–Al–Hf system at 1830 °F (1000 °C).

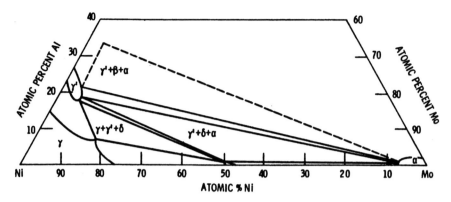

Fig. A8. Ni–Al–Mo system at 1900 °F (1038 °C).

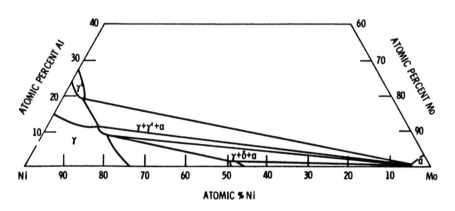

Fig. A9. Ni–Al–Mo system at 2140 °F (1170 °C).

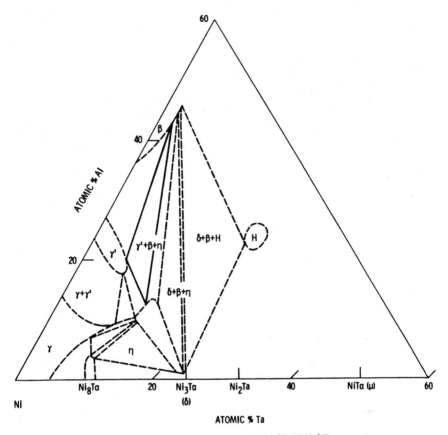

Fig. A10. Ni–Al–Ta system at 1830 °F (1000 °C).

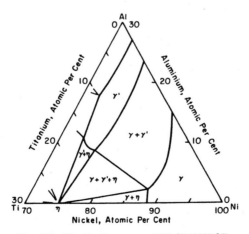

Fig. A11. Ni–Al–Ti system at 1830 °F (1000 °C).

569

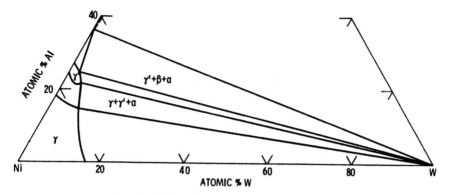

Fig. A12. Ni–Al–W system at 2280 °F (1250 °C).

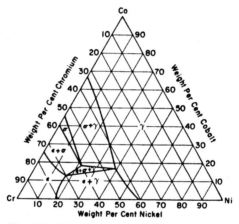

Fig. A13. Ni–Co–Cr system at 2200 °F (1200 °C).

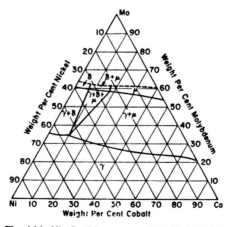

Fig. A14. Ni–Co–Mo system at 2200 °F (1200 °C).

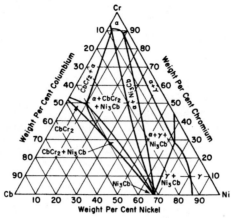

Fig. A15. Ni–Cr–Cb system at 2010 °F (1100 °C).

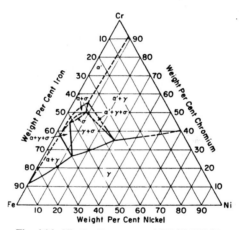

Fig. A16. Ni–Cr–Fe system at 1650 °F (900 °C).

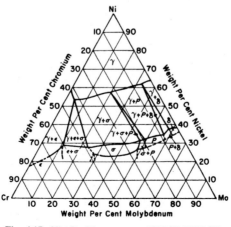

Fig. A17. Ni–Cr–Mo system at 2200 °F (1200 °C).

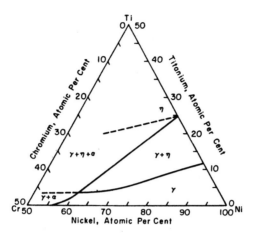

Fig. A18. Ni–Cr–Ti system at 1830 °F (1000 °C).

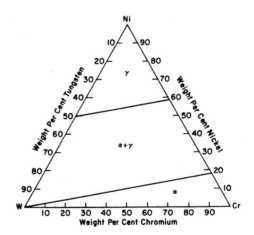

Fig. A19. Ni–Cr–W system at 2200 °F (1200 °C).

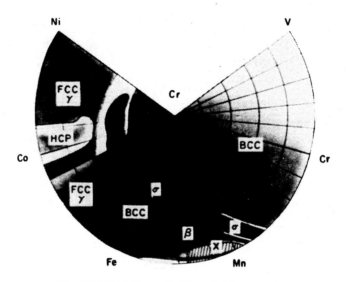

Fig. A20. Polar diagram for Cr vs. first long period.

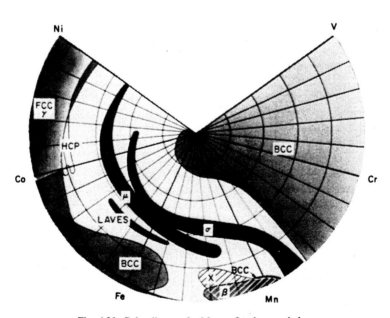

Fig. A21. Polar diagram for Mo vs. first long period.

ACKNOWLEDGMENTS

The assistance of Dr. M. J. Donachie, Jr. (Pratt & Whitney Aircraft) and Professor P. Nash (Illinois Institute of Technology) in preparing this compilation is gratefully acknowledged.

REFERENCES

1. C. T. Sims and W. C. Hagel, *The Superalloys*, Wiley, New York, 1972, p. 577.
2. A. Taylor and R. W. Floyd, *J. Inst. Met.*, **81**, 451 (1952–1953).
3. S. M. Merchant and M. R. Notis, *Mat. Sci. Eng.*, **66**, p. 47 (1984).
4. E. L. Kamen and P. A. Beck, NACA Technical Note 2603, 1952, p. 40.
5. S. P. Rideout and P. A. Beck, NACA Technical Report 1122, 1953.
6. D. S. Duvall, Pratt & Whitney Aircraft Report (AMRDL) 68-060, 1968.
7. P. Nash and D. R. F. West, *Met. Sci.*, **15**, 350 (1981).
8. D. B. Miracle et al., *Met. Trans. A*, **15A**, 483 (1984).
9. P. Nash and D. R. F. West, *Met. Sci.*, **13**, 673 (1979).
10. A. Taylor and R. W. Floyd, *J. Inst. Met.*, **81**, 31 (1952–1953).
11. P. Nash, S. Fielding and D. R. F. West, *Met. Sci.*, **17**, 194 (1983).
12. W. D. Manly and P. A. Beck, NACA Technical Note 2602, February 1952.
13. J. J. English, DMIC Report 183, Battelle Memorial Institute, Columbus, OH, 1963, p. 99.
14. R. E. Lismer, L. Pryce and K. W. Andrews, *J. Iron Steel Inst.*, **171**, 56 (1952).
15. A. Taylor and R. W. Floyd, *J. Inst. Met.*, **80**, 586 (1951–1952).
16. J. J. English, DMIC Report 152, Battelle Memorial Institute, Columbus, OH, 1961, p. 196.

Appendix B

Superalloy Data

Compiled by T. P. GABB and ROBERT L. DRESHFIELD

NASA Lewis Research Center, Cleveland, Ohio

This section contains chemical compositions, stress rupture, tensile, and physical properties of selected superalloys. These selected nickel-, cobalt-, and iron-base alloys are believed to be in general use, of historical interest, or recently developed with use pending. The data represent nominal values in their most common condition. The mechanical properties of directionally processed materials are presented for the longitudinal direction unless otherwise noted. As has been discussed elsewhere in the book, some alloys may be heat treated or thermomechanically processed to achieve significant changes in mechanical behavior. The data is not intended and should not be used for design purposes; rather, it is intended to allow study and comparison among the various alloys.

The 1000-h stress rupture curves are displayed for each class of alloys on separate plots. For clarity, not all alloys tabulated have been graphically displayed; however, it is suggested that the behavior of the alloys can be inferred by referring to alloys with similar properties.

The physical properties (dynamic modulus of elasticity and coefficient of thermal expansion) have been averaged for groups of alloys because of their similarity.

ACKNOWLEDGMENTS

We wish to thank Cabot Corporation, Cannon-Muskegon Corporation, Carpenter Technology Corporation, Cytemp Specialty Steel Division, General Electric Company,

Inco Alloys International, Carl Lund, Pratt & Whitney Aircraft Company, and Special Metals Corporation for their assistance, particularly in providing materials property data. The authors also acknowledge the assistance of Garrett Turbine Engine Company, TRW, and Westinghouse in identifying alloys of interest.

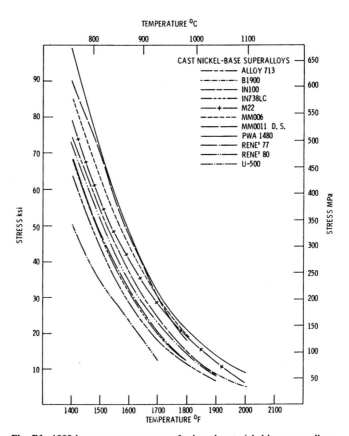

Fig. B1. 1000-h stress rupture curves of selected cast nickel-base superalloys.

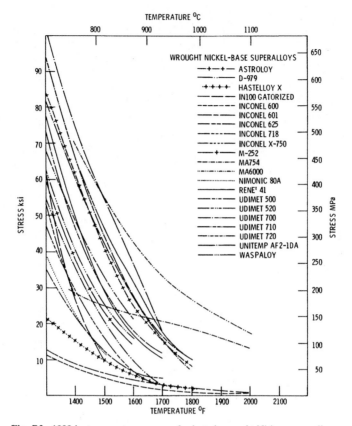

Fig. B2. 1000-h stress rupture curves of selected wrought Ni-base superalloys.

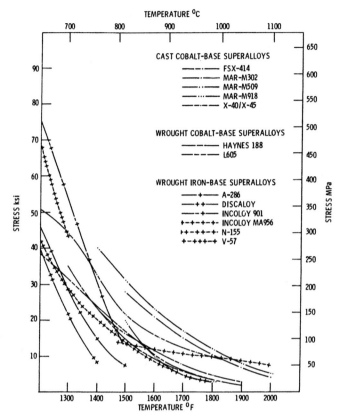

Fig. B3. 1000-h stress rupture curves of selected cast and wrought Co-base and wrought Fe-base superalloys.

Table B1. Nominal Chemistry and Density of Cast Nickel- and Cobalt-Base Alloys

Alloy	Ni	Cr	Co	Mo	W	Ta	Cb	Al	Ti	Fe	Mn	Si	C	B	Zr	Others	Density lb/in.³	Density g/cm³
								Ni-Base Alloys										
Alloy 713C	74	12.5	0.0	4.2	0.0	0.0	2.0	6.1	0.8	0.0	0.0	0.0	0.12	0.012	0.10		0.286	7.9
Alloy 713LC	75	12.0	0.0	4.5	0.0	0.0	2.0	5.9	0.6	0.0	0.0	0.0	0.05	0.010	0.10		0.289	8.0
B-1900	64	8.0	10.0	6.0	0.0	4.0	0.0	6.0	1.0	0.0	0.0	0.0	0.10	0.015	0.10		0.297	8.2
C-1023	58	15.5	10.0	8.5	0.0	0.0	0.0	4.2	3.6	0.0	0.0	0.0	0.16	0.006	0.00		—	—
CMSX-2	66	8.0	4.6	0.6	7.9	5.8	0.0	5.6	0.9	0.0	0.0	0.0	0.00	0.000	0.00		0.311	8.6
GMR-235	63	15.5	0.0	5.3	0.0	0.0	0.0	3.0	2.0	10.0	0.3	0.6	0.15	0.060	0.00		0.291	8.0
IN-100	60	10.0	15.0	3.0	0.0	0.0	0.0	5.5	4.7	0.0	0.0	0.0	0.18	0.014	0.06	1.0V	0.280	7.7
IN-731	67	9.5	10.0	2.5	0.0	0.0	0.0	5.5	4.6	0.0	0.0	0.0	0.18	0.015	0.06	1.0V	0.280	7.7
IN-738LC	61	16.0	8.5	1.7	2.6	1.7	0.9	3.4	3.4	0.0	0.0	0.0	0.11	0.010	0.05		0.283	8.1
IN-939	48	22.5	19.0	0.0	2.0	1.4	1.0	1.9	3.7	0.0	0.0	0.0	0.15	0.009	0.09		0.295	8.2
IN-792	61	12.4	9.0	1.9	3.8	3.9	0.0	3.1	4.5	0.0	0.0	0.0	0.12	0.020	0.10		0.298	8.3
M22	71	5.7	0.0	2.0	11.0	3.0	0.0	6.3	0.0	0.0	0.0	0.0	0.13	0.000	0.60		0.312	8.6
MM-002 (RR-7080)	61	9.0	10.0	0.0	10.0	2.5	0.0	5.5	1.5	0.0	0.0	0.0	0.14	0.015	0.05	1.5Hf	—	—
MM-004 (IN-713+Hf)	74	12.0	0.0	4.5	0.0	0.0	2.0	5.9	0.6	0.0	0.0	0.0	0.05	0.015	0.05	1.3Hf	—	—
MM-005 (René 125+Hf)	59	8.5	10.0	2.0	8.0	3.8	0.0	4.8	2.5	0.0	0.0	0.0	0.11	0.015	0.05	1.4Hf	0.308	8.5

Table B1. (*Continued*)

Alloy	Ni	Cr	Co	Mo	W	Ta	Cb	Al	Ti	Fe	Mn	Si	C	B	Zr	Others	Density lb/in.^{3}	Density g/cm^3
MM-006 (MAR-M 246+Hf)	63	9.0	10.0	2.5	10.0	1.5	0.0	5.5	1.5	0.0	0.0	0.0	0.14	0.015	0.05	1.8Hf	0.311	8.6
MM-009 (MAR-M 200+Hf)	59	9.0	10.0	0.0	12.5	0.0	1.0	5.0	2.0	0.0	0.0	0.0	0.14	0.015	0.05	1.8Hf	0.311	8.6
MM-0011 (MAR-M 247)	60	8.3	10.0	0.7	10.0	3.0	0.0	5.5	1.0	0.0	0.0	0.0	0.14	0.015	0.05	1.5Hf	0.308	8.5
MAR-M 421	61	15.8	9.5	2.0	3.8	0.0	2.0	4.3	1.8	0.0	0.0	0.0	0.14	0.015	0.05		0.292	8.1
PWA 1480	63	10.0	5.0	0.0	4.0	12.0	0.0	5.0	1.5	0.0	0.0	0.0	0.00	0.000	0.00		0.313	8.7
René 77	58	14.6	15.0	4.2	0.0	0.0	0.0	4.3	3.3	0.0	0.0	0.0	0.07	0.016	0.04		0.286	7.9
René 80	60	14.0	9.5	4.0	4.0	0.0	0.0	3.0	5.0	0.0	0.0	0.0	0.17	0.015	0.03		0.295	8.2
SEL	51	15.0	22.0	4.5	0.0	0.0	0.0	4.4	2.4	0.0	0.0	0.0	0.08	0.015	0.00		0.290	8.0
SEL-15	58	11.0	14.5	6.5	1.5	0.0	0.5	5.4	2.5	0.0	0.0	0.0	0.07	0.015	0.00		0.289	8.0
SRR-99	66	9.0	5.0	0.0	9.5	2.9	0.7	5.5	1.8	0.0	0.0	0.0	0.03	0.000	0.00		0.307	8.5
TRW-NASA VIA	61	6.1	7.5	2.0	5.8	9.0	0.5	5.4	1.0	0.0	0.0	0.0	0.13	0.020	0.13	0.4Hf, 0.5Re	0.317	8.8
Udimet 500	52	18.0	19.0	4.2	0.0	0.0	0.0	3.0	3.0	0.0	0.0	0.0	0.07	0.007	0.05		0.290	8.0
UDM56	64	16.0	5.0	1.5	6.0	0.0	0.0	4.5	2.0	0.0	0.0	0.0	0.02	0.070	0.03	0.5V	0.295	8.2
Co-Base Alloys																		
FSX-414	10	29.0	52.0	0.0	7.5	0.0	0.0	0.0	0.0	1.0	0.0	0.0	0.25	0.010	0.00		0.300	8.3
MAR-M 302	0	21.5	58.0	0.0	10.0	9.0	0.0	0.0	0.0	0.0	0.0	0.0	0.85	0.005	0.20		0.333	9.2
MAR-M 509	10	23.5	55.0	0.0	7.0	3.5	0.0	0.0	0.2	0.0	0.0	0.0	0.60	0.000	0.50		0.320	8.9
WI-52	0	21.0	63.0	0.0	11.0	0.0	2.0	0.0	0.0	2.0	0.3	0.3	0.45	0.000	0.00		0.321	8.9
X-40/X-45	10	25.5	54.0	0.0	7.5	0.0	0.0	0.0	0.0	0.0	0.7	0.7	0.50	0.000	0.00		0.311	8.6

Table B2. Ultimate Tensile Strengths of Cast Nickel- and Cobalt-Base Superalloys and Thermal Expansion Coefficients of Nickel-, Cobalt-, and Iron-Base Superalloys

Cast Alloy	Ultimate Tensile Strength, ksi (MPa)					Condition of Test Material[a]
	70 °F (21 °C)	1200 °F (650 °C)	1400 °F (760 °C)	1600 °F (870 °C)	1800 °F (980 °C)	
Nickel Base						
1 Alloy 713C	123 (850)	126 (870)	136 (940)	105 (725)	68 (470)	As cast
2 Alloy 713LC	130 (895)	157 (1080)	138 (955)	109 (750)	68 (470)	As cast
3 B-1900	141 (975)	147 (1015)	138 (955)	115 (795)	80 (550)	As cast
4 CMSX-2[a]	172 (1185)	—	188 (1295)	148 (1020)	—	2400 °F/3/GFQ+1800 °F/5/AC+1600 °F/20/AC
5 GMR-235	103 (710)	96 (660)	—	—	—	As cast
6 IN-100	147 (1015)	161 (1110)	155 (1070)	128 (885)	82 (565)	As cast
7 IN-731	121 (835)	130 (895)	133 (915)	109 (750)	76 (525)	As cast
8 IN-738LC	150 (1035)	—	—	—	—	2050 °F/2/AC+1550 °F/24/AC
9 IN-792	170 (1170)	158 (1090)	144 (990)	122 (840)	—	2050 °F/2/AC+1550 °F/24/AC
10 IN-939	152 (1050)	143 (985)	133 (915)	93 (640)	47 (325)	2120 °F/4/RAC+1830 °F/6/RAC+1650 °F/24/AC+1290 °F/16/AC
11 Inconel 718 Cast	149 (1025)	128 (885)	—	—	—	2000 °F/1-2/AC+1700 °F/1/OQ+1325 °F/8/SC+1150 °F/8/AC
12 M-22	106 (730)	121 (835)	132 (910)	128 (885)	79 (545)	As cast

(Table continues on next page.)

Table B2. (*Continued*)

Cast Alloy	Ultimate Tensile Strength, ksi (MPa)					Condition of Test Material[d]
	70 °F (21 °C)	1200 °F (650 °C)	1400 °F (760 °C)	1600 °F (870 °C)	1800 °F (980 °C)	
13 MM-002	150 (1035)	155 (1070)	150 (1035)	120 (825)	80 (550)	1600 °F/20/AC
14 MM-004	145 (1000)	145 (1000)	130 (895)	95 (655)	55 (380)	As cast
15 MM-005	155 (1070)	155 (1070)	155 (1070)	125 (860)	80 (550)	—
16 MM-006	160 (1105)	155 (1070)	155 (1070)	125 (860)	82 (565)	1550 °F/50/AC
17 MM-009	150 (1035)	155 (1070)	150 (1035)	120 (825)	78 (540)	1600 °F/50/AC
18 MM-0011	150 (1035)	155 (1070)	150 (1070)	120 (825)	80 (550)	1550 °F/50/AC
19 MM-0011[b]	157 (1085)	—	164 (1135)	138 (950)	—	2250 °F/2/GFQ+1800 °F/5/AC+ 1600 °F/20/AC
20 MM-0011[c]	123 (850)	130 (895)	131 (905)	123 (850)	86 (595)	2230 °F/2/GFQ+1800 °F/5/AC+ 1600 °F/20/AC
21 MAR-M 421	157 (1080)	140 (965)	138 (955)	109 (750)	55 (380)	2100 °F/2+1950 °F/4+1400 °F/16
22 PWA 1480[a]	—	165 (1140)	164 (1130)	144 (995)	99 (685)	2350 °F/4/AC+1975 °F/4/AC+ 1600 °F/32/AC
23 René 77	148 (1020)	152 (1050)	136 (940)	—	—	2125 °F/4/AC+1975 °F/4/AC+ 1700 °F/24/AC+1400 °F/16/AC
24 René 80	149 (1030)	149 (1030)	144 (995)	102 (705)	—	2225 °F/2/HeQ+2000 °F/4/HeQ+ 1925 °F/4/FC+1550 °F/16/AC
25 SEL	148 (1020)	132 (910)	127 (875)	—	—	1400 °F/16/AC

26 SEL-15	154 (1060)	160 (1100)	158 (1090)	122 (840)	—	1400 °F/4/AC
27 TRW-NASA VIA	152 (1050)	165 (1140)	159 (1100)	126 (870)	86 (595)	As cast
28 U-500	135 (930)	128 (885)	124 (855)	96 (662)	19 (130)	2100 °F/4/AC+1975 °F/4/AC+ 1400 °F/16/AC
29 UDM 56	137 (945)	138 (950)	137 (945)	103 (710)	—	2125 °F/2/AC+1600 °F/16/AC
Cobalt Base						
1 FSX-414	107 (740)	70 (485)	58 (400)	45 (310)	—	2100 °F/4/FC to 1800 °F/FC to 1000 °F/AC
2 MAR-M 302	135 (930)	114 (785)	102 (705)	65 (450)	40 (275)	As cast
3 MAR-M 509	114 (785)	81 (560)	83 (600)	51 (350)	31 (205)	As cast
4 WI-52	109 (750)	107 (740)	88 (605)	60 (415)	40 (275)	As cast
5 X-40/X-45	108 (745)	75 (515)	70 (485)	47 (325)	29 (200)	As cast

Mean Coefficient of Thermal Expansion (70 °F to Temperature) $10^6/°F$

	1000 °F (540 °C)	1200 °F (650 °C)	1400 °F (760 °C)	1600 °F (870 °C)	1800 °F (980 °C)
1 Typical for Ni-base superalloys	7.9	8.1	8.4	8.8	9.2
2 Typical for Co-base superalloys	8.2	8.5	8.8	9.0	9.2
3 Typical for Fe-base superalloys	9.3	9.5	9.9	10.0	10.1

[a] Single crystal [001].
[b] Directionally solidified, longitudinal.
[c] Directionally solidified, transverse.
[d] GFQ, gas furnace quench; AC, air cool; RAC, rapid air cool; SC, slow cool; HeQ, helium quench; FC, furnace cool; OQ, oil quench.

Table B3. Tensile Yield Strengths and Elongations of Cast Nickel- and Cobalt-Base Superalloys and Dynamic Modulus of Elasticity of Nickel-, Iron-, and Cobalt-Base Superalloys

Cast Alloy	Yield Strength at 0.2% Offset, ksi (MPa)					Tensile Elongation (%)				
	70 °F (21 °C)	1200 °F (650 °C)	1400 °F (760 °C)	1600 °F (870 °C)	1800 °F (980 °C)	70 °F (21 °C)	1200 °F (650 °C)	1400 °F (760 °C)	1600 °F (870 °C)	1800 °F (980 °C)
Nickel Base										
1 Alloy 713C	107 (740)	104 (720)	108 (745)	72 (495)	44 (305)	8	7	6	14	20
2 Alloy 713LC	109 (750)	114 (785)	110 (760)	84 (580)	41 (285)	15	11	11	12	22
3 B-1900	120 (830)	134 (925)	117 (810)	101 (695)	60 (415)	8	6	4	4	7
4 CMSX-2[a]	165 (1135)	—	181 (1245)	125 (860)	—	10	—	17	20	18
5 GMR-235	93 (640)	82 (565)	—	—	43 (295)	3	3	—	—	—
6 IN-100	123 (850)	129 (890)	125 (860)	101 (695)	54 (370)	9	6	7	6	6
7 IN-731	105 (725)	108 (745)	112 (775)	88 (610)	52 (360)	7	5	5	4	7
8 IN-738LC	130 (895)	—	—	—	—	7	—	—	—	—
9 IN-792	154 (1060)	—	144 (995)	96 (665)	—	4	7	4	8	—
10 IN-939	116 (800)	101 (695)	92 (635)	58 (400)	30 (205)	5	7	7	18	25
11 Inconel 718 Cast	122 (840)	104 (715)	—	—	—	9	9	—	—	—
12 M-22	99 (685)	111 (765)	112 (775)	98 (675)	52 (360)	6	5	5	5	6
13 MM-002	120 (825)	120 (825)	125 (860)	80 (550)	50 (345)	7	7	5	7	12
14 MM-004	110 (760)	110 (760)	90 (620)	60 (415)	35 (240)	11	10	6	15	20
15 MM-005	120 (825)	120 (825)	125 (860)	80 (550)	50 (345)	5	7	5	7	12
16 MM-006	125 (860)	125 (860)	125 (860)	80 (550)	50 (345)	6	7	7	10	14
17 MM-009	120 (825)	120 (825)	125 (860)	80 (550)	50 (345)	5	5	5	7	10
18 MM-0011	120 (825)	125 (860)	125 (860)	80 (550)	50 (345)	8	7	7	6	10
19 MM-0011[b]	124 (855)	—	134 (925)	103 (710)	—	12	—	9	13	—
20 MM-0011[c]	106 (730)	110 (760)	111 (765)	107 (740)	67 (460)	12	11	9	13	27
21 MAR-M 421	130 (895)	125 (860)	120 (830)	80 (550)	36 (250)	5	4	3	6	6

Alloy										
22 PWA 1480[a]	130 (895)	132 (910)	131 (905)	106 (730)	72 (495)	4	6	8	12	20
23 René 77	115 (795)	104 (720)	100 (690)	—	—	7	12	—	—	—
24 René 80	124 (855)	105 (725)	104 (720)	77 (530)	—	5	8	10	12	—
25 SEL	131 (905)	115 (795)	115 (795)	—	—	6	8	7	—	—
26 SEL-15	130 (895)	125 (865)	118 (815)	93 (640)	—	9	6	5	6	—
27 TRW-NASA VIA	136 (940)	137 (945)	137 (945)	112 (775)	—	4	4	5	3	—
28 U-500	118 (815)	102 (705)	102 (705)	87 (600)	—	13	18	9	9	—
29 UDM 56	123 (850)	108 (745)	105 (725)	76 (525)	—	3	4	5	5	—
Cobalt Base										
1 FSX-414	64 (440)	31 (215)	28 (195)	24 (165)	—	11	15	18	23	—
2 MAR-M 302	100 (690)	70 (485)	65 (450)	42 (290)	24 (165)	3	3	3	8	16
3 MAR-M 509	85 (585)	55 (380)	50 (345)	42 (290)	24 (165)	5	5	5	10	16
4 WI-52	85 (585)	58 (400)	50 (345)	40 (275)	28 (195)	5	8	9	11	20
5 X-40/X-45	76 (525)	38 (260)	—	—	—	9	12	10	16	31

	Dynamic Modulus of Elasticity 10^6 psi (GPa)					
	70 °F (21 °C)	1000 °F (540 °C)	1200 °F (650 °C)	1400 °F (760 °C)	1600 °F (870 °C)	1800 °F (980 °C)
1 Typical for equiaxed Ni-based superalloys	28.9 (199)	26.0 (179)	24.9 (172)	23.9 (165)	22.6 (156)	21.1 (145)
2 Typical for DS (long.) and SC [001] Ni-based superalloys	18.5 (128)	16.0 (110)	15.3 (105)	14.4 (99)	13.5 (93)	12.6 (87)
3 Typical for Co-based superalloys	32.7 (225)	—	25.8 (178)	23.9 (165)	22.5 (155)	19.8 (137)
4 Typical for Fe-based superalloys	29.3 (202)	24.1 (166)	22.0 (151)	—	—	—

[a] SC, single crystal [001].
[b] DS, directionally solidified superalloys; longitudinal.
[c] DS, directionally solidified superalloys; transverse.

Table B4.   1000-h Rupture Strengths of Cast Nickel- and Cobalt-Base Superalloys

Cast Alloy	1000-h Rupture Strength, ksi (MPa)			
	1400 °F (760 °C)	1600 °F (870 °C)	1800 °F (980 °C)	2000 °F (1090 °C)
		Nickel base		
Alloy 713C	65 (450)	28 (195)	13 (90)	—
Alloy 713LC	60 (415)	30 (205)	13 (90)	—
B-1900	—	37 (255)	15 (105)	5 (35)
CMSX-2[a]	—	50 (345)	25 (170)	13 (90)
GMR-235	—	26 (180)	11 (75)	—
IN-100	75 (515)	37 (255)	15 (105)	—
IN-731	—	—	15 (105)	—
IN-738LC	—	—	12 (85)	—
IN-792	79 (545)	38 (260)	15 (105)	—
IN-939	62 (425)	28 (195)	9 (60)	—
M-22	79 (545)	41 (285)	19 (130)	6 (40)
MM-002	85 (565)	44 (305)	18 (125)	—
MM-004	70 (485)	30 (205)	13 (90)	—
MM-005	82 (565)	44 (305)	17 (115)	—
MM-006	85 (585)	44 (305)	18 (125)	—
MM-009	85 (585)	44 (305)	18 (125)	—
MM-0011	85 (585)	44 (305)	18 (125)	—
MM-0011[b]	94 (650)	47 (325)	18 (125)	—
MAR-M 421	66 (455)	32 (220)	15 (105)	—
PWA 1480[a]	99 (680)	49 (335)	21 (145)	9 (65)
René 77	—	32 (220)	9 (60)	—
René 80	—	35 (240)	15 (105)	—
SEL	—	25 (170)	7 (50)	—
SEL-15	—	43 (295)	11 (75)	—
TRW-NASA VIA	85 (585)	44 (305)	20 (140)	—
U-500	50 (345)	24 (165)	—	—
UDM 56	73 (505)	39 (270)	18 (125)	—
		Cobalt base		
FSX-414	24 (165)	12 (85)	5 (35)	—
MAR-M 302	—	20 (140)	10 (70)	3 (20)
MAR-M 509	40 (275)	20 (140)	12 (85)	5 (35)
WI-52	—	21 (145)	10 (70)	—
X-40/X-45	33 (230)	16 (110)	10 (70)	—

[a] Single crystal [001].
[b] Directionally solidified, longitudinal.

Table B5. Nominal Chemistry and Density of Wrought Nickel-, Cobalt-, and Iron-Base Alloys

Alloy	Ni	Cr	Co	Mo	W	Ta	Cb	Al	Ti	Fe	Mn	Si	C	B	Zr	Other	Density lb/in³	Density gm/cc
								Nickel-Base Alloys										
Astroloy	55.0	15.0	17.0	5.3	0.0	0.0	0.0	4.0	3.5	0.0	0.0	0.0	0.06	0.030	0.00		0.286	7.9
Cabot 214	75.0	16.0	0.0	0.0	0.0	0.0	0.0	4.5	0.0	2.5	0.0	0.0	0.00	0.000	0.00	0.01Y	0.291	8.1
D-979	45.0	15.0	0.0	4.0	0.0	0.0	0.0	1.0	3.0	27.0	0.3	0.2	0.05	0.010	0.00		0.296	8.2
Hastelloy C-22	51.6	21.5	2.5	13.5	4.0	0.0	0.0	0.0	0.0	5.5	1.0	0.1	0.01	0.000	0.00	0.3V	0.314	8.7
Hastelloy C-276	0.0	15.5	2.5	16.0	3.7	0.0	0.0	0.0	0.0	5.5	1.0	0.1	0.01	0.000	0.00	0.3V	0.321	8.9
Hastelloy G-30	42.7	29.5	2.0	5.5	2.5	0.0	0.8	0.0	0.0	15.0	1.0	1.0	0.03	0.000	0.00	2.0Cu	0.297	8.2
Hastelloy S	67.0	15.5	0.0	14.5	0.0	0.0	0.0	0.3	0.0	1.0	0.5	0.4	0.00	0.009	0.00	0.05La	0.316	8.8
Hastelloy X	47.0	22.0	1.5	9.0	0.6	0.0	0.0	0.0	0.0	18.5	0.5	0.5	0.10	0.000	0.00		0.297	8.2
Haynes 230	57.0	22.0	0.0	2.0	14.0	0.0	0.0	0.3	0.0	0.0	0.5	0.4	0.10	0.000	0.00	0.02La	0.319	8.8
IN-100																		
GATORIZE	55.8	12.4	18.5	3.2	0.0	0.0	0.0	5.0	4.3	0.0	0.0	0.0	0.07	0.020	0.06	0.8V	0.284	8.1
Inconel 600	76.0	15.5	0.0	0.0	0.0	0.0	0.0	0.0	0.0	8.0	0.5	0.2	0.08	0.000	0.00		0.304	8.4
Inconel 601	60.5	23.0	0.0	0.0	0.0	0.0	0.0	1.4	0.0	14.1	0.5	0.2	0.05	0.000	0.00		0.291	8.1
Inconel 617	54.0	22.0	12.5	9.0	0.0	0.0	0.0	1.0	0.3	0.0	0.0	0.0	0.07	0.000	0.00		0.302	8.4
Inconel 625	61.0	21.5	0.0	9.0	0.0	0.0	3.6	0.2	0.2	2.5	0.2	0.2	0.05	0.000	0.00		0.305	8.4
Inconel 706	41.5	16.0	0.0	0.0	0.0	0.0	2.9	0.2	1.8	40.0	0.2	0.2	0.03	0.000	0.00		0.292	8.1
Inconel 718	52.5	19.0	0.0	3.0	0.0	0.0	5.1	0.5	0.9	18.5	0.2	0.2	0.04	0.000	0.00		0.297	8.2

(Table continues on next page.)

Table B5. (*Continued*)

Alloy	Ni	Cr	Co	Mo	W	Ta	Cb	Al	Ti	Fe	Mn	Si	C	B	Zr	Other	Density lb/in³	Density gm/cc
Inconel MA 754	78.0	20.0	0.0	0.0	0.0	0.0	0.0	0.3	0.5	0.0	0.0	0.0	0.05	0.000	0.00	$0.6Y_2O_3$	0.300	8.3
Inconel MA 6000	69.0	15.0	0.0	2.0	4.0	2.0	0.0	4.5	2.5	0.0	0.0	0.0	0.05	0.010	0.15	$2.5Y_2O_3$	0.293	8.1
Inconel X750	73.0	15.5	0.0	0.0	0.0	0.0	1.0	0.7	2.5	7.0	0.5	0.2	0.04	0.000	0.00		0.298	8.3
M-252	55.0	20.0	10.0	10.0	0.0	0.0	0.0	1.0	2.6	0.0	0.5	0.5	0.15	0.005	0.00		0.298	8.3
Nimonic 75	76.0	19.5	0.0	0.0	0.0	0.0	0.0	0.0	0.4	3.0	0.3	0.3	0.10	0.000	0.00		0.302	8.4
Nimonic 80A	76.0	19.5	0.0	0.0	0.0	0.0	0.0	1.4	2.4	0.0	0.3	0.3	0.06	0.003	0.06		0.295	8.2
Nimonic 90	59.0	19.5	16.5	0.0	0.0	0.0	0.0	1.5	2.5	0.0	0.3	0.3	0.07	0.003	0.06		0.296	8.2
Nimonic 105	53.0	15.0	20.0	5.0	0.0	0.0	0.0	4.7	1.2	0.0	0.3	0.3	0.13	0.005	0.10		0.289	8.0
Nimonic 115	60.0	14.3	13.2	0.0	0.0	0.0	0.0	4.9	3.7	0.0	0.0	0.0	0.15	0.160	0.04		0.284	7.9
Nimonic 263	51.0	20.0	20.0	5.9	0.0	0.0	0.0	0.5	2.1	0.0	0.4	0.3	0.06	0.001	0.02		.302	8.4
Nimonic PE.16	43.0	16.5	1.0	1.1	0.0	0.0	0.0	1.2	1.2	33.0	0.1	0.1	0.05	0.020	0.00		0.29	8.0
Nimonic PK.33	56.0	18.5	14.0	7.0	0.0	0.0	0.0	2.0	2.0	0.3	0.1	0.1	0.05	0.030	0.00		0.297	8.2
René 41	55.0	19.0	11.0	10.0	0.0	0.0	0.0	1.5	3.1	0.0	0.0	0.0	0.09	0.005	0.00		0.298	8.3
René 95	61.0	14.0	8.0	3.5	3.5	0.0	3.5	3.5	2.5	0.0	0.0	0.0	0.15	0.010	0.05		0.297	8.2
TD Nickel	98.0	0.0	0.0	0.0	0.0	0.0	0.0	0.0	0.0	0.0	0.0	0.0	0.00	0.000	0.00	$2.0ThO_2$	0.322	8.9
Udimet 500	54.0	18.0	18.5	4.0	0.0	0.0	0.0	2.9	2.9	0.0	0.0	0.0	0.08	0.006	0.05		0.290	8.0
Udimet 520	57.0	19.0	12.0	6.0	1.0	0.0	0.0	2.0	3.0	0.0	0.0	0.0	0.05	0.005	0.00		0.292	8.1
Udimet 700	55.0	15.0	17.0	5.0	0.0	0.0	0.0	4.0	3.5	0.0	0.0	0.0	0.06	0.030	0.00		0.286	7.9
Udimet 710	55.0	18.0	15.0	3.0	1.5	0.0	0.0	2.5	5.0	0.0	0.0	0.0	0.07	0.020	0.00		0.292	8.1

Udimet 720	55.0	17.9	14.7	3.0	1.3	0.0	0.0	2.5	5.0	0.0	0.0	0.0	0.03	0.033	0.03		0.292	8.1
Unitemp AF2-1DA	59.0	12.0	10.0	3.0	6.0	1.5	0.0	4.6	3.0	1.0	0.0	0.0	0.35	0.014	0.10		0.299	8.3
Unitemp AF2-1DA6	60.0	12.0	10.0	2.7	6.5	1.5	0.0	4.0	2.8	0.0	0.0	0.0	0.04	0.015	0.10		0.301	8.3
Waspaloy	58.0	19.5	13.5	4.3	0.0	0.0	0.0	1.3	3.0	0.0	0.0	0.0	0.08	0.006	0.00		0.296	8.2
Cobalt-Base Alloys																		
Haynes 188	22.0	22.0	39.2	0.0	14.0	0.0	0.0	0.0	0.0	3.0	0.0	0.0	0.10	0.000	0.00		0.330	9.1
L-605	10.0	20.0	52.9	0.0	15.0	0.0	0.0	0.0	0.0	0.0	0.0	0.0	0.05	0.000	0.00		0.330	9.1
MAR-M 918	20.0	20.0	52.5	0.0	0.0	7.5	0.0	0.0	0.0	0.0	0.0	0.0	0.05	0.000	0.10		0.320	8.9
MP35N	35.0	20.0	35.0	10.0	0.0	0.0	0.0	0.0	0.0	0.0	0.0	0.0	0.00	0.000	0.00		0.304	8.4
MP159	25.5	19.0	35.7	7.0	0.0	0.0	0.6	0.2	3.0	9.0	0.0	0.0	0.00	0.000	0.00		0.301	8.3
Iron-Base Alloys																		
A-286	26.0	15.0	0.0	1.3	0.0	0.0	0.0	0.2	2.0	54.0	1.3	0.5	0.05	0.015	0.00		0.286	7.9
Discaloy	26.0	13.5	0.0	2.7	0.0	0.0	0.0	0.1	1.7	54.0	0.9	0.8	0.04	0.005	0.00		0.288	8.0
Haynes 556	20.0	22.0	20.0	3.0	2.5	0.9	0.1	0.3	0.0	29.0	1.5	0.4	0.10	0.000	0.00	0.2N, 0.02La	0.297	8.3
Alloy 901	42.5	12.5	0.0	5.7	0.0	0.0	0.0	0.2	2.8	36.0	0.1	0.1	0.05	0.015	0.00		0.297	8.2
Incoloy 903	38.0	0.0	15.0	0.0	0.0	0.0	3.0	0.7	1.4	41.0	0.0	0.0	0.00	0.000	0.00		0.294	8.1
Incoloy 909	38.0	0.0	13.0	0.0	0.0	0.0	4.7	0.0	1.5	42.0	0.0	0.4	0.01	0.001	0.00		0.296	8.2
Incoloy MA 956	0.0	20.0	0.0	0.0	0.0	0.0	0.0	4.5	0.5	74.0	0.0	0.0	0.00	0.000	0.00	$0.5Y_2O_3$	0.26	7.2
N-155	20.0	21.0	20.0	3.0	2.5	0.0	1.0	0.0	0.0	30.0	1.5	0.5	0.15	0.000	0.00	0.15N	0.296	8.2
V-57	27.0	14.8	0.0	1.3	0.0	0.0	0.0	0.3	3.0	52.0	0.3	0.7	0.08	0.010	0.00		0.287	7.9

Table B6. Ultimate Tensile Strengths of Wrought Nickel-, Cobalt-, and Iron-Base Superalloys

Wrought Alloy	Form	Ultimate Tensile Strength					Condition of Test Material[a]
		70 °F (21 °C)	1000 °F (540 °C)	1200 °F (650 °C)	1400 °F (760 °C)	1600 °F (870 °C)	
				Nickel Base			
1 Astroloy	Bar	205 (1415)	180 (1240)	190 (1310)	168 (1160)	112 (775)	2000 °F/4/OQ+1600 °F/8/AC+1800 °F/4/AC+1200 °F/24/AC+1400 °F/8/AC
2 Cabot 214	Bar	133 (915)	104 (715)	98 (675)	84 (560)	64 (440)	2050 °F
3 D-979	Bar	204 (1410)	188 (1295)	160 (1105)	104 (720)	50 (345)	1900 °F/1/OQ+1550 °F/6/AC+1300 °F/16/AC
4 Hastelloy C-22	Sht.	116 (800)	91 (625)	85 (585)	76 (525)	—	2050 °F/RQ
5 Hastelloy G-30	Sht.	100 (690)	71 (490)	—	—	—	2150 °F/RAC-WQ
6 Hastelloy S	Bar	130 (845)	112 (775)	105 (720)	84 (575)	50 (340)	1950 °F/AC
7 Hastelloy X	Sht.	114 (785)	94 (650)	83 (570)	63 (435)	37 (255)	2150 °F/1/RAC
8 Haynes 230	Bar	126 (870)	105 (720)	98 (675)	84 (575)	56 (385)	2250 °F/AC
9 IN-100	Gat.	229 (1580)	217 (1500)	200 (1380)	159 (1095)	—	
10 Inconel 600	Bar	96 (660)	81 (560)	65 (450)	38 (260)	20 (140)	2050 °F/2/AC
11 Inconel 601	Sht.	107 (740)	105 (725)	76 (525)	42 (290)	23 (160)	2100 °F/1/AC
12 Inconel 617	Bar	107 (740)	84 (580)	82 (565)	64 (440)	40 (275)	2150 °F/AC
13 Inconel 617	Sht.	112 (770)	86 (590)	86 (590)	68 (470)	45 (310)	2150 °F/0.2/AC
14 Inconel 625	Bar	140 (965)	132 (910)	121 (835)	80 (550)	40 (275)	2100 °F/1/WQ
15 Inconel 706	Bar	190 (1310)	166 (1145)	150 (1035)	105 (725)	—	1800 °F/1/AC+1550 °F/3/AC+1325 °F/8/FC+1150 °F/8/AC
16 Inconel 718	Bar	208 (1435)	185 (1275)	178 (1228)	138 (950)	49 (340)	1800 °F/1/AC+1325 °F/8/FC+1150 °F/18/AC
17 Inconel 718 Dir. Age	Bar	222 (1530)	196 (1350)	179 (1235)	—	—	1325 °F/8/SC+1150 °F/8/AC
18 Inconel 718 Super	Bar	196 (1350)	174 (1200)	164 (1130)	—	—	1700 °F/1/AC+1325 °F/8/SC+1150 °F/8/AC
19 Inconel MA 754	Bar (L)	140 (965)	110 (760)	87 (600)	50 (345)	36 (250)	2400 °F/1/AC
20 Inconel MA 6000	Bar (L)	188 (1295)	168 (1155)	158 (1090)	142 (975)	107 (740)	2250 °F/0.5/AC+1750 °F/2/AC+1550 °F/24/AC

21 Inconel X750	Bar	174 (1200)	152 (1050)	136 (940)	—	—	2100 °F/2/AC+1550 °F/24/AC+1300 °F/20/AC
22 M-252	Bar	180 (1240)	178 (1230)	168 (1160)	137 (945)	74 (510)	1900 °F/4/AC+1400 °F/16/AC
23 Nimonic 75	Bar	108 (745)	98 (675)	78 (540)	45 (310)	22 (150)	1925 °F/1/AC
24 Nimonic 80A	Bar	145 (1000)	127 (875)	115 (795)	87 (600)	45 (310)	1975 °F/8/AC+1300 °F/16/AC
25 Nimonic 90	Bar	179 (1235)	156 (1075)	136 (940)	95 (655)	48 (330)	1975 °F/8/AC+1300 °F/16/AC
26 Nimonic 105	Bar	171 (1180)	164 (1130)	159 (1095)	135 (930)	96 (660)	2100 °F/4/AC+1940 °F/16/AC+1560 °F/16/AC
27 Nimonic 115	Bar	180 (1240)	158 (1090)	163 (1125)	157 (1085)	120 (830)	2175 °F/1.5/AC+2010 °F/6/AC
28 Nimonic 263	Sht.	141 (970)	116 (800)	112 (770)	94 (650)	40 (280)	2100 °F/0.2/WQ+1470 °F/8/AC
29 Nimonic PE.16	Bar	128 (885)	107 (740)	96 (660)	74 (510)	31 (215)	1900 °F/4/AC+1470 °F/2/AC+1290 °F/16/AC
30 Nimonic PK.33	Sht.	171 (1180)	145 (1000)	145 (1000)	128 (885)	74 (510)	2010–2040 °F/0.25/AC+1560 °F/4/AC
31 René 41	Bar	206 (1420)	203 (1400)	194 (1340)	160 (1105)	90 (620)	1950 °F/4/AC+1400 °F/16/AC
32 René 95	Bar	235 (1620)	224 (1550)	212 (1460)	170 (1170)	—	1650 °F/24/ +2025 °F/1/ OQ+1350 °F/64/AC
33 TD Nickel	Bar	100 (690)	45 (310)	38 (260)	33 (230)	28 (195)	1800–2000 °F/0.25–2/AC
34 Udimet 500	Bar	190 (1310)	180 (1240)	176 (1215)	151 (1040)	93 (640)	1975 °F/4/AC+1550 °F/24/AC+1400 °F/16/AC
35 Udimet 520	Bar	190 (1310)	180 (1240)	170 (1175)	105 (725)	75 (515)	2025 °F/4/AC+1550 °F/24/AC+1400 °F/16/AC
36 Udimet 700	Bar	204 (1410)	185 (1275)	180 (1240)	150 (1035)	100 (690)	2150 °F/4/AC+1975 °F/4/AC+1550 °F/24/AC+1400 °F/16/AC
37 Udimet 710	Bar	172 (1185)	167 (1150)	187 (1290)	148 (1020)	102 (705)	2150 °F/4/AC+1975 °F/4/AC+1550 °F/24/AC+1400 °F/16/AC
38 Udimet 720	Bar	228 (1570)	—	211 (1455)	211 (1455)	167 (1150)	2035 °F/2/AC+1975 °F/4/OQ+1200 °F/24/AC+1400 °F/8/AC
39 Unitemp AF2-1DA6	Bar	226 (1560)	215 (1480)	203 (1400)	187 (1290)	—	2100 °F/4/AC+1400 °F/16/AC

(Table continues on next page.)

Table B6. (*Continued*)

Wrought Alloy	Form	Ultimate Tensile Strength					Condition of Test Material[a]
		70 °F (21 °C)	1000 °F (540 °C)	1200 °F (650 °C)	1400 °F (760 °C)	1600 °F (870 °C)	
40 Waspaloy	Bar	185 (1275)	170 (1170)	162 (1115)	94 (650)	40 (275)	1975 °F/4/AC+1550 °F/24/AC+1400 °F/16/AC
				Cobalt Base			
1 Haynes 188	Sht.	139 (960)	107 (740)	103 (710)	92 (635)	61 (420)	2150 °F/1/RAC
2 L-605	Sht.	146 (1005)	116 (800)	103 (710)	66 (455)	47 (325)	2250 °F/1/RAC
3 MAR-M 918	Sht.	130 (895)	—	—	—	—	2175 °F/4/AC
4 MP35N	Bar	294 (2025)	—	—	—	—	53% CW+1050 °F/4/AC
5 MP159	Bar	275 (1895)	227 (1565)	223 (1540)	—	—	48% CW+1225 °F/4/AC
				Iron Base			
1 A-286	Bar	146 (1005)	131 (905)	104 (720)	64 (440)	—	1800 °F/1/OQ+1325 °F/16/AC
2 Alloy 901	Bar	175 (1205)	149 (1030)	139 (960)	105 (725)	—	2000 °F/2/WQ+1450 °F/2/AC+1325 °F/24/AC
3 Discaloy	Bar	145 (1000)	125 (865)	104 (720)	70 (485)	—	1850 °F/2/OQ+1350 °F/20/AC+1200 °F/20/AC
4 Haynes 556	Sht.	118 (815)	93 (645)	85 (590)	69 (470)	48 (330)	2150 °F/AC
5 Incoloy 903	Bar	190 (1310)	—	145 (1000)	—	—	1550 °F/1/WQ+1325 °F/8/FC+1150 °F/8/AC
6 Incoloy 909	Bar	190 (1310)	168 (1160)	149 (1025)	89 (615)	—	1800 °F/1/WQ+1325 °F/8/FC+1150 °F/8/AC
7 Incoloy MA956	Sht. (L)	94 (645)	54 (370)	33 (230)	23 (160)	18 (125)	2375 °F/1/AC
8 N-155	Bar	118 (815)	94 (650)	79 (545)	62 (428)	38 (260)	2150 °F/1/WQ+1500 °F/4/AC
9 V-57	Bar	170 (1170)	145 (1000)	130 (895)	90 (620)	—	1800 °F/2-4/OQ+1350 °F/16/AC

[a] OQ, oil quench; AC, air cool; Sht, sheet; Dir. Age, Direct Age; L, longitudinal; RQ, rapid quench; RAC-WQ, rapid air quench-water quench; gat, gatorized; FC, furnace cool; CW, cold worked; SC, slow cool.

Table B7. Tensile Yield Strengths and Elongations of Wrought Nickel-, Cobalt-, and Iron-Based Superalloys

Wrought Alloy	Form	Yield Strength at 0.2% Offset, ksi (MPa)					Tensile Elongation (%)				
		70 °F (21 °C)	1000 °F (540 °C)	1200 °F (650 °C)	1400 °F (760 °C)	1600 °F (870 °C)	70 °F (21 °C)	1000 °F (540 °C)	1200 °F (650 °C)	1400 °F (760 °C)	1600 °F (870 °C)
		Nickel Base									
1 Astroloy	Bar	152 (1050)	140 (965)	140 (965)	132 (910)	100 (690)	16	16	18	21	25
2 Cabot 214	Bar	81 (560)	74 (510)	73 (505)	72 (495)	45 (310)	38	19	14	9	11
3 D-979	Bar	146 (1005)	134 (925)	129 (890)	95 (655)	44 (305)	15	15	21	17	18
4 Hastelloy C-22	Sht.	59 (405)	40 (275)	36 (250)	35 (240)	—	57	61	65	63	—
5 Hastelloy G-30	Sht.	46 (315)	25 (170)	—	—	—	64	75	—	—	—
6 Hastelloy S	Bar	65 (455)	49 (340)	47 (320)	45 (310)	32 (220)	49	50	56	70	47
7 Hastelloy X	Sht.	52 (360)	42 (290)	40 (275)	38 (260)	26 (180)	43	45	37	37	50
8 Haynes 230	Bar	56 (385)	41 (280)	38 (265)	38 (265)	39 (270)	48	56	55	48	64
9 IN-100 Gat.	Bar	164 (1125)	159 (1095)	161 (1110)	148 (1015)	—	22	22	21	14	—
10 Inconel 600	Bar	41 (285)	32 (220)	30 (205)	26 (180)	6 (40)	45	41	49	70	80
11 Inconel 601	Sht.	66 (455)	51 (350)	45 (310)	32 (220)	8 (55)	40	34	33	78	128
12 Inconel 617	Bar	43 (295)	29 (200)	25 (170)	26 (180)	28 (195)	70	68	75	84	118
13 Inconel 617	Sht.	50 (345)	33 (230)	32 (220)	33 (230)	30 (205)	55	62	61	59	73
14 Inconel 625	Bar	71 (490)	60 (415)	61 (420)	60 (415)	40 (275)	50	50	34	45	125
15 Inconel 706	Bar	146 (1005)	132 (910)	125 (860)	96 (660)	—	20	19	24	32	—
16 Inconel 718	Bar	172 (1185)	154 (1065)	148 (1020)	107 (740)	48 (330)	21	18	19	25	88
17 Inconel 718 Dir. Age	Bar	198 (1365)	171 (1180)	158 (1090)	—	—	16	15	23	—	—

(Table continues on next page.)

593

Table B7. (*Continued*)

Wrought Alloy	Form	Yield Strength at 0.2% Offset, ksi (MPa)					Tensile Elongation (%)				
		70 °F (21 °C)	1000 °F (540 °C)	1200 °F (650 °C)	1400 °F (760 °C)	1600 °F (870 °C)	70 °F (21 °C)	1000 °F (540 °C)	1200 °F (650 °C)	1400 °F (760 °C)	1600 °F (870 °C)
18 Inconel 718 Super	Bar	160 (1105)	148 (1020)	139 (960)	—	—	16	18	14	—	—
19 Inconel MA 754	Bar (L)	85 (585)	75 (515)	69 (475)	40 (275)	31 (215)	21	19	25	34	32
20 Inconel MA 6000	Bar (L)	186 (1285)	147 (1010)	127 (875)	113 (780)	76 (525)	4	6	6	6	9
21 Inconel X750	Bar	118 (815)	105 (725)	103 (710)	—	—	27	26	10	—	—
22 M-252	Bar	122 (840)	111 (765)	108 (745)	104 (720)	70 (485)	16	15	11	10	18
23 Nimonic 75	Bar	41 (285)	29 (200)	29 (200)	23 (160)	13 (90)	40	40	46	67	68
24 Nimonic 80A	Bar	90 (620)	77 (530)	80 (550)	73 (505)	38 (260)	39	37	21	17	30
25 Nimonic 90	Bar	117 (810)	105 (725)	99 (685)	78 (540)	38 (260)	33	28	14	12	23
26 Nimonic 105	Bar	120 (830)	112 (775)	111 (765)	107 (740)	71 (490)	16	22	24	25	27
27 Nimonic 115	Bar	125 (865)	115 (795)	118 (815)	116 (800)	80 (550)	27	18	23	24	16
28 Nimonic 263	Sht.	84 (580)	70 (485)	70 (485)	67 (460)	26 (180)	39	42	27	21	25
29 Nimonic PE.16	Bar	77 (530)	70 (485)	70 (485)	54 (370)	20 (140)	37	26	30	42	80
30 Nimonic PK.33	Sht.	113 (780)	105 (725)	105 (725)	97 (670)	61 (420)	30	30	26	18	24
31 René 41	Bar	154 (1060)	147 (1020)	145 (1000)	136 (940)	80 (550)	14	14	14	11	19
32 René 95	Bar	190 (1310)	182 (1255)	177 (1220)	160 (1100)	—	15	12	14	15	—
33 TD Nickel	Bar	80 (550)	43 (295)	36 (250)	31 (215)	26 (180)	25	14	12	11	9

No. / Alloy	Form										
34 Udimet 500	Bar	122 (840)	115 (795)	110 (760)	106 (730)	72 (495)	32	28	28	39	20
35 Udimet 520	Bar	125 (860)	120 (825)	115 (795)	105 (725)	75 (520)	21	20	17	15	20
36 Udimet 700	Bar	140 (965)	130 (895)	124 (855)	120 (825)	92 (635)	17	16	16	20	27
37 Udimet 710	Bar	132 (910)	123 (850)	125 (860)	118 (815)	92 (635)	7	10	15	25	29
38 Udimet 720	Bar	173 (1195)	—	164 (1130)	152 (1050)	—	13	—	17	9	—
39 Unitemp AF2-1DA6	Bar	147 (1015)	151 (1040)	148 (1020)	144 (995)	—	20	19	18	16	—
40 Waspaloy	Bar	115 (795)	105 (725)	100 (690)	98 (675)	75 (520)	25	23	34	28	35
Cobalt Base											
1 Haynes 188	Sht.	70 (485)	44 (305)	44 (305)	42 (290)	38 (260)	56	70	61	43	73
2 L-605	Sht.	67 (460)	36 (250)	35 (240)	38 (260)	35 (240)	64	59	35	12	35
3 MAR-M 918	Sht.	130 (895)	—	—	—	—	48	—	—	—	—
4 MP35N	Bar	235 (1620)	—	—	—	—	10	—	—	—	—
5 MP159	Bar	265 (1825)	217 (1495)	205 (1415)	—	—	8	8	7	—	—
Iron Base											
1 A-286	Bar	105 (725)	88 (605)	88 (605)	62 (430)	—	25	19	13	19	—
2 Alloy 901	Sht.	130 (895)	113 (780)	110 (760)	92 (635)	—	14	14	13	19	—
3 Discaloy	Bar	106 (730)	94 (650)	91 (630)	62 (430)	—	19	16	19	—	—
4 Haynes 556	Sht.	60 (410)	35 (240)	33 (225)	32 (220)	29 (195)	48	54	52	49	53
5 Incoloy 903	Bar	160 (1105)	—	130 (895)	—	—	14	—	18	—	—
6 Incoloy 909	Bar	148 (1020)	137 (945)	126 (870)	78 (540)	—	16	14	24	34	—
7 Incoloy MA956	Sht. (L)	80 (555)	41 (285)	25 (170)	20 (140)	17 (115)	10	20	20	14	9
8 N-155	Bar	58 (400)	49 (340)	43 (295)	36 (250)	25 (175)	40	33	32	32	33
9 V-57	Bar	120 (830)	110 (760)	108 (745)	70 (485)	—	26	19	22	34	—

Table B8. 1000-h Rupture Strengths of Wrought Nickel-, Cobalt-, and Iron-Base Superalloys

Wrought Alloy	Form	1000-h Rupture Strength, ksi (MPa)			
		1200 °F (650 °C)	1400 °F (760 °C)	1600 °F (870 °C)	1800 °F (980 °C)
Nickel base					
Astroloy	Bar	112 (770)	62 (425)	25 (170)	8 (55)
Cabot 214		—	—	4 (30)	2 (15)
D-979	Bar	75 (515)	36 (250)	10 (70)	—
Hastelloy S	Bar	—	13 (90)	4 (25)	—
Hastelloy X	Sht.	31 (215)	15 (105)	6 (40)	2 (15)
Haynes 230		—	18 (125)	8 (55)	2 (15)
IN-100 Gat.	Bar	120 (825)	28 (195)	—	—
Inconel 600	Bar	—	—	4 (30)	2 (15)
Inconel 601	Sht.	28 (195)	9 (60)	4 (30)	2 (15)
Inconel 617	Bar	52 (360)	24 (165)	9 (60)	4 (30)
Inconel 617	Sht.	—	23 (160)	9 (60)	4 (30)
Inconel 625	Bar	54 (370)	23 (160)	7 (50)	3 (20)
Inconel 706	Bar	84 (580)	—	—	—
Inconel 718	Bar	86 (595)	28 (195)	—	—
Inconel 718 Dir. Age	Bar	59 (405)	—	—	—
Inconel 718 Super	Bar	87 (600)	—	—	—
Inconel MA 754	Bar (L)	37 (255)	29 (200)	23 (160)	19 (130)
Inconel MA 6000	Bar (L)	—	71 (490)	43 (295)	27 (185)
Inconel X750	Bar	68 (470)	—	7 (50)	—
M-252	Bar	82 (565)	39 (270)	14 (95)	—
Nimonic 75	Bar	25 (170)	7 (50)	1 (5)	—
Nimonic 80A	Bar	61 (420)	23 (160)	—	—
Nimonic 90	Bar	66 (455)	30 (205)	9 (60)	—
Nimonic 105	Bar	—	48 (330)	19 (130)	4 (30)
Nimonic 115	Bar	—	61 (420)	27 (185)	10 (70)
Nimonic PE.16	Bar	50 (345)	22 (150)	—	—
Nimonic PK.33	Sht.	95 (655)	45 (310)	13 (90)	—
René 41	Bar	102 (705)	50 (345)	17 (115)	—
René 95	Bar	125 (860)	—	—	—
TD Nickel	Bar	26 (180)	21 (145)	16 (110)	12 (85)
Udimet 500	Bar	110 (760)	47 (325)	18 (125)	—
Udimet 520	Bar	85 (585)	50 (345)	22 (150)	—
Udimet 700	Bar	102 (705)	62 (425)	29 (200)	8 (55)
Udimet 710	Bar	126 (870)	67 (460)	29 (200)	10 (70)
Udimet 720	Bar	97 (670)	—	—	—
Unitemp AF2-1DA6	Bar	128 (885)	52 (360)	—	—
Waspaloy	Bar	89 (615)	42 (290)	16 (110)	—
Cobalt base					
Haynes 188	Sht.	—	24 (165)	10 (70)	4 (30)
L-605	Sht.	39 (270)	24 (165)	11 (75)	4 (30)
MAR-M 918	Sht.	—	9 (60)	3 (20)	1 (5)

(Table continues on next page.)

Table B8. (*Continued*)

Wrought Alloy	Form	1000-h Rupture Strength, ksi (MPa)			
		1200 °F (650 °C)	1400 °F (760 °C)	1600 °F (870 °C)	1800 °F (980 °C)
Iron-base					
A-286	Bar	46 (315)	15 (105)	—	—
Alloy 901	Sht.	76 (525)	30 (205)	—	—
Discaloy	Bar	40 (275)	9 (60)	—	—
Haynes 556	Sht.	40 (275)	18 (125)	8 (55)	3 (20)
Incoloy 903	Bar	74 (510)	—	—	—
Incoloy 909	Bar	50 (345)	—	—	—
Incoloy MA956	Sht. (L)	—	16 (110)	12 (85)	10 (65)
N-155	Bar	43 (295)	20 (140)	10 (70)	3 (20)
V-57	Bar	70 (485)	—	—	—

Appendix C

Registered Trademarks

René 41	Allvac Metals Company (a Teledyne Company)
Hastelloy	Cabot Corporation
Haynes	
René	General Electric Company
Vitallium	Howmet Corporation
Inconel	The International Nickel Company
Incoloy	
Nimonic	
Monel	
MA	
APKL	
MAR-M	Martin Marietta Corporation
Udimet	Special Metals, Inc.
Gatorize	United Aircraft Company
MERL	
Waspaloy	
Unitemp	Universal-Cyclops Steel Corporation
Discaloy	Westinghouse Corporation
Nicrotung	
REP	Whittaker Corporation
CMSX	Cannon-Muskegon Corporation
Multiphase	Standard Pressed Steel Co.

INDEXES

Get thorough insight into the index, by which the whole book is governed.
Jonathan Swift (1667–1745)

Alloy Index

This index lists alloys that are specifically cited in text, figures, or tables. Those in Appendix B (pp. 576–597) are not given index citations owing to their ease of accessibility.

Author Index

This index lists all pages of text citation, but it omits page references to bibliographies that are easily accessible by way of text entries.

605

Subject Index